AF676666

Cellular Mechanisms of Cardiovascular Disease

Cellular Mechanisms of Cardiovascular Disease

Editors

Tânia Martins-Marques
Gonçalo F. Coutinho
Attila Kiss

MDPI

Basel • Beijing • Wuhan • Barcelona • Belgrade • Novi Sad • Cluj • Manchester

Editors

Tânia Martins-Marques	Gonçalo F. Coutinho	Attila Kiss
University of Coimbra	University of Coimbra	Medical University of Vienna
Coimbra, Portugal	Coimbra, Portugal	Vienna, Austria

Editorial Office
MDPI
St. Alban-Anlage 66
4052 Basel, Switzerland

This is a reprint of articles from the Special Issue published online in the open access journal *Biomedicines* (ISSN 2227-9059) (available at: https://www.mdpi.com/journal/biomedicines/special_issues/Cellular_Cardiovascular).

For citation purposes, cite each article independently as indicated on the article page online and as indicated below:

Lastname, A.A.; Lastname, B.B. Article Title. *Journal Name* **Year**, *Volume Number*, Page Range.

ISBN 978-3-0365-8988-6 (Hbk)
ISBN 978-3-0365-8989-3 (PDF)
doi.org/10.3390/books978-3-0365-8989-3

© 2023 by the authors. Articles in this book are Open Access and distributed under the Creative Commons Attribution (CC BY) license. The book as a whole is distributed by MDPI under the terms and conditions of the Creative Commons Attribution-NonCommercial-NoDerivs (CC BY-NC-ND) license.

Contents

Contents

biomedicines MDPI

Editorial

Editorial of the Special Issue: Cellular Mechanisms of Cardiovascular Disease

Tânia Martins-Marques [1,2,3,*], Gonçalo Coutinho [1,2,3,4] and Attila Kiss [5]

1 Univ Coimbra, Coimbra Institute for Clinical and Biomedical Research (iCBR), Faculty of Medicine, 3000-548 Coimbra, Portugal; goncalofcoutinho@gmail.com
2 Univ Coimbra, Center for Innovative Biomedicine and Biotechnology (CIBB), 3004-504 Coimbra, Portugal
3 Clinical Academic Centre of Coimbra (CACC), 3004-561 Coimbra, Portugal
4 University Hospital and Center of Coimbra, Cardiothoracic Surgery Department, 3000-075 Coimbra, Portugal
5 Ludwig Boltzmann Institute for Cardiovascular Research, Center for Biomedical Research and Translational Surgery, Medical University of Vienna, 1090 Vienna, Austria; attila.kiss@meduniwien.ac.at
* Correspondence: tania.m.marques@fmed.uc.pt

Cardiovascular diseases (CVD) remain the major cause of mortality and disability worldwide, having contributed to 19.1 million deaths in 2020. Although cardiovascular outcomes have significantly improved due to early diagnosis, improved and timely treatment, the prevalence of CVD is expected to increase in the coming years, namely due to population aging and other comorbidities highlighting the pressing need to identify novel biomarkers and disease-modifying treatments [1]. Therefore, a major future challenge in cardiovascular medicine lies in understanding the precise molecular basis of cardiac and vascular remodeling, which is the focus of this Special Issue.

Across the spectrum of CVD, numerous cellular and molecular changes have been reported to contribute to cardiac dysfunction [2]. In two review papers by Frąk et al., and Franczyk et al., the molecular mechanisms underlying the development and progression of CVD are comprehensively overviewed, focusing on the role of endothelial dysfunction, inflammation and oxidative stress as major drivers of atherosclerosis, coronary artery disease, hypertension and coronary artery spasm [3,4]. Both papers provide a relevant basis to identify potential diagnostic and therapeutic targets across different CVD.

Consistent with the impact of oxidative stress in CVD development, Islam et al. demonstrate that the H_2S prodrug, SG-1002, confers protection against oxidative damage and hypertrophy in vitro [5]. At a mechanistic level, the authors suggest that the effects of SG-1002 are mediated by an increase in the H_2S levels and expression of antioxidant proteins, with a concomitant decrease in the expression of atrial natriuretic peptide (ANP) and brain natriuretic peptide (BNP).

As the primary trigger for atherosclerosis, endothelial dysfunction has been a matter of intense research. In this Special Issue, Motoji et al. demonstrate that the administration of *Candida albicans* water-soluble fraction (CAWS) to an apolipoprotein-E-deficient (ApoE$^{-/-}$) mice model induces a phenotype of Kawasaki disease (KD)-like vasculitis, exacerbating the formation of aortic plaque lesions [6]. Importantly, the authors show that administration of statins limited atherosclerosis and inflammatory cell infiltration, suggesting that statin therapy can be used to prevent cardiovascular events in KD patients.

Grounded on the important link between diet and vascular health, da Silva et al. investigated the impact of a ketogenic regimen in ApoE$^{-/-}$ mice [7]. They demonstrated that a ketogenic diet promoted systemic inflammation and atherosclerotic plaque burden, which was accompanied by tissue-specific changes in aquaporins' expression in the liver and adipose tissue. However, further studies are required to establish the link between nutritional ketosis and the regulatory role of aquaporins in inflammation during atherogenesis.

In addition to the mechanisms underlying endothelial dysfunction, it is important to have a significantly better knowledge of the pathways leading to the destabilization

Citation: Martins-Marques, T.; Coutinho, G.; Kiss, A. Editorial of the Special Issue: Cellular Mechanisms of Cardiovascular Disease. *Biomedicines* **2023**, *11*, 2494. https://doi.org/10.3390/biomedicines11092494

Received: 29 August 2023
Accepted: 4 September 2023
Published: 8 September 2023

Copyright: © 2023 by the authors. Licensee MDPI, Basel, Switzerland. This article is an open access article distributed under the terms and conditions of the Creative Commons Attribution (CC BY) license (https://creativecommons.org/licenses/by/4.0/).

and rupture of atherosclerotic plaques that ultimately cause myocardial infarction (MI). In this Special Issue, Puylaert et al. identified Gasdermin D as a major player in pyroptotic cell death in atherosclerotic lesions [8], and demonstrated that Gasdermin D deficiency did not prevent plaque formation in the thoracic aorta of ApoE$^{-/-}$ mice but delayed the progression of lesions in the brachiocephalic artery, with a concomitant decrease in plaque necrosis. Although additional evidence is required, this study suggests that Gasdermin D constitutes a valuable therapeutic target to increase atherosclerotic plaque stability.

The role of neutrophil extracellular traps (NETs) in atherothrombosis and plaque destabilization is thoroughly summarized in a systematic review by Nappi et al. [9]. The authors focus not only on the potential of using NETs as biomarkers of CVD, including acute coronary syndromes and peripheral arterial disease, but also on the therapeutic implications of targeting NETs in CVD.

Purinergic signaling plays a major role not only in regulating normal cardiovascular function, but also during pathological contexts, namely upon calcification of blood vessels and cardiac valves, which is the focus of the study carried out by Klauzen et al. [10]. In this paper, it is reported that valve interstitial cells (VIC) and valve endothelial cells (VEC) from stenotic human heart valves display significant changes in the expression levels of purinergic genes, which is associated with the pathological calcification of VIC, likely representing a valuable target in the search for an anti-calcification therapy.

The clinical benefits of using sodium–glucose cotransporter-2 inhibitors (SGLT-2i) have been shown and these stand a major therapeutic option for the treatment of heart failure (HF) independent of left ventricular ejection fraction (LVEF) and diabetes. However, the cardioprotective mechanisms of SGLT2i are still not fully known. In this Special Issue, Sullivan et al. focus on the importance of SGLT-2i in reducing edema in pre-clinical models of HF with reduced ejection fraction (HFrEF), highlighting the promising role of this strategy in improving the quality of life and reducing the number of hospitalizations of HFrEF patients [11].

It has been widely demonstrated that advanced glycation end-products (AGEs) contribute to a variety of microvascular and macrovascular complications. In this Special Issue, Heber et al. demonstrate that the levels of plasma methylglyoxal, a major AGE precursor, are increased in acute MI (AMI) patients following reperfusion via primary percutaneous coronary intervention [12]. Importantly, higher methylglyoxal levels within 24 h after AMI were associated with poorer left ventricular function after 4 days, suggesting that methylglyoxal constitutes a relevant and potential therapeutic target for post-MI remodeling and cardiac dysfunction.

In order to identify novel therapies for AMI, Docshin et al. characterized the mechanisms of early activation of regenerative processes in post-infarction cardiac tissue [13]. In this study, they demonstrated that cardiac mesenchymal cells isolated after MI in rats have an increased proliferative capacity, and a higher expression of the *Bmp2/Runx2* and Notch signaling genes, which have an important role upon cardiogenesis, paving the way for future studies aiming to ascertain its impact upon myocardial recovery in vivo.

Besides environmental and genetic causative factors, the epigenetic regulation of CVD-related genes is associated with the development and progression of multiple CVD [14]. In this Special Issue, Tarazón et al. observed a genome-wide hypomethylation in cardiac tissue samples of ischemic cardiomyopathy patients, which was related to changes in the DNMT3B system, the main DNA methyltransferase [15]. Overall, this study provides additional evidence to strengthen the potential of targeting epigenetic key enzymes for CVD treatment.

First described as an important epigenetic modification, by regulating chromatin structure and histone activity, protein acetylation is now known as a broader post-translational modification involved in the control of cell metabolism and enzymatic activity. In this Special Issue, Dubois-Deruy et al. discuss the role of lysine acetyltransferases and deacetylases in the pathophysiology of obesity, type 2 diabetes and HF, as well as the preclinical evidence demonstrating the cardioprotective potential of the pharmacological modulation of cardiac acetylation [16].

In addition to classical risk factors, CVD burden is now recognized to be aggravated due to the growing population of cancer survivors treated with cardiotoxic cancer therapies, which is discussed in two review papers collected in this Special Issue. Franczyk et al. consider evidence relating the use of receptor tyrosine kinase inhibitors (RTKIs) in treatment of renal cell carcinoma to the development of hypertension, myocardial ischemia and HF, highlighting the importance of defining proper surveillance strategies for high-risk patients [17]. Rocca et al. focus on mitochondrial dynamics as key determinants of anticancer drug-dependent cardiotoxicity, which constitute relevant therapeutic targets to prevent cardiomyocyte dysfunction or loss in cancer survivors [18].

Besides anticancer drugs, therapeutic regimens used for systemic disorders have also been reported to increase HF-associated hospitalizations, such as the dipeptidyl-peptidase-4 (DPP4) inhibitor saxagliptin, widely used to control type 2 diabetes. In the paper by Vörös et al., the mechanisms underlying saxagliptin cardiotoxicity are investigated [19]. The authors demonstrate that the levels of DPP4 and its substrate neuropeptide Y (NPY) are decreased in failing human hearts. In vitro experiments using fibroblast/cardiomyocyte co-cultures suggest that saxagliptin may interfere with NPY-mediated cardiac tissue remodeling in HF, but further studies are warranted to confirm this association.

Overall, this Special Issue brings together relevant cell-based and pre-clinical studies identifying novel mechanisms underlying cardiac and vascular remodeling and dysfunction, paving the way for the identification of more efficient diagnostic and therapeutic tools for a wide range of CVD.

Conflicts of Interest: The authors declare no conflict of interest.

References

1. Townsend, N.; Kazakiewicz, D.; Wright, F.L.; Timmis, A.; Huculeci, R.; Torbica, A.; Gale, C.P.; Achenbach, S.; Weidinger, F. Epidemiology of cardiovascular disease in Europe. *Nat. Rev. Cardiol.* **2021**, *19*, 133–143. [CrossRef] [PubMed]
2. Martins-Marques, T.; Hausenloy, D.J.; Sluijter, J.P.G.; Leybaert, L.; Girao, H. Intercellular Communication in the Heart: Therapeutic Opportunities for Cardiac Ischemia. *Trends Mol. Med.* **2020**, *27*, 248–262. [CrossRef] [PubMed]
3. Frąk, W.; Wojtasińska, A.; Lisińska, W.; Młynarska, E.; Franczyk, B.; Rysz, J. Pathophysiology of Cardiovascular Diseases: New Insights into Molecular Mechanisms of Atherosclerosis, Arterial Hypertension, and Coronary Artery Disease. *Biomedicines* **2022**, *10*, 1938. [CrossRef] [PubMed]
4. Franczyk, B.; Dybiec, J.; Frąk, W.; Krzemińska, J.; Kućmierz, J.; Młynarska, E.; Szlagor, M.; Wronka, M.; Rysz, J. Cellular Mechanisms of Coronary Artery Spasm. *Biomedicines* **2022**, *10*, 2349. [CrossRef] [PubMed]
5. Islam, R.K.; Donnelly, E.; Donnarumma, E.; Hossain, F.; Gardner, J.D.; Islam, K.N. H2S Prodrug, SG-1002, Protects against Myocardial Oxidative Damage and Hypertrophy In Vitro via Induction of Cystathionine β-Synthase and Antioxidant Proteins. *Biomedicines* **2023**, *11*, 612. [CrossRef] [PubMed]
6. Motoji, Y.; Fukazawa, R.; Matsui, R.; Nagi-Miura, N.; Miyagi, Y.; Itoh, Y.; Ishii, Y. Kawasaki Disease-like Vasculitis Facilitates Atherosclerosis, and Statin Shows a Significant Antiatherosclerosis and Anti-Inflammatory Effect in a Kawasaki Disease Model Mouse. *Biomedicines* **2022**, *10*, 1794. [CrossRef] [PubMed]
7. da Silva, I.V.; Gullette, S.; Florindo, C.; Huang, N.K.; Neuberger, T.; Ross, A.C.; Soveral, G.; Castro, R. The Effect of Nutritional Ketosis on Aquaporin Expression in Apolipoprotein E-Deficient Mice: Potential Implications for Energy Homeostasis. *Biomedicines* **2022**, *10*, 1159. [CrossRef] [PubMed]
8. Puylaert, P.; Van Praet, M.; Vaes, F.; Neutel, C.H.G.; Roth, L.; Guns, P.J.; De Meyer, G.R.Y.; Martinet, W. Gasdermin D Deficiency Limits the Transition of Atherosclerotic Plaques to an Inflammatory Phenotype in ApoE Knock-Out Mice. *Biomedicines* **2022**, *10*, 1171. [CrossRef] [PubMed]
9. Nappi, F.; Bellomo, F.; Avtaar Singh, S.S. Worsening Thrombotic Complication of Atherosclerotic Plaques Due to Neutrophils Extracellular Traps: A Systematic Review. *Biomedicines* **2023**, *11*, 113. [CrossRef] [PubMed]
10. Klauzen, P.; Semenova, D.; Kostina, D.; Uspenskiy, V.; Malashicheva, A. Purinergic Signaling in Pathologic Osteogenic Differentiation of Aortic Valve Interstitial Cells from Patients with Aortic Valve Calcification. *Biomedicines* **2023**, *11*, 307. [CrossRef] [PubMed]
11. Sullivan, R.D.; McCune, M.E.; Hernandez, M.; Reed, G.L.; Gladysheva, I.P. Suppression of Cardiogenic Edema with Sodium–Glucose Cotransporter-2 Inhibitors in Heart Failure with Reduced Ejection Fraction: Mechanisms and Insights from Pre-Clinical Studies. *Biomedicines* **2022**, *10*, 2016. [CrossRef] [PubMed]

12. Heber, S.; Haller, P.M.; Kiss, A.; Jäger, B.; Huber, K.; Fischer, M.J.M. Association of Plasma Methylglyoxal Increase after Myocardial Infarction and the Left Ventricular Ejection Fraction. *Biomedicines* **2022**, *10*, 605. [CrossRef] [PubMed]
13. Docshin, P.M.; Karpov, A.A.; Mametov, M.V.; Ivkin, D.Y.; Kostareva, A.A.; Malashicheva, A.B. Mechanisms of Regenerative Potential Activation in Cardiac Mesenchymal Cells. *Biomedicines* **2022**, *10*, 1283. [CrossRef] [PubMed]
14. Shi, Y.; Zhang, H.; Huang, S.; Yin, L.; Wang, F.; Luo, P.; Huang, H. Epigenetic regulation in cardiovascular disease: Mechanisms and advances in clinical trials. *Signal Transduct. Target. Ther.* **2022**, *7*, 200. [CrossRef] [PubMed]
15. Tarazón, E.; Pérez-Carrillo, L.; Giménez-Escamilla, I.; García-Manzanares, M.; Martínez-Dolz, L.; Portolés, M.; Roselló-Lletí, E. DNMT3B System Dysregulation Contributes to the Hypomethylated State in Ischaemic Human Hearts. *Biomedicines* **2022**, *10*, 866. [CrossRef] [PubMed]
16. Dubois-Deruy, E.; El Masri, Y.; Turkieh, A.; Amouyel, P.; Pinet, F.; Annicotte, J.S. Cardiac Acetylation in Metabolic Diseases. *Biomedicines* **2022**, *10*, 1834. [CrossRef] [PubMed]
17. Franczyk, B.; Rysz, J.; Ławiński, J.; Ciałkowska-Rysz, A.; Gluba-Brzózka, A. Cardiotoxicity of Selected Vascular Endothelial Growth Factor Receptor Tyrosine Kinase Inhibitors in Patients with Renal Cell Carcinoma. *Biomedicines* **2023**, *11*, 181. [CrossRef] [PubMed]
18. Rocca, C.; De Francesco, E.M.; Pasqua, T.; Granieri, M.C.; De Bartolo, A.; Gallo Cantafio, M.E.; Muoio, M.G.; Gentile, M.; Neri, A.; Angelone, T.; et al. Mitochondrial Determinants of Anti-Cancer Drug-Induced Cardiotoxicity. *Biomedicines* **2022**, *10*, 520. [CrossRef] [PubMed]
19. Vörös, I.; Onódi, Z.; Tóth, V.É.; Gergely, T.G.; Sághy, É.; Görbe, A.; Kemény, Á.; Leszek, P.; Helyes, Z.; Ferdinandy, P.; et al. Saxagliptin Cardiotoxicity in Chronic Heart Failure: The Role of DPP4 in the Regulation of Neuropeptide Tone. *Biomedicines* **2022**, *10*, 1573. [CrossRef] [PubMed]

Disclaimer/Publisher's Note: The statements, opinions and data contained in all publications are solely those of the individual author(s) and contributor(s) and not of MDPI and/or the editor(s). MDPI and/or the editor(s) disclaim responsibility for any injury to people or property resulting from any ideas, methods, instructions or products referred to in the content.

biomedicines

MDPI

Review

Pathophysiology of Cardiovascular Diseases: New Insights into Molecular Mechanisms of Atherosclerosis, Arterial Hypertension, and Coronary Artery Disease

Weronika Frąk, Armanda Wojtasińska, Wiktoria Lisińska, Ewelina Młynarska *, Beata Franczyk and Jacek Rysz

Department of Nephrology, Hypertension and Family Medicine, Medical University of Lodz, ul. Zeromskiego 113, 90-549 Lodz, Poland

* Correspondence: emmlynarska@gmail.com; Tel.: +48-(042)-6393750

Abstract: Cardiovascular diseases (CVDs) are disorders associated with the heart and circulatory system. Atherosclerosis is its major underlying cause. CVDs are chronic and can remain hidden for a long time. Moreover, CVDs are the leading cause of global morbidity and mortality, thus creating a major public health concern. This review summarizes the available information on the pathophysiological implications of CVDs, focusing on coronary artery disease along with atherosclerosis as its major cause and arterial hypertension. We discuss the endothelium dysfunction, inflammatory factors, and oxidation associated with atherosclerosis. Mechanisms such as dysfunction of the endothelium and inflammation, which have been identified as critical pathways for development of coronary artery disease, have become easier to diagnose in recent years. Relatively recently, evidence has been found indicating that interactions of the molecular and cellular elements such as matrix metalloproteinases, elements of the immune system, and oxidative stress are involved in the pathophysiology of arterial hypertension. Many studies have revealed several important inflammatory and genetic risk factors associated with CVDs. However, further investigation is crucial to improve our knowledge of CVDs progression and, more importantly, accelerate basic research to improve our understanding of the mechanism of pathophysiology.

Keywords: cardiovascular disease; atherosclerosis; arterial hypertension; coronary artery disease; inflammation; matrix metalloproteinases; oxidative stress; vascular endothelium dysfunction; genetic factor

Citation: Frąk, W.; Wojtasińska, A.; Lisińska, W.; Młynarska, E.; Franczyk, B.; Rysz, J. Pathophysiology of Cardiovascular Diseases: New Insights into Molecular Mechanisms of Atherosclerosis, Arterial Hypertension, and Coronary Artery Disease. *Biomedicines* **2022**, *10*, 1938. https://doi.org/10.3390/biomedicines10081938

Academic Editors: Tânia Martins-Marques, Gonçalo F. Coutinho and Attila Kiss

Received: 30 June 2022
Accepted: 6 August 2022
Published: 10 August 2022

Publisher's Note: MDPI stays neutral with regard to jurisdictional claims in published maps and institutional affiliations.

Copyright: © 2022 by the authors. Licensee MDPI, Basel, Switzerland. This article is an open access article distributed under the terms and conditions of the Creative Commons Attribution (CC BY) license (https://creativecommons.org/licenses/by/4.0/).

1. Introduction

Cardiovascular diseases (CVDs) are a group of disorders of the heart and blood vessels [1]. They are a set of heterogeneous diseases whose underlying cause of development is most often atherosclerosis [2]. CVDs are chronic diseases that gradually evolve throughout life and remain asymptomatic for a long time [3]. Moreover, CVDs are the leading cause of morbidity and mortality in patients worldwide [4]. In Europe, CVDs are responsible for 45% of deaths [5], thus being particularly important for public health. Atherosclerosis, coronary artery disease (CAD), and arterial hypertension (AH) are the leading causes of CVDs [6].

Atherosclerosis is the main cause of cardiovascular-related death worldwide [7]. It is a thickening and hardening of the arterial wall, accompanies aging, and is related to major adverse impact on the cardiovascular system and various other diseases [8]. Elevated plasma cholesterol level (>150 mg/dL) is a major cause of the development of atherosclerosis [9].

CAD is a common heart condition in which we can observe the narrowing or blockage of major blood vessels—coronary arteries. CAD is caused primarily by plaque formation within the intima of the vessel wall [10], with plaque being defined as a fatty material growing inside intima along with a severe inflammation, especially if the inflammation is chronic. This in turn causes difficulties in supplying the cardiomyocytes with enough blood, oxygen, and nutrients [11]. As a result, atherosclerotic plaque may erode or rupture, initially resulting in thrombosis and then a closure of the vessel, leading to myocardial infarction, stroke, limb ischemia, and death [12]. The other factors causing this condition are a diseased endothelium, low-grade inflammation, and lipid accumulation [13].

AH is one of the most common CVDs. AH causes few or no symptoms, but is an important risk factor for a myocardial infarction, stroke, renal failure, and peripheral vascular disease [14]. The diagnosis of AH, in accordance with the most significant guidelines, is diagnosed when a person's systolic blood pressure (SBP) in the office or clinic is ≥140 mm Hg and/or their diastolic blood pressure (DBP) is ≥90 mm Hg following a repeated examination [15]. CVDs are caused by multiple factors. Some of them are unvarying, such as age, gender, and genetic background, whereas others could be variable and, therefore, reduced (smoking, physical inactivity, poor dietary habits, elevated BP, type 2 diabetes, dyslipidemia, and obesity) [16].

In this review, we summarize the available evidence of the pathophysiological implications of CVDs, focusing on atherosclerosis, CAD, and AH. It is particularly essential to explain the mechanisms of their formation and progression.

2. Coronary Artery Disease and Atherosclerosis

CAD and atherosclerosis are discussed first, due to their broad subject matter and mutual implication. The pathophysiological basis of these diseases is constantly being researched in order to finally find the reasons behind their formation and what factors additionally exacerbate the ongoing processes and may directly contribute to their induction. Understanding the exact course of the entire pathophysiological and pathogenic process will allow for accurate prevention and diagnosis of both diseases. Cardiovascular diseases are classified as civilization diseases [17]; hence, it is important to create accurate and effective algorithms that will help doctors during their work. Those that already exist require further improvement and improvement, due to the continuous discoveries and introduction of new pharmacotherapeutic solutions.

In our review, we focus on endothelial dysfunction as one of the first and most important causes of the processes leading to CAD and atherosclerosis [18]. In the case of both diseases, these processes constitute a starting point for further research and implications in the course of the disease development. This brings us to the remainder of the paper, i.e., inflammatory processes involving diseased tissue and oxidative factors.

We also do not forget about the genetic basis; dynamically developing research shows completely new faces of these diseases known to us and makes it possible to reflect on the real causes of their diversity and changeable forms, as well as create new possibilities for better diagnostics and considering whether the current judgement that the lifestyle and drugs mainly allow controlling the disease-causing process. They allow the use of new solutions and deepen our knowledge in this subject.

2.1. *Endothelial Dysfunction in Atherosclerosis*

In arterial vasculature there are areas (branch points, bifurcations, and major curvatures—arterial geometry), which are much more prone to atherosclerotic lesions [19]. The mechanical forces such as a turbulent flow, which is related to the geometry and shape of vessels, also influence the endothelial cell [19].

Atherosclerosis occurs in these regions as a result of differences in flow, which is present at sites of low shear stress, turbulence, and oscillating flow. We do not maintain that these factors cause atherosclerosis but rather that they "prime the soil" in which lesions start to develop [20]. Endothelial cells are exposed to various degrees and types of shear stress, which have influence on their shape, intracellular signaling, and gene expression [21]. In the regular state, which is the quiescent state of the endothelium, nitric oxide (NO) is produced in order to bind to cysteine groups in NF-κB and the mitochondria, which inhibit cellular processes. Moreover, the endothelial layer is covered by a glycocalyx, which is layer of proteoglycans and extracellular matrix components, involved in transendothelial transport, e.g., of lipoproteins, which can be lost or reduced in inflammation due to plasminogen activator inhibitor [21]. In homeostasis, endothelial cells prevent platelet activation, blood clotting, and leukocyte adherence by secreting substances such as NO, prostacyclin, t-PA, and antithrombin III [21]. When inflammation occurs there is an increase in the number of adhesion molecules (E-selectin, ICAM, and VCAM), which participate in the infiltration of leukocytes through the endothelial layer [22]. With leukocytes, lipoproteins penetrate the endothelium, and they are trapped in the subendothelial space and oxidatively modified. Endothelial dysfunction in modern cardiovascular medicine is described as changes in the production and availability of endothelial-derived NO, prostacyclin, and endothelin, as well as their impact on vascular reactivity. In this case, reactive oxygen species (ROS) such as H_2O_2 reach the regulatory molecules, which leads to the activation of the cells [23]. The endothelial membrane is permeable to compounds such as NO and H_2O_2, which causes the activation of the transcription factors and protease. Moreover, production of endothelial ROS may be triggered by inflammation and cells that participate in this process, such as leukocytes and growth factors. Other mechanisms that cause endothelial dysfunction are the formation of peroxynitrite, NO synthase uncoupling, prostacyclin formation inhibition, endothelin expression stimulation, and reduced NO signaling due to the inhibition of soluble guanylate cyclase activity [24]. All these mechanisms promote a vasoconstrictive and procoagulant milieu. Moreover, endothelial cells can make a transition to the mesenchymal cells [21]. Consequently, the extracellular matrix deposits between cells and dysregulates the junctional proteins (e.g., occludin and claudin-5), which lose cell–cell contact [21]. As a result, the endothelium loses its integrity with media, and this causes a higher activity in that field. Then, the changes extend beyond NO metabolism reactivity, including increased level of permeability for lipoprotein, oxydation, leukocyte adhesion and accumulation, and altered extracellular matrix metabolism, with all of these accumulating in the arterial wall [19]. In this way, macrophages, cholesterol, and inflammatory cells access the media, and atherosclerosis begins. Moreover, the mechanical forces such as a turbulent flow, which is related to the geometry and shape of vessels, also influence the endothelial cell [19].

2.2. *Inflammatory and Oxidising Factors in Atherosclerosis*

Following the mechanisms of endothelial dysfunction, we can see that inflammatory factors also play a huge role in the development of this pathology. Atherosclerosis is characterized by the retention of lipids and inflammatory cells such as macrophages, T lymphocytes, and mast cells in damaged arterial wall, the intima [25]. Modified lipids activate inflammatory cells in the intima, producing chemokines and cytokines such as tumor necrosis factor (TNF-alpha), interleukin -1, -4, and -6, and interferon-gamma, which activate other leukocytes, endothelial cells, and adhesion molecules, especially vascular cell adhesion molecule-1 (VCAM), intercellular adhesion molecule-1 (ICAM), and E-selectin, on the endothelial surface. These, in turn, recruit other inflammatory cells. As a result, the monocyte-derived macrophages release enzymes in order to modify the lipoproteins. These modified lipoproteins become atherosclerotic plaques. Then, macrophages absorb and build in the cholesterol-rich lipoproteins from LDL, as well as secrete pro-oxidant substances, which contribute to the process of atherosclerosis: ROS and RNS (reactive nitrogen species). These are the same compounds that participate in the endothelial dysfunction, which aggravates the condition of the endothelium, indicating that the process is self-perpetuating. This is not desirable, because this damage of the cellular functions of biomolecules (such as proteins, carbohydrates, and lipids) can result in lipid peroxidation and LDL oxidation. As we know, oxidized phospholipids trigger inflammation, because of the extensive binding to the Toll-like receptors, which can activate the transcription factors nuclear factor-κB (NF-κB) cytokines which trigger proinflammation; hence, oxLDL is called a clinical marker of plaque inflammation [26]. OxLDL irritates endothelial cells, increasing the production of adhesion molecules. ROS and RNS convert the LDL-C to the OX-LDL, which are built into the intimal layer. The inner layer also migrates the muscle cells from the media and proliferates itself. When all these processes combine, the atherosclerotic plaque is created, featuring fiber tissue, muscle cells, and many inflammatory cells. Accelerated cell turnover is likely to lead to an enhanced macromolecular permeability, increasing lipid uptake in the regions with a disturbed flow. This in turn would lead to the atherosclerotic phenotype expression of VEGF increasing in response to low shear stress, leading to greater endothelial permeability [27]. In addition, in the vessels, hyperglycemia promotes the overproduction of ROS by the mitochondrial electron transport chain. Excess superoxide leads to DNA strand breakage and activation of nuclear poly ADP ribose polymerase (PARP). These processes inhibit glyceraldehyde-3-phosphate dehydrogenase (GAPDH), shunting the early glycolytic intermediates into the pathogenic signaling pathways during inflammation [28].

2.3. Epigenetic Factors in Atherosclerosis

The first major epigenetic mechanism that contributes to the complexity of atherosclerosis is DNA methylation, which is catalyzed by DNA methyltransferase 1 (DNMT1) and 3b (DNMT3b) DNA methylation [27]. Moreover, DNA demethylation is also an important mechanism explaining the pathogenesis of atherosclerosis. During studies on mice, the *TET2* overexpression significantly reduced atherosclerotic lesion formation, likely by oxidatively demethylating 5meC to 5hmC in the endothelial vessel wall. This *TET2*-induced rescue occurs via the upregulation of autophagy, as *TET2* overexpression decreases the methylation level of the promoters of autophagic flux-related genes [29]. Moreover, the loss of *TET2* functions in hematopoietic cells and myeloid cells enhanced atherosclerosis in mice, as shown in [30]. Bone marrow was transplanted from control mice into an atherosclerosis-prone *Ldlr−/−* recipient mice, and a diet high in cholesterol was introduced. After 5, 9, and 13 weeks on the diet, the recipients of *Tet2−/−* marrow had 2.0-fold, 1.7-fold, and 1.4-fold larger lesions in the arteries. DNA methylation and histone acetylation are key processes in regulating the expression of inflammatory cytokines and chemokines in atherosclerosis. This involves methylation at the C5 position of cytosine residues in a *CpG* dinucleotide context, exerted by DNA methyltransferases (DNMTs). DNMTs are capable of both methylation and demethylation, making the modification reversible, but those modifications are reserved for the Tet methylcytosine dioxygenases (TET1, 2, and 3) [29]. In addition, studies have shown that other genes, *DNMT3A*, *JAK2*, and *ASXL1*, which mutated, increase the risk of incident coronary heart disease 12-fold in *JAK2* V617F and 1.7-fold to 2.0-fold in other genes mentioned above [30]. Mutations in the *DNMT3A* and *TET2* influence DNA methylation, whereas those in the *ASXL1* alter histone modifications, thereby influencing clonal expansion of hematopoietic stem cells [31]. This phenomenon was found to influence risk of CAD [32].

2.4. Endothelial Dysfunction in CAD

The vascular endothelium is the layer of cells lying under the epithelium lining the inside of the vessel and the muscular layer, which is a boundary between the circulating blood and the vascular wall. Its cells are specialized in maintaining vascular homeostasis, which is crucial for the proper functioning of organs, especially the heart. Through its role in signal transduction and as a source of many vasoactive substances, it is their key regulator. The vascular endothelium reacts to physical and chemical stimuli through the release of autocrine and paracrine vasoactive agents. Factors of endothelial origin regulate surface tension and cell adhesion, including platelet activation and leukocyte adhesion, smooth muscle cell proliferation, and vascular wall inflammation. The endothelium is considered to be a strong indicator of cardiovascular function and fitness. Its dysfunction is considered to be the earliest marker of atherosclerosis and, in effect, CAD.

First, we focus on the critical process of the NO signaling pathway [33]. Nitric oxide is a gas with relatively small particles that strongly dilates blood vessels and has additional anti-inflammatory and antioxidant properties [34]. It is synthesized by three distinct subtypes of the NO synthase (NOS) enzyme, each with unique expression patterns and functional properties: neuronal NOS (nNOS, NOS1), inducible NOS (iNOS, NOS2), and endothelial NOS (eNOS, NOS3). Broadly, these proteins catalyze the production of NO and l-citrulline from L-arginine and O_2, using electrons donated from dihydronicotinamide adenine dinucleotide phosphate (NADPH). Finally, in the presence of heme and tetrahydrobiopterin (BH_4), NOS monomers form homodimers, which are capable of using the donated NADPH electrons to catalyze the two-step oxidation of L-arginine to L-citrulline and NO. The expression of the nicotinamide adenine dinucleotide phosphate oxidase in the vessel wall with the consequent overproduction of NO has been proposed as an initial step in the chronic dysregulation of normal NO production by eNOS, which is characteristic of the monomeric forms of eNOS. In effect, eNOS produces a superoxide, rather than NO. The superoxide reacts with NO to form peroxynitrite, which in turn increases the uncoupling of eNOS and further superoxide production [35,36]. This again promotes eNOS uncoupling and is itself a mediator of effects [37]. When uncoupled, eNOS switches from its oxygenase-induced oxidant excess, and then exerts a deleterious effect on the endothelial and vascular function.

In the second step, NOS catalyzes the oxidation of N^{ω}-hydroxy-L-arginine to L-citrulline, thereby releasing NO [38,39]. For example, nNOS and eNOS are highly dependent on Ca^{2+}-activated CaM for homodimerization and activity, whereas iNOS is minimally dependent on calcium concentration. These nuances have critical functional effects. NADPH oxidase (NOX) is another element that has received much attention as a key element in the development of vascular dysfunction. NOX has the primary function of producing reactive oxygen species (ROS) and is considered the main source of the ROS production in endothelial cells. The enzymatic production of NO by eNOS is critical in mediating the endothelial function, and oxidative stress can cause eNOS dysregulation and endothelial dysfunction [40].

Taking this into account, the endothelial dysfunction is directly related to a decreased production and sensitivity of cells to NO. As a result, we have an effective disturbance in the functioning of the entire vessel and its homeostasis, which leads to an observation of prothrombotic and proinflammatory phenomena, along with lower susceptibility of the blood vessel wall.

Another factor of interest is phospholipase A2 and its influence on the endothelial dysfunction. Lp-PLA2 (lipoprotein-bound phospholipase A2), also known as platelet-activating acetylhydrolase, is a vascular-specific inflammatory enzyme mainly produced by macrophages, lymphocytes, and foam cells in the atherosclerotic plaques. The circulation of Lp-PLA2 is mainly associated with apolipoprotein B-containing lipoproteins and, therefore, closely related to low-density lipoproteins (LDLs). Lp-PLA2 can trigger proinflammatory and proatherogenic properties in the vascular wall. The enzyme hydrolyzes oxidized phospholipids on the LDL particles in the intima of the artery, producing two highly inflammatory mediators with proinflammatory and atherosclerotic effects. Elevated levels of Lp-PLA2 correlate with arterial stiffness in patients with a stable CAD (Figure 1), regardless of the risk factors and pharmacotherapy [41].

2.5. Inflammation in CAD

Systemic vasculitis is a term referring to a group of diseases characterized by inflammation and fibrinoid necrosis of blood vessel walls. The underlying pathogenesis involves many mechanisms such as cell-mediated inflammation, immune complex (IC)-mediated inflammation, and ANCA-mediated inflammation, For example, inflammation in GCA is mostly a T-cell-driven process in which dendritic cells present antigens in blood vessel walls. These T-cells activate other inflammatory cells such as monocytes and macrophages, which, as a result, release proinflammatory cytokines (interleukin-1, interleukin-6, and interferon-γ). Inflammation in diseases such as polyarteritis nodosa and cryoglobulinemia is driven by IC deposition (antibody-mediated IC formation, microaneurysms) [42], whereas Wegener's granulomatosis, Churg–Strauss syndrome, microscopic polyangiitis, and necrotizing glomerulonephritis result from interactions between antibodies and enzymes within inflammatory cells, which is typical for ANCA-related vasculitis [43]. As a consequence of these processes, manufactured antibodies and immune complexes attach to the inner layer of blood vessels. This is a cause of aggravated ET-1 release, which, in a positive feedback loop, recruits more monocytes and macrophages. Vasculitis, as a very diverse group of diseases, has many cellular mechanisms that we mentioned above and affects different arteries and veins, in terms of both distribution and size [44]. The size of vessels is also crucial in terms of factors sustaining inflammation. Vasculitis of small veins is related to involvement of the endothelium, necrose (which is associated with ANCA), damage of the arterial walls, aneurysm formation, and hemorrhage. It is also characterized by leukocytoclasia, which is the excessive accumulation of neutrophils. Moreover, in vasculitis of large systemic vessels, the response to the inflammation leads to thickening of the intima, restenosis, and overall remodeling. These changes are manifested in the malfunctioning of blood vessels and subsequent events such as infarction and hemorrhage [44].

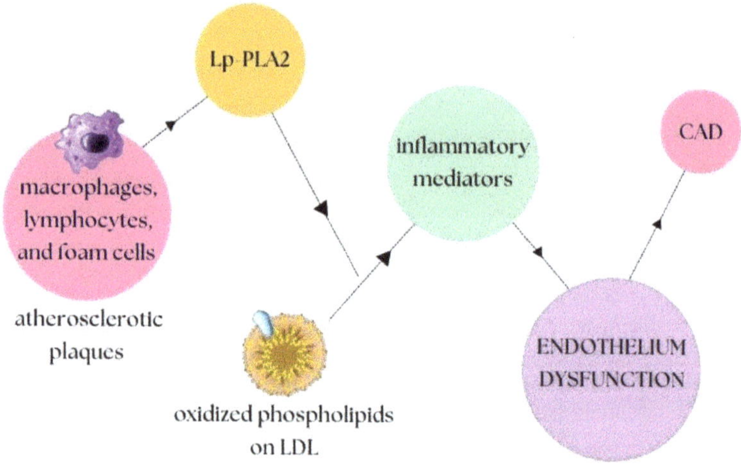

Figure 1. Lp-PLA2-dependent activation cycle [40]. The macrophages, lymphocytes, and foam cells present in the atherosclerotic plaques have an influence on the increased level of Lp-PLA2, which in turn catalyzes a reaction that, in the presence of oxidized phospholipids on LDL, is a direct contribution to the secretion of an increased amount of inflammatory mediators, which in turn leads to endothelial dysfunction and further CAD.

2.6. Genetic Background in CAD

In recent years, there has been substantial research showing correlations between genetic factors and endothelial function and dysfunction, which in turn are associated with an increased risk of developing CAD. In our work, we selected genes that, in our opinion, have gained attention in recent years and have shown a significant impact on the development of CAD. However, it should be remembered that, since 2007, in addition to all the genes known so far that have a confirmed impact on CAD, new research has resulted in nearly 60 distinct genetic loci [45].

As mentioned before, atherosclerosis underlies CAD. Previous studies have shown that it has an important but poorly understood and defined genetic component [41,46]. An earlier genome-wide association study (GWAS) identified many loci associated with an increased risk of CAD. A recent study by Redouane Aherrahrou, Liang Guo, and others accurately described the characteristics of atherosclerosis that are associated with the migration and proliferation processes in the vascular smooth muscle (VSMC) cells. The phenotypic variability shown was of great importance here; more specifically, four loci directly related to the atherosclerosis in VSMC were identified. Moreover, as many as 79 out of 163 loci associated with CAD are associated with one of the VSMC phenotypes [47].

One of them is the chromosome 1q41 locus, which harbors *MIA3* protein. The G allele of the lead risk SNP rs67180937 is associated with a lower VSMC *MIA3* expression and a lower proliferation. The *MIA3* protein, which plays a role in the secretion of collagen, is the likely cause of the genetic basis of CAD at its 1q141 locus. Silencing this protein resulted in a reduction in the VSMC profiling, with a lentivirus used for this purpose. This protein also significantly impacts the thickness of the fibrous cap of the atherosclerotic plaque, both in humans and in mice [48].

Next one, *JCAD* (junctional cadherin 5-associated, also known as *KIAA1462*, encoding a junctional protein associated with CAD) is one of more than 160 GWAS-identified genes [49]. *JCAD* is a protein that binds cells in the endothelium and is responsible for the regulation of pathological angiogenesis, less frequently than its development [50]. CAD depletion also increased EC apoptosis and reduced EC proliferation, migration, and angiogenesis [51]. *JCAD* depletion inhibits the activation of the YAP/TAZ pathway and the expression of further proatherogenic genes including CTGF and Cyr61. Proteomic studies suggest that *JCAD* regulates the YAP/TAZ activation by interacting with the actin-binding protein TRIOBP, thus stabilizing stress fiber formation. In addition, endothelial *JCAD* expression was increased in murine and human atherosclerotic plaques [52].

Moreover, the SIRT1 protein (Sirtuin 1) plays an important role in regulating the physiological mechanisms taking place in the cell, consequently influencing the mechanisms against CAD. SIRT1 is a cardioprotective molecule due to it regulating the expression of eNOS. It regulates angiogenesis, and it protects the endothelium against dysfunctional changes and damage to the heart muscle resulting from a reduced perfusion and ischemia. Suppression of the SIRT1 causes monocyte affinity due to endothelial dysfunction [53]. The gene encoding transcription factor *TCF21* has been linked to CAD risk by GWAS in multiple racial/ethnic groups. *TCF21* antagonizes the MYOCD/SRF pathway through multiple mechanisms, further establishing a role for this CAD-associated gene in smooth muscle cells [54]. Two genes associated with high susceptibility of atherosclerotic plaque were also found. These two genes changed at different timepoints after myocardial infarction, and both had the lowest prognosis of heart failure when expressed at low levels. *TLR2* and *CD14* are closely associated with the worsening of CAD, the instability of atherosclerotic plaques, and the prognosis of heart failure after myocardial infarction [55]. Other studies revealed an atheroprotective role of *SVEP1*. The deficiency of wildtype *SVEP1* increased the endothelial *CXCL1* expression, leading to an enhanced recruitment of proinflammatory leukocytes from blood to plaque. Consequently, elevated vascular inflammation resulted in an enhanced plaque progression in the *SVEP1* deficiency (Figure 2) [56]. An intronic region of the disintegrin and metalloproteinase with thrombospondin motifs-7 (*ADAMTS7*) is one of the two crucial factors that can reduce the wildtype *SVEP1* [57]. *ADAMTS7* is one of the many proteins involved in the remodeling of the blood vessel walls due to the properties that dissolve their substrate proteins [57]. Increased expression of this gene is associated with an increased proteolytic activity and a migration of smooth muscle cells, which favors the abovementioned remodeling properties, as demonstrated on the smooth muscle cells in mice [58].

Two more genes have attracted our attention: *NOS3* and *GUCY1A3* and their common variants. Their presence significantly contributes to a reduction in the level of BP, which is known to be one of the important factors in reducing the occurrence of CAD [59]. Loss-of-function mutations in GUCY1A3 and CCT7 result in a reduced level of α1 subunit of soluble guanylyl cyclase (α1-sGC), as well as β1-sGC protein content, and they impair the soluble guanylyl cyclase activity, which was correlated with risk of myocardial infarction in a large family enriched for premature CAD^2 [60].

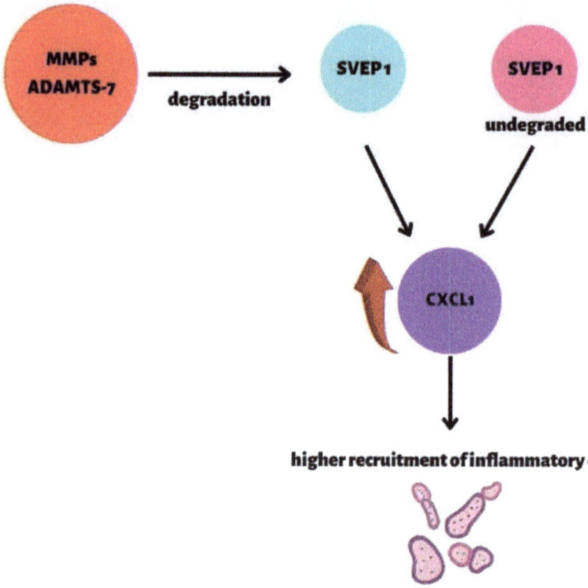

Figure 2. The role of *SVEP1* in atherosclerosis and CAD [56]. *CXCL1* expression is dependent on the presence of *SVEP1*. Only in the presence of *SVEP1* is the *CXCL1* expression silenced. Two factors have a significant influence on the expression of *CXCL1*: MMPs and ADAMTS- 7 (specific metalloproteinases), which can reduce the wildtype *SVEP1*; mutant *SVEP1* (*SVEP1_p.D2702G* missense variant). Consequently, the secretion and recruitment of inflammatory cells are increased.

The following genes, when inactivated, show increased risk of CAD: *LDRL* (low-density lipoprotein receptor), *APOA5* (apolipoprotein A-V) [61], and *LPL* (lipoprotein lipase) [45]. Abnormal mutations in *APOA5* increase the risk of CAD due to this gene coding a protein responsible for the increased activity of *LPL* [61]. The opposite effect is shown by mutations in the *APOC3* and *ANGPTL4* genes, which are responsible for the inhibition of lipoprotein lipase [62,63].

3. Arterial Hypertension

3.1. Matrix Metalloproteinases in AH

The remodeling of the vascular extracellular matrix (ECM) during hypertensive impairment has been associated with the involvement of matrix metalloproteinases (MMPs) and their tissue inhibitors (TIMPs). Several MMPs and TIMPs might participate in the vascular remodeling associated with the AH. Matrix metalloproteinases (MMPs) are a family of zinc-dependent endopeptidases, involved in many physiological and pathological processes, particularly tissue repair and modulation, cellular differentiation, cell mobility, angiogenesis, and cell proliferation, migration, and apoptosis. The dysfunction of MMP activity leads to a progression of various pathologies, tissue destruction, fibrosis, and matrix weakening [64]. It is worth mentioning that MMPs are also involved in the development of hypertension-mediated damage to the vascularity, heart, and kidneys, leading to organ failure and cardiovascular complications (Figure 3) [65]. Endogenous tissue inhibitors of MMPs (TIMPs) are crucial in restraining the degradation of ECM. Furthermore, the pathological processes in the vessel wall might be provoked by an excessive amount of MMPs, due to an imbalance between MMPs and TIMPs [66].

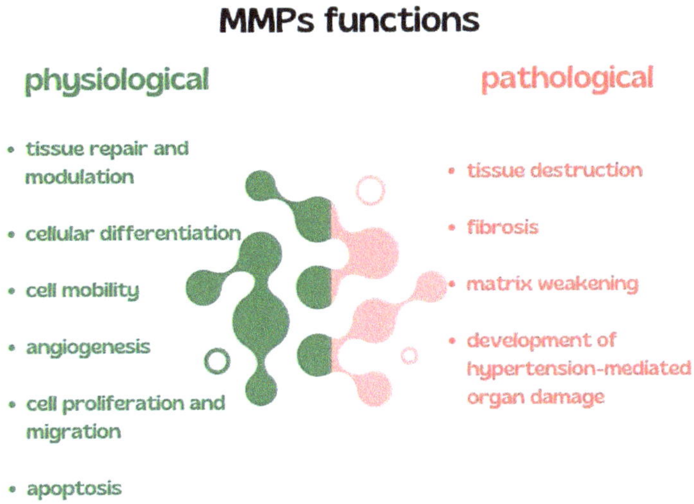

Figure 3. Summary of physiological and pathological functions of MMPs. MMPs, matrix metalloproteinases.

The results of a meta-analysis conducted by Marchesi et al. showed increased plasma levels and activities of MMP-2, MMP-9, and TIMP-1 among hypertensive patients [67]. Furthermore, Bisogni et al. found that a high BP along with the development and progression of CVDs correlated with increased levels of MMP-2 and MMP-9. It is also worth mentioning that MMP-9 is involved in the vascular remodeling, thus leading to the perpetuation of the elevated BP [65]. Additionally, Kostov et al. reported elevated MMP-1 levels and increased collagen degradation in patients with AH [68]. Thus, the impaired vascular remodeling and aggravated proteolytic activity might result from an MMP/TIMP imbalance in vascularity, principally in the intima and media of the vessel wall [16]. On the contrary, a study by Basu et al. attempted to demonstrate a crucial role of the TIMP-3, which is the protection of the arterial ECM in response to Ang II [69]. Furthermore, Pushpakumar et al. observed worsening of kidney functions and renovascular remodeling due to an increased activity of MMP-9 among hypertensive mice with TIMP2 deficiency [70]. These alterations in clinical AH suggest an important role for MMPs in AH. What is interesting, the determination of the MMPs and TIMPs in the serum can be used as a noninvasive approach to diagnosing and monitoring the structural changes in the cardiovascular system in AH [71].

Alterations in the structure, along with the degradation of several components of the vascular wall, including ECM, are critical during AH. The ECM is critical for maintaining the homeostasis in vasculature. It is thought to support stability and vascular cell behaviors [72]. A study by Cui et al. indicated that MMPs cause a degradation of the ECM proteins such as collagen or elastin; therefore, MMPs disturb the structural integrity of the vascular wall and lead to a decreased elasticity of the vascular wall [73].

Vascular ECM remodeling during AH applies effects on the functional and structural alterations in the vascular smooth muscle cells (VSMCs). Wang et al. demonstrated that MMPs promote VSMC growth and proliferation by activating growth factors such as insulin growth factor-1, transforming growth factor-h, and heparin-bound epidermal growth factor, in addition to promoting interactions between VSMC and these growth factors [16].

Martínez et al. found that, in rats, MMP-2 may regulate BP by destroying the vasodilator peptides adrenomedullin (AM) 1–52, 8–52, and 11–52 (Figure 4). AM plays a role in the regulation of BP. MMP-2 cleaves the vasodilator peptide AM (1–52, 8–52, and 11–52) into smaller peptides AM (11–22), which act as the vasoconstrictors. MMP-2 activity may be related to the development of AH, both by reducing the levels of the potent vasodilator AM (1–52, 8–52, and 11–52) and by generating hypertensive molecules [74].

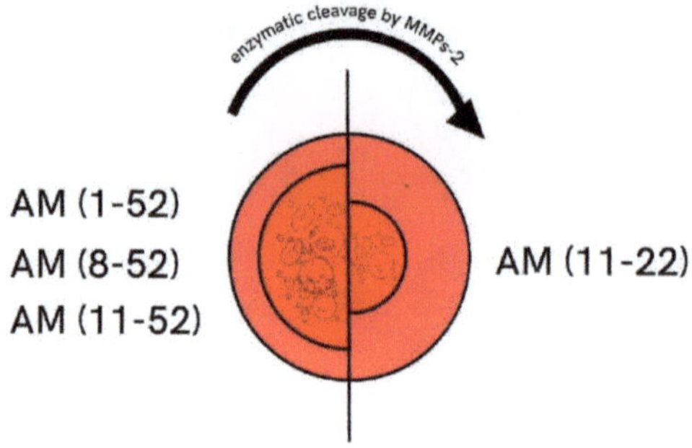

Figure 4. MMP-2 cleavage of AM. Adrenomedullin (AM 11–52) is a peptide that affects vessels and leads to its vasodilation. MMP-2 regulates BP by reducing the AM (1–52, 8–52, and 11–52) into smaller peptides AM (11–22), which act as vasoconstrictors. Therefore, MMP-2 may worsen vascular function and increase BP in hypertensive patients. Numbers in brackets are the peptide length before and after the MMP2 cleavage. MMP-2, matrix metalloproteinase 2; AM, adrenomedullin; BP, blood pressure.

In addition, the MMPs may play a role in hypertensive complications such as intracranial hemorrhage. A study by Wakisaka et al. found increased levels of MMP-9 in the endothelial cells and ECM of cerebral vessels that are considered to be associated with a spontaneous intracranial hemorrhage in the hypertensive rat models [75].

MMPs/TIMPs are involved in the regulation of the ECM metabolism, which plays a significant role with regard to the maintenance of tissue integrity [72].

3.2. *Immune System in AH*

Both innate and adaptive immune cells have been implicated in the development of AH. The majority of studies correlated the presence of AH with elevated levels of circulating inflammatory markers, cytokines, and antibodies. The findings by Mirhafez et al. showed an association between the concentrations of several cytokines and AH (an increased level of IL-6, IL-1β, IL-1α, IL-18, IL-2, IL-8, TNF-α, IFN-γ, C-reactive protein (CRP), and MCP-1, and a decreased level of IL-10) among patients with AH [76].

Sesso et al. evaluated the relationship between IL-6 and CRP and the risk of developing AH in a nested case–control study of 400 women. The results showed that IL-6 was indifferently associated with while CRP was firmly related to the AH risk [77]. Furthermore, CRP has been found to be a mediator in the development of endothelial dysfunction, vascular stiffness, and elevated BP [78]. Kong et al. investigated CRP gene polymorphisms. The results showed that plasma levels of CRP could predict the development of AH. On the contrary, the relationship between genotype and CRP levels was not associated with a change in AH risk [79]. Lima et al. demonstrated that IL-10 inhibits the pressor activity of Ang II and vascular dysfunction associated with AH, while also modulating the RhoA/Rho kinase pathway. Strategies to increase IL-10 levels during AH may enhance the benefits provided by regular treatments [80]. Peng et al. established in mice models that elevated levels of IL-4 are linked to the development of with cardiomyopathy, as a result of an angiotensin II-induced cardiac damage [81].

Macrophages are the main effector cells of the innate immune system, and numerous studies indicate their role in the pathogenesis of AH. Monocyte levels are elevated in hypertensives in comparison to normotensives [82]. Interestingly Ang II-preactivated circulating monocytes in hypertensive patients might lead to subendothelial infiltration and thereafter augment the risk of arteriosclerotic complications [83].

Macrophages have a role in mediating hypertensive end-organ damage. In one study, perivascular macrophages were associated with the neurovascular and cognitive dysfunction induced by AH [84]. Shen et al. demonstrated that microglia, which are the resident immune cells in the brain, are the main molecular factors in mediating a neuroinflammation and modulating neuronal excitation, which contributes to an increased BP [85].

Barbaro et al. identified a novel mechanism via which excess sodium contributes to inflammation and AH. The result showed a pathway of sodium entrance into dendritic cells. Therefore, high-salt-treated DCs produce the prohypertensive cytokines IL-17 and INF-γ [86].

Antigen-presenting cells (APCs) are involved in the evolution of the inflammation associated with AH. Hevia et al. showed that APCs are essential for the development of AH, as the deletion of APCs produces rapid changes in the BP in mice with angiotensin II plus a high-salt diet. Additionally, the APCs activate the intrarenal renin/angiotensin system components and take part in the modulation of the natriuresis and tubular sodium transporters [87].

The NOD-like receptor protein 3 (NLRP3) inflammasome participates in the development of AH [88]. Krishnan et al. investigated MCC950—a recently-identified inhibitor of NLRP3 activity. The result showed that, in mice with established AH, MCC950 lowered BP and decreased renal inflammation, together with reduced fibrosis and kidney dysfunction [89].

Some studies investigated the neuro-immune axis associated with AH. Abboud et al. indicated that excessive sympathetic activity and reduced parasympathetic activity have a major role in pathological processes [90]. Furthermore, Harwani et al. demonstrated anti-inflammatory nicotinic/cholinergic modulation of the innate immune system in rats. Interestingly, they revealed proinflammatory innate immune responses in the hypertensive rats, prior to the development of AH [91].

Evidence from human and animal studies strongly suggests an association between inflammation and AH. Most recent findings in this field add to the growing body of evidence suggesting that AH is an inflammatory disease, while also drawing attention to the new possibilities for treating AH.

3.3. Oxidative Stress in AH

Disruption of redox signaling is a common pathophysiological mechanism observed in AH. Oxidative stress, resulting in either enhanced ROS production or decreases in antioxidant defense, is associated with an elevated BP, endothelial dysfunction, and vascular remodeling [92]. Oxidative stress and inflammatory responses act cooperatively in the pathogenesis of AH [93]. In the vascular system, the major source of ROS production is NOX, whose expression is increased in hypertensive conditions [94]. Dysregulation of enzymes such as NADPHNOX, nitric oxide synthase (NOS), xanthine oxidase, mitochondrial enzymes, or superoxide dismutase (SOD), which generate $\cdot O_2^-$, H_2O_2, and $\cdot OH$ together with reduced levels of antioxidants, results in increased formation of ROS within the vasculature. ROS contribute to vascular injury by promoting VSMC growth, extracellular matrix protein deposition, activation of matrix MMPs, inflammation, endothelial dysfunction, and increased vascular tone [95]. Furthermore, Crowley et al. indicated a novel mechanism via which oxidative stress promotes an inflammation in the vascular wall. The chaperone protein cyclophilin A (CypA), secreted from VSMCs due to ROS stimulation, leads to the recruitment of the inflammatory cells within the vasculature. Additionally, CypA triggers activation of MMPs, thus exaggerating vascular injury [93].

A variety of studies have indicated the role of NOX enzymes in the vascular remodeling during AH. Dikalova et al. indicated that an overexpression in the vascular smooth muscle of NOX1 exacerbates the hypertrophic and hypertensive responses to Ang II and increases the superoxide production in mice. Therefore, NOX1 participates in the development of cardiovascular pathologies [96]. Interestingly, Nosalski et al. investigated the pharmacological inhibition of NOX1/NOX4 in rats. The result showed that pharmacological inhibition of NOX4, elevated BP, increased accumulation of immune cell accumulation, and increased perivascular collagen deposition led to an accelerated vascular aging among both normotensive and hypertensive rats. Interestingly, the NOX1 inhibition did not affect the development of AH [97]. Murdoch et al. evaluated the overexpression of Nox2 and its association with AH. Thus, NOX2 contributes to the vascular remodeling and endothelial dysfunction, in addition to being involved in the pathophysiology of AH [98]. On the contrary, NOX4 promotes the protection of the vascular system in situations of increased stress induced by ischemia or angiotensin II [99]. NOX4 enhances the H_2O_2 production; thus, it has valuable effect on vasodilator function and BP [100].

A large body of evidence has shown that the *nuclear factor erythroid factor 2-related factor* 2(*Nrf2*) is involved in the AH pathophysiology. Tanase et al. investigated *Nrf2* in oxidative stress and its role in AH. *Nrf2* is a critical redox-sensitive transcription factor, functioning as a target nuclear receptor against oxidative stress, and it is a major component of the redox homeostasis of cells [101]. Chronic oxidative stress inhibits *Nrf2* activity and function [102]. It is worth mentioning that Farooqui and colleagues demonstrated the development of AH and renal function impairment due to an *Nrf2* inhibition in mice. The result showed that, in mice treated with a pro-oxidant—L-buthionine sulfoximine (10 mmol/L in drinking water) and an *Nrf2* inhibitor—ML385 (10 mg/kg body weight/day, intraperitoneally), oxidative stress, renal functional impairment, inflammation, and elevated BP were revealed [103]. Additionally, microRNA (miR)-140-5p exaggerates AH and oxidative stress in mouse models. Liu et al. established that downregulation of miR-140-5p reduced oxidative stress and ROS levels by activating the protein expression of *Nrf2* [104]. Furthermore, Biernacki et al. investigated changes in oxidative metabolism and apoptosis in the hearts of the hypertensive rats. The result showed that inhibition of lipolysis by fatty-acid amide hydrolase inhibitors can increase the enzymatic activity and nonenzymatic antioxidant activity in rats, whereas *Nrf2* expression is suppressed [105]. Therefore, treatments involving the upregulation of *Nrf2* expression might be promising. Nonetheless, further research is needed on the therapeutic potential of *Nrf2*.

It has become clear that inflammation, ROS, and BP elevation are significant in the pathophysiology of AH.

4. Conclusions

In this review, we focused on the important molecular aspects of three cardiovascular diseases: atherosclerosis, CAD, and AH. In atherosclerosis, we paid attention to the role of the oxidative processes, their pathomechanism, and the influence on the development of the disease. Important epigenetic factors, particularly those that have been highlighted in research in recent years, were also highlighted. In CAD, we focused on the endothelial dysfunction and the most important factors leading to it, as well as its new aspects. The most recent discoveries in the field of genetic background and the discovery of the role of many genes that directly or indirectly contribute to the increased risk of CAD were also considered to be of interest.

In AH, the most noteworthy aspects were the matrix metalloproteinases and the functioning of the immune system and its dysfunctions. Here, we also drew attention to the influence of oxidative stress.

These findings might shed new light on the cellular mechanisms of CVDs and prospective targets for the prevention and treatment in the near future. However, the major scientific achievements of recent years and the many new discoveries and mechanisms still require careful attention and additional studies.

Author Contributions: Conceptualization, E.M., B.F. and J.R.; methodology, W.F., A.W., W.L. and E.M.; software, E.M.; validation, E.M.; formal analysis, W.F., A.W., W.L., E.M., B.F. and J.R.; investigation, W.F., A.W. and W.L.; resources, E.M., B.F. and J.R.; data curation, E.M.; writing—original draft preparation, W.F., A.W. and W.L.; writing—review and editing, E.M.; visualization, W.F., A.W., W.L. and E.M.; supervision, E.M., B.F. and J.R.; project administration, E.M.; funding acquisition, E.M., B.F. and J.R. All authors have read and agreed to the published version of the manuscript.

Funding: This research received no external funding.

Institutional Review Board Statement: Not applicable.

Informed Consent Statement: Not applicable.

Data Availability Statement: The data used in this article were sourced from materials mentioned in the references.

Conflicts of Interest: The authors declare no conflict of interest.

References

1. Pagidipati, N.J.; Gaziano, T.A. Estimating Deaths from Cardiovascular Disease: A Review of Global Methodologies of Mortality Measurement. *Circulation* **2013**, *127*, 749–756. [CrossRef] [PubMed]
2. Wong, N.D.; Budoff, M.J.; Ferdinand, K.; Graham, I.M.; Michos, E.D.; Reddy, T.; Shapiro, M.D.; Toth, P.P. Atherosclerotic cardiovascular disease risk assessment: An American Society for Preventive Cardiology clinical practice statement. *Am. J. Prev. Cardiol.* **2022**, *10*, 100335. [CrossRef] [PubMed]
3. Sonja, F.; Nola, I.A. Management of Measurable Variable Cardiovascular Disease' Risk Factors. *Curr. Cardiol. Rev.* **2018**, *14*, 153–163. [CrossRef]
4. Amini, M.; Zayeri, F.; Salehi, M. Trend analysis of cardiovascular disease mortality, incidence, and mortality-to-incidence ratio: Results from global burden of disease study 2017. *BMC Public Health* **2021**, *21*, 401. [CrossRef] [PubMed]
5. Sanz, M.; Del Castillo, A.M.; Jepsen, S.; Juanatey, J.R.G.; D'Aiuto, F.; Bouchard, P.; Chapple, I.; Dietrich, T.; Gotsman, I.; Graziani, F.; et al. Periodontitis and cardiovascular diseases: Consensus report. *J. Clin. Periodontol.* **2020**, *47*, 268–288. [CrossRef] [PubMed]
6. Cheung, A.K.; Sarnak, M.J.; Yan, G.; Dwyer, J.T.; Heyka, R.J.; Rocco, M.V.; Teehan, B.P.; Levey, A.S. Atherosclerotic cardiovascular disease risks in chronic hemodialysis patients. *Kidney Int.* **2000**, *58*, 353–362. [CrossRef]
7. Mahmoud, R.; Setorki, M.; Doudi, M.; Baradaran, A.; Nasri, H. Atherosclerosis: Process, Indicators, Risk Factors and New Hopes. *Int. J. Prev. Med.* **2014**, *5*, 927–946.
8. Mitchell, G.F.; Powell, J.T. Arteriosclerosis. *Arter. Thromb. Vasc. Biol.* **2020**, *40*, 1025–1027. [CrossRef]
9. Helvaci, M.R.; Tonyali, O.; Abyad, A. The Safest Value of Plasma Triglycerides. *World Fam. Med. J./Middle East J. Fam. Med.* **2019**, *17*, 22–27. [CrossRef]
10. Shao, C.; Wang, J.; Tian, J.; Tang, Y.-D. Coronary Artery Disease: From Mechanism to Clinical Practice. In *Coronary Artery Disease: Therapeutics and Drug Discovery*; Advances in Experimental Medicine and Biology; Springer: Singapore, 2020; Volume 1177, pp. 1–36. [CrossRef]
11. Badimon, L.; Padró, T.; Vilahur, G. Atherosclerosis, platelets and thrombosis in acute ischaemic heart disease. *Eur. Heart J. Acute Cardiovasc. Care* **2012**, *1*, 60–74. [CrossRef] [PubMed]
12. Zannad, F.; Bauersachs, R. Rivaroxaban: A New Treatment Paradigm in the Setting of Vascular Protection? *Thromb. Haemost.* **2018**, *118*, S12–S22. [CrossRef] [PubMed]
13. Libby, P.; Ridker, P.M.; Hansson, G.K. Progress and challenges in translating the biology of atherosclerosis. *Nature* **2011**, *473*, 317–325. [CrossRef] [PubMed]
14. Brouwers, S.; Sudano, I.; Kokubo, Y.; Sulaica, E.M. Arterial hypertension. *Lancet* **2021**, *398*, 249–261. [CrossRef]
15. Unger, T.; Borghi, C.; Charchar, F.; Khan, N.A.; Poulter, N.R.; Prabhakaran, D.; Ramirez, A.; Schlaich, M.; Stergiou, G.S.; Tomaszewski, M.; et al. 2020 International Society of Hypertension Global Hypertension Practice Guidelines. *Hypertension* **2020**, *75*, 1334–1357. [CrossRef]
16. Wang, X.; Khalil, R.A. Matrix Metalloproteinases, Vascular Remodeling, and Vascular Disease. *Adv. Pharmacol.* **2018**, *81*, 241–330. [CrossRef]
17. Kopp, W. How Western Diet and Lifestyle Drive The Pandemic Of Obesity And Civilization Diseases. *Diabetes Metab. Syndr. Obes. Targets Ther.* **2019**, *12*, 2221–2236. [CrossRef]
18. Medina-Leyte, D.; Zepeda-García, O.; Domínguez-Pérez, M.; González-Garrido, A.; Villarreal-Molina, T.; Jacobo-Albavera, L. Endothelial Dysfunction, Inflammation and Coronary Artery Disease: Potential Biomarkers and Promising Therapeutical Approaches. *Int. J. Mol. Sci.* **2021**, *22*, 3850. [CrossRef]
19. Gimbrone, M.A., Jr.; García-Cardeña, G. Endothelial Cell Dysfunction and the Pathobiology of Atherosclerosis. *Circ. Res.* **2016**, *118*, 620–636. [CrossRef]
20. VanderLaan, P.; Reardon, C.A.; Getz, G.S. Site Specificity of Atherosclerosis: Site-Selective Responses to Atherosclerotic Modulators. *Arter. Thromb. Vasc. Biol.* **2004**, *24*, 12–22. [CrossRef]
21. Mussbacher, M.; Schossleitner, K.; Kral-Pointner, J.B.; Salzmann, M.; Schrammel, A.; Schmid, J.A. More than Just a Monolayer: The Multifaceted Role of Endothelial Cells in the Pathophysiology of Atherosclerosis. *Curr. Atheroscler. Rep.* **2022**, *24*, 483–492. [CrossRef] [PubMed]
22. Michiels, C. Endothelial cell functions. *J. Cell. Physiol.* **2003**, *196*, 430–443. [CrossRef] [PubMed]
23. Deanfield, J.E.; Halcox, J.P.; Rabelink, T.J. Endothelial function and dysfunction: Testing and clinical relevance. *Circulation* **2007**, *115*, 1285–1295. [CrossRef]
24. Gutiérrez, E.; Flammer, A.J.; Lerman, L.O.; Elízaga, J.; Lerman, A.; Fernández-Avilés, F. Endothelial dysfunction over the course of coronary artery disease. *Eur. Heart J.* **2013**, *34*, 3175–3181. [CrossRef]
25. Bäck, M.; Yurdagul, A., Jr.; Tabas, I.; Öörni, K.; Kovanen, P.T. Inflammation and its resolution in atherosclerosis: Mediators and therapeutic opportunities. *Nat. Rev. Cardiol.* **2019**, *16*, 389–406. [CrossRef] [PubMed]

26. Xia, Z.; Gu, M.; Jia, X.; Wang, X.; Wu, C.; Guo, J.; Zhang, L.; Du, Y.; Wang, J. Integrated DNA methylation and gene expression analysis identifies SLAMF7 as a key regulator of atherosclerosis. *Aging* **2018**, *10*, 1324–1337. [CrossRef]
27. Marchio, P.; Guerra-Ojeda, S.; Vila, J.M.; Aldasoro, M.; Victor, V.M.; Mauricio, M.D. Targeting Early Atherosclerosis: A Focus on Oxidative Stress and Inflammation. *Oxid. Med. Cell Longev.* **2019**, *2019*, 8563845. [CrossRef]
28. Wolf, D.; Ley, K. Immunity and Inflammation in Atherosclerosis. *Circ. Res.* **2019**, *124*, 315–327. [CrossRef]
29. Du, W.; Xu, A.; Huang, Y.; Cao, J.; Zhu, H.; Yang, B.; Shao, X.; He, Q.; Ying, M. The role of autophagy in targeted therapy for acute myeloid leukemia. *Autophagy* **2021**, *17*, 2665–2679. [CrossRef]
30. Jaiswal, S.; Natarajan, P.; Silver, A.J.; Gibson, C.J.; Bick, A.G.; Shvartz, E.; McConkey, M.; Gupta, N.; Gabriel, S.; Ardissino, D.; et al. Clonal Hematopoiesis and Risk of Atherosclerotic Cardiovascular Disease. *N. Engl. J. Med.* **2017**, *377*, 111–121. [CrossRef]
31. Asada, S.; Kitamura, T. Clonal hematopoiesis and associated diseases: A review of recent findings. *Cancer Sci.* **2021**, *112*, 3962–3971. [CrossRef] [PubMed]
32. Patel, A.P.; Natarajan, P. Completing the genetic spectrum influencing coronary artery disease: From germline to somatic variation. *Cardiovasc. Res.* **2019**, *115*, 830–843. [CrossRef] [PubMed]
33. Tousoulis, D.; Kampoli, A.-M.; Tentolouris, C.; Papageorgiou, N.; Stefanadis, C. The Role of Nitric Oxide on Endothelial Function. *Curr. Vasc. Pharmacol.* **2012**, *10*, 4–18. [CrossRef]
34. Gresele, P.; Momi, S.; Guglielmini, G. Nitric oxide-enhancing or -releasing agents as antithrombotic drugs. *Biochem. Pharmacol.* **2019**, *166*, 300–312. [CrossRef]
35. Yang, T.; Zhang, X.-M.; Tarnawski, L.; Peleli, M.; Zhuge, Z.; Terrando, N.; Harris, R.A.; Olofsson, P.S.; Larsson, E.; Persson, A.E.G.; et al. Dietary nitrate attenuates renal ischemia-reperfusion injuries by modulation of immune responses and reduction of oxidative stress. *Redox Biol.* **2017**, *13*, 320–330. [CrossRef] [PubMed]
36. Ma, L.; Hu, H.L.; Feng, X.; Wang, S. Nitrate and Nitrite in Health and Disease. *Aging Dis.* **2018**, *9*, 938–945. [CrossRef]
37. Meza, C.A.; La Favor, J.D.; Kim, D.-H.; Hickner, R.C. Endothelial Dysfunction: Is There a Hyperglycemia-Induced Imbalance of NOX and NOS? *Int. J. Mol. Sci.* **2019**, *20*, 3775. [CrossRef]
38. Förstermann, U.; Sessa, W.C. Nitric oxide synthases: Regulation and function. *Eur. Heart J.* **2012**, *33*, 829–837. [CrossRef]
39. Carlstrom, M.; Montenegro, M.F. Therapeutic value of stimulating the nitrate-nitrite-nitric oxide pathway to attenuate oxidative stress and restore nitric oxide bioavailability in cardiorenal disease. *J. Intern. Med.* **2019**, *285*, 2–18. [CrossRef]
40. Mourouzis, K.; Siasos, G.; Oikonomou, E.; Zaromitidou, M.; Tsigkou, V.; Antonopoulos, A.; Bletsa, E.; Stampouloglou, P.; Vlasis, K.; Vavuranakis, M.; et al. Lipoprotein-associated phospholipase A2 levels, endothelial dysfunction and arterial stiffness in patients with stable coronary artery disease. *Lipids Health Dis.* **2021**, *20*, 12. [CrossRef]
41. Franzén, O.; Ermel, R.; Cohain, A.; Akers, N.K.; Di Narzo, A.; Talukdar, H.A.; Foroughi-Asl, H.; Giambartolomei, C.; Fullard, J.F.; Sukhavasi, K.; et al. Cardiometabolic risk loci share downstream cis- and trans-gene regulation across tissues and diseases. *Science* **2016**, *353*, 827–830. [CrossRef] [PubMed]
42. Guillevin, L.; Dörner, T. Vasculitis: Mechanisms involved and clinical manifestations. *Arthritis Res. Ther.* **2007**, *9*, S9. [CrossRef] [PubMed]
43. Jennette, J.C.; Thomas, D.B.; Falk, R.J. Microscopic polyangiitis (*microscopic polyarteritis*). *Semin. Diagn. Pathol.* **2001**, *18*, 3–13. [PubMed]
44. Maugeri, N.; Rovere-Querini, P.; Baldini, M.; Sabbadini, M.G.; Manfredi, A.A. Translational Mini-Review Series on Immunology of Vascular Disease: Mechanisms of vascular inflammation and remodelling in systemic vasculitis. *Clin. Exp. Immunol.* **2009**, *156*, 395–404. [CrossRef]
45. Khera, A.V.; Won, H.-H.; Peloso, G.M.; O'Dushlaine, C.; Liu, D.; Stitziel, N.O.; Natarajan, P.; Nomura, A.; Emdin, C.A.; Gupta, N.; et al. Association of Rare and Common Variation in the Lipoprotein Lipase Gene with Coronary Artery Disease. *JAMA* **2017**, *317*, 937–946. [CrossRef]
46. Eric, G.R.; Segrè, A.V.; van de Bunt, M.; Wen, X.; Xi, H.S.; Hormozdiari, F.; Ongen, H. Using an atlas of gene regulation across 44 human tissues to inform complex disease- and trait-associated variation. *Nat. Genet.* **2018**, *50*, 956–967. [CrossRef]
47. Aherrahrou, R.; Guo, L.; Nagraj, V.P.; Aguhob, A.A.; Hinkle, J.; Chen, L.; Soh, J.Y.; Lue, D.; Alencar, G.F.; Boltjes, A.; et al. Genetic Regulation of Atherosclerosis-Relevant Phenotypes in Human Vascular Smooth Muscle Cells. *Circ. Res.* **2020**, *127*, 1552–1565. [CrossRef]
48. Cherepanova, O.A.; Gomez, D.; Shankman, L.S.; Swiatlowska, P.; Williams, J.; Sarmento, O.F.; Alencar, G.F.; Hess, D.L.; Bevard, M.H.; Greene, E.S.; et al. Activation of the pluripotency factor OCT4 in smooth muscle cells is atheroprotective. *Nat. Med.* **2016**, *22*, 657–665. [CrossRef]
49. Nikpay, M.; Goel, A.; Won, H.H.; Hall, L.M.; Willenborg, C.; Kanoni, S.; Saleheen, D.; Kyriakou, T.; Nelson, C.P.; Hopewell, J.C.; et al. A comprehensive 1000 Genomes–based genome-wide association meta-analysis of coronary artery disease. *Nat. Genet.* **2015**, *47*, 1121–1130. [CrossRef]
50. Hara, T.; Monguchi, T.; Iwamoto, N.; Akashi, M.; Mori, K.; Oshita, T.; Okano, M.; Toh, R.; Irino, Y.; Shinohara, M.; et al. Targeted Disruption of JCAD (Junctional Protein Associated with Coronary Artery Disease)/KIAA1462, a Coronary Artery Disease–Associated Gene Product, Inhibits Angiogenic Processes In Vitro and In Vivo. *Arter. Thromb. Vasc. Biol.* **2017**, *37*, 1667–1673. [CrossRef]

51. Jones, P.D.; Kaiser, M.A.; Najafabadi, M.G.; Koplev, S.; Zhao, Y.; Douglas, G.; Kyriakou, T.; Andrews, S.; Rajmohan, R.; Watkins, H.; et al. *JCAD*, a Gene at the 10p11 Coronary Artery Disease Locus, Regulates Hippo Signaling in Endothelial Cells. *Arter. Thromb. Vasc. Biol.* **2018**, *38*, 1711–1722. [CrossRef] [PubMed]
52. Xu, S.; Xu, Y.; Liu, P.; Zhang, S.; Liu, H.; Slavin, S.; Kumar, S.; Koroleva, M.; Luo, J.; Wu, X.; et al. The novel coronary artery disease risk gene JCAD/KIAA1462 promotes endothelial dysfunction and atherosclerosis. *Eur. Heart J.* **2019**, *40*, 2398–2408. [CrossRef] [PubMed]
53. Askin, L.; Tibilli, H.; Tanriverdi, O.; Turkmen, S. The Relationship between Coronary Artery Disease and SIRT1 Protein. *North. Clin. Istanb.* **2020**, *7*, 631–635. [CrossRef] [PubMed]
54. Nagao, M.; Lyu, Q.; Zhao, Q.; Wirka, R.C.; Bagga, J.; Nguyen, T.; Cheng, P.; Kim, J.B.; Pjanic, M.; Miano, J.M.; et al. Coronary Disease-Associated Gene *TCF21* Inhibits Smooth Muscle Cell Differentiation by Blocking the Myocardin-Serum Response Factor Pathway. *Circ. Res.* **2020**, *126*, 517–529. [CrossRef]
55. Qi, B.; Chen, J.-H.; Tao, L.; Zhu, C.-M.; Wang, Y.; Deng, G.-X.; Miao, L. Integrated Weighted Gene Co-expression Network Analysis Identified That TLR2 and CD40 Are Related to Coronary Artery Disease. *Front. Genet.* **2021**, *11*, 613744. [CrossRef]
56. Winkler, M.J.; Müller, P.; Sharifi, A.M.; Wobst, J.; Winter, H.; Mokry, M.; Ma, L.; Van Der Laan, S.W.; Pang, S.; Miritsch, B.; et al. Functional investigation of the coronary artery disease gene SVEP1. *Basic Res. Cardiol.* **2020**, *115*, 67. [CrossRef]
57. Bengtsson, E.; Hultman, K.; Dunér, P.; Asciutto, G.; Almgren, P.; Orho-Melander, M.; Melander, O.; Nilsson, J.; Hultgårdh-Nilsson, A.; Gonçalves, I. ADAMTS-7 is associated with a high-risk plaque phenotype in human atherosclerosis. *Sci. Rep.* **2017**, *7*, 3753. [CrossRef]
58. Pu, X.; Xiao, Q.; Kiechl, S.; Chan, K.; Ng, F.L.; Gor, S.; Poston, R.N.; Fang, C.; Patel, A.; Senver, E.C.; et al. ADAMTS7 Cleavage and Vascular Smooth Muscle Cell Migration Is Affected by a Coronary-Artery-Disease-Associated Variant. *Am. J. Hum. Genet.* **2013**, *92*, 366–374. [CrossRef]
59. Emdin, C.A.; Khera, A.V.; Klarin, D.; Natarajan, P.; Zekavat, S.; Nomura, A.; Haas, M.E.; Aragam, K.; Ardissino, D.; Wilson, J.G.; et al. Phenotypic Consequences of a Genetic Predisposition to Enhanced Nitric Oxide Signaling. *Circulation* **2018**, *137*, 222–232. [CrossRef]
60. Jeanette, E.; Stark, K.; Esslinger, U.B.; Rumpf, P.M.; Koesling, D.; de Wit, C.; Kaiser, F.J.; Braunholz, D.; Medack, A. Dysfunctional nitric oxide signalling increases risk of myocardial infarction. *Nature* **2013**, *504*, 432–436. [CrossRef]
61. Do, R.; Project, N.E.S.; Stitziel, N.; Won, H.-H.; Jørgensen, A.B.; Duga, S.; Merlini, P.A.; Kiezun, A.; Farrall, M.; Goel, A.; et al. Exome sequencing identifies rare LDLR and APOA5 alleles conferring risk for myocardial infarction. *Nature* **2015**, *518*, 102–106. [CrossRef] [PubMed]
62. Jørgensen, A.B.; Frikke-Schmidt, R.; Nordestgaard, B.G.; Tybjærg-Hansen, A. Loss-of-Function Mutations in *APOC3* and Risk of Ischemic Vascular Disease. *N. Engl. J. Med.* **2014**, *371*, 32–41. [CrossRef] [PubMed]
63. Dewey, F.E.; Gusarova, V.; O'Dushlaine, C.; Gottesman, O.; Trejos, J.; Hunt, C.; Van Hout, C.V.; Habegger, L.; Buckler, D.; Lai, K.-M.V.; et al. Inactivating Variants in ANGPTL4 and Risk of Coronary Artery Disease. *N. Engl. J. Med.* **2016**, *374*, 1123–1133. [CrossRef] [PubMed]
64. Laronha, H.; Caldeira, J. Structure and Function of Human Matrix Metalloproteinases. *Cells* **2020**, *9*, 1076. [CrossRef] [PubMed]
65. Bisogni, V.; Cerasari, A.; Pucci, G.; Vaudo, G. Matrix Metalloproteinases and Hypertension-Mediated Organ Damage: Current Insights. *Integr. Blood Press. Control* **2020**, *13*, 157–169. [CrossRef]
66. Raffetto, J.D.; Khalil, R.A. Matrix metalloproteinases and their inhibitors in vascular remodeling and vascular disease. *Biochem. Pharmacol.* **2008**, *75*, 346–359. [CrossRef]
67. Marchesi, C.; Dentali, F.; Nicolini, E.; Maresca, A.M.; Tayebjee, M.H.; Franz, M.; Guasti, L.; Venco, A.; Schiffrin, E.; Lip, G.Y.; et al. Plasma levels of matrix metalloproteinases and their inhibitors in hypertension: A Systematic Review and Meta-Analysis. *J. Hypertens.* **2012**, *30*, 3–16. [CrossRef]
68. Kostov, K.; Blazhev, A. Changes in Serum Levels of Matrix Metalloproteinase-1 and Tissue Inhibitor of Metalloproteinases-1 in Patients with Essential Hypertension. *Bioengineering* **2022**, *9*, 119. [CrossRef]
69. Basu, R.; Lee, J.; Morton, J.S.; Takawale, A.; Fan, D.; Kandalam, V.; Wang, X.; Davidge, S.T.; Kassiri, Z. TIMP3 is the primary TIMP to regulate agonist-induced vascular remodelling and hypertension. *Cardiovasc. Res.* **2013**, *98*, 360–371. [CrossRef]
70. Pushpakumar, S.; Kundu, S.; Pryor, T.; Givvimani, S.; Lederer, E.; Tyagi, S.C.; Sen, U. Angiotensin-II induced hypertension and renovascular remodelling in tissue inhibitor of metalloproteinase 2 knockout mice. *J. Hypertens.* **2013**, *31*, 2270–2281. [CrossRef]
71. Cabral-Pacheco, G.A.; Garza-Veloz, I.; La Rosa, C.C.-D.; Ramirez-Acuña, J.M.; Perez-Romero, A.B.; Guerrero-Rodriguez, J.F.; Martinez-Avila, N.; Martinez-Fierro, M.L. The Roles of Matrix Metalloproteinases and Their Inhibitors in Human Diseases. *Int. J. Mol. Sci.* **2020**, *21*, 9739. [CrossRef] [PubMed]
72. Cai, Z.; Gong, Z.; Li, Z.; Li, L.; Kong, W. Vascular Extracellular Matrix Remodeling and Hypertension. *Antioxid. Redox Signal.* **2021**, *34*, 765–783. [CrossRef]
73. Cui, N.; Hu, M.; Khalil, R.A. Biochemical and Biological Attributes of Matrix Metalloproteinases. *Prog. Mol. Biol. Transl. Sci.* **2017**, *147*, 1–73. [CrossRef]
74. Martínez, A.; Oh, H.-R.; Unsworth, E.J.; Bregonzio, C.; Saavedra, J.M.; Stetler-Stevenson, W.; Cuttitta, F. Matrix metalloproteinase-2 cleavage of adrenomedullin produces a vasoconstrictor out of a vasodilator. *Biochem. J.* **2004**, *383*, 413–418. [CrossRef]
75. Wakisaka, Y.; Chu, Y.; Miller, J.; Rosenberg, G.A.; Heistad, D.D. Spontaneous Intracerebral Hemorrhage during Acute and Chronic Hypertension in Mice. *J. Cereb. Blood Flow Metab.* **2010**, *30*, 56–69. [CrossRef] [PubMed]

76. Mirhafez, S.R.; Mohebati, M.; Disfani, M.F.; Karimian, M.S.; Ebrahimi, M.; Avan, A.; Eslami, S.; Pasdar, A.; Rooki, H.; Esmaeili, H.; et al. An imbalance in serum concentrations of inflammatory and anti-inflammatory cytokines in hypertension. *J. Am. Soc. Hypertens.* **2014**, *8*, 614–623. [CrossRef] [PubMed]
77. Sesso, H.D.; Wang, L.; Buring, J.E.; Ridker, P.M.; Gaziano, J.M. Comparison of Interleukin-6 and C-Reactive Protein for the Risk of Developing Hypertension in Women. *Hypertension* **2007**, *49*, 304–310. [CrossRef] [PubMed]
78. Hage, F. C-reactive protein and Hypertension. *J. Hum. Hypertens.* **2014**, *28*, 410–415. [CrossRef]
79. Kong, H.; Qian, Y.-S.; Tang, X.-F.; Zhang, J.; Gao, P.-J.; Zhang, Y.; Zhu, D.-L. C-reactive protein (CRP) gene polymorphisms, CRP levels and risk of incident essential hypertension: Findings from an observational cohort of Han Chinese. *Hypertens. Res.* **2012**, *35*, 1019–1023. [CrossRef]
80. Lima, V.V.; Zemse, S.M.; Chiao, C.-W.; Bomfim, G.F.; Tostes, R.C.; Webb, R.C.; Giachini, F.R. Interleukin-10 limits increased blood pressure and vascular RhoA/Rho-kinase signaling in angiotensin II-infused mice. *Life Sci.* **2016**, *145*, 137–143. [CrossRef]
81. Peng, H.; Sarwar, Z.; Yang, X.-P.; Peterson, E.L.; Xu, J.; Janic, B.; Rhaleb, N.; Carretero, O.A.; Rhaleb, N.-E. Profibrotic Role for Interleukin-4 in Cardiac Remodeling and Dysfunction. *Hypertension* **2015**, *66*, 582–589. [CrossRef] [PubMed]
82. Parissis, J.T.; Korovesis, S.; Giazitzoglou, E.; Kalivas, P.; Katritsis, D. Plasma profiles of peripheral monocyte-related inflammatory markers in patients with arterial hypertension. Correlations with plasma endothelin-1. *Int. J. Cardiol.* **2002**, *83*, 13–21. [CrossRef]
83. Dörffel, Y.; Lätsch, C.; Stuhlmüller, B.; Schreiber, S.; Scholze, S.; Burmester, G.R.; Scholze, J. Preactivated Peripheral Blood Monocytes in Patients with Essential Hypertension. *Hypertension* **1999**, *34*, 113–117. [CrossRef] [PubMed]
84. Faraco, G.; Sugiyama, Y.; Lane, D.; Garcia-Bonilla, L.; Chang, H.; Santisteban, M.; Racchumi, G.; Murphy, M.; Van Rooijen, N.; Anrather, J.; et al. Perivascular macrophages mediate the neurovascular and cognitive dysfunction associated with hypertension. *J. Clin. Investig.* **2016**, *126*, 4674–4689. [CrossRef] [PubMed]
85. Shen, X.Z.; Li, Y.; Li, L.; Shah, K.H.; Bernstein, K.; Lyden, P.D.; Shi, P. Microglia Participate in Neurogenic Regulation of Hypertension. *Hypertension* **2015**, *66*, 309–316. [CrossRef]
86. Barbaro, N.R.; Foss, J.D.; Kryshtal, D.O.; Tsyba, N.; Kumaresan, S.; Xiao, L.; Mernaugh, R.L.; Itani, H.A.; Loperena, R.; Chen, W.; et al. Dendritic Cell Amiloride-Sensitive Channels Mediate Sodium-Induced Inflammation and Hypertension. *Cell Rep.* **2017**, *21*, 1009–1020. [CrossRef]
87. Hevia, D.; Araos, P.; Prado, C.; Luppichini, E.F.; Rojas, M.; Alzamora, R.; Cifuentes-Araneda, F.; Gonzalez, A.A.; Amador, C.A.; Pacheco, R.; et al. Myeloid CD11c $^{+}$ Antigen-Presenting Cells Ablation Prevents Hypertension in Response to Angiotensin II Plus High-Salt Diet. *Hypertension* **2018**, *71*, 709–718. [CrossRef]
88. Krishnan, S.M.; Dowling, J.; Ling, Y.H.; Diep, H.; Chan, C.T.; Ferens, D.; Kett, M.M.; Pinar, A.; Samuel, C.S.; Vinh, A.; et al. Inflammasome activity is essential for one kidney/deoxycorticosterone acetate/salt-induced hypertension in mice. *J. Cereb. Blood Flow Metab.* **2016**, *173*, 752–765. [CrossRef]
89. Krishnan, S.M.; Ling, Y.H.; Huuskes, B.M.; Ferens, D.M.; Saini, N.; Chan, C.T.; Diep, H.; Kett, M.M.; Samuel, C.S.; Kemp-Harper, B.K.; et al. Pharmacological inhibition of the NLRP3 inflammasome reduces blood pressure, renal damage, and dysfunction in salt-sensitive hypertension. *Cardiovasc. Res.* **2019**, *115*, 776–787. [CrossRef]
90. Abboud, F.M.; Harwani, S.; Chapleau, M. Autonomic Neural Regulation of the Immune System: Implications for Hypertension and Cardiovascular Disease. *Hypertension* **2012**, *59*, 755–762. [CrossRef]
91. Harwani, S.C.; Chapleau, M.W.; Legge, K.L.; Ballas, Z.K.; Abboud, F.M. Neurohormonal Modulation of the Innate Immune System Is Proinflammatory in the Prehypertensive Spontaneously Hypertensive Rat, a Genetic Model of Essential Hypertension. *Circ. Res.* **2012**, *111*, 1190–1197. [CrossRef] [PubMed]
92. Rodrigo, R.; González, J.; Paoletto, F. The role of oxidative stress in the pathophysiology of hypertension. *Hypertens. Res.* **2011**, *34*, 431–440. [CrossRef] [PubMed]
93. Crowley, S.D. The Cooperative Roles of Inflammation and Oxidative Stress in the Pathogenesis of Hypertension. *Antioxid. Redox Signal.* **2014**, *20*, 102–120. [CrossRef] [PubMed]
94. Poznyak, A.V.; Grechko, A.V.; Orekhova, V.A.; Khotina, V.; Ivanova, E.A.; Orekhov, A.N. NADPH Oxidases and Their Role in Atherosclerosis. *Biomedicines* **2020**, *8*, 206. [CrossRef] [PubMed]
95. Touyz, R.M.; Schiffrin, E.L. Reactive oxygen species in vascular biology: Implications in hypertension. *Histochem. Cell Biol.* **2004**, *122*, 339–352. [CrossRef]
96. Dikalova, A.; Clempus, R.; Lassègue, B.; Cheng, G.; McCoy, J.; Dikalov, S.; San Martin, A.; Lyle, A.; Weber, D.S.; Weiss, D.; et al. Nox1 Overexpression Potentiates Angiotensin II-Induced Hypertension and Vascular Smooth Muscle Hypertrophy in Transgenic Mice. *Circulation* **2005**, *112*, 2668–2676. [CrossRef]
97. Nosalski, R.; Mikolajczyk, T.; Siedlinski, M.; Saju, B.; Koziol, J.; Maffia, P.; Guzik, T. Nox1/4 inhibition exacerbates age dependent perivascular inflammation and fibrosis in a model of spontaneous hypertension. *Pharmacol. Res.* **2020**, *161*, 105235. [CrossRef]
98. Murdoch, C.E.; Alom-Ruiz, S.P.; Wang, M.; Zhang, M.; Walker, S.; Yu, B.; Brewer, A.; Shah, A.M. Role of endothelial Nox2 NADPH oxidase in angiotensin II-induced hypertension and vasomotor dysfunction. *Basic Res. Cardiol.* **2011**, *106*, 527–538. [CrossRef]
99. Schroeder, K.; Zhang, M.; Benkhoff, S.; Mieth, A.; Pliquett, R.; Kosowski, J.; Kruse, C.; Luedike, P.; Michaelis, U.R.; Weissmann, N.; et al. Nox4 Is a Protective Reactive Oxygen Species Generating Vascular NADPH Oxidase. *Circ. Res.* **2012**, *110*, 1217–1225. [CrossRef]

100. Ray, R.; Murdoch, C.E.; Wang, M.; Santos, C.X.; Zhang, M.; Alom-Ruiz, S.; Anilkumar, N.; Ouattara, A.; Cave, A.C.; Walker, S.J.; et al. Endothelial Nox4 NADPH Oxidase Enhances Vasodilatation and Reduces Blood Pressure In Vivo. *Arter. Thromb. Vasc. Biol.* **2011**, *31*, 1368–1376. [CrossRef]
101. Tanase, D.M.; Apostol, A.G.; Costea, C.F.; Tarniceriu, C.C.; Tudorancea, I.; Maranduca, M.A.; Floria, M.; Serban, I.L. Oxidative Stress in Arterial Hypertension (HTN): The Nuclear Factor Erythroid Factor 2-Related Factor 2 (Nrf2) Pathway, Implications and Future Perspectives. *Pharmaceutics* **2022**, *14*, 534. [CrossRef] [PubMed]
102. Dovinova, I.; Kvandová, M.; Balis, P.; Gresova, L.; Majzunova, M.; Horakova, L.; Chan, J.Y.; Barancik, M. The role of Nrf2 and PPARgamma in the improvement of oxidative stress in hypertension and cardiovascular diseases. *Physiol. Res.* **2020**, *69*, S541–S553. [CrossRef] [PubMed]
103. Farooqui, Z.; Mohammad, R.S.; Lokhandwala, M.F.; Banday, A.A. Nrf2 inhibition induces oxidative stress, renal inflammation and hypertension in mice. *Clin. Exp. Hypertens.* **2021**, *43*, 175–180. [CrossRef] [PubMed]
104. Liu, Q.; Ren, K.; Liu, S.; Li, W.; Huang, C.; Yang, X. MicroRNA-140-5p aggravates hypertension and oxidative stress of atherosclerosis via targeting Nrf2 and Sirt2. *Int. J. Mol. Med.* **2019**, *43*, 839–849. [CrossRef] [PubMed]
105. Biernacki, M.; Łuczaj, W.; Jarocka-Karpowicz, I.; Ambrożewicz, E.; Toczek, M.; Skrzydlewska, E. The Effect of Long-Term Administration of Fatty Acid Amide Hydrolase Inhibitor URB597 on Oxidative Metabolism in the Heart of Rats with Primary and Secondary Hypertension. *Molecules* **2018**, *23*, 2350. [CrossRef]

biomedicines

Review

Cellular Mechanisms of Coronary Artery Spasm

Beata Franczyk, Jill Dybiec, Weronika Frąk, Julia Krzemińska, Joanna Kućmierz, Ewelina Młynarska *, Magdalena Szlagor, Magdalena Wronka and Jacek Rysz

Department of Nephrology, Hypertension and Family Medicine, Medical University of Lodz, ul. Żeromskiego 113, 90-549 Łódź, Poland
* Correspondence: emmlynarska@gmail.com; Tel.: +48-(042)-6393750

Abstract: Coronary artery spasm (CAS) is a reversible phenomenon caused by spontaneous excessive vascular smooth muscle contractility and vascular wall hypertonicity, which results in partial or complete closure of the lumen of normal or atherosclerotic coronary arteries. The clinical picture of CAS includes chest discomfort which is similar in quality to that of stable effort angina. Mechanisms underlying the development of CAS are still unclear. CAS certainly is a multifactorial disease. In this review, we paid attention to the role of the main pathophysiologic mechanisms in CAS: endothelial dysfunction, chronic inflammation, oxidative stress, smooth muscle hypercontractility, atherosclerosis and thrombosis, and mutations leading to deficient aldehyde dehydrogenase 2 (ALDH2) activity. These findings might shed novel insight on the underlying mechanisms and identify potential diagnostic and therapeutic targets for cardiovascular diseases in the future.

Keywords: coronary artery spasm; cellular mechanism; endothelial dysfunction; oxidative stress; smooth muscle hypercontractility; inflammation; atherosclerosis; thrombosis; variant angina

Citation: Franczyk, B.; Dybiec, J.; Frąk, W.; Krzemińska, J.; Kućmierz, J.; Młynarska, E.; Szlagor, M.; Wronka, M.; Rysz, J. Cellular Mechanisms of Coronary Artery Spasm. *Biomedicines* **2022**, *10*, 2349. https://doi.org/10.3390/biomedicines10102349

Academic Editors: Tânia Martins-Marques, Gonçalo F. Coutinho and Attila Kiss

Received: 29 August 2022
Accepted: 17 September 2022
Published: 21 September 2022

Publisher's Note: MDPI stays neutral with regard to jurisdictional claims in published maps and institutional affiliations.

Copyright: © 2022 by the authors. Licensee MDPI, Basel, Switzerland. This article is an open access article distributed under the terms and conditions of the Creative Commons Attribution (CC BY) license (https://creativecommons.org/licenses/by/4.0/).

1. Introduction

Coronary artery spasm (CAS) is a reversible phenomenon caused by spontaneous excessive vascular smooth muscle contractility and vascular wall hypertonicity, which results in partial or complete closure of the lumen of normal or atherosclerotic coronary arteries [1,2]. The concept of CAS was first postulated by Prinzmetal et al. who described angina that occurs at rest or during regular daily activities which could not be explained by the increased oxygen demand of the myocardium [3]. Prevalence of CAS is diverse between countries: in the Japanese, 24.3%, followed by the Taiwanese, 19.3%, and Caucasian, 7.5%, populations [2]. Among patients aged 40 to 70 years, CAS is more common in men than in women [1]; however, it is mostly a disease of middle- and older-aged men and post-menopausal women [4].

The clinical picture of CAS includes chest discomfort which is similar in quality to that of stable effort angina [4]. A typical CAS attack is transient, often lasts only a few seconds, and is unpredictable; however, it arises particularly from midnight to early morning [4,5]. It occurs at rest and is a vague sensation of compression in the precordium or upper abdomen with radiation mostly to the neck, jaw, and left shoulder [6]. Angina may be accompanied by cold sweats, syncope, and a lowering of blood pressure [6]. Kishida et al. revealed that 82% (872 of 1062 episodes) of CAS episodes were asymptomatic and that syncope occurred in 12.5% (30 of 240 patients) of patients with CAS [7].

The main risk factors for CAS are smoking, age, high-sensitivity C-reactive protein (hs-CRP), hypertension, LDL cholesterol, and diabetes mellitus [1,2].

Mechanisms underlying the development of CAS are still unclear. CAS certainly is a multifactorial disease [5]. The main pathophysiologic mechanisms in CAS are dysfunction of the autonomic nervous system, endothelial dysfunction, chronic inflammation, oxidative stress, smooth muscle hypercontractility, atherosclerosis and thrombosis, and mutations leading to deficient aldehyde dehydrogenase 2 (ALDH2) activity [1].

In patients with CAS due to dysfunction of the endothelium, deficiency in nitric oxide (NO) is observed [8]. One of the damaging factors for the endothelium is oxidative stress. Free radicals degrade NO which results in artery spasms. Cigarette smoking is one of the main risk factors that intensify this process [9]. In addition, it is critical for people suffering from CAS to cease smoking, as the inflammation in the body, an essential part of smoking, triggers coronary spasms [10–13].

Fortunately, it has been reported that antioxidants such as vitamin C or E can restore disturbed arterial reactivity [14,15]. Furthermore, estrogen, a hormone responsible for enhancing NO synthase activity, can be considered as a protective factor. It has been shown that high estrogen levels (typical of the pre-menopausal period) were associated with a lower frequency of ischemic episodes [16].

One of the triggers of CAS is also vascular smooth muscle cell hyperreactivity. The excessive intracellular influx of calcium ions, disturbances in the functioning of calcium channels, and malfunctioning of ATP-sensitive potassium channels may result in the occurrence of coronary artery spasms [17–19]. RhoA/Rho-kinase (ROK) activity and a number of neurotransmitters are also involved in the pathogenesis of hypercontractility leading to CAS [2,20–23].

Another important risk factor is also deficiency of magnesium, an endogenous calcium channel antagonist [1,4,5,24]. Intravenous magnesium administration is beneficial in patients with CAS [5].

The main pathogenetic mechanisms of CAS are presented in Figure 1.

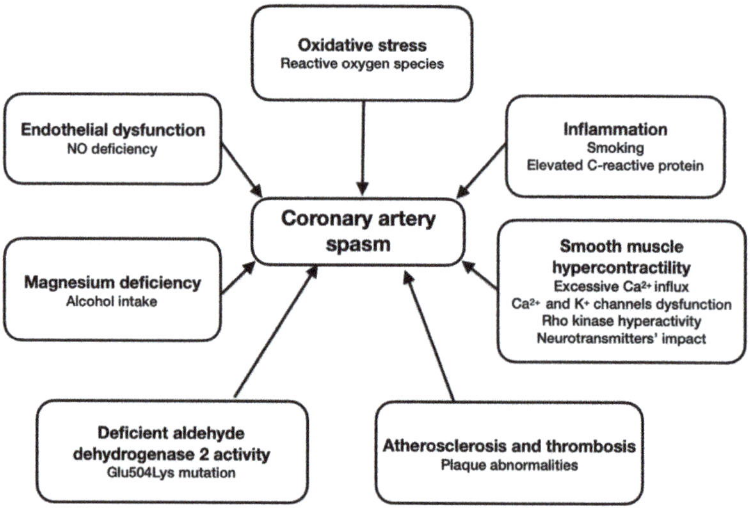

Figure 1. Pathogenetic mechanisms of CAS.

Lifestyle change and elimination of risk factors, as well as adherence to prescribed pharmacotherapy, form the basis of the management of CAS and reduce the risk of further episodes in the future [25]. The pharmacotherapy of CAS is based on the use of calcium channel blockers (CCBs) and/or nitrates. In exceptional cases, invasive therapies can be used [26].

Diagnosis of vasospastic angina may be problematic [5]. The primary role in the initial evaluation of a patient with an attack of a coronary artery spasm is to perform an electrocardiography (ECG) [5]. However, coronary angiography with a provocation test is considered the gold standard for diagnosing the disease. There are also other modern imaging tests such as intravascular ultrasound (IVUS) and optical coherence tomography (OCT), which are much more accurate but less commonly used [1].

2. Endothelial Dysfunction

The vascular endothelium is known as a regulatory organ that is essential for the proper function of the cardiovascular system. Due to its ability to produce biologically active substances, the endothelium is a significant factor that maintains homeostasis. Moreover, it is crucial for the fluidity of blood because of its anticoagulant, fibrinolytic, and antithrombotic properties. Therefore, its dysfunction plays an important role in the pathomechanism of blood vessel alterations [27].

One of the multiple roles of a normal functional endothelium is the production of NO. This compound is responsible for vasodilatation by suppressing vasoconstrictors such as angiotensin II and endothelium I [1]. Furthermore, NO deficiency may intensify their synthesis [28,29].

In young healthy people, acetylcholine (ACh) induces an increase in coronary artery diameter by releasing NO [8]. Nevertheless, there are other endothelium-dependent vasodilators such as serotonin, histamine, or ergonovine which are also a virtue of nitric oxide-releasing mechanisms [30]. On the other hand, a dysfunctional endothelium is characterized by the deficiency of NO. Consequently, in subjects with coronary atherosclerosis, intracoronary infusion of ACh results in spasms [8]. The difference between those two groups proved useful in the diagnosis of CAS. Injections of ACh are used as a provocative test [4]. However, it has been reported that coronary hyperconstriction induced by ACh involves all coronary segments. Spasms caused by ergonovine or serotonin concern only the given coronary site or segment which is similar to spontaneous spasms. Due to this fact, ACh may not be an appropriate option [10].

Nonetheless, nitrates via conversion into NO in vivo are independent of endothelial mechanisms. Synthesis of NO from L-arginine can be inhibited by L-monomethyl-arginine (L-NMMA) [30]. Kugiyama et al. [31] conducted a study in which they infused L-NMMA into coronary arteries in 21 patients with coronary spastic angina (CSA) and in 28 control patients. A coronary spasm was induced by ACh. Administration of L-NMMA in the control group resulted in a decrease in the basal diameter of a coronary artery but ended up with no effect in the other group. Moreover, the dilator response to nitroglycerin was significantly higher in patients with CSA. This is a result of the super-sensitivity of spasm arteries to nitroglycerin. It might be due to the deficiency of endogenous NO activity.

According to Kawano's research [16], variation in estrogen levels is strictly connected with the frequency of myocardial ischemia. As it is known, estrogen is responsible for enhancing NO synthase activity [32]. Due to the similarity between the endothelial function of a brachial and coronary artery [33], flow-mediated dilation of the brachial artery was assessed in the study. This magnitude is mostly based on endothelium-derived nitric oxide [34], and as Kawano showed, it is related to the variation in estradiol levels during the menstrual cycle. The ischemic episodes occurred more frequently with low estrogen levels and less frequently at high estrogen levels. This is why CAS occurs more often in post-menopausal women. However, no similar association was demonstrated in progesterone levels.

3. Oxidative Stress

Oxidative stress is the state of imbalance between the action of reactive oxygen species (ROS) and the biological ability to dispose of reactive intermediates or to repair the damage. Due to enzyme activity, the reducing environment in cells is maintained. Disruption of this mechanism can contribute to the production of free radicals and peroxides. Free radicals are defined as substances that have one or more unpaired electrons. This feature makes them highly reactive and allows them to donate their electrons to other molecules. Consequently, it leads to chain reactions and then oxidative damage [35].

Studies showed that oxidative stress plays an important role in the pathogenesis of endothelial dysfunction [36,37]. ROS are responsible for the degradation of NO, thus stimulating vasoconstriction and causing endothelial damage [30].

Thioredoxin is a ubiquitous enzyme, and one of its functions is cytoprotection against oxidative stress [38]. Miyamoto et al. [39] reported that plasma levels of thioredoxin were increased in subjects with coronary spastic angina. Furthermore, it was shown that higher thioredoxin levels were strictly connected with a more frequent occurrence of anginal attacks. It can be concluded that the high activity of the disease is associated with intensified oxidative stress.

Smoking has been recognized as one of the major risk factors for coronary spasms [9]. Cigarette smoke is the source of a large number of free radicals causing the degradation of NO [40]. It has been reported that the number of smokers was significantly higher in the coronary spastic angina group than in the chest pain syndrome group [39]. Oxidative activity of estrogen protects pre-menopausal women from CAS, but this does not apply to those who smoke [16]. Motoyama et al. [14] concluded that vitamin C can improve impaired endothelium-dependent vasodilation in chronic smokers. Serum levels of vitamin C were lower in smokers than in nonsmokers. Additionally, plasma levels of thiobarbituric-acid-reactive substances (TBARS) were remarkably higher in those addicted to cigarettes. TBARS are known as an indicator of oxidative stress. However, the infusion of vitamin C resulted in a decrease in TBARS levels in smokers but did not change the levels in nonsmokers. These results are presented in Table 1.

Table 1. Association between particular indicators and risk of CSA.

References	Study Design	Year	All Patients	Examined Indicator	Conclusions
Motoyama et al. [14]	Clinical trial	1997	40	vitamin C	Decreased vitamin C levels in smokers.
Motoyama et al. [14]	Clinical trial	1997	40	TBARS	Increased TBARS levels in smokers.
Miyamoto et al. [39]	Clinical trial	2004	170	thioredoxin	Increased thioredoxin levels in subjects with CSA.
Miwa et al. [41]	Clinical trial	1996	103	vitamin E	Decreased vitamin E levels in subjects with CSA.

TBARS, thiobarbituric-acid-reactive substances; CSA, coronary spastic angina.

Not only vitamin C is helpful for patients with coronary spastic angina. Studies have also shown that vitamin E is able to restore disturbed arterial reactivity. Miwa et al. [41] concluded that plasma vitamin E levels were markedly lower in subjects suffering from active variant angina than in those without coronary spasms. This finding suggests a connection between oxidative stress and CAS. In Motoyama's research [15], the effect of the oral administration of 300 mg/day of vitamin E on endothelium-dependent vasodilation was examined. It was shown that supplementation of vitamin E resulted in improvement in flow-dependent vasodilation. Furthermore, this management also caused a decrease in plasma TBARS levels.

The mechanism of either vitamin C and E is based on increasing the availability of intracellular reduced glutathione (GSH) and thiols [42]. Glutathione is mainly responsible for protection from oxidative stress and the prevention of nitric oxide inactivation. It has been reported that intracoronary infusion of GSH can restore the proper function of the endothelium [43] and suppress constrictor response to ACh in epicardial coronary arteries [44].

4. Inflammation

The evidence based on many studies [4,10,11,45], as well as clinical settings, suggests an association between CAS and inflammation, especially connected with Rho-kinase regulation [10,11]. CAS indicates an association with chronic inflammation by elevated biomarkers such as hs-CRP [2,12,45], interleukin-6, peripheral leukocytes, monocytes [2], and soluble CD40 ligands [12]. Moreover, adhesive molecules, such as P-selection, are elevated in patients with CAS [46]. A chronic low-grade inflammatory state may lead to

CAS through RhoA/Rho-kinase pathway activation and a reduction in endothelial NO activity [11]. C-reactive protein, a sensitive marker of inflammation, suppresses endothelial NO activity and activates RhoA signaling [47] and Rho-kinase activity in white blood cells, which is a prognostic factor for the severity of CAS, correlating with the Il-6 level in the plasma [2]. Moreover, another inflammatory marker, an increased mastocyte level, has been reported in patients with CAS [48]. Cigarette smoking, a major risk factor for coronary artery spasms, is connected with low-grade inflammation and an increased hs-CRP level [12,13]. Thus, it confirms that even minor elevations of the hs-CRP level in serum are essentially and independently associated with coronary spasms. Furthermore, a recent study suggested that coronary spasms are associated with inflammation of coronary adventitia and perivascular adipose tissue [49].

5. Smooth Muscle Hypercontractility

The activity of vascular smooth muscles, contraction and relaxation, is regulated by the phosphorylation and dephosphorylation of the myosin light chain (MLC). Physiologically, phosphorylation is induced by an increase in the intracellular concentration of calcium ions, which, being in a complex with calmodulin, activate myosin light chain kinase leading to phosphorylation of MLC. In coronary artery spasms, excessive contraction of the smooth muscles of the coronary vessels occurs in response to an increase in the intracellular Ca^{2+} influx [4].

Elevated expression of L-type Ca^{2+} channels and an increase in Ca^{2+} entry into vascular smooth muscle cells (VSMCs) through the channels may also initiate the spasm [50]. Moreover, a Ca^{2+} influx through the alpha1H Ca^{2+} system is crucial to coronary arteries' relaxation. The deficiency of α1HT-type calcium channels inhibits the relaxing effect of ACh [17], which may contribute to the pathogenesis of coronary artery spasms.

Phospholipase C overactivity, which is dependent on Ca^{2+}, may also cause CAS through enhanced contraction of VSMCs [51].

ROK and RhoA, being VSMC contractility regulators, are involved in the pathogenesis of coronary artery spasms. Properly, the Rho-kinase metabolic pathway modulates the level of MLC phosphorylation by the inhibition of myosin phosphatase.

The hyper-reactivity of Rho kinase in smooth muscle cells promotes its contraction by sensitizing the myosin light chain to calcium ions, as well as indirectly increasing the phosphorylation of this chain, promoting vasoconstriction. A study on animal models showed that hydroxyfasudil, the Rho kinase inhibitor, prevented dose-dependent excessive coronary contractions, supporting the role of Rho kinase in the pathogenesis of CAS [52,53].

It is essential to mention that inflammation may be also a trigger for vascular smooth muscle cell hyperactivity. The activity of Rho kinase in the coronary artery can be increased by the proinflammatory mediator, interleukin 1β (Il-1β) [10], which confirms the participation of the inflammatory process in the generation of excessive smooth muscle contractions in CAS.

Coronary artery spasms may also be the result of a defect in the endothelial enzyme responsible for the production of NO, which is one of the key mediators inducing vasodilation. Rho kinase can inhibit the production of NO [54], and the absence of this relaxing factor might result in CAS. Treatment that shows the appropriate effect in this situation is statins, which increase the activity of endothelial NO, decrease ROS, and suppress the RhoA/ROCK pathway [2].

ATP-sensitive potassium channels (KATP) are responsible for the regulation of vascular tone. KATP channels are made up of two types of subunits: the lumen-forming subunit (usually Kir 6.2) and the sulfonylurea receptor (SUR). Studies [18,19] suggest that the SUR2 KATP channel is a crucial regulator of episodic vasomotor activity, and the loss of function of SUR2 KATP channels is of key importance for the proper function of the coronary vessels. The loss of KATP activity in the smooth muscles of the coronary vessels correlates with the occurrence of excessive contraction of the coronary artery. Based on these conclusions, it can be confirmed that there is a relationship between malfunctioning KATP channels and the occurrence of Prinzmetal variant angina.

Transmitters such as serotonin [20], dopamine [21], or histamine [22] may also be important factors influencing the VSMCs involved in the pathogenesis of CAS. Direct injection of ACh, a parasympathetic neurotransmitter of the nervous system, into the coronary artery also causes excessive contraction [23]. Focal administration of Il-1β may damage the inner membrane of the coronary vessels, sensitizing these places to the coronary administration of serotonin and histamine, which may lead to vasoconstriction [55].

There are multiple different pathways in which vascular smooth muscle hypercontractility may cause coronary artery spasms, but their connotation remains to be elucidated.

6. Atherosclerosis and Thrombosis

Atherosclerosis is characterized by the chronic accumulation of cholesterol-rich plaque in the arteries and is linked to a broad spectrum of cardiovascular diseases [56]. The disturbance of vascular endothelial structure and its function has also a major role in the pathogenesis of atherosclerosis [57,58].

In recent years, there has been substantial research showing correlations between atherosclerosis and vasomotor dysfunction. Although normal vessels are not excluded, CAS occurs more frequently among arteries with atherosclerotic segments [59]. Pelligrini et al. found that patients with CAS have more advanced atherosclerosis and a higher prevalence of vulnerable plaques, compared to those without CAS [60]. Furthermore, many studies have revealed that the presence of atherosclerosis together with CAS is associated with worse patient outcomes [59]. It is also worth mentioning that the CAS could trigger the rapture of stable plaque. In this way, coronary thrombosis and myocardial infarction (MI) might occur [61]. These findings by Yamagishi et al. suggest that atherosclerosis exists at the location of the focal vasospasm even in the absence of coronary disease on the angiography. Therefore, the development of focal vasospasms is linked to the presence of atherosclerotic lesions [62].

Spasms and atherosclerosis most likely have similar etiological pathways, such as endothelial dysfunction and arterial remodeling [63]. However, in accordance with recent research by Morita et al., atherosclerosis and CAS pathophysiologies might differ. Cigarette smoking, low diastolic blood pressure, and systemic low-grade inflammation are risk factors for CAS. On the contrary, the predictors for coronary atherosclerotic stenosis are age, diabetes mellitus, low HDL cholesterol, hypertension or high systolic blood pressure, and uric acid [64]. Interestingly, in the arterial vasculature, the atherosclerotic lesions develop at different locations where the spasm occurs. Moreover, recent studies showed that percutaneous coronary intervention (PCI) with stenting due to atherosclerotic stenosis does not contribute to the recurrence of CAS and that spasms often occur diffusely in the distal segments of the stented lesion [4].

CAS and atherosclerosis play an essential role in the pathogenesis of coronary heart diseases [65].

We also do not forget that coronary thrombosis is recognized as one of the causes of acute coronary syndromes including acute myocardial infarction, unstable angina, and sudden ischemic death [66,67]. Intimal tear, intimal erosion, and microthrombi are the three main abnormal morphologic findings of optical coherence tomography in patients with vasospasm-induced acute coronary syndrome as compared to individuals with chronic stable variant angina [68]. Kitano et al. established that spastic segments with focal spasms were more prone to have intracoronary thrombi [69]. Furthermore, intracoronary thrombi were found at all spastic segments, despite the presence of atherosclerotic lesions. Teragawa et al. indicated that CAS plays a significant role in thrombogenicity [70]. However, there were no significant differences between the spastic section with the intracoronary thrombi and the spastic segment in terms of the rate and severity of plaque formation as determined by CAS [70].

Patients with CAS have an increased risk of rapid plaque progression and ischemic events because coronary artery spasms trigger local thrombus formation and extensive inflammatory response [71]. Interestingly, patients with coronary spasms have increased plasma levels of hs-CRP and P-selection. Platelets are activated after attacks of CAS. Further, thrombogenesis is intensified [5]. Therefore, we can conclude that CAS might induce thrombosis in the coronary circulation.

7. Deficient Aldehyde Dehydrogenase 2 Activity

Susceptibility to disease is often genetically determined and is specific to a population [72]. This is also the case with CSA, which is a common disease among East Asians [73]. A mutation in the ALDH2 gene is believed to be the cause of this condition; moreover, it is likely that nearly 1 billion people worldwide carry the mutation, most of whom are Easterners [74]. It has been suggested that the risk of coronary heart disease and myocardial infarction may be increased due to the Glu504Lys variant in the ALDH2 gene, which is responsible for reducing the ability of ALDH2 to metabolize acetaldehyde [72]. East Asians are at higher risk for this variant, which is quite common in this population—it occurs in up to 40% of East Asians, while it is not observed in other parts of the world [73]. The mutation in question is a point mutation involving the substitution of glutamic acid for lysine and occurs on chromosome 12q24.2, leading to a mutant, dominant allele (A) [75]. Homozygotes carrying the defective variant are most at risk for enzyme deficiency, while heterozygotes have moderate enzyme deficiency [73]. It is also suggested that carriers of the A allele have a 48% risk of coronary artery disease (CAD) compared to those without the mutation [75].

Alcohol consumption has been proven to be one of the risk factors for the development of coronary heart disease [72]. After consuming ethanol, we can divide its metabolism into two stages. The first involves the conversion of alcohol to acetaldehyde with the involvement of alcohol dehydrogenases ADH; then, in the second stage, this product is oxidized to acetic acid via aldehyde dehydrogenases [72,74,75]. It is believed that acetaldehyde has 30 times more toxicity than ethanol; moreover, it forms free radicals that can react with DNA [74]. It is aldehyde dehydrogenase 2 that is thought to be the main enzyme that oxidizes the harmful aldehyde [72,76]. However, the action of ALDH2 is not limited to removing toxic acetaldehyde from the body but also aldehydes formed by lipid peroxidation [73]. Unfortunately, ALDH2 activity can be reduced by the presence of the 504lys variant, which leads to an increase in harmful acetaldehyde in the blood after ethanol ingestion due to impaired metabolism. As a result, homozygotes have an 18-fold higher concentration of noxious aldehyde, while heterozygotes have a 5-fold higher concentration of noxious aldehyde compared to wild-type homozygotes. The risk of CAD is likely related to the amount of alcohol consumed, with large amounts of alcohol having a higher risk [75]. It has also been suggested that high levels of acetaldehyde may affect circulation and blood pressure, thereby increasing the risk of CAD [72]. What is more, patients with ALDH2 deficiency are at risk for many other diseases such as esophageal and gastric cancer, cirrhosis, Alzheimer's disease, and osteoporosis [74].

Mizuno Y et al. [73] investigated the relationship between smoking and the presence of a defective ALDH2 variant in the pathogenesis of CSA. The study showed that Asians with defective ALDH2*2 alleles have a higher risk of CSA. It was also noted that the genetic factor interacts with and exacerbates the deleterious effects of smoking on vasoconstriction, and the joint effect of the two factors interacts more strongly than each factor alone by increasing reactive aldehydes. It has been pointed out that reactive aldehydes may be a target for prevention and treatment in people at risk or suffering from CSA [73]. On the other hand, Li Y et al. [76] conducted a meta-analysis in which they evaluated the association between the G487A polymorphism of the ALDH2 gene and CAD in the Chinese population. They found a positive correlation of this gene variant with susceptibility to CAD [76]. Gu J. et al. [72] also conducted a meta-analysis in which they examined the relationship between the ALDH2 Glu504Lys polymorphism and the risk of CAD or myocardial infarction among the Asian population. Interestingly, it was shown that the Chinese and Korean population with the 504lys variant has a higher risk of developing CAD and MI which was not observed in the Japanese population [72]. Zhang L. et al. [75] evaluated the relationship between ALDH2 polymorphism and the risk of CAD. They analyzed 11 population-based studies, which included Chinese, Korean, and Japanese. Based on the study, they suggested that a defective dominant A allele is associated with lower concentrations of high-density lipoprotein C, which may also influence high CAD risk [75]. Another topic of interest was addressed by Fujioka K. et al. [74] who evaluated whether administration of the dietary supplement ESSENTIAL AD2 affects acetaldehyde levels in ALDH2-deficient subjects after alcohol consumption. They studied 12 subjects who were heterozygotes for mutations in the ALDH2 gene. Interestingly, after 28 days of supplement therapy, a reduction in serum acetaldehyde levels was observed after alcohol consumption. A reduction in liver enzymes was also observed during the study [74]. The Table 2 analyzes some of the studies discussed [72–76].

Table 2. Association between mutation in the ALDH2 gene and CAD risk.

Authors	**Gu et al. [72]**	**Mizuno et al. [73]**	**Fujioka et al. [74]**	**Zhang et al. [75]**	**Li et al. [76]**
Year	2013	2017	2019	2015	2018
Study design	Meta-analysis	Clinical trial	Clinical trial	Meta-analysis	Meta-analysis
All patients	6762	410	12	8366	5644
Aim of the study	Evaluation of the association between the Glu504Lys mutation in the ALDH2 gene and the risk of CAD and myocardial infarction.	Evaluation of the correlation between smoking and the presence of ALDH2 gene mutation and their impact on CAD risk.	Evaluation of the effect of Essential AD2 supplementation by ALDH2-deficient subjects on blood acetaldehyde levels.	Evaluation of the correlation between polymorphisms of ALDH2 and CAD.	The effect of the G487A ALDH2 mutation on the occurrence of CAD.
Conclusions	The Glu504Lys mutation in the ALDH2 gene in the Asian population is associated with a high risk of myocardial infarction and CAD. The polymorphism's correlation with high risk of the above diseases is particularly evident in the Korean and Chinese populations but not in the Japanese population.	The synergism of smoking and ALDH2 gene mutation has a greater impact on CAD risk than either factor alone by increasing the amount of reactive aldehydes.	Daily supplementation of essential AD2 by ALDH2-deficient individuals resulted in lower blood levels of acetaldehyde.	The presence of a dominant A allele in the genotype of individuals with the mutation is associated with a higher risk of CAD. Patients with a mutation in the ALDH2 gene consume significantly less alcohol than patients without the A allele.	The G487A of ALDH2 mutation significantly increases the risk of CAD.

ALDH2, aldehyde dehydrogenase 2; CAD, coronary artery disease.

8. The Role of Magnesium

Magnesium levels in the body may be related to the occurrence of CAS. Its deficiency is one of the factors causing coronary vasospasms [1,4,5,24]. Magnesium, as an endogenous calcium antagonist [4,5], causes blockage of calcium channels, and therefore coronary smooth muscle contraction does not occur [24,26]. Compulsive alcohol consumption can also lead to angina attacks in a magnesium-dependent mechanism. This is due to magnesium deficiency caused by excessive urinary excretion of magnesium [24].

The basis of CAS is an increase in intracellular calcium ion levels and increased sensitivity to it. The increased sensitivity to calcium ions is due to the increased activity of the RhoA/ROCK pathway. It is also influenced by a decrease in NO release. As a result, coronary smooth muscle contraction occurs [5,24].

Magnesium supplementation is crucial in patients with low magnesium levels [1,4,5]. Beneficial effects of magnesium supply in patients suffering from CAS were reported to consist of a reduction in coronary vasospasms and spasm attacks and alleviation of hyperventilation-induced angina attacks [5,24,26].

An analysis of a case report by Popow et al. [77] describing coronary vasospasms caused by hypocalcemia (0.69 mmol/L; norm: 1.13–1.29 mmol/L) and hypomagnesemia (0.52 mmol/L; norm: 0.7–1.0 mmol/L) led to similar conclusions. The severe pain accompanying the pathology subsided after the infusion of calcium and magnesium. Although this is a rare cause of CAS, it should be taken into account during the differential diagnosis [77].

9. Diagnosis of Vasospastic Angina

The diagnosis of vasospastic angina (VSA) can be problematic [5]. Taking a clinical history and performing an ECG are the first step in the initial evaluation of a patient with an acute attack [78]. Patient symptoms are uncharacteristic and include transient retrosternal pain lasting a few seconds, occurring most often at rest, usually from midnight to the early morning hours [5]. ECG monitoring is important during an attack in the outpatient setting; however, it does not provide evidence of coronary artery spasms, and an attack may not occur at this time [5,79]. Changes in the ECG most often include the appearance of a peak T-wave; less often, we may see a lowering or elevation of the ST segment, as well as the appearance of a negative T-wave or U-wave [26,79]. These abnormalities are observed when the main coronary artery contracts partially or completely, while a mild spasm may not cause any changes in examination [26]. ECG-based diagnosis is often problematic because in many cases, several minutes after the episode occurs, parameters normalize [80]. In addition to the aforementioned changes during systole, we can observe various types of arrhythmias including supraventricular tachyarrhythmias, atrioventricular blocks, or ventricular tachycardias [26]. It is noteworthy that ventricular arrhythmias occur more frequently in patients with a shorter-lasting episode of ischemia, characterized by low severity [80].

Coronary angiography with provocation testing is considered the most convincing and reliable test for diagnosing VSA [1]. However, provocation testing is justified when we suspect VSA in a patient, but our suspicions have not been definitively confirmed [78]. The provocation test involves an intracoronary injection of vasoconstricting agents, among which ergonovine and acetylcholine are the most commonly used [1,26]. The test assesses the percentage of vessel lumen reduction, which can be 50%, 70%, 75%, or 90% [1,79]. Provocative tests have the ability to induce contraction in a given coronary vessel over a well-defined period of time, providing the opportunity to document and monitor this phenomenon in a laboratory setting [5]. Indications for a provocative test can be divided into strong, good, and controversial, and Figure 2 presents some of them [81].

Unfortunately, there are also several contraindications to the provocation test, some of which are listed in Table 3 [1].

Table 3. Contraindications to the provocation test [1].

Some of the Contraindications to Performing an Intracoronary Provocation Test
Pregnancy
Severe hypertension
Significant left coronary artery trunk stenosis
Advanced heart failure
Severe aortic stenosis

The diagnosis of the disease can also be made when fast-acting nitrate administered sublingually results in the resolution of ECG changes [5,79]. It is also possible when nitroglycerin causes a rapid relief of symptoms of the disease; however, in addition, at least one condition shown in Figure 3 must be met [5].

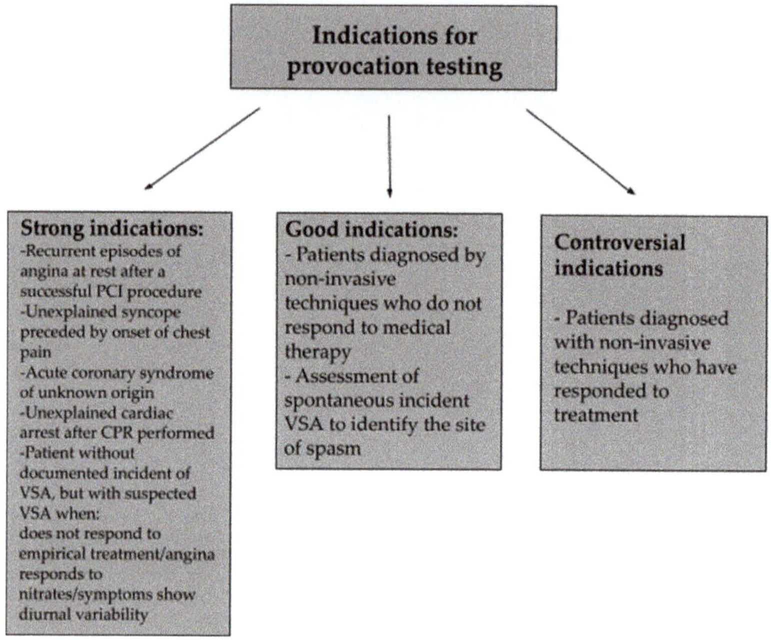

Figure 2. Some of the indications for the provocative test including their division [81].

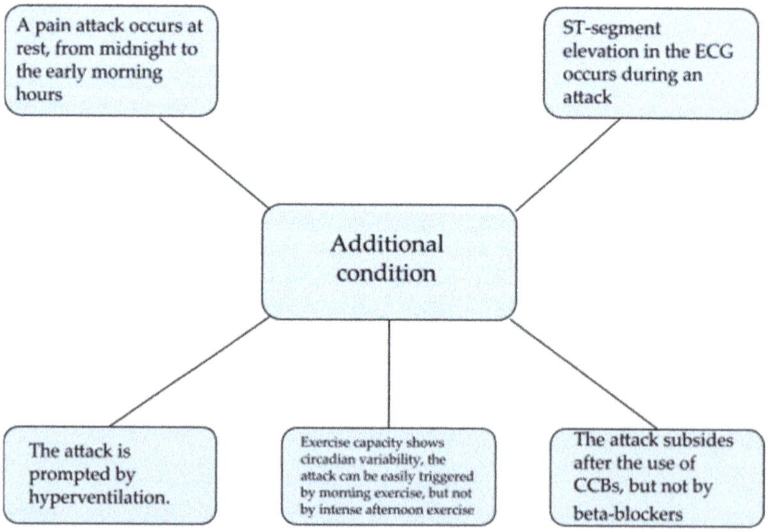

Figure 3. Additional criteria necessary for the diagnosis of angina pectoris when symptoms are relieved by nitroglycerin. Meeting one of the listed criteria is sufficient [5]. ECG, electrocardiography; CCBs, calcium channel blockers.

The use of cardiac biomarkers released during coronary artery contraction, such as creatinine kinase or troponins I and C, can be helpful in the diagnosis, but their levels are not always elevated [79]. If there are indications, a non-invasive exercise test can be performed, but only less than 30% of patients will have ischemic changes in the ECG resulting from exercise-induced vasospasms, and the rest of the patients will have a normal result [78]. Modern coronary artery imaging methods, such as IVUS and OCT, also appear to be useful in detailed diagnosis. IVUS visualizes the vessel's thickened intima-media and even small atherosclerotic plaques at the site of the focal vasospasm; moreover, it detects lesions that may not be visible on angiography. OCT, on the other hand, allows imaging of structural changes such as erosions at the site of the coronary artery spasm [1].

Particularly noteworthy are the international diagnostic criteria presented by the International Study Group for Vasomotor Disorders in Coronary Artery Disease (COVADIS), which are helpful in the diagnosis of VSA. According to this group, diagnosis should be based on three basic pillars, among which are: (I) typical clinical signs, such as a spontaneous incident of nitrate-responsive angina, (II) ECG-documented myocardial ischemia during the spontaneous incident in the form of ST-segment elevation/decrease or new negative U-waves in at least two adjacent leads, and (III) documentation of coronary artery spasms [81].

10. Management of Coronary Artery Spasm

The basis of management of CAS is lifestyle changes and the elimination of risk factors [1,5,26]. It is recommended to stop smoking, consuming alcohol [1,4,5,26], and using substances such as cocaine [1,26]. Moreover, it is important to avoid emotional stress [1,4,5,26] and exercise in the early morning [1,4].

Another component of conservative treatment is pharmacotherapy. In the case of an acute CAS attack, nitrate is used sublingually or orally in spray form. Nitroglycerin or isosorbide dinitrate (ISDN) is converted to NO in vivo. Usage of nitrates is explained by the high sensitivity of the contracted coronary arteries to nitrates and the deficiency of endogenous NO [4,5]. Since these are short-acting drugs, it is necessary to use further preparations, which are CCBs [4]. CCBs are first-line treatment [1,26]. Both dihydropyridine and non-dihydropyridine CCBs have been shown to be effective in reducing recurrent angina [26]. However, some studies report that the use of non-dihydropyridine CCBs results in an almost complete reduction in CAS recurrence [1]. It is important that CCBs should be taken before sleep [4,5,82]. In a more selective (dihydropyridine) or less selective (non-dihydropyridine) way, CCBs act on the L-type calcium channels of myocytes in the vessels by inhibiting calcium influx. Thus, they reduce vascular resistance and cause coronary artery relaxation [83]. Long-acting nitrates are other drugs used to reduce the risk of angina [26], whereas the superiority of one drug over the other has not been demonstrated [1]. Since treatment with long-acting nitrates is accompanied by the phenomenon of tolerance, it is recommended to dose the drug in such a way that an 8-hour break is maintained [4,5,83]. Long-acting nitrates can be used both as monotherapy and as an adjunct to treatment with CCBs [26]. Underlying the action mechanism of nitrates is mitochondrial denitrification, which occurs in the vessel wall with the involvement of aldehyde dehydrogenase. As a result of these transformations, NO is produced. Hence, vasodilation occurs. The effect of nitrates varies with the dose used. Small doses cause a reduction in venous return and preload. In contrast, high-dose nitrates, similarly to CCBs, result in a decrease in afterload and thus a decrease in the heart's oxygen demand while improving oxygen supply to the myocardium [83].

Statins are a group of drugs that cause CAS reduction and improve the prognosis of patients. The effect of statins is possible due to their properties that cause inhibition of the RhoA/ROCK pathway and an increase in NO activity [4]. They are an important adjunct to CAS pharmacotherapy [26]. Inhibitors of the RhoA/ROCK pathway, such as fasudil, a Rho-kinase inhibitor, may prove beneficial in the treatment of CAS because of the contraction-reducing properties of vascular smooth muscle cells (VSMCs). However, further studies are needed [4,5,26].

The use of aspirin in high doses, i.e., >325 mg per day, inhibits prostacyclin formation. This results in vasoconstriction; therefore, such treatment is contraindicated in CAS patients. In contrast, the use of low-dose aspirin, i.e., <100 mg per day, blocks the production of thromboxane A2, resulting in vasodilation; however, there are no conclusive reports on this topic. The use of low-dose aspirin in CAS patients is still a matter of debate [1,26].

Another drug with positive effects in CAS patients is Nicorandil. This drug causes coronary artery dilatation as a result of its nitrate- and potassium-channel-activating properties [1,4,5,83]. Nicorandil is recommended for patients with refractory CAS [26]. The use of magnesium in CAS patients and its mode of action has been discussed above, while antioxidants or estrogens have beneficial effects by improving endothelial function and reducing nitrate tolerance [4].

The use of alpha-1 adrenergic receptor antagonists is controversial, and it is thought that they may be a component of treatment when dealing with CAS refractory to conventional treatment [26]. Patients with CAS may benefit by taking magnesium or antioxidants (vitamins C and E) [1,4,5,26]. Meanwhile, in post-menopausal women, it is recommended to take estrogen, especially for patients with refractory CAS [4,5].

Drugs contraindicated in CAS include beta-blockers [1,26,82] but also, among others, catecholamines, parasympathetic stimulants, and ergot alkaloids [5,26]. They have vasoconstrictive effects and cause coronary vasospasms [1,5,82]. An exception is nebivolol, which owes its selectivity to β1 receptors and its ability to produce NO [1].

A summary of pharmacotherapy is shown in Table 4 [1,4,5,26,83].

Table 4. Pharmacological treatment of CAS [1,4,5,26,83].

Medication	Mechanism	Effect
CCBs	Inhibition of L-type calcium channels of myocytes in vessels	Reduction of vascular resistance and relaxation of coronary arteries
nitrates	NO production as a result of transformations occurring in the vessel wall	Vasodilatation
statins	Inhibition of RhoA/ROCK pathway, increase in NO activity, decrease in ROS	CAS inhibition
aspirin	Blockage of thromboxane A2 production (dose < 100 mg per day)	Vasodilatation
Rho-kinase inhibitors	Inhibition of RhoA/ROCK pathway	Decreased contraction of VSMCs
nicorandil	Activation of nitrates and potassium channel	Dilation of the coronary arteries

CCBs, calcium channel blockers.

Besides the non-invasive CAS treatment methods outlined, there are also invasive methods. Patients with atherosclerotic lesions may benefit from PCI and coronary artery bypass graft (CABG) [1]. Another method is implantation of a cardioverter-defibrillator (ICD). It is designed to prevent ventricular arrhythmias that can result in sudden cardiac death. However, it is suggested that this method of treatment should only be used in selected patients [26].

A study by Lin et al. [84] shows another potential treatment option for refractory CAS—sympathectomy. The study showed that sympathectomy, compared to traditional CAS treatment, was more effective in protecting against a syndrome of episodes of a major adverse cardiac event and death. However, the authors emphasize that further studies are needed [84].

11. Conclusions

Coronary arteries can contract and relax through several mechanisms. Hence, coronary constriction is not always pathological. Nevertheless, in some diseases, coronary constriction becomes more predominant and results in symptoms of a wide variety of heart and vasculature diseases.

In this review, we focused on the important molecular aspects of CAS. We paid attention to the role of endothelial dysfunction and oxidative stress, their pathomechanism, and their influence on the development of cardiovascular diseases. We also drew attention to the influence of smooth muscle hypercontractility, as an excessive intracellular influx of calcium ions, disturbances in the functioning of calcium channels, and malfunctioning of ATP-sensitive potassium channels may result in the occurrence of CAS. Moreover, the most recent discoveries have proven that inflammation plays a critical role in modulating all stages of CAS. The important role of atherosclerosis and thrombosis was also highlighted. Deficiency of aldehyde dehydrogenase 2 activity and magnesium contributes to CAS and was also considered to be of interest.

These findings might shed novel insight on the underlying mechanisms and identify potential diagnostic and therapeutic targets for cardiovascular diseases in the future.

Author Contributions: Conceptualization: B.F., E.M. and J.R.; methodology: J.D., W.F., J.K. (Julia Krzemińska), J.K. (Joanna Kućmierz), M.S., M.W. and E.M.; software: E.M.; validation: B.F., E.M. and J.R.; formal analysis: J.D., W.F., J.K. (Julia Krzemińska), J.K. (Joanna Kućmierz), M.S., M.W. and E.M.; investigation: J.D., W.F., J.K. (Julia Krzemińska), J.K. (Joanna Kućmierz), M.S. and M.W.; resources: B.F., E.M. and J.R.; data curation: E.M.; writing—original draft preparation: E.M.; writing—review and editing: E.M.; visualization: J.D., W.F., J.K. (Julia Krzemińska), J.K. (Joanna Kućmierz), M.S., M.W. and E.M.; supervision: B.F., E.M. and J.R.; project administration: B.F., E.M. and J.R.; funding acquisition: B.F. and J.R. All authors have read and agreed to the published version of the manuscript.

Funding: This research received no external funding.

Institutional Review Board Statement: Not applicable.

Informed Consent Statement: Not applicable.

Data Availability Statement: The data used in this article are sourced from materials mentioned in the References section.

Conflicts of Interest: The authors declare no conflict of interest.

References

1. Matta, A.; Bouisset, F.; Lhermusier, T.; Campelo-Parada, F.; Elbaz, M.; Carrié, D.; Roncalli, J. Coronary Artery Spasm: New Insights. *J. Interv. Cardiol.* **2020**, *2020*, 5894586. [CrossRef] [PubMed]
2. Hung, M.-J.; Hu, P.; Hung, M.-Y. Coronary Artery Spasm: Review and Update. *Int. J. Med. Sci.* **2014**, *11*, 1161–1171. [CrossRef] [PubMed]
3. Prinzmetal, M.; Kennamer, R.; Merliss, R.; Wada, T.; Bor, N. Angina Pectoris I. A Variant Form of Angina Pectoris: Preliminary Report. *Am. J. Med.* **1959**, *27*, 375–388. [CrossRef]
4. Yasue, H.; Mizuno, Y.; Harada, E. Coronary Artery Spasm—Clinical Features, Pathogenesis and Treatment. *Proc. Jpn. Acad. Ser. B Phys. Biol. Sci.* **2019**, *95*, 53–66. [CrossRef]
5. Yasue, H.; Nakagawa, H.; Itoh, T.; Harada, E.; Mizuno, Y. Coronary Artery Spasm–Clinical Features, Diagnosis, Pathogenesis, and Treatment. *J. Cardiol.* **2008**, *51*, 2–17. [CrossRef]
6. Hung, M.-Y.; Kounis, N.G.; Lu, M.-Y.; Hu, P. Myocardial Ischemic Syndromes, Heart Failure Syndromes, Electrocardiographic Abnormalities, Arrhythmic Syndromes and Angiographic Diagnosis of Coronary Artery Spasm: Literature Review. *Int. J. Med. Sci.* **2020**, *17*, 1071–1082. [CrossRef]
7. Kishida, H.; Tada, Y.; Fukuma, N.; Saitoh, T.; Kusama, Y.; Sano, J. Significant Characteristics of Variant Angina Patients with Associated Syncope. *Jpn. Heart J.* **1996**, *37*, 317–326. [CrossRef]
8. Yasue, H.; Matsuyama, K.; Matsuyama, K.; Okumura, K.; Morikami, Y.; Ogawa, H. Responses of Angiographically Normal Human Coronary Arteries to Intracoronary Injection of Acetylcholine by Age and Segment. Possible Role of Early Coronary Atherosclerosis. *Circulation* **1990**, *81*, 482–490. [CrossRef]
9. Sugiishi, M.; Takatsu, F. Cigarette Smoking Is a Major Risk Factor for Coronary Spasm. *Circulation* **1993**, *87*, 76–79. [CrossRef]
10. Shimokawa, H. Cellular and Molecular Mechanisms of Coronary Artery Spasm: Lessons from Animal Models. *Jpn. Circ. J.* **2000**, *64*, 1–12. [CrossRef]
11. Itoh, T.; Mizuno, Y.; Harada, E.; Yoshimura, M.; Ogawa, H.; Yasue, H. Coronary Spasm Is Associated with Chronic Low-Grade Inflammation. *Circ. J.* **2007**, *71*, 1074–1078. [CrossRef] [PubMed]
12. Hubert, A.; Seitz, A.; Pereyra, V.M.; Bekeredjian, R.; Sechtem, U.; Ong, P. Coronary Artery Spasm: The Interplay Between Endothelial Dysfunction and Vascular Smooth Muscle Cell Hyperreactivity. *Eur. Cardiol.* **2020**, *15*, e12. [CrossRef] [PubMed]
13. Yasue, H.; Hirai, N.; Mizuno, Y.; Harada, E.; Itoh, T.; Yoshimura, M.; Kugiyama, K.; Ogawa, H. Low-Grade Inflammation, Thrombogenicity, and Atherogenic Lipid Profile in Cigarette Smokers. *Circ. J.* **2006**, *70*, 8–13. [CrossRef]

14. Motoyama, T.; Kawano, H.; Kugiyama, K.; Hirashima, O.; Ohgushi, M.; Yoshimura, M.; Ogawa, H.; Yasue, H. Endothelium-Dependent Vasodilation in the Brachial Artery Is Impaired in Smokers: Effect of Vitamin C. *Am. J. Physiol.* **1997**, *273*, H1644–H1650. [CrossRef] [PubMed]
15. Motoyama, T.; Kawano, H.; Kugiyama, K.; Hirashima, O.; Ohgushi, M.; Tsunoda, R.; Moriyama, Y.; Miyao, Y.; Yoshimura, M.; Ogawa, H.; et al. Vitamin E Administration Improves Impairment of Endothelium-Dependent Vasodilation in Patients with Coronary Spastic Angina. *J. Am. Coll. Cardiol.* **1998**, *32*, 1672–1679. [CrossRef]
16. Kawano, H.; Ogawa, H. Endothelial Dysfunction and Coronary Artery Spasm. *Curr. Drug Targets Cardiovasc. Haematol. Disord.* **2004**, *4*, 23–33. [CrossRef]
17. Chen, C.-C.; Lamping, K.G.; Nuno, D.W.; Barresi, R.; Prouty, S.J.; Lavoie, J.L.; Cribbs, L.L.; England, S.K.; Sigmund, C.D.; Weiss, R.M.; et al. Abnormal Coronary Function in Mice Deficient in Alpha1H T-Type Ca^{2+} Channels. *Science* **2003**, *302*, 1416–1418. [CrossRef]
18. Kakkar, R.; Ye, B.; Stoller, D.A.; Smelley, M.; Shi, N.-Q.; Galles, K.; Hadhazy, M.; Makielski, J.C.; McNally, E.M. Spontaneous Coronary Vasospasm in KATP Mutant Mice Arises From a Smooth Muscle–Extrinsic Process. *Circ. Res.* **2006**, *98*, 682–689. [CrossRef]
19. Chutkow, W.A.; Pu, J.; Wheeler, M.T.; Wada, T.; Makielski, J.C.; Burant, C.F.; McNally, E.M. Episodic Coronary Artery Vasospasm and Hypertension Develop in the Absence of Sur2 KATP Channels. *J. Clin. Investig.* **2002**, *110*, 203–208. [CrossRef]
20. McFadden, E.P.; Clarke, J.G.; Davies, G.J.; Kaski, J.C.; Haider, A.W.; Maseri, A. Effect of Intracoronary Serotonin on Coronary Vessels in Patients with Stable Angina and Patients with Variant Angina. *N. Engl. J. Med.* **1991**, *324*, 648–654. [CrossRef]
21. Crea, F.; Chierchia, S.; Kaski, J.C.; Davies, G.J.; Margonato, A.; Miran, D.O.; Maseri, A. Provocation of Coronary Spasm by Dopamine in Patients with Active Variant Angina Pectoris. *Circulation* **1986**, *74*, 262–269. [CrossRef] [PubMed]
22. Ginsburg, R.; Bristow, M.R.; Kantrowitz, N.; Baim, D.S.; Harrison, D.C. Histamine Provocation of Clinical Coronary Artery Spasm: Implications Concerning Pathogenesis of Variant Angina Pectoris. *Am. Heart J.* **1981**, *102*, 819–822. [CrossRef]
23. Yasue, H.; Horio, Y.; Nakamura, N.; Fujii, H.; Imoto, N.; Sonoda, R.; Kugiyama, K.; Obata, K.; Morikami, Y.; Kimura, T. Induction of Coronary Artery Spasm by Acetylcholine in Patients with Variant Angina: Possible Role of the Parasympathetic Nervous System in the Pathogenesis of Coronary Artery Spasm. *Circulation* **1986**, *74*, 955–963. [CrossRef] [PubMed]
24. Kusama, Y.; Kodani, E.; Nakagomi, A.; Otsuka, T.; Atarashi, H.; Kishida, H.; Mizuno, K. Variant Angina and Coronary Artery Spasm: The Clinical Spectrum, Pathophysiology, and Management. *J. Nippon Med. Sch.* **2011**, *78*, 4–12. [CrossRef]
25. Kondo, T.; Terada, K. Coronary-Artery Vasospasm. *N. Engl. J. Med.* **2017**, *376*, e52. [CrossRef]
26. Picard, F.; Sayah, N.; Spagnoli, V.; Adjedj, J.; Varenne, O. Vasospastic Angina: A Literature Review of Current Evidence. *Arch. Cardiovasc. Dis.* **2019**, *112*, 44–55. [CrossRef]
27. Rubanyi, G.M. The Role of Endothelium in Cardiovascular Homeostasis and Diseases. *J. Cardiovasc. Pharm.* **1993**, *22* (Suppl. 4), S1–S14. [CrossRef]
28. Boulanger, C.; Lüscher, T.F. Release of Endothelin from the Porcine Aorta. Inhibition by Endothelium-Derived Nitric Oxide. *J. Clin. Investig.* **1990**, *85*, 587–590. [CrossRef]
29. Takemoto, M.; Egashira, K.; Usui, M.; Numaguchi, K.; Tomita, H.; Tsutsui, H.; Shimokawa, H.; Sueishi, K.; Takeshita, A. Important Role of Tissue Angiotensin-Converting Enzyme Activity in the Pathogenesis of Coronary Vascular and Myocardial Structural Changes Induced by Long-Term Blockade of Nitric Oxide Synthesis in Rats. *J. Clin. Investig.* **1997**, *99*, 278–287. [CrossRef]
30. Moncada, S.; Palmer, R.M.; Higgs, E.A. Nitric Oxide: Physiology, Pathophysiology, and Pharmacology. *Pharm. Rev.* **1991**, *43*, 109–142.
31. Kugiyama, K.; Yasue, H.; Okumura, K.; Ogawa, H.; Fujimoto, K.; Nakao, K.; Yoshimura, M.; Motoyama, T.; Inobe, Y.; Kawano, H. Nitric Oxide Activity Is Deficient in Spasm Arteries of Patients with Coronary Spastic Angina. *Circulation* **1996**, *94*, 266–271. [CrossRef] [PubMed]
32. Mendelsohn, M.E.; Karas, R.H. The Protective Effects of Estrogen on the Cardiovascular System. *N. Engl. J. Med.* **1999**, *340*, 1801–1811. [CrossRef] [PubMed]
33. Anderson, T.J.; Uehata, A.; Gerhard, M.D.; Meredith, I.T.; Knab, S.; Delagrange, D.; Lieberman, E.H.; Ganz, P.; Creager, M.A.; Yeung, A.C. Close Relation of Endothelial Function in the Human Coronary and Peripheral Circulations. *J. Am. Coll. Cardiol.* **1995**, *26*, 1235–1241. [CrossRef]
34. Joannides, R.; Haefeli, W.E.; Linder, L.; Richard, V.; Bakkali, E.H.; Thuillez, C.; Lüscher, T.F. Nitric Oxide Is Responsible for Flow-Dependent Dilatation of Human Peripheral Conduit Arteries in Vivo. *Circulation* **1995**, *91*, 1314–1319. [CrossRef] [PubMed]
35. Hayyan, M.; Hashim, M.A.; AlNashef, I.M. Superoxide Ion: Generation and Chemical Implications. *Chem. Rev.* **2016**, *116*, 3029–3085. [CrossRef]
36. Diaz, M.N.; Frei, B.; Vita, J.A.; Keaney, J.F. Antioxidants and Atherosclerotic Heart Disease. *N. Engl. J. Med.* **1997**, *337*, 408–416. [CrossRef]
37. Gryglewski, R.J.; Palmer, R.M.; Moncada, S. Superoxide Anion Is Involved in the Breakdown of Endothelium-Derived Vascular Relaxing Factor. *Nature* **1986**, *320*, 454–456. [CrossRef]
38. Nakamura, H.; Nakamura, K.; Yodoi, J. Redox Regulation of Cellular Activation. *Annu. Rev. Immunol.* **1997**, *15*, 351–369. [CrossRef]

39. Miyamoto, S.; Kawano, H.; Sakamoto, T.; Soejima, H.; Kajiwara, I.; Hokamaki, J.; Hirai, N.; Sugiyama, S.; Yoshimura, M.; Yasue, H.; et al. Increased Plasma Levels of Thioredoxin in Patients with Coronary Spastic Angina. *Antioxid. Redox Signal.* **2004**, *6*, 75–80. [CrossRef]
40. Ota, Y.; Kugiyama, K.; Sugiyama, S.; Ohgushi, M.; Matsumura, T.; Doi, H.; Ogata, N.; Oka, H.; Yasue, H. Impairment of Endothelium-Dependent Relaxation of Rabbit Aortas by Cigarette Smoke Extract–Role of Free Radicals and Attenuation by Captopril. *Atherosclerosis* **1997**, *131*, 195–202. [CrossRef]
41. Miwa, K.; Miyagi, Y.; Igawa, A.; Nakagawa, K.; Inoue, H. Vitamin E Deficiency in Variant Angina. *Circulation* **1996**, *94*, 14–18. [CrossRef] [PubMed]
42. Meister, A. Glutathione-Ascorbic Acid Antioxidant System in Animals. *J. Biol. Chem.* **1994**, *269*, 9397–9400. [CrossRef]
43. Kugiyama, K.; Ohgushi, M.; Motoyama, T.; Hirashima, O.; Soejima, H.; Misumi, K.; Yoshimura, M.; Ogawa, H.; Sugiyama, S.; Yasue, H. Intracoronary Infusion of Reduced Glutathione Improves Endothelial Vasomotor Response to Acetylcholine in Human Coronary Circulation. *Circulation* **1998**, *97*, 2299–2301. [CrossRef] [PubMed]
44. Kugiyama, K.; Miyao, Y.; Sakamoto, T.; Kawano, H.; Soejima, H.; Miyamoto, S.; Yoshimura, M.; Ogawa, H.; Sugiyama, S.; Yasue, H. Glutathione Attenuates Coronary Constriction to Acetylcholine in Patients with Coronary Spastic Angina. *Am. J. Physiol. Heart Circ. Physiol.* **2001**, *280*, H264–H271. [CrossRef] [PubMed]
45. Lewis, J.R.; Kisilevsky, R.; Armstrong, P.W. Prinzmetal's Angina, Normal Coronary Arteries and Pericarditis. *Can Med. Assoc. J.* **1978**, *119*, 36–39. [PubMed]
46. Kaikita, K.; Ogawa, H.; Yasue, H.; Sakamoto, T.; Suefuji, H.; Sumida, H.; Okumura, K. Soluble P-Selectin Is Released into the Coronary Circulation after Coronary Spasm. *Circulation* **1995**, *92*, 1726–1730. [CrossRef]
47. Jialal, I.; Devaraj, S.; Singh, U. C-Reactive Protein and the Vascular Endothelium. *J. Am. Coll. Cardiol.* **2006**, *47*, 1379–1381. [CrossRef]
48. Forman, M.B.; Oates, J.A.; Robertson, D.; Robertson, R.M.; Roberts, L.J.; Virmani, R. Increased Adventitial Mast Cells in a Patient with Coronary Spasm. *N. Engl. J. Med.* **1985**, *313*, 1138–1141. [CrossRef]
49. Ohyama, K.; Matsumoto, Y.; Takanami, K.; Ota, H.; Nishimiya, K.; Sugisawa, J.; Tsuchiya, S.; Amamizu, H.; Uzuka, H.; Suda, A.; et al. Coronary Adventitial and Perivascular Adipose Tissue Inflammation in Patients With Vasospastic Angina. *J. Am. Coll. Cardiol.* **2018**, *71*, 414–425. [CrossRef]
50. Kuga, T.; Shimokawa, H.; Hirakawa, Y.; Kadokami, Y.; Arai, Y.; Fukumoto, Y.; Kuwata, K.; Kozai, T.; Egashira, K.; Takeshita, A. Increased Expression of L-Type Calcium Channels in Vascular Smooth Muscle Cells at Spastic Site in a Porcine Model of Coronary Artery Spasm. *J. Cardiovasc. Pharm.* **2000**, *35*, 822–828. [CrossRef]
51. Nakano, T.; Osanai, T.; Tomita, H.; Sekimata, M.; Homma, Y.; Okumura, K. Enhanced Activity of Variant Phospholipase C-Δ1 Protein (R257H) Detected in Patients With Coronary Artery Spasm. *Circulation* **2002**, *105*, 2024–2029. [CrossRef] [PubMed]
52. Shimokawa, H.; Seto, M.; Katsumata, N.; Amano, M.; Kozai, T.; Yamawaki, T.; Kuwata, K.; Kandabashi, T.; Egashira, K.; Ikegaki, I.; et al. Rho-Kinase-Mediated Pathway Induces Enhanced Myosin Light Chain Phosphorylations in a Swine Model of Coronary Artery Spasm. *Cardiovasc. Res.* **1999**, *43*, 1029–1039. [CrossRef]
53. Kandabashi, T.; Shimokawa, H.; Miyata, K.; Kunihiro, I.; Kawano, Y.; Fukata, Y.; Higo, T.; Egashira, K.; Takahashi, S.; Kaibuchi, K.; et al. Inhibition of Myosin Phosphatase by Upregulated Rho-Kinase Plays a Key Role for Coronary Artery Spasm in a Porcine Model With Interleukin-1β. *Circulation* **2000**, *101*, 1319–1323. [CrossRef] [PubMed]
54. Büssemaker, E.; Pistrosch, F.; Förster, S.; Herbrig, K.; Gross, P.; Passauer, J.; Brandes, R.P. Rho Kinase Contributes to Basal Vascular Tone in Humans: Role of Endothelium-Derived Nitric Oxide. *Am. J. Physiol. Heart Circ. Physiol.* **2007**, *293*, H541–H547. [CrossRef] [PubMed]
55. Shimokawa, H.; Ito, A.; Fukumoto, Y.; Kadokami, T.; Nakaike, R.; Sakata, M.; Takayanagi, T.; Egashira, K.; Takeshita, A. Chronic Treatment with Interleukin-1 Beta Induces Coronary Intimal Lesions and Vasospastic Responses in Pigs in Vivo. The Role of Platelet-Derived Growth Factor. *J. Clin. Investig* **1996**, *97*, 769–776. [CrossRef]
56. Botts, S.R.; Fish, J.E.; Howe, K.L. Dysfunctional Vascular Endothelium as a Driver of Atherosclerosis: Emerging Insights Into Pathogenesis and Treatment. *Front. Pharm.* **2021**, *12*, 787541. [CrossRef]
57. Xu, S.; Ilyas, I.; Little, P.J.; Li, H.; Kamato, D.; Zheng, X.; Luo, S.; Li, Z.; Liu, P.; Han, J.; et al. Endothelial Dysfunction in Atherosclerotic Cardiovascular Diseases and Beyond: From Mechanism to Pharmacotherapies. *Pharm. Rev.* **2021**, *73*, 924–967. [CrossRef]
58. Frąk, W.; Wojtasińska, A.; Lisińska, W.; Młynarska, E.; Franczyk, B.; Rysz, J. Pathophysiology of Cardiovascular Diseases: New Insights into Molecular Mechanisms of Atherosclerosis, Arterial Hypertension, and Coronary Artery Disease. *Biomedicines* **2022**, *10*, 1938. [CrossRef]
59. Shin, D.I.; Baek, S.H.; Her, S.H.; Han, S.H.; Ahn, Y.; Park, K.-H.; Kim, D.-S.; Yang, T.-H.; Choi, D.-J.; Suh, J.-W.; et al. The 24-Month Prognosis of Patients With Positive or Intermediate Results in the Intracoronary Ergonovine Provocation Test. *JACC Cardiovasc. Interv.* **2015**, *8*, 914–923. [CrossRef]
60. Pellegrini, D.; Konst, R.; van den Oord, S.; Dimitriu-Leen, A.; Mol, J.-Q.; Jansen, T.; Maas, A.; Gehlmann, H.; van Geuns, R.-J.; Elias-Smale, S.; et al. Features of Atherosclerosis in Patients with Angina and No Obstructive Coronary Artery Disease. *EuroIntervention* **2022**, *18*, e397–e404. [CrossRef]
61. Slavich, M.; Patel, R.S. Coronary Artery Spasm: Current Knowledge and Residual Uncertainties. *Int. J. Cardiol. Heart Vasc.* **2016**, *10*, 47–53. [CrossRef] [PubMed]

62. Yamagishi, M.; Miyatake, K.; Tamai, J.; Nakatani, S.; Koyama, J.; Nissen, S.E. Intravascular Ultrasound Detection of Atherosclerosis at the Site of Focal Vasospasm in Angiographically Normal or Minimally Narrowed Coronary Segments. *J. Am. Coll. Cardiol.* **1994**, *23*, 352–357. [CrossRef]
63. Takagi, Y.; Yasuda, S.; Takahashi, J.; Tsunoda, R.; Ogata, Y.; Seki, A.; Sumiyoshi, T.; Matsui, M.; Goto, T.; Tanabe, Y.; et al. Clinical Implications of Provocation Tests for Coronary Artery Spasm: Safety, Arrhythmic Complications, and Prognostic Impact: Multicentre Registry Study of the Japanese Coronary Spasm Association. *Eur. Heart J.* **2013**, *34*, 258–267. [CrossRef] [PubMed]
64. Morita, S.; Mizuno, Y.; Harada, E.; Nakagawa, H.; Morikawa, Y.; Saito, Y.; Katoh, D.; Kashiwagi, Y.; Yoshimura, M.; Murohara, T.; et al. Differences and Interactions between Risk Factors for Coronary Spasm and Atherosclerosis–Smoking, Aging, Inflammation, and Blood Pressure. *Intern. Med.* **2014**, *53*, 2663–2670. [CrossRef]
65. Vidal-Perez, R.; Abou Jokh Casas, C.; Agra-Bermejo, R.M.; Alvarez-Alvarez, B.; Grapsa, J.; Fontes-Carvalho, R.; Rigueiro Veloso, P.; Garcia Acuña, J.M.; Gonzalez-Juanatey, J.R. Myocardial Infarction with Non-Obstructive Coronary Arteries: A Comprehensive Review and Future Research Directions. *World J. Cardiol.* **2019**, *11*, 305–315. [CrossRef]
66. Abbate, R.; Cioni, G.; Ricci, I.; Miranda, M.; Gori, A.M. Thrombosis and Acute Coronary Syndrome. *Thromb. Res.* **2012**, *129*, 235–240. [CrossRef]
67. Lin, C.S.; Penha, P.D.; Zak, F.G.; Lin, J.C. Morphodynamic Interpretation of Acute Coronary Thrombosis, with Special Reference to Volcano-like Eruption of Atheromatous Plaque Caused by Coronary Artery Spasm. *Angiology* **1988**, *39*, 535–547. [CrossRef]
68. Park, H.-C.; Shin, J.H.; Jeong, W.K.; Choi, S.I.; Kim, S.-G. Comparison of Morphologic Findings Obtained by Optical Coherence Tomography in Acute Coronary Syndrome Caused by Vasospasm and Chronic Stable Variant Angina. *Int. J. Cardiovasc. Imaging* **2015**, *31*, 229–237. [CrossRef]
69. Kitano, D.; Takayama, T.; Sudo, M.; Kogo, T.; Kojima, K.; Akutsu, N.; Nishida, T.; Haruta, H.; Fukamachi, D.; Kawano, T.; et al. Angioscopic Differences of Coronary Intima between Diffuse and Focal Coronary Vasospasm: Comparison of Optical Coherence Tomography Findings. *J. Cardiol.* **2018**, *72*, 200–207. [CrossRef]
70. Teragawa, H.; Orita, Y.; Oshita, C.; Uchimura, Y. Intracoronary Thrombogenicity in Patients with Vasospastic Angina: An Observation Using Coronary Angioscopy. *Diagnostics* **2021**, *11*, 1632. [CrossRef]
71. Nishi, T.; Kume, T.; Yamada, R.; Okamoto, H.; Koto, S.; Yamashita, M.; Ueno, M.; Kamisaka, K.; Sasahira, Y.; Enzan, A.; et al. Layered Plaque in Organic Lesions in Patients With Coronary Artery Spasm. *J. Am. Heart Assoc.* **2022**, *11*, e024880. [CrossRef] [PubMed]
72. Gu, J.-Y.; Li, L.-W. ALDH2 Glu504Lys Polymorphism and Susceptibility to Coronary Artery Disease and Myocardial Infarction in East Asians: A Meta-Analysis. *Arch. Med. Res.* **2014**, *45*, 76–83. [CrossRef] [PubMed]
73. Mizuno, Y.; Hokimoto, S.; Harada, E.; Kinoshita, K.; Yoshimura, M.; Yasue, H. Variant Aldehyde Dehydrogenase 2 (ALDH2*2) in East Asians Interactively Exacerbates Tobacco Smoking Risk for Coronary Spasm—Possible Role of Reactive Aldehydes. *Circ. J.* **2016**, *81*, 96–102. [CrossRef] [PubMed]
74. Fujioka, K.; Gordon, S. Effects of "Essential AD2" Supplement on Blood Acetaldehyde Levels in Individuals Who Have Aldehyde Dehydrogenase (ALDH2) Deficiency. *Am. J.* **2019**, *26*, 583–588. [CrossRef]
75. Zhang, L.L.; Wang, Y.Q.; Fu, B.; Zhao, S.L.; Kui, Y. Aldehyde Dehydrogenase 2 (ALDH2) Polymorphism Gene and Coronary Artery Disease Risk: A Meta-Analysis. *Genet. Mol. Res.* **2015**, *14*, 18503–18514. [CrossRef]
76. Li, Y.; Wang, H.; Wu, J.; Kim, H.J.; Yang, X.; Geng, H.; Gong, G. ALDH2 Gene G487A Polymorphism and Coronary Artery Disease: A Meta-analysis Including 5644 Participants. *J. Cell. Mol. Med.* **2018**, *22*, 1666–1674. [CrossRef] [PubMed]
77. Popow, M.; Kochanowski, J.; Krakowian, M. Coronary Spasm Secondary to Hypocalcaemia and Hypomagnesaemia. *Kardiol. Pol. (Pol. Heart J.)* **2015**, *73*, 57. [CrossRef] [PubMed]
78. Rodriguez Ziccardi, M.; Hatcher, J.D. Prinzmetal Angina. In *StatPearls*; StatPearls Publishing: Treasure Island, FL, USA, 2022.
79. Swarup, S.; Patibandla, S.; Grossman, S.A. Coronary Artery Vasospasm. In *StatPearls*; StatPearls Publishing: Treasure Island, FL, USA, 2022.
80. de Luna, A.B.; Cygankiewicz, I.; Baranchuk, A.; Fiol, M.; Birnbaum, Y.; Nikus, K.; Goldwasser, D.; Garcia-Niebla, J.; Sclarovsky, S.; Wellens, H.; et al. Prinzmetal Angina: ECG Changes and Clinical Considerations: A Consensus Paper. *Ann. Noninvasive Electrocardiol.* **2014**, *19*, 442–453. [CrossRef] [PubMed]
81. Beltrame, J.F.; Crea, F.; Kaski, J.C.; Ogawa, H.; Ong, P.; Sechtem, U.; Shimokawa, H.; Bairey Merz, C.N. Coronary Vasomotion Disorders International Study Group (COVADIS) International Standardization of Diagnostic Criteria for Vasospastic Angina. *Eur. Heart J.* **2017**, *38*, 2565–2568. [CrossRef]
82. Song, J.-K. Coronary Artery Vasospasm. *Korean Circ. J.* **2018**, *48*, 767–777. [CrossRef]
83. Balla, C.; Pavasini, R.; Ferrari, R. Treatment of Angina: Where Are We? *Cardiology* **2018**, *140*, 52–67. [CrossRef] [PubMed]
84. Lin, Y.; Liu, H.; Yu, D.; Wu, M.; Liu, Q.; Liang, X.; Pang, X.; Chen, K.; Luo, L.; Dong, S. Sympathectomy versus Conventional Treatment for Refractory Coronary Artery Spasm. *Coron. Artery Dis.* **2019**, *30*, 418–424. [CrossRef] [PubMed]

biomedicines

MDPI

Article

H_2S Prodrug, SG-1002, Protects against Myocardial Oxidative Damage and Hypertrophy In Vitro via Induction of Cystathionine β-Synthase and Antioxidant Proteins

Rahib K. Islam [1,†], Erinn Donnelly [1,†], Erminia Donnarumma [2], Fokhrul Hossain [1], Jason D. Gardner [1] and Kazi N. Islam [3,*]

1 Departments of Pharmacology and Experimental Medicine, Genetics, and Physiology, Louisiana State University Health Sciences Center, 1901 Perdido St., New Orleans, LA 70112, USA
2 Mitochondrial Biology Group, Institute Pasteur, CNRS UMR 3691, 75015 Paris, France
3 Agricultural Research Development Program, College of Engineering, Science, Technology and Agriculture, Central State University, 1400 Brush Row Road, Wilberforce, OH 45384, USA
* Correspondence: kislam@centralstate.edu; Tel.: +1-937-376-6635
† These authors contributed equally to this work.

Abstract: Endogenously produced hydrogen sulfide (H_2S) is critical for cardiovascular homeostasis. Therapeutic strategies aimed at increasing H_2S levels have proven cardioprotective in models of acute myocardial infarction (MI) and heart failure (HF). The present study was undertaken to investigate the effects of a novel H_2S prodrug, SG-1002, on stress induced hypertrophic signaling in murine HL-1 cardiac muscle cells. Treatment of HL-1 cells with SG-1002 under serum starvation without or with H_2O_2 increased the levels of H_2S, H_2S producing enzyme, and cystathionine β-synthase (CBS), as well as antioxidant protein levels, such as super oxide dismutase1 (SOD1) and catalase, and additionally decreased oxidative stress. SG-1002 also decreased the expression of hypertrophic/HF protein markers such as atrial natriuretic peptide (ANP), brain natriuretic peptide (BNP), galectin-3, TIMP1, collagen type III, and TGF-β1 in stressed HL-1 cells. Treatment with SG-1002 caused a significant induction of cell viability and a marked reduction of cellular cytotoxicity in HL-1 cells under serum starvation incubated without or with H_2O_2. Experimental results of this study suggest that SG-1002 attenuates myocardial cellular oxidative damage and/or hypertrophic signaling via increasing H_2S levels or H_2S producing enzymes, CBS, and antioxidant proteins.

Keywords: H_2S; SG-1002; oxidative stress; reactive oxygen species (ROS); cardiovascular diseases; gasotransmitters; antioxidants; cardiomyocytes

Citation: Islam, R.K.; Donnelly, E.; Donnarumma, E.; Hossain, F.; Gardner, J.D.; Islam, K.N. H_2S Prodrug, SG-1002, Protects against Myocardial Oxidative Damage and Hypertrophy In Vitro via Induction of Cystathionine β-Synthase and Antioxidant Proteins. *Biomedicines* **2023**, *11*, 612. https://doi.org/10.3390/biomedicines11020612

Academic Editors: Tânia Martins-Marques, Gonçalo F. Coutinho and Attila Kiss

Received: 1 October 2022
Revised: 15 February 2023
Accepted: 16 February 2023
Published: 18 February 2023

Copyright: © 2023 by the authors. Licensee MDPI, Basel, Switzerland. This article is an open access article distributed under the terms and conditions of the Creative Commons Attribution (CC BY) license (https://creativecommons.org/licenses/by/4.0/).

1. Introduction

Sulfur has a critical role in protein structure/function and redox status/signaling in all living organisms. Although a sulfur containing molecule, hydrogen sulfide (H_2S), is now recognized as a central player in physiology and pathobiology, the full scope and depth of sulfur metabolome's impact on human health and longevity has been vastly underestimated and is only starting to be grasped. Since many pathological conditions have been related to abnormally low levels of H_2S in blood and/or tissues, and are amenable to treatment by H_2S supplementation, development of safe and efficacious H_2S donors deserves to be undertaken with a sense of urgency; these prodrugs also hold the promise of becoming widely used for disease prevention and as antiaging agents [1]. One such prodrug is an SG-1002, a precursor to a natural-occurring molecule, H_2S, for which deficits have been shown to exist in a number of serious diseases including cardiovascular disease (CVD), type II diabetes, cancer, hypertension, etc. [1].

DATS (diallyl trisulfide), DBTS (dibutenyl trisulfide), TC-2153 (benzopentathiepin 8-trifluoromethylv-1,2,3,4,5-benzopentathiepin-6-amine hydrochloride), and SG-1002 have the potential as pharmacological therapeutic agents that collectively could prove to be useful in the treatment myriad of disease conditions related to oxidative stress and cellular damage inflicted by reactive oxygen species (ROS) including most aging related diseases [1]. In contrast to other donors, SG-1002 is an H_2S prodrug and only operates through H_2S signaling. Other donors elicit pharmacologic effects that are partially through H_2S.

Among the known H_2S prodrugs, SG-1002 is unique in that it lacks a carbon-based scaffold. The absence of a carbon-based scaffold has various advantages for SG-1002 as a medicinal agent. As the number of atoms in the scaffold increases, a given dose contains decreasing amounts of pharmacophore; further, toxicity is frequently scaffold dependent. Additionally, it is not a therapeutic targeted-based agent, it disobeys drug likeness laws, it has a 100% prodrug-to-H2S conversion efficiency, and it is bioactivated. SG-1002 is an α-sulfur rich microcrystalline material that is water insoluble and has traces of ionic chemicals (sodium sulfate, sodium thiosulfate, and sodium polythionates) that have a substantial influence on its physicochemical behavior [1].

Furthermore, SG-1002 is orally active, which is a great benefit as a medicinal substance. The oral route of administration increases patient compliance because it avoids more intrusive methods like injections and/or infusions, which often require patients to visit hospitals or health facilities. Additionally, SG-1002 has entered clinical studies where safety and early efficacy data for two distinct purposes have been established. It is also effective in numerous disease models.

To the best of our knowledge only in one case (SG-1002) has safety been demonstrated in a formal Phase 1 clinical study [2], so to realize the therapeutic potential of these four agents, it will be necessary to invest considerable resources to carry out the required clinical trials. Encouraging results in animal models have been obtained with SG-1002 in heart failure (HF), atherosclerosis, ischemic damage, and Duchenne muscular dystrophy [1]. Based on previously published articles on the roles of SG-1002 in several disease models, the present study was undertaken to determine the effects of SG-1002 on oxidative stress induced hypertrophic signaling in HL-1 cells. The HL-1 cell line is derived from an AT-1 mouse atrial cardiomyocyte tumor and has several key advantages over other types of cardiac cells. It can be recovered from frozen and passaged indefinitely while still maintaining its differentiated biochemical and morphological features, as well as the ability to contract. Because of these properties, HL-1 cardiomyocytes can be used to model the effects of HF, hypertrophy and oxidative stress in vitro [3].

Once thought of only as a toxic gas, H_2S now belongs to a class of compounds known as gasotransmitters. H_2S along with carbon monoxide (CO) and nitric oxide (NO) are endogenously produced gases that are required for cardiovascular homeostasis [4]. Previous studies have shown that NO is highly cytoprotective and that maintaining NO bioavailability is protective against the development and progression of HF [5–7]. Similarly, to H_2S, NO levels are reduced in HF patients compared with healthy controls [5]. Endothelial nitric oxide synthase (eNOS) is elevated in the presence of H_2S leading to the increased levels of circulating NO [4,8]. Similarly, NO therapy was found to increase the levels of H_2S in murine HF model [9]. Endogenously produced H_2S exerts a variety of cytoprotective actions in vivo, by acting as an antioxidant and promoting Nrf2 and NRF-1 signaling [10–12], and thereby augmenting NO-mediated signaling [4].

H_2S is of particular interest as a possible agent to combat HF because of its ability to promote vasodilation and its anti-inflammatory [13], antioxidant [14] and anti-apoptotic [15] properties. Previous studies have shown that genetic overexpression of the H_2S producing enzyme cystathionine γ-lyase (CSE) protects against HF while deficiency exacerbates it [8]. H_2S is a potent antioxidant that can eliminate free radicals and prevent new reactive oxygen species (ROS) from forming, which are particularly detrimental in myocardial infarction/reperfusion (MI/R) injuries [14]. Additionally, H_2S has recently been implicated as a potential treatment against hypertrophic signaling [16], which is responsible for the pathological remodeling associated with HF.

Three enzymes are responsible for the endogenous production of H_2S: cystathionine β-synthase (CBS), CSE, and 3-mercaptopyruvate sulfur transferase (3-MST). H_2S (and its metabolites) are found in most organ systems, including the heart, liver, kidney, brain, nervous system, lung, airway tissues, gastrointestinal tract, reproductive organs, skeletal muscle, pancreas, synovial joints, connective tissue, cochlea, and adipose tissues [17]. Impaired H_2S generation as a result of CSE dysfunction has been cited as a significant contributor to pathology in numerous disease states [4,18,19]. Furthermore, H_2S is critically important to maintaining homeostasis in the cardiovascular system, both in the heart and in the circulation [20,21]. In preclinical models, H_2S therapy attenuates disease severity by antioxidant activity, promoting angiogenesis, modulating mitochondrial function, reducing inflammation, upregulating antioxidant gene programs, inhibiting cell death, and attenuating fibrosis [8,20,22,23].

In mammals, CSE is predominately responsible for the manufacture of H_2S in the cardiovascular system, while CBS is present in greater quantities in the central nervous system [24]. 3-MST is responsible for manufacturing 90% of the H_2S in the brain [24]. A study by Islam et al. has shown that NO therapy protects cardiac function in the myocardial ischemia (MI) murine model via induction of H_2S/H_2S producing enzymes (CSE and CBS) and antioxidant levels [9]. Through the use of an exogenous H_2S donor, SG-1002, we attempted to identify a mechanism for the decrease in stress-induced cardiomyocyte hypertrophic signaling. Furthermore, we also sought to ascertain whether this prodrug may promote expression of CBS, CSE, and 3-MST as well as antioxidant proteins, enabling cells to create more endogenous H_2S. This is the first report demonstrating the beneficial effects of SG-1002 on stress/hypertrophic murine cardiomyocytes.

2. Materials and Methods

2.1. Cell Culture

Murine HL-1 cardiac muscle cells were maintained in 10% fetal bovine serum (FBS) containing Claycomb media followed by serum starvation (1% FBS) media for 24 h. Serum starved/stressed cells were treated for 1 h with either 10 μM SG-1002 (Received from Dr. David J. Lefer, LSU, New Orleans), or H_2O_2 (500 μM), or hydroxylamine (HA) 10 (μM), or a hypertrophic agonist such as endothelin-1 (ET-1) (10 nM), or in combination. Treated cells were analyzed for mRNA levels (by RTqRT-PCR), H_2S levels, oxidative stress, and related assays.

2.2. Real Time Quantitative Reverse Transcription Polymerase Chain Reaction (RTqRT-PCR)

RNA was isolated from the stressed HL-1 cells treated without or with 10 μM of SG-1002, or H_2O_2 (500 μM) or HA (10 μM) or in combination for 1 h in 1% serum (FBS) containing media. 1 μg of RNA was used for the synthesis of cDNA. TaqMan primers from Life Technology Co, Carlsbad, CA, USA. for ANP, BNP, TGF β1, collagen type III, galectin 3, TIMP1, SOD-1, catalase, CSE, CBS, and 3-MST were used to amplify q-PCR. 18s rRNA was used as a house keeping gene.

2.3. Measurement of H_2S

Stressed HL-1 cells treated without or with 10 μM of SG-1002, or H_2O_2 (500 μM) or in combination for 1 h in 1% serum (FBS) containing media. The levels of H_2S were determined in culture media samples by using gas chromatography chemiluminescence method as described previously [9].

2.4. Immunoblot Assay

The polyclonal antibody for SOD1, catalase, and the monoclonal antibody for GAPDH were purchased from Santa Cruz Biotechnology, Inc., Santa Cruz, CA. Murine HL-1 cardiac muscle cells were maintained in 10% fetal bovine serum (FBS) containing Claycomb media followed by serum starvation (1% FBS) media for 24 h. Serum starved/stressed cells were treated for 1 h with either 10 μM SG-1002. Serum starved/stressed HL-1 cells were treated with or without 10 μM of SG-1002 for 1 h in 1% serum containing media followed protein extraction. The level of SOD1, catalase and GAPDH were determined by using immunoblot analysis [10]. Total protein samples were prepared from treated HL-1 cardiomyocytes using RIPA lysis buffer and quantified using BCA protein assay kits from Pierce, Inc. Waltham, MA, USA. 15 μg of protein samples were loaded in each well and separated were then separated via electrophoresis on a 4–20% Mini-PROTEAN TGX Precast Gel (Bio-Rad Laboratories, Inc., Hercules, CA, USA) and transferred to a 0.45 μM nitrocellulose membrane (Bio-Rad Laboratories, Inc.). Membranes were blocked for at least two hours with Odyssey Blocking Buffer (Li-Cor, Lincoln, NE, USA), diluted 1:1 with PBS followed by probing with primary (such as SOD1, Catalase, and GAPDH) antibodies for 3 h at 4° and fluorescence conjugated secondary antibodies secondary antibodies for 1.5 h at room temperature. Proteins bands were visualized and scanned/quantified by Li-Cor/Odyssey infrared imaging system. GAPDH was employed as a loading control.

2.5. Measurement of Advanced Oxidative Protein Products (AOPP)

All reagents or chemicals used in our experiments were purchased from Sigma-Aldrich. Advanced oxidative protein products AOPP Assay Kit was obtained from Abcam (ab242295). AOPP is a simple, reproducible, and consistent system for the detection of advanced oxidation protein products in plasma, lysates, and tissue homogenates. This kit includes a chloramine standard and an AOPP Human Serum Albumin conjugate for use as a positive control. Oxidative stress was measured by determining the levels of AOPP in cultured HL-1 cells after incubation without or with either 10 μM of SG-1002 or 500 μM of H_2O_2 or in combination in 1% FBS containing media. AOPP was determined using a spectrophotometric method. Samples were incubated with glacial acetic acid and the absorbance was read at 340 nm. Chloramine T with potassium iodide was used as calibrator.

2.6. Cell Proliferation Assay

Cell viability was determined utilizing a Vybrant® MTT (3-(4,5-Dimethylthiazol-2-Yl)-2,5-Diphenyltetrazolium Bromide) cell proliferation assay kit (Thermo Fisher Scientific, Waltham, MA, USA). Cells were maintained on a 96-well tissue culture plate (Falcon, Corning, NY, USA) in 10% serum and cultured for 24 h in serum starvation (1%) prior to treatment. Cells were then pretreated with DMSO or SG-1002 for 30 min, followed by treatment without or with 10 μM SG-1002 or 500 μM of H_2O_2, or in combination for one hour. 10 μL of the 12 mM MTT stock solution was loaded into each well and the plate was incubated in the dark at 37 °C for 2 h. Following the incubation period, 175 μL of the media was removed, and 50 μL of DMSO was added to the remaining 25 μL in each well. The plate was then incubated at 37 °C for 10 min and the optical density was determined using a spectrophotometer at 540 nm.

2.7. Lactate Dehydrogenase (LDH) Cytotoxicity Assay

Stressed HL-1 cells were incubated with either 10 μM of SG-1002, or H_2O_2 (500 μM) and or in combination for 1 h in 1% serum containing media. Treated HL-1 cellular cytotoxicity was determined by using the assay kits obtained from Thermo Fisher Scientific.

2.8. Measurement of ROS

Cellular ROS were measured using a cellular ROS/superoxide detection assay kit (Abcam, Cambridge, UK). Cells were maintained on a 96-well tissue culture plate (Falcon, Corning, NY, USA) in 10% FBS and cultured for 24 h in serum starvation (1% FBS) prior to treatment. Cells were then pretreated with DMSO or SG-1002 for 30 min, followed by treatment with 500 μM of H_2O_2, and/or 10 μM SG-1002 or in combination for one hour in combination in 1% serum containing media. After treatment, cells were loaded with 100 μL/well of ROS superoxide detection solution and incubated for one hour in the dark at 37 °C. The plate was then read from the bottom at excitation 485/20 and emission 528/20.

2.9. Statistical Analysis

All data in this study were expressed as the mean ± SE from at least three independent experiments from given n sizes. Statistical significance between two groups was determined using the two-tailed Student's t test. Statistical significance of multiple treatments was determined by one-way analysis of variance followed by the Bonferroni post hoc test when appropriate. A p value of <0.05 was considered significant.

3. Results

3.1. Time Course and Dose Responses of SG-1002 on H_2S Production

HL-1 cardiac muscle cells were incubated either with different concentrations of SG-1002 for 1 h (Figure 1A) or with 10 μM of SG-1002 for several time points (Figure 1B) in 1% serum containing media followed by the measurement of H_2S levels in culture media. As can be seen 10 and 20 μM SG-1002 was able to significantly increase H_2S levels in HL-1 cells (Figure 1A). It was also observed that treatment with 10 μM of SG-1002 produced more H_2S in 1 h as compared with other time points. This result suggests that SG-1002 induces the production of H_2S.

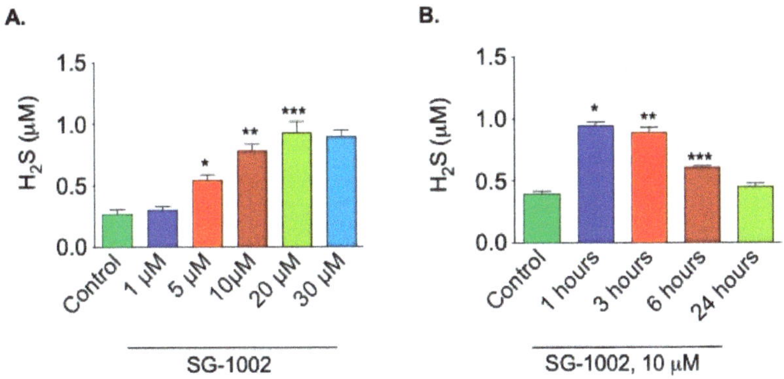

Figure 1. Time course and dose responses of SG-1002 on H_2S production. HL-1 cardiac muscle cells were incubated either with different concentrations of SG-1002 for 1 h (**A**) or with 10 mM of SG-1002 for several time periods (**B**) in 1% serum containing media followed by the measurement of H_2S levels in culture media. *, **, and ***. $p < 0.05$ versus (vs.) control (n = 4).

3.2. Induction of CBS by SG-1002 in Cultured HL-1 Cardiac Muscle Cells

Treatment with SG-1002 significantly increased ($p < 0.05$) the mRNA expression of CBS in HL-1 cardiac muscle cells cultured for 1 h in 1% serum, as compared to the control. mRNA expression of CSE and 3-MST were also slightly up-regulated compared to the control, though not significantly (Figure 2A–C). We were also interested in whether SG-1002 can induce CBS expression in the presence of an inhibitor of CBS, such as hydroxylamine (HA). We observed that SG-1002 induces CBS mRNA expression even in the presence of HA as compared with HA alone (Figure 2D).

3.3. Inhibition of Oxidative Stress by SG-1002 in Cultured HL-1 Cardiac Muscle Cells

Hl-1 cells treated with SG-1002 had significantly lower levels of oxidative stress when compared to the controls. AOPP and ROS levels were measured in 4 treated groups of HL-1 cells following 24 h of serum starvation: DMSO vehicle control, 10 μM SG-1002, DMSO and 500 μM H_2O_2 (to further stress the cells), or SG-1002 and 500 μM H_2O_2. The cells treated with SG-1002 alone had the lowest levels of oxidative stress (AOPP and ROS), followed by the DMSO-treated control group (Figure 3A,B). The H_2O_2 group had the highest levels of oxidative stress and SG-1002 was able to antagonize the effects of H_2O_2 (Figure 3A,B).

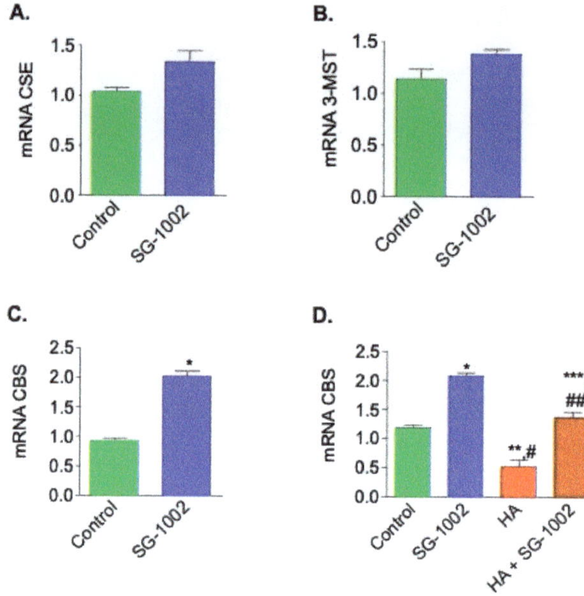

Figure 2. Effects of SG-1002 on mRNA Expression of CBS, CSE, and 3-MST in HL-1 cells. HL-1 cells were incubated with specific reagents as described in Materials and Methods for 1 h in 1% serum containing media. cDNA was prepared from RNA obtained from treated HL-1 cells followed by analysis of mRNA of CSE (**A**), 3-MST (**B**), and CBS (**C**,**D**) using TaqMan PCR assay system. *, $p < 0.05$ vs. control, (**A–C**), (*t* test) (n = 4); *, and **, $p < 0.05$ vs. control; #, $p < 0.05$ SG-1002 vs. HA; ##, $p < 0.05$ HA vs. SG-1002 +HA; ***, $p < 0.05$ SG-1002 vs. SG 1002 + HA (D) (Anova) (n = 3).

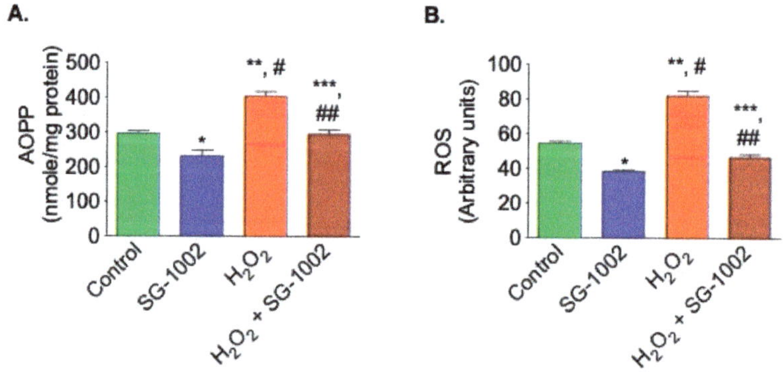

Figure 3. Reduction of oxidative stress in HL-1 cells by SG-1002. Oxidative stress was measured by determining the levels of AOPP (**A**) and ROS (**B**) in cultured HL-1 cells after tread with specific reagents as described in Materials and Methods. The experiment was repeated at least three times to verify the reproducibility. *, and **, $p < 0.05$ vs. control; #, $p < 0.05$ SG-1002 vs. H_2O_2; ##, $p < 0.05$ H_2O_2 vs. SG-1002 + H_2O_2; ***, $p < 0.05$ SG-1002 vs. SG-1002 + H_2O_2 (Anova) (n = 4).

3.4. Effects of SG-1002 on SOD1 and Catalase Levels in HL-1 Cells

It was then of interest to determine the levels of antioxidative enzyme genes expression in HL-1 cells were treated without or with either 10 μM of SG-1002, or H_2O_2 (500 μM) or in combination for 1 h in 1% serum containing media. cDNA was prepared from RNA obtained from cultured HL-1 cells followed by analysis of mRNA of SOD1 (Figure 4A) and catalase (Figure 4B) using the TaqMan PCR assay system. Treatment of HL-1 cells with SG-1002 significantly increased mRNA expression of SOD1 and catalase. Furthermore,

SG-1002 also increased the expression of both enzymes even under oxidative stress induced by H_2O_2 (Figure 4A,B).

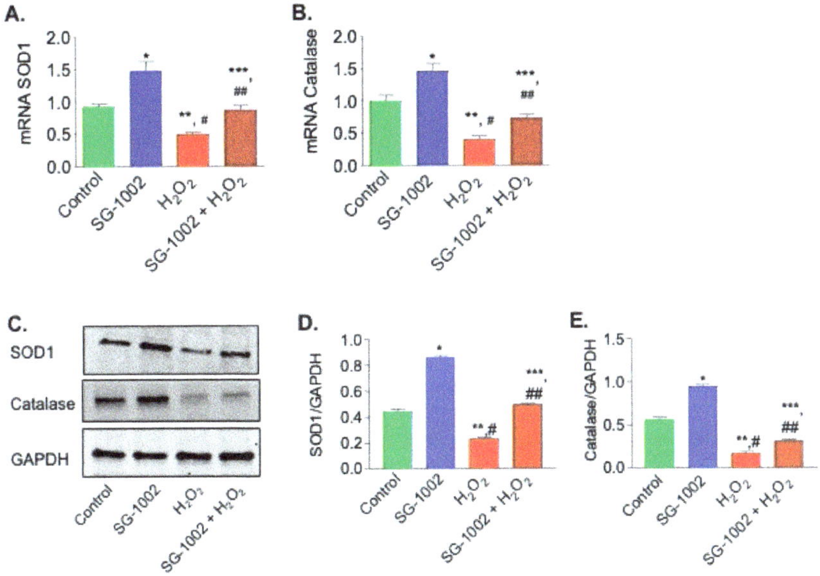

Figure 4. Effects of SG-1002 on mRNA and proteins levels of SOD1 and catalase in HL-1 cells. cDNA was prepared from RNA obtained from treated HL-1 cells followed by analysis of mRNA of SOD1 (**A**) and catalase (**B**) using TaqMan PCR assay system. SOD1 and catalase levels were determined in the protein extracts obtained from treated HL-1 cells by using immunoblot analysis. (**C**) represents blots for SOD1, catalase and GAPDH, (**D**,**E**) represent the quantitation of blots in (**C**). * and **, $p < 0.05$ vs. control; #, $p < 0.05$ SG-1002 vs. H_2O_2; ##, $p < 0.05$ H_2O_2 vs. SG-1002 + H_2O_2; ***, $p < 0.05$ SG 1002 vs. SG 1002 + H_2O_2 (Anova), (n = 4).

Then we were interested in determining the effects of SG-1002 on the protein levels of SOD1 and catalase by utilizing immunoblot analysis in stressed HL1 cells. We found that in stressed HL1 cells, SG-1002 significantly increased SOD1 and catalase levels as compared with the control (Figure 4C–E). This data suggests that SG-1002 protects HL-1 cells from oxidative damage via inducing antioxidant proteins.

3.5. Effects of SG-1002 on H_2S Production and CBS mRNA Expression When Cells Were Cultured under Stress

It was also of interest to examine whether SG-1002 is able to induce production of H_2S and H_2S producing enzyme CBS, when cells were treated with H_2O_2. As can be seen here under oxidative stress, SG-1002 also significantly induced the production of H_2S as well as CBS mRNA in HL-1 cells treated with SG-1002 plus H_2O_2 as compared with H_2O_2 alone (Figure 5A,B).

3.6. Effects of SG -1002 on the Expression of HF Biomarkers in HL-1 Cells

It is well known the expression of both ANP and BNP are elevated during hypertrophy or HF. Therefore, it was of interest to determine the levels of these gene expression in H_2O_2 and ET-1-treated HL-1 cells. As can be seen in Figure 6, SG-1002 significantly inhibited H_2O_2 or ET-1 induced ANP (Figure 6A) and BNP (Figure 6B) in HL-1 cells.

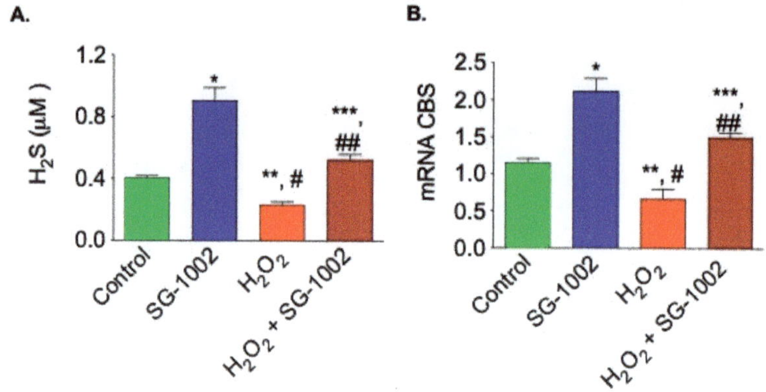

Figure 5. Effects of SG-1002 on H_2S production and CBS mRNA expression when cells were cultured under stress. HL-1 cells were incubated with specific reagents as described in Materials and Methods for 1 h in 1% serum containing media followed by the measurement of H_2S levels in culture media (**A**) and analysis of CBS mRNA (**B**) using the TaqMan PCR assay system. *, and **, $p < 0.05$ vs. control; #, $p < 0.05$ SG-1002 vs. H_2O_2; ##, $p < 0.05$ H_2O_2 vs. SG-1002 + H_2O_2; ***, $p < 0.05$ SG-1002 vs. SG 1002 + H_2O_2 (Anova) (n = 3).

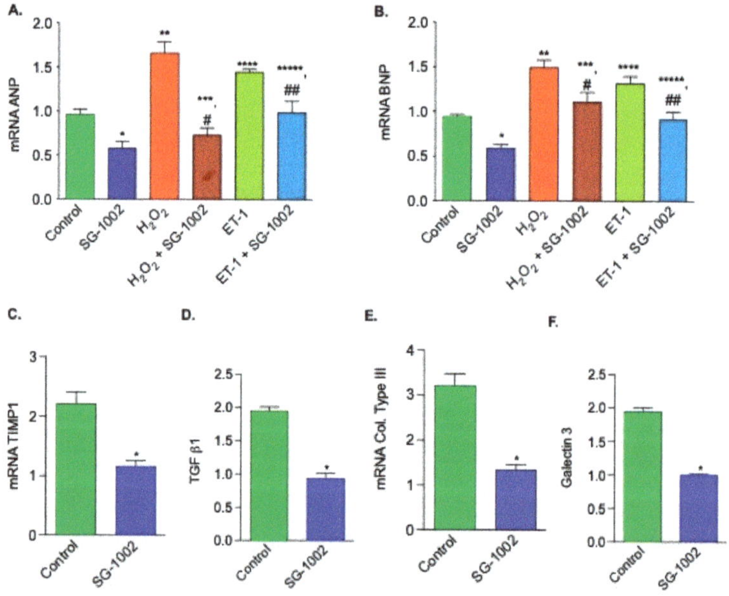

Figure 6. Effects of SG -1002 on the expression of HF markers in HL-1 cells. cDNA was prepared from RNA obtained from treated HL-1 cells as described in Materials and Methods followed by analysis of mRNA of ANP (**A**), BNP (**B**), TIMP1 (**C**), TGF b1 (**D**), collagen (col) type III (**E**), and galectin 3 (**F**) using the TaqMan PCR assay system. *, **, and ****, $p < 0.05$ vs. control; #, $p < 0.05$ SG-1002 vs. H_2O_2 + SG-1002; ***, $p < 0.05$ H_2O_2 vs. SG-1002 + H_2O_2; ##, $p < 0.05$ ET-1 vs. SG-1002 + ET1; *****, $p < 0.05$, SG-1002 vs. ET1 + SG-1002 (Anova) (**A**,**B**) (n = 4); *t* test (**C**–**F**) (n = 3).

We also tested the effects of SG-1002 on mRNA expression of other HF biomarkers, for example, TIMP1, TGF b1, collagen type III, and galectin 3. mRNAs of all these upregulated biomarkers in stressed HL1 cells are markedly decreased by SG-1002 (Figure 6C–F). Inhibition of hypertrophic genes expression in stressed HL-1 cells by SG-1002 suggests that SG-1002 plays important roles in the protection of HL-1 cells from oxidative damage.

3.7. SG-1002 Decreases Oxidative Stress-Induced Cellular Death and Cytotoxicity in Muscle Cells

Next, the effects of SG-1002 on cells viability and cytotoxicity were measured using HL1 cells treated without or with either 10 µM of SG-1002, or H_2O_2 (500 µM) and or in combination for 1 h in 1% serum containing media. Cell viability (Figure 7A) and cytotoxicity (Figure 7B) were determined by using MTT and LDH cytotoxicity assays, respectively.

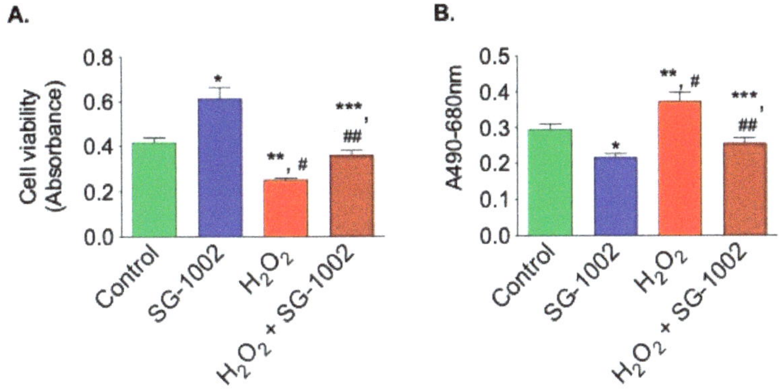

Figure 7. SG-1002 decreases oxidative stress-induced cellular death and cytotoxicity in HL-1 cardiac muscle cells. HL-1 cells were incubated with specific reagents as described in Materials and Methods. Cell viability (**A**) and cytotoxicity (**B**) were determined by using MTT and LDH cytotoxicity assays, respectively. * and **, $p < 0.05$ vs. control; #, $p < 0.05$ SG-1002 vs. H_2O_2; ##, $p < 0.05$ H_2O_2 vs. SG-1002 + H2O2; ***, $p < 0.05$ SG-1002 vs. SG 1002 + H_2O_2 (Anova) (n = 3).

Treatment with SG-1002 alone markedly improves cell viability when compared with controls. Additionally, when the cells were further stressed with 500 µM H_2O_2, treatment with SG-1002 was able to significantly ($p < 0.05$) improve cell viability as determined by MTT assay (Figure 7A). Cytotoxicity was measured by using lactate dehydrogenase (LDH) cytotoxicity assay (Figure 7B). Treatment with SG-1002 significantly lowered levels of LDH when compared to controls when HL-1 cells were under oxidative stress (Figure 7B).

The current data unambiguously show that SG-1002 reduces myocardial cellular oxidative damage and/or hypertrophic signaling by elevating H_2S levels, CBS, the enzyme that produces H_2S, and antioxidant proteins.

4. Discussion

DATS, DBTS, and SG-1002 have the potential to be used prophylactically for a variety of purposes, including the promotion of immunity, prevention or slowing the onset of chronic-degenerative diseases, and protecting the vital organs from damage induced by other drugs such as paracetamol, corticosteroids, chemotherapy agents, etc. SG-1002 has an advantage in this context, namely its efficient conversion into H_2S without causing halitosis and body odor. Novel, long acting, and controllable H_2S-based therapeutics (i.e., H_2S donors, prodrugs, and H_2S enzyme activators) may represent valuable candidates for drug development. However, challenges to their use in clinical practice remain. Currently, H_2S donors have poor pharmacokinetics with a very short half-life and uncontrolled release [20]. Two commercially available inorganic salts, Na_2S and NaHS, rapidly increase free H_2S concentration, but this increase is short-lived (seconds). These compounds also have a narrow therapeutic dose window leading to potential toxicity [25]. Because it is not possible to supplement H_2S levels in a sustained manner, use of these compounds for treating chronic disorders, such as HF, is not clinically feasible. Additionally, given the toxicity of supraphysiological H_2S levels, it is critical that novel H_2S agents with favorable pharmacokinetic profiles are developed specifically for clinical application. Based on the published literature regarding the importance of H_2S on cardiac homeostasis and the benefit of SG-1002 in several disease models, the present study was undertaken to determine the effects of this prodrug, SG-1002, on oxidative stress induced hypertrophic signaling in stressed murine cardiomyocytes (HL-1 cells). In this study, we investigated a mechanism of the reduction in stress induced cardiomyocyte hypertrophic signaling through the use of an exogenous H_2S donor, SG-1002. Our aim was to determine whether this prodrug could induce expression of H_2S producing enzymes (CBS, CSE and 3-MST) as well as antioxidant protection, which would allow cells to produce more endogenous H_2S.

Low levels of H_2S in blood or tissue has been correlated with the onset of disease onset of disease states related to oxidative cell damage, chronic inflammation [26], immune dysfunction [26,27], endoplasmic reticulum (ER) stress [28,29], dysregulation of mitochondrial bioenergetics [30], and hyperproliferation of cells or viruses [31], suggesting causal links that are actively being investigated. Furthermore, an inverse link between illness progression and H_2S in blood and/or tissues has been established in some cases [32–34]. So called "H_2S-poor" disease states that are amenable to correction by H_2S donors include [35,36] aging, ischemia, cardiac hypertrophy, HF, liver disease (cirrhosis, steatosis), hypertension, atherosclerosis, endothelial dysfunction, diabetic complications, preeclampsia, Alzheimer's disease (AD), and Huntington's disease (HD).

It has been reported that oxidative stress results in cardiac contractile dysfunction and is a key player in the transition from compensatory cardiac hypertrophy to decompensated HF [37]. H_2S is known to reduce oxidative stress in two ways: by direct inactivation of oxidant species and via upregulation of endogenous antioxidant defenses [38,39]. In our study we observed treatment of stressed HL-1 cells with H_2S prodrug, SG-1002, decreased ROS and AOPP (oxidative stress markers) and increased SOD1 and catalase via induction of CBS/H_2S. SG-1002 may also attenuate oxidative stress via a third mechanism: recoupling of eNOS. Uncoupled eNOS (as seen in HF) results in decreased NO generation, and leads to excessive peroxynitrite concentrations [40].

L-cysteine is converted into H_2S via CSE. In vivo studies have shown that H_2S protects against acute MI/R injury [39]. In a recent study, wild-type mice and CSE-knock out (KO) mice underwent transverse aortic constriction (TAC) and then received either SG-1002 or vehicle. CSE-KO mice developed worsened cardiac remodeling and function compared to control mice. Interestingly, CSE-KO mice that received SG-1002 exhibited improved cardiac remodeling and function compared to vehicle treated mice [11]. SG-1002 has also been examined in preclinical studies of HF [8]. SG-1002 attenuates left ventricular (LV) remodeling and dysfunction in a pressure overload model of HF [8]. Administration of SG-1002 significantly decreased disease markers and increased both H_2S and NO levels [8]. A subsequent phase 1 clinical trial was performed using SG-1002 for a 21-day period in healthy and HF subjects. In this trial it was observed that SG-1002 significantly augments circulating H_2S levels in HF patients and healthy subjects. SG-1002 was also found to be well tolerated and safe at all doses tested [2].

In a model of acute limb ischemia, the proangiogenic effects of SG-1002 were evaluated (ALI). Pigs were given either a placebo or SG-1002 for the 35-day research after they had the intravascular occlusion technique to induce ALI. Moreover, SG-1002 also preserved existing capillaries in ischemic limbs to a 1.6 times greater extent than in pigs that received placebo [41].

Terminally ill children between 18 month and 14 years of age presented with osteosarcoma, hydrocephalus with cancerous tumor, medulloblastoma, squamous cell carcinoma, and acute lymphoblastic leukemia (all refractory to chemotherapy and/or radio therapy) were treated at doses between 1200 mg and 3600 mg SG-1002 daily, no adverse reaction was reported. Additionally, improvements in fatigue, inflammation, pain, headache, cardiac function, blood sugar regulation and reduced tumor volume were reported [1]. Bibli et al. found that treatment of human umbilical vein endothelial cells with SG-1002 induced protein S-sulfhydration and protection of membrane lipids from peroxidation [42]; these in vitro findings are consistent with the significant increases in H_2S levels of blood and tissues attained when SG-1002 is orally administered to mice [8], swine [41], and humans [2].

Another recent study investigated the effects of SG-1002 on homocysteine induced cardiac remodeling and dysfunction. $CBS^{+/-}$ and $CBS^{+/+}$ (WT) mice treated with SG-1002 in their chow. At baseline, $CBS^{+/-}$ mice showed an increased afterload (increased end systolic pressure with conserved stroke volume). This phenotype is abolished with SG-1002 treatment by reducing end systolic pressure and, at the same time, significantly increasing the end-diastolic volume. Additionally SG-1002 was found to prevent the $CBS^{+/-}$ mice from developing pathological cardiac remodeling [43]. This work established that increasing H_2S levels with SG-1002 in the setting of HF increases cardiac mitochondrial content/function and improves cardiac function via AMPK activation [44]. Another recent study showed the impact of SG-1002 on atherosclerosis in mice. In this study, the left carotid artery was partially ligated. Mice that received SG-1002 had significantly reduced plaque formation, indicating a potential anti-atherosclerotic effect of SG-1002 [45].

Many previous studies have established that oxidative stress/ROS exacerbate HF while antioxidants protect against it [46]. Certain ROS, such as hydrogen peroxide (H_2O_2), act as signaling molecules [47] for the immune system and helps recruit white blood cells to initiate healing to damaged tissues [48]. They are usually eliminated via interactions with superoxide dismutase, catalase, glutathione peroxidase, and peroxiredoxins [49]. Similarly, in our current study it was observed that overproduction of ROS and AOPP in stressed HL-1 cells were scavenged by catalase and SOD1 induced by SG-1002 treatment. In many chronic conditions, more ROS are produced than can be eliminated, causing an imbalance between the antioxidant and oxidant systems. This imbalance is toxic and causes damage to many organelles including the mitochondria. In the setting of HF, this imbalance can cause changes in normal autophagy pathways [50], worsening arteriosclerosis [51], and persistent low levels of systemic inflammation [52].

A study found that mice that had undergone MI/R had smaller infarct sizes and improved LV function when treated with a H_2S donor [53]. Our previous study in HF murine model has shown that NO mediated induction of H_2S producing enzymes and/or H_2S causing the activation of antioxidant gene regulated transcription factor, Nrf2 resulting in increased antioxidant protection of cells from damage [9]. Present study shows that treatment of stressed HL1 cells with a H_2S donor, SG-1002, increased the levels of the H_2S-producing enzyme, CBS, as well as antioxidant proteins, SOD1 and catalase, leading to the protection of cells from damage. This suggests that SG-1002 works at least in part by increasing expression of antioxidant systems. H_2S is a powerful antioxidant compound that is capable of scavenging free radicals in vitro [54]. Similarly, in the present study it was observed that, when stressed cardiomyocytes were treated with SG-1002, they had lower levels of oxidative stress because of the induction of H_2S/ or H2S producing enzymes such as CBS. We also noted that the cells that were further stressed with H_2O_2 and treated with SG-1002 had much lower levels of oxidative stress compared to the H_2O_2 alone group, but higher levels than both the SG-1002 alone and control group, suggesting that high levels of pre-existing oxidative stress impair its effect.

While some studies point to apoptosis after MI/R being cardioprotective, others point to it being a detrimental to LV function. Changes in normal cell death and apoptosis are known to occur in HF [55]. We assessed the effects of SG-1002 on cell death and cytotoxicity in stressed HL-1 cells. The assays we used included MTT and LDH cytotoxicity, both of which provide information about the amount but not the pathways that cause cell death and cytotoxicity. In order to determine which pro-survival pathways are being affected by treatment with SG-1002, future studies may investigate increases in genes and proteins related to improved survival such as beclin-1, or decreases in pro-apoptotic markers, such as bax, caspase-3, or some other related genes.

Another way sulfide donor is said to have cardioprotective effects is by attenuating hypertrophy. A previous study found that mice treated with SG-1002 had less cardiac enlargement, preserved LV function, and less fibrosis after TAC when compared to mice that received the vehicle [8]. Our previous study has shown that the elevated levels of hypertrophic genes (HF indicators) such as ANP and BNP in HF murine models were markedly decreased by nitric oxide mediated production of H_2S or H_2S producing enzymes [9]. Similarly, in present cellular study it was observed that SG-1002 treatment of stressed HL-1 cells significantly reduced expression of HF biomarkers. Interestingly, an antioxidant property of SG-1002 was clearly observed when H_2O_2 induced oxidative stress in HL-1 was antagonized by SG-1002. A study showed that exogenous H_2S not only prevented hypertrophy in neonatal rat cardiac ventricular myocytes (NRCMs), but also improved the viability of the hypertrophic cells, possibly by altering glucose metabolism [56].

Our results are consistent with the research on SG-1002's impact on apoptosis, oxidative stress, and hypertrophy. Future research should examine how SG-1002 impacts other clinical HF markers, notably pro-fibrotic markers. Other studies have already found H_2S can reduce organ fibrosis, though the mechanism of how it produces this effect is unknown [57]. In addition, other studies have found that another possible mechanism of H_2S's protective effect is through the mitochondria. H_2S/CSE has been shown to decrease methylation of mitochondrial transcription factor A (TFAM) [58], which regulates mtDNA copy number and mitochondrial biogenesis [59]. H_2S can also improve the function of mitochondria under hypoxic conditions. Under normal conditions, H_2S can reduce ATP production by inhibiting cytochrome c oxidase [60]. However, under conditions where oxygen concentrations are low (such as in I/R injuries), CSE can be translocated to mitochondria, where it produces H_2S, which helps to preserve the generation of ATP [61]. Furthermore, Elrod et al. found that mice subjected to MI/R injury and treated with an H_2S donor at the time of reperfusion had preserved mitochondrial function and membrane integrity, as well as smaller infarcts, less fractional shortening, and preserved ejection fraction when compared to vehicle-treated mice [53]. With the evidence of mitochondrial involvement growing, future studies may benefit from examining the effects of SG-1002 on mitochondrial gene expression and function.

Based on our previous findings in murine models (9), our present study may support a similar mechanism for under stressed cardiomyocytes which indicates that an increased production of H_2S or H_2S producing enzyme by SG-1002 may activate nuclear factor erythroid 2-related factor 2 (Nrf2), in turn causing the elevation of antioxidant gene expression/antioxidants levels and decreasing the levels of oxidative stress resulting in protection of cardiac cells from damage leading to the inhibition of myocardial hypertrophy/HF (Figure 8).

A small (n = 18) phase 1 clinical trial found SG-1002 to be a safe and well-tolerated drug in both healthy controls and HF patients [2]. SG-1002 is emerging as a potential H_2S pro-drug due to its safety, mode of administration, and unique ability to efficiently generate H_2S with no byproducts in a slow and sustained, dose and enzyme independent manner. These features position SG-1002 as the H_2S donor of choice when studying biological systems in vivo. However, its negligible solubility in water makes it a poor choice for in vitro experiments. Larger clinical studies are being designed to further test the safety and efficacy of SG-1002 for the treatment of HF/hypertrophy and CVD. Our study and others have shown that H_2S prodrugs like SG-1002 increase the levels of H_2S producing enzymes and antioxidants while decreasing hypertrophic gene expression and oxidative stress, making it a promising novel therapeutic drug.

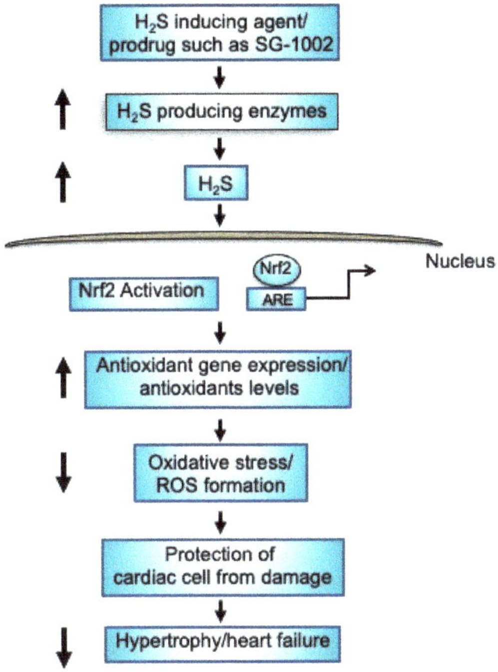

Figure 8. Mechanisms of protection of cardiac cells by H_2S producing agents such as prodrug SG-1002. Treatment of SG-1002 induces H_2S producing enzymes resulting in increased production of H_2S. Increased levels of H_2S activates nuclear Nrf2 as well as causing elevation of antioxidant gene expression/antioxidants levels resulting in protection of cardiac cell from damage leading to the inhibition of hypertrophy/HF.

5. Conclusions

Data from the current study may provide a mechanism for the effects of SG-1002 on stressed cardiomyocytes. This induces the production of a H_2S/H_2S producing enzyme, CBS, resulting in induction of antioxidant gene expression/antioxidants levels and reduction of oxidative stress levels leading to the protection of cardiomyocytes from damage. SG-1002, via a similar mechanism, is probably involved in an in vivo system where it may protect myocardial hypertrophy/HF. SG-1002 is a leading H_2S donor candidate due to its safety, oral activity, and unique ability to efficiently generate H_2S with no byproducts in a slow and sustained manner, independent for both dose and enzyme. As found in a small phase 1 clinical trial, SG-1002 was considered to be a safe and well-tolerated drug in both healthy controls and HF patients. Therefore, it is necessary to study the effects of SG-1002 in cardiovascular diseases in detail through larger clinical trials.

Author Contributions: Conceptualization, methodology, software, validation, formal analysis, data curation, writing, and original draft preparation, R.K.I.; methodology, software, validation, formal analysis, E.D. (Erinn Donnelly), E.D. (Erminia Donnarumma) and F.H.; review, editing, and visualization, J.D.G.; Methodology, software, data curation, validation, investigation, resources, review and editing, visualization, project administration, and funding acquisition, K.N.I. All authors have read and agreed to the published version of the manuscript.

Funding: This work was supported by Evans Allen Research, NI201445XXXXG018 Accession # 1022014.

Institutional Review board Statement: Not applicable.

Informed Consent Statement: Not applicable.

Data Availability Statement: The data presented in this study are available on request from the corresponding author in accordance with the state regulations and appropriate laws.

Conflicts of Interest: The authors declare no conflict of interest.

References

1. Gojon, G.; Morales, G.A. SG1002 and Catenated Divalent Organic Sulfur Compounds as Promising Hydrogen Sulfide Prodrugs. *Antioxid. Redox Signal.* **2020**, *33*, 1010–1045. [CrossRef]
2. Polhemus, D.J.; Li, Z.; Pattillo, C.B.; Gojon, G., Sr.; Gojon, G., Jr.; Giordano, T.; Krum, H. A novel hydrogen sulfide prodrug, SG1002, promotes hydrogen sulfide and nitric oxide bioavailability in heart failure patients. *Cardiovasc. Ther.* **2015**, *33*, 216–226. [CrossRef] [PubMed]
3. Claycomb, W.C.; Lanson, N.A., Jr.; Stallworth, B.S.; Egeland, D.B.; Delcarpio, J.B.; Bahinski, A.; Izzo, N.J., Jr. HL-1 cells: A cardiac muscle cell line that contracts and retains phenotypic characteristics of the adult cardiomyocyte. *Proc. Natl. Acad. Sci. USA* **1998**, *95*, 2979–2984. [CrossRef]
4. King, A.L.; Bhushan, S.; Otsuka, H.; Kondo, K.; Nicholson, C.K.; Bradley, J.M.; Islam, K.N.; Calvert, J.W.; Tao, Y.-X.; Dugas, T.R.; et al. Hydrogen sulfide cytoprotective signaling is endothelial nitric oxide synthase-nitric oxide dependent. *Proc. Natl. Acad. Sci. USA* **2014**, *111*, 3182–3187. [CrossRef]
5. Bhushan, S.; Kondo, K.; Polhemus, D.J.; Otsuka, H.; Nicholson, C.K.; Tao, Y.X.; Huang, H.; Georgiopoulou, V.V.; Murohara, T.; Calvert, J.W.; et al. Nitrite therapy improves left ventricular function during heart failure via restoration of nitric oxide-mediated cytoprotective signaling. *Circ. Res.* **2014**, *114*, 1281–1291. [CrossRef]
6. Cai, H.; Harrison, D.G. Endothelial dysfunction in cardiovascular diseases: The role of oxidant stress. *Circ. Res.* **2000**, *87*, 840–844. [CrossRef] [PubMed]
7. Elrod, J.W.; Duranski, M.R.; Langston, W.; Greer, J.J.; Tao, L.; Dugas, T.R.; Kevil, C.G.; Champion, H.C.; Lefer, D.J. eNOS gene therapy exacerbates hepatic ischemia-reperfusion injury in diabetes: A role for eNOS uncoupling. *Circ. Res.* **2006**, *99*, 78–85. [CrossRef]
8. Kondo, K.; Bhushan, S.; King, A.L.; Prabhu, S.D.; Hamid, T.; Koenig, S.; Murohara, T.; Predmore, B.L.; Gojon, G., Sr.; Gojon, G., Jr.; et al. H_2S protects against pressure overload-induced heart failure via upregulation of endothelial nitric oxide synthase. *Circulation* **2013**, *127*, 1116–1127. [CrossRef] [PubMed]
9. Donnarumma, E.; Bhushan, S.; Bradley, J.M.; Otsuka, H.; Donnelly, E.L.; Lefer, D.J.; Islam, K.N. Nitrite Therapy Ameliorates Myocardial Dysfunction via H2S and Nuclear Factor-Erythroid 2-Related Factor 2 (Nrf2)-Dependent Signaling in Chronic Heart Failure. *J. Am. Heart Assoc.* **2016**, *5*, e003551. [CrossRef] [PubMed]
10. Ling, K.; Zhou, W.; Guo, Y.; Hu, G.; Chu, J.; Xie, F.; Li, Y.; Wang, W. H2S attenuates oxidative stress via Nrf2/NF-kappaB signaling to regulate restenosis after percutaneous transluminal angioplasty. *Exp. Biol. Med.* **2021**, *246*, 226–239. [CrossRef] [PubMed]
11. Calvert, J.W.; Elston, M.; Nicholson, C.K.; Gundewar, S.; Jha, S.; Elrod, J.W.; Ramachandran, A.; Lefer, D.J. Genetic and pharmacologic hydrogen sulfide therapy attenuates ischemia-induced heart failure in mice. *Circulation* **2010**, *122*, 11–19. [CrossRef] [PubMed]
12. Peake, B.F.; Nicholson, C.K.; Lambert, J.P.; Hood, R.L.; Amin, H.; Amin, S.; Calvert, J.W. Hydrogen sulfide preconditions the db/db diabetic mouse heart against ischemia-reperfusion injury by activating Nrf2 signaling in an Erk-dependent manner. *Am. J. Physiol. Heart Circ. Physiol.* **2013**, *304*, H1215–H1224. [CrossRef] [PubMed]
13. Li, L.; Rossoni, G.; Sparatore, A.; Lee, L.C.; Del Soldato, P.; Moore, P.K. Anti-inflammatory and gastrointestinal effects of a novel diclofenac derivative. *Free Radic. Biol. Med.* **2007**, *42*, 706–719. [CrossRef] [PubMed]
14. Streeter, E.; Ng, H.H.; Hart, J.L. Hydrogen sulfide as a vasculoprotective factor. *Med. Gas Res.* **2013**, *3*, 9. [CrossRef]
15. Kamat, P.K.; Kalani, A.; Tyagi, S.C.; Tyagi, N. Hydrogen Sulfide Epigenetically Attenuates Homocysteine-Induced Mitochondrial Toxicity Mediated Through NMDA Receptor in Mouse Brain Endothelial (bEnd3) Cells. *J. Cell. Physiol.* **2015**, *230*, 378–394. [CrossRef] [PubMed]
16. Zhang, L.; Wang, Y.; Li, Y.; Li, L.; Xu, S.; Feng, X.; Liu, S. Hydrogen Sulfide (H2S)-Releasing Compounds: Therapeutic Potential in Cardiovascular Diseases. *Front. Pharmacol.* **2018**, *9*, 1066. [CrossRef]
17. Predmore, B.L.; Lefer, D.J.; Gojon, G. Hydrogen sulfide in biochemistry and medicine. *Antioxid. Redox Signal.* **2012**, *17*, 119–140. [CrossRef]
18. Gallagher, P.M.; Meleady, R.; Shields, D.C.; Tan, K.S.; McMaster, D.; Rozen, R.; Evans, A.; Graham, I.M.; Whitehead, A.S. Homocysteine and risk of premature coronary heart disease. Evidence for a common gene mutation. *Circulation* **1996**, *94*, 2154–2158. [CrossRef]
19. Wang, K.; Ahmad, S.; Cai, M.; Rennie, J.; Fujisawa, T.; Crispi, F.; Baily, J.; Miller, M.R.; Cudmore, M.; Hadoke, P.W.; et al. Dysregulation of hydrogen sulfide producing enzyme cystathionine gamma-lyase contributes to maternal hypertension and placental abnormalities in preeclampsia. *Circulation* **2013**, *127*, 2514–2522. [CrossRef]
20. Polhemus, D.J.; Lefer, D.J. Emergence of hydrogen sulfide as an endogenous gaseous signaling molecule in cardiovascular disease. *Circ. Res.* **2014**, *114*, 730–737. [CrossRef]
21. Kolluru, G.K.; Shen, X.; Bir, S.C.; Kevil, C.G. Hydrogen sulfide chemical biology: Pathophysiological roles and detection. *Nitric Oxide* **2013**, *35*, 5–20. [CrossRef]

22. Polhemus, D.; Kondo, K.; Bhushan, S.; Bir, S.C.; Kevil, C.G.; Murohara, T.; Lefer, D.J.; Calvert, J.W. Hydrogen sulfide attenuates cardiac dysfunction after heart failure via induction of angiogenesis. *Circ. Heart Fail.* **2013**, *6*, 1077–1086. [CrossRef] [PubMed]
23. Kimura, Y.; Kimura, H. Hydrogen sulfide protects neurons from oxidative stress. *FASEB J.* **2004**, *18*, 1165–1167. [CrossRef]
24. Ahmad, A.; Sattar, M.A.; Rathore, H.A.; Khan, S.A.; Lazhari, M.I.; Afzal, S.; Hashmi, F.; Abdullah, N.A.; Johns, E.J. A critical review of pharmacological significance of Hydrogen Sulfide in hypertension. *Indian J. Pharmacol.* **2015**, *47*, 243–247.
25. Caliendo, G.; Cirino, G.; Santagada, V.; Wallace, J.L. Synthesis and biological effects of hydrogen sulfide (H2S): Development of H2S-releasing drugs as pharmaceuticals. *J. Med. Chem.* **2010**, *53*, 6275–6286. [CrossRef]
26. Li, L.; Moore, P.K. Could hydrogen sulfide be the next blockbuster treatment for inflammatory disease? *Expert Rev. Clin. Pharmacol.* **2013**, *6*, 593–595. [CrossRef]
27. Miller, T.W.; Wang, E.A.; Gould, S.; Stein, E.V.; Kaur, S.; Lim, L.; Amarnath, S.; Fowler, D.H.; Roberts, D.D. Hydrogen sulfide is an endogenous potentiator of T cell activation. *J. Biol. Chem.* **2012**, *287*, 4211–4221. [CrossRef]
28. Kabil, O.; Vitvitsky, V.; Banerjee, R. Sulfur as a signaling nutrient through hydrogen sulfide. *Annu. Rev. Nutr.* **2014**, *34*, 171–205. [CrossRef]
29. Wu, J.; Pan, W.; Wang, C.; Dong, H.; Xing, L.; Hou, J.; Fang, S.; Li, H.; Yang, F.; Yu, B. H2S attenuates endoplasmic reticulum stress in hypoxia-induced pulmonary artery hypertension. *Biosci. Rep.* **2019**, *39*, BSR20190304. [CrossRef]
30. Szczesny, B.; Modis, K.; Yanagi, K.; Coletta, C.; Le Trionnaire, S.; Perry, A.; Wood, M.E.; Whiteman, M.; Szabo, C. AP39, a novel mitochondria-targeted hydrogen sulfide donor, stimulates cellular bioenergetics, exerts cytoprotective effects and protects against the loss of mitochondrial DNA integrity in oxidatively stressed endothelial cells in vitro. *Nitric Oxide* **2014**, *41*, 120–130. [CrossRef]
31. Alshorafa, A.K.; Guo, Q.; Zeng, F.; Chen, M.; Tan, G.; Tang, Z.; Yin, R. Psoriasis is associated with low serum levels of hydrogen sulfide, a potential anti-inflammatory molecule. *Tohoku J. Exp. Med.* **2012**, *228*, 325–332. [CrossRef] [PubMed]
32. Chen, Y.H.; Yao, W.Z.; Geng, B.; Ding, Y.L.; Lu, M.; Zhao, M.W.; Tang, C.S. Endogenous hydrogen sulfide in patients with COPD. *Chest* **2005**, *128*, 3205–3211. [CrossRef] [PubMed]
33. Jiang, H.L.; Wu, H.C.; Li, Z.L.; Geng, B.; Tang, C.S. Changes of the new gaseous transmitter H2S in patients with coronary heart disease. *Di Yi Jun Yi Da Xue Xue Bao* **2005**, *25*, 951–954.
34. Kovacic, D.; Glavnik, N.; Marinsek, M.; Zagozen, P.; Rovan, K.; Goslar, T.; Mars, T.; Podbregar, M. Total plasma sulfide in congestive heart failure. *J. Card Fail.* **2012**, *18*, 541–548. [CrossRef]
35. Szabo, C.; Papapetropoulos, A. International Union of Basic and Clinical Pharmacology. CII: Pharmacological Modulation of H2S Levels: H2S Donors and H2S Biosynthesis Inhibitors. *Pharmacol. Rev.* **2017**, *69*, 497–564. [CrossRef] [PubMed]
36. Zivanovic, J.; Kouroussis, E.; Kohl, J.B.; Adhikari, B.; Bursac, B.; Schott-Roux, S.; Petrovic, D.; Miljkovic, J.L.; Thomas-Lopez, D.; Jung, Y.; et al. Selective Persulfide Detection Reveals Evolutionarily Conserved Antiaging Effects of S-Sulfhydration. *Cell Metab.* **2019**, *30*, 1152–1170.e13. [CrossRef] [PubMed]
37. Dhalla, A.K.; Hill, M.F.; Singal, P.K. Role of oxidative stress in transition of hypertrophy to heart failure. *J. Am. Coll. Cardiol.* **1996**, *28*, 506–514. [CrossRef] [PubMed]
38. Jha, S.; Calvert, J.W.; Duranski, M.R.; Ramachandran, A.; Lefer, D.J. Hydrogen sulfide attenuates hepatic ischemia-reperfusion injury: Role of antioxidant and antiapoptotic signaling. *Am. J. Physiol. Heart Circ. Physiol.* **2008**, *295*, H801–H806. [CrossRef]
39. Calvert, J.W.; Jha, S.; Gundewar, S.; Elrod, J.W.; Ramachandran, A.; Pattillo, C.B.; Kevil, C.G.; Lefer, D.J. Hydrogen sulfide mediates cardioprotection through Nrf2 signaling. *Circ. Res.* **2009**, *105*, 365–374. [CrossRef]
40. Kuzkaya, N.; Weissmann, N.; Harrison, D.G.; Dikalov, S. Interactions of peroxynitrite, tetrahydrobiopterin, ascorbic acid, and thiols: Implications for uncoupling endothelial nitric-oxide synthase. *J. Biol. Chem.* **2003**, *278*, 22546–22554. [CrossRef]
41. Rushing, A.M.; Donnarumma, E.; Polhemus, D.J.; Au, K.R.; Victoria, S.E.; Schumacher, J.D.; Li, Z.; Jenkins, J.S.; Lefer, D.J.; Goodchild, T.T. Effects of a novel hydrogen sulfide prodrug in a porcine model of acute limb ischemia. *J. Vasc. Surg.* **2019**, *69*, 1924–1935. [CrossRef] [PubMed]
42. Bibli, S.I.; Hu, J.; Leisegang, M.S.; Wittig, J.; Zukunft, S.; Kapasakalidi, A.; Fisslthaler, B.; Tsilimigras, D.; Zografos, G.; Filis, K.; et al. Shear stress regulates cystathionine gamma lyase expression to preserve endothelial redox balance and reduce membrane lipid peroxidation. *Redox Biol.* **2020**, *28*, 101379. [CrossRef]
43. Shen, X.; Carlstrom, M.; Borniquel, S.; Jadert, C.; Kevil, C.G.; Lundberg, J.O. Microbial regulation of host hydrogen sulfide bioavailability and metabolism. *Free Radic. Biol. Med.* **2013**, *60*, 195–200. [CrossRef]
44. Shimizu, Y.; Polavarapu, R.; Eskla, K.L.; Nicholson, C.K.; Koczor, C.A.; Wang, R.; Lewis, W.; Shiva, S.; Lefer, D.J.; Calvert, J.W. Hydrogen sulfide regulates cardiac mitochondrial biogenesis via the activation of AMPK. *J. Mol. Cell. Cardiol.* **2018**, *116*, 29–40. [CrossRef]
45. Bibli, S.I.; Hu, J.; Sigala, F.; Wittig, I.; Heidler, J.; Zukunft, S.; Tsilimigras, D.I.; Randriamboavonjy, V.; Wittig, J.; Kojonazarov, B.; et al. Cystathionine gamma Lyase Sulfhydrates the RNA Binding Protein Human Antigen R to Preserve Endothelial Cell Function and Delay Atherogenesis. *Circulation* **2019**, *139*, 101–114. [CrossRef]
46. Rosenbaugh, E.G.; Savalia, K.K.; Manickam, D.S.; Zimmerman, M.C. Antioxidant-based therapies for angiotensin II-associated cardiovascular diseases. *Am. J. Physiol. Regul. Integr. Comp. Physiol.* **2013**, *304*, R917–R928. [CrossRef]
47. Veal, E.A.; Day, A.M.; Morgan, B.A. Hydrogen peroxide sensing and signaling. *Mol. Cell* **2007**, *26*, 1–14. [CrossRef]
48. Niethammer, P.; Grabher, C.; Look, A.T.; Mitchison, T.J. A tissue-scale gradient of hydrogen peroxide mediates rapid wound detection in zebrafish. *Nature* **2009**, *459*, 996–999. [CrossRef]

49. Islam, K.N.; Takahashi, M.; Higashiyama, S.; Myint, T.; Uozumi, N. Fragmentation of ceruloplasmin following non-enzymatic glycation reaction. *J. Biochem.* **1995**, *118*, 1054–1060. [CrossRef] [PubMed]
50. Li, B.; Chi, R.F.; Qin, F.Z.; Guo, X.F. Distinct changes of myocyte autophagy during myocardial hypertrophy and heart failure: Association with oxidative stress. *Exp. Physiol.* **2016**, *101*, 1050–1063. [CrossRef] [PubMed]
51. Libby, P. Inflammation in Atherosclerosis. *Nature* **2002**, *420*, 868–874. [CrossRef]
52. Ledwaba, L.; Tavel, J.A.; Khabo, P.; Maja, P.; Qin, J.; Sangweni, P.; Liu, X.; Follmann, D.; Metcalf, J.A.; Orsega, S.; et al. Pre-ART levels of inflammation and coagulation markers are strong predictors of death in a South African cohort with advanced HIV disease. *PLoS ONE* **2012**, *7*, e24243. [CrossRef] [PubMed]
53. Elrod, J.W.; Calvert, J.W.; Morrison, J.; Doeller, J.E.; Kraus, D.W.; Tao, L.; Jiao, X.; Scalia, R.; Kiss, L.; Szabo, C.; et al. Hydrogen sulfide attenuates myocardial ischemia-reperfusion injury by preservation of mitochondrial function. *Proc. Natl. Acad. Sci. USA* **2007**, *104*, 15560–15565. [CrossRef] [PubMed]
54. Ahmad, A.; Sattar, M.Z.; Rathore, H.A.; Hussain, A.I.; Khan, S.A.; Fatima, T.; Afzal, S.; Abdullah, N.A.; Johns, E.J. Antioxidant Activity and Free Radical Scavenging Capacity of L-Arginine and NaHS: A Comparative In Vitro Study. *Acta Pol. Pharm. Drug Res.* **2015**, *72*, 245–252.
55. Gottlieb, R.A.; Burleson, K.O.; Kloner, R.A.; Babior, B.M.; Engler, R.L. Reperfusion Injury Induces Apoptosis in Rabbit Cardiomyoctes. *J. Clin. Investig.* **1994**, *94*, 1621–1628. [CrossRef]
56. Liang, M.; Jin, S.; Wu, D.D.; Wang, M.J.; Zhu, Y.C. Hydrogen sulfide improves glucose metabolism and prevents hypertrophy in cardiomyocytes. *Nitric Oxide* **2015**, *46*, 114–122. [CrossRef]
57. Zhang, S.; Pan, C.; Zhou, F.; Yuan, Z.; Wang, H.; Cui, W.; Zhang, G. Hydrogen Sulfide as a Potential Therapeutic Target in Fibrosis. *Oxid. Med. Cell. Longev.* **2015**, *2015*, 593407. [CrossRef]
58. Li, S.; Yang, G. Hydrogen Sulfide Maintains Mitochondrial DNA Replication via Demethylation of TFAM. *Antioxid. Redox Signal.* **2015**, *23*, 630–642. [CrossRef]
59. Clayton, D.A. Replication and Transcription of Vertebrate Mitochondrial DNA. *Annu. Rev. Cell Biol.* **1991**, *7*, 453–478. [CrossRef]
60. Blackstone, E.; Morrison, M.; Roth, M.B. H2S induces a suspended animation-like state in mice. *Science* **2005**, *308*, 518. [CrossRef]
61. Fu, M.; Zhang, W.; Wu, L.; Yang, G.; Li, H.; Wang, R. Hydrogen sulfide (H2S) metabolism in mitochondria and its regulatory role in energy production. *Proc. Natl. Acad. Sci. USA* **2012**, *109*, 2943–2948. [CrossRef] [PubMed]

Disclaimer/Publisher's Note: The statements, opinions and data contained in all publications are solely those of the individual author(s) and contributor(s) and not of MDPI and/or the editor(s). MDPI and/or the editor(s) disclaim responsibility for any injury to people or property resulting from any ideas, methods, instructions or products referred to in the content.

biomedicines

MDPI

Article

Kawasaki Disease-like Vasculitis Facilitates Atherosclerosis, and Statin Shows a Significant Antiatherosclerosis and Anti-Inflammatory Effect in a Kawasaki Disease Model Mouse

Yusuke Motoji [1], Ryuji Fukazawa [2,*], Ryosuke Matsui [2], Noriko Nagi-Miura [3], Yasuo Miyagi [1], Yasuhiko Itoh [2] and Yosuke Ishii [1]

1 Department of Cardiovascular Surgery, Graduate School of Medicine, Nippon Medical School, Tokyo 113-8603, Japan; yusuke-motoji@nms.ac.jp (Y.M.); show@nms.ac.jp (Y.M.); yosuke-i@nms.ac.jp (Y.I.)
2 Department of Pediatrics, Graduate School of Medicine, Nippon Medical School, Tokyo 113-8603, Japan; r-matsui@nms.ac.jp (R.M.); yasuhiko@nms.ac.jp (Y.I.)
3 Laboratory for Immunopharmacology of Microbial Products, Tokyo University of Pharmacy and Life Sciences, Hachioji 192-0392, Japan; miuranno@toyaku.ac.jp
* Correspondence: oraora@nms.ac.jp; Tel.: +81-3-3822-2131

Abstract: Kawasaki disease (KD) is an acute form of systemic vasculitis that may promote atherosclerosis in adulthood. This study examined the relationships between KD, atherosclerosis, and the long-term effects of HMG-CoA inhibitors (statins). Candida albicans water-soluble fraction (CAWS) was injected intraperitoneally into 5-week-old male apolipoprotein-E-deficient (Apo E-/-) mice to create KD-like vasculitis. Mice were divided into 4 groups: the control, CAWS, CAWS+statin, and late-statin groups. They were sacrificed at 6 or 10 weeks after injection. Statin was started after CAWS injection in all groups except the late-statin group, which was administered statin internally 6 weeks after injection. Lipid plaque lesions on the aorta were evaluated with Oil Red O. The aortic root and abdominal aorta were evaluated with hematoxylin and eosin staining and immunostaining. CAWS vasculitis significantly enhanced aortic atherosclerosis and inflammatory cell invasion into the aortic root and abdominal aorta. Statins significantly inhibited atherosclerosis and inflammatory cell invasion, including macrophages. CAWS vasculitis, a KD-like vasculitis, promoted atherosclerosis in Apo E-/- mice. The long-term oral administration of statin significantly suppressed not only atherosclerosis but also inflammatory cell infiltration. Therefore, statin treatment may be used for the secondary prevention of cardiovascular events during the chronic phase of KD.

Keywords: Kawasaki disease; vasculitis; atherosclerosis; statin

Citation: Motoji, Y.; Fukazawa, R.; Matsui, R.; Nagi-Miura, N.; Miyagi, Y.; Itoh, Y.; Ishii, Y. Kawasaki Disease-like Vasculitis Facilitates Atherosclerosis, and Statin Shows a Significant Antiatherosclerosis and Anti-Inflammatory Effect in a Kawasaki Disease Model Mouse. *Biomedicines* **2022**, *10*, 1794. https://doi.org/10.3390/biomedicines10081794

Academic Editors: Tânia Martins-Marques, Gonçalo F. Coutinho and Attila Kiss

Received: 13 June 2022
Accepted: 22 July 2022
Published: 26 July 2022

Publisher's Note: MDPI stays neutral with regard to jurisdictional claims in published maps and institutional affiliations.

Copyright: © 2022 by the authors. Licensee MDPI, Basel, Switzerland. This article is an open access article distributed under the terms and conditions of the Creative Commons Attribution (CC BY) license (https://creativecommons.org/licenses/by/4.0/).

1. Introduction

Kawasaki disease (KD) is an acute inflammatory syndrome predominantly affecting infants, causing systemic vasculitis [1]. In Japan, the total number of patients with KD has exceeded 360,000 thus far. Of these patients, 150,000 were over 20 years of age, and 15,000 were over 40 years (from 1964 to 31 December 2016) [2,3]. Among adult patients with KD, it has been reported that vasculitis might induce the early progression of atherosclerosis [4–6]. The pathological image of chronic KD cardiovascular sequelae is mainly composed of sclerotic lesions after vasculitis, which is pathologically different from usual atherosclerotic lesions [6,7]. However, it has been demonstrated that chronic vascular endothelial cell damage persists in patients with KD due to the tissue infiltration of inflammatory cells such as macrophage and activation of the immune system by releasing inflammatory cytokines [5,6,8]. Endothelial dysfunction could be said to share a common ground with ordinary atherosclerosis [5,6,8,9].

We have reported findings of prolonged vascular endothelial dysfunction and vascular aging in a pediatric specimen of a human KD coronary artery aneurysm. These findings were similar to early-stage features in adult atherosclerotic lesions [10]. Arditi et al., using an animal model of KD-like vasculitis caused by Lactobacillus casei cell wall bacterial components (LCWE), also reported that complications of dyslipidemia, a risk factor of atherosclerosis, and KD could increase the risk of early-onset atherosclerosis [11]. Currently, several small case studies have reported that HMG-CoA inhibitors (statins) have anti-inflammatory effects and improve vascular endothelial cell function [12,13]. Consequently, Japanese and U.S. treatment guidelines for the remote stage of patients with KD and coronary artery aneurysms recommend the clinical use of statins to prevent atherosclerosis and the restorative effect of coronary artery damage [2,14]. However, there is little evidence for oral statin administration as the basic and clinical mechanisms of the effects of statin have not been elucidated. We have studied the inflammatory response and management of acute KD using the Candida albicans water-soluble fraction (CAWS) model of KD-like vasculitis [15]. CAWS is a substance similar to Candida albicans extracted from the feces of patients with KD. CAWS can be injected into animal models to cause coronary arteritis, vasculitis, and myocarditis similar to KD [16,17]. This study aimed to examine whether a pre-existing KD vasculitis accelerated the development of atherosclerosis in young adulthood and how statins might have therapeutic effects on the prevention and development of atherosclerosis. We created a KD-like vasculitis model of atherosclerosis with CAWS in apolipoprotein-E-deficient (Apo E-/-) mice, an animal model of atherosclerosis, and evaluated whether CAWS vasculitis was a factor promoting atherosclerosis development in the remote stage. We also investigated the therapeutic effect of statins on atherosclerosis caused by KD-like vasculitis.

2. Materials and Methods

The study was performed in accordance with the Guide for the Care and Use of Laboratory Animals published by the US National Institutes of Health (NIH publication number 85-23, revised 1996). The study protocol was approved by the Animal Care and Use Committee of Nippon Medical School (approval number: 2020-029).

1) Animals

Five-week-old male C.KOR/StmSlc-ApoE^{shl} mice were purchased from Sankyo Labo Service Co., Ltd. (Tokyo, Japan). All mice were maintained under specific pathogen-free conditions according to the guidelines for animal care of the National Institute of Infectious Diseases in Tokyo [18]. To prevent any gender effects, we used only male mice. Water and food were available ad libitum. Mice were fed a normal diet until 7 weeks of age and were exchanged to a high-fat diet containing 0.15% cholesterol after 7 weeks of age.

2) Preparation of CAWS and Statin

CAWS was prepared from the C. albicans strain NBRC1385 according to a previously reported method [19]. Briefly, 5 L of C-limiting medium was maintained in a glass incubator for two days at 27 °C, while air was supplied at a rate of 5 L/min, and the mixture was swirled at 400 rpm. Following culture, an equal volume of ethanol was added. After standing the mixture overnight, the precipitate was collected and dissolved in 250 mL of distilled water. Ethanol was also added. Subsequently, the mixture was allowed to stand overnight again. The precipitate obtained was collected and dried with acetone to obtain CAWS.

The HMG-CoA inhibitor, Atorvastatin Calcium Hydrate (atorvastatin), was provided by Sankyo Ltd. (Tokyo, Japan). Atorvastatin was crushed and dissolved in 0.5 w/v% Methylcellulose 400 Solution, Sterilized (WACO FUJIFILM CORPORATION, Tokyo, Japan). Statin was started orally at the same time as the high-fat diet. The latter was at a dose of 10 mg/kg/day from 7 weeks of age. The oral dose for mice was determined according to the clinical dose in humans. To eliminate errors in statin feeding due to individual taking differences, statins were administered directly orally daily using an oral sonde.

3) Experimental Procedures

Mice were divided into the following four groups for comparison (Figure 1). Mice were sacrificed at 6 (11 weeks of age) and 10 (15 weeks of age) weeks after CAWS injections. The life cycle of mice is assumed to be much faster than that of humans, and male mice generally reach sexual maturity at 8 weeks of age. In terms of human age, 5 weeks of age corresponds to childhood, 7 weeks to 10–15 years of age, 11 weeks to adolescence to middle age, and 15 weeks to middle age to early adulthood.

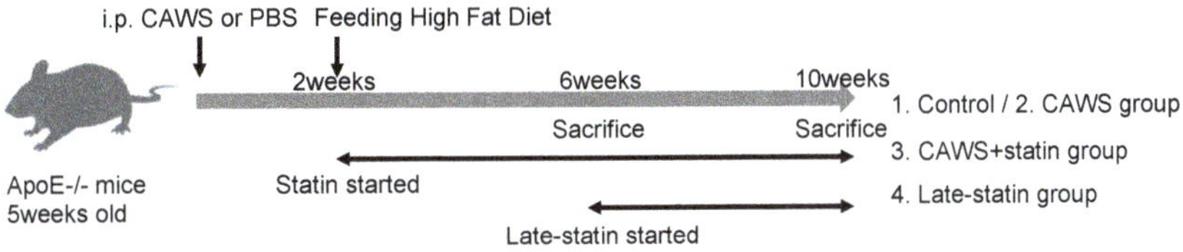

Figure 1. Experimental procedure. CAWS (4 mg for 5 consecutive days; control group received PBS instead) was injected intraperitoneally into 5-week-old Apo E-/- mice to induce KD-like vasculitis (the control group received phosphate-buffered saline instead). After 2 weeks of CAWS administration, the diet of all mice was changed to a high-fat diet to promote atherosclerosis. Mice were divided into four groups: 1. control group, 2. CAWS group, 3. CAWS+statin group, and 4. late-statin group. Statins (10 mg/kg/day) were dissolved in 0.5 mL of 0.5% Methylcellulose solution and then administered orally with a sonde. The control group was administered methylcellulose solution without statin. The CAWS and CAWS+statin groups received statins daily from 2 weeks after CAWS administration to the end of the experiment. The Late-statin group received statins daily from 6 weeks after CAWS administration to the end of the experiment. After 6 and 10 weeks of CAWS administration, mice were sacrificed, and the samples were collected.

(1) Control group: Five-week-old Apo E-/- mice were injected intraperitoneally with phosphate-buffered saline (PBS) instead of CAWS. Instead of statins, the same volume of methylcellulose solution was orally administered to mice daily in the same manner.

(2) CAWS group: CAWS (4 mg/mouse) was injected intraperitoneally into 5-week-old Apo E-/- mice for 5 consecutive days, as done in the study of Hashimoto et al. [15]. Instead of statin, mice were orally injected with a methylcellulose solution daily in the same manner.

(3) CAWS+statin group: Mice were orally administered statin daily from 2 weeks after CAWS administration (from 7 weeks of age) until the last day of the experiment.

(4) Late-statin group: Mice were started on statin orally from 6 (11 weeks of age) weeks after CAWS administration until the last day of the experiment. Since statins have no indication for infancy in clinical practice, we designed a model in which statins are administered from young adulthood.

4) Assessment of atherosclerotic lesions in the aorta

Mice were anesthetized, and the aortas were excised from the aortic arch to the iliac bifurcation. Whole aortas en face were prepared and stained with Oil red O. Lesion areas were quantified and analyzed using the hybrid cell count system (KEYENCE) and KEYENCE BZX analyzer (Osaka, Japan). Image analysis was performed by a trained observer blinded to treatment [20]. The lipid-stained plaque area in the aorta en face preparations was expressed as a percent of the aortic surface area [11].

5) Assessment of the area of aortic root horizontal transection

Aortic roots were embedded in paraffin and identified in serial sections (5 µm), and stained with hematoxylin-eosin (HE) as described in our early publication [15]. The abdominal aorta below the renal artery was harvested and embedded in paraffin. Serial sections (5 µm) were prepared similarly and stained with HE. The area (mm^2) of inflammatory cell infiltration was measured, and the ratio of the inflammatory area to the total tissue area on the aortic root was calculated.

6) Evaluation of macrophage cell and TGFβ receptor expression in the aortic root

Sections of the aortic root were analyzed immunohistochemically for the presence of macrophage cells and transforming growth factor (TGF) β receptor expression. Immunostaining was performed to identify macrophage cells and their fractions, which represent inflammatory cells. Immunostaining for TGFβ receptors was also performed because it has been reported that TGFβ signaling is involved in vascular remodeling in KD vasculitis [21]. Accordingly, the following sheep anti-rabbit antibodies were used: anti-Galectin 3 (MAC-2) antibody (1/250, 60 min RT, Abcam; ab76245, UK), a specific marker for macrophages; anti-CD80 (Abcam; ab215166, UK), a marker specific for M1; anti-Mannose receptor (CD206) antibody (1/10,000, 30 min RT, Abcam; ab64693, UK), a marker specific for M2; and anti-TGFβ receptor II antibody (1/500, 60 min RT, Abcam; ab186838, UK). The sections were further treated with the secondary antibodies and developed using HRP-conjugated DAB substrate (Abcam; ab236446, UK).

7) Evaluation of macrophage cells in the abdominal aorta

For abdominal aorta specimens, an anti-galectin 3 (MAC-2) antibody (1/250, 60 min RT, Abcam; ab76245, UK) was used to stain macrophages. Similarly, HRP-conjugated DAB substrate (Abcam; ab236446, UK) was used for staining.

8) Serological Evaluation

Serum samples collected were stored at −20 Celsius degrees until analysis.

Sandwich enzyme-linked immunosorbent assays were used to detect the plasma levels of high-sensitivity C-reactive protein (hs-CRP) (MyBioSource, San Diego, CA, USA; hs-CRP elisa kit, MBS262829) and a low-density lipoprotein/very low-density lipoproteins (LDL/VLDL) (Abcam, San Francisco, CA, USA; cholesterol Assay Kit-HDL and LDL/VLDL, ab65390) according to the manufacturers' instructions.

9) Statistical Analysis

Statistical data were expressed as median (upper and lower quartiles) or mean ± standard deviation. Statistical analyses were performed using JMP statistical software version 16 (SAS Institute Inc., Cary, NC, USA). The Kruskal-Wallis test was used to analyze statistical differences among groups. When significance was detected, the Wilcoxon test was used as a post-hoc test to compare values between both groups. A p-value < 0.05 was considered statistically significant. In all correlation analyses, Spearman's rank correlation was used, and Spearman's coefficients are denoted by ρ.

3. Results

1) Oil red O staining of the entire aorta (Figure 2)

All mice treated with CAWS administration had accelerated aortic plaque lesion formation. While the plaque in the ascending aorta and descending thoracic aorta was relatively mild, extensive plaque was observed in the abdominal aorta and below. The aortic plaque coverage ratio was significantly higher in the CAWS group than in the control group at both 6 and 10 weeks after CAWS administration (6 weeks: 5.2 ± 3.2% vs. 20.6 ± 5.9% [p = 0.037]; 10 weeks: 6.5 ± 2.4% vs. 33.0 ± 7.6% [p = 0.012]; Table 1 and Figure 2a,b).

Compared to the CAWS group, the CAWS + statin group had a decreasing trend after 6 weeks of CAWS administration and a significant decrease 10 weeks after CAWS administration (6 weeks: 20.6 ± 5.9% vs. 12.7 ± 6.1% [p = 0.111]; 10 weeks: 33.0 ± 7.6% vs. 17.5 ± 3.9% [p = 0.012]; Table 1 and Figure 2a,b).

Furthermore, the aortic plaque coverage ratio in the late-statin group (21.0 ± 4.0%) was significantly lower than that in the CAWS group at 10 (p = 0.022) but equivalent to 6 weeks (p = 0.676).

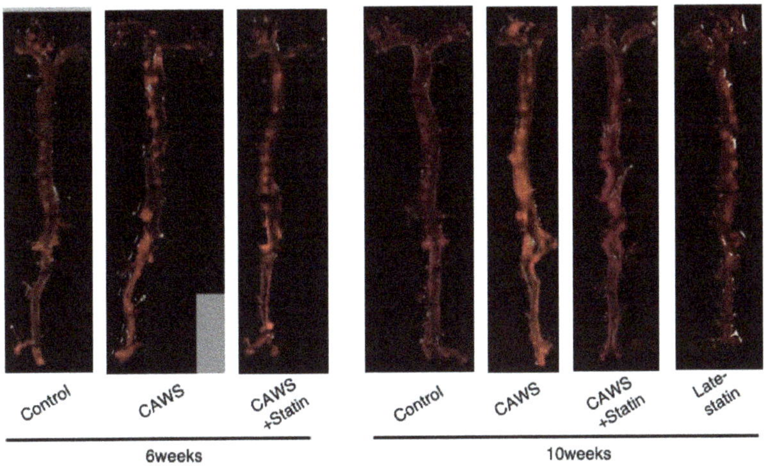

(a)

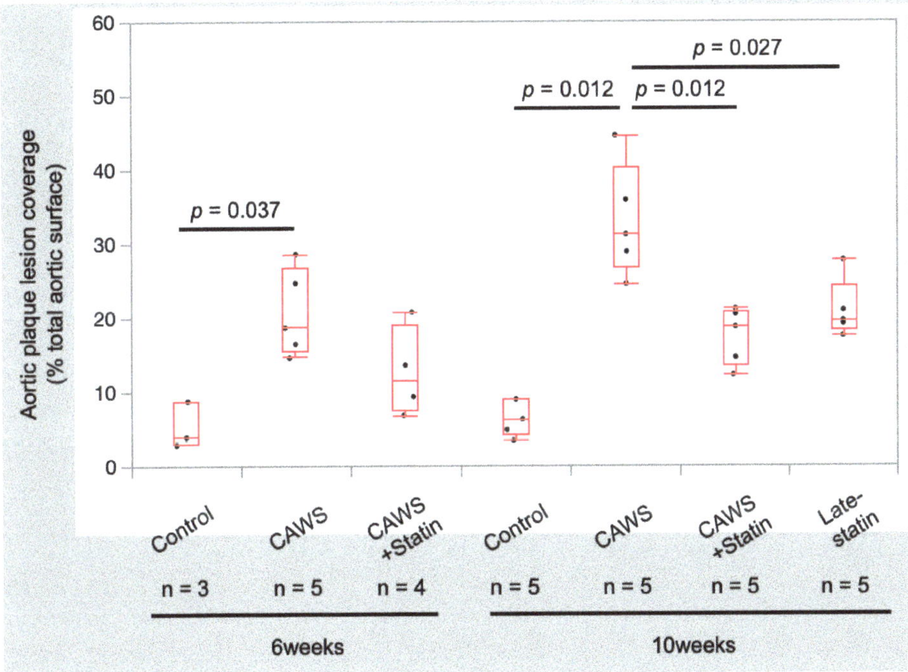

(b)

Figure 2. Histological findings of aortic plaque lesion coverage. The paraformaldehyde-fixed aorta was stained with Oil Red O. Quantification of aortic plaque lesion coverage area ratios on the entire aortic surface was performed in Apo E-/- mice. The plaque area was significantly higher in the CAWS mice than in statin-treated mice. Interestingly, the late-statin group showed a similar reduction in the aortic plaque area as the statin group. Total aortic plaque coverage ratios were calculated and compared between groups. Area ratios were expressed as percentages. CAWS significantly promoted atherosclerosis at both 6 and 10 weeks after administration. In both the late-statin and CAWS+statin groups at 10 weeks, statins significantly inhibited atherosclerosis. The inhibitory effects were comparable between both groups. (**a**) Quantification of aortic plaque lesion coverage area ratios on the entire aortic surface. (**b**) Total aortic plaque coverage ratios.

Table 1. Summary of the data at 6 and 10 weeks after CAWS administration.

6 weeks after CAWS administration	Control	CAWS	CAWS+statin		Late-statin	
Body weight (g)	28.4 ± 1.2 (*n* = 7)	28.9 ± 2.4 (*n* = 9)	27.1 ± 1.4 * (*n* = 8)	**		
Spleen weight (g)	0.14 ± 0.02 (*n* = 7)	0.23 ± 0.02 * (*n* = 9)	0.22 ± 0.04 (*n* = 8)	**		
Spleen/body weight ratio (%)	0.5 ± 0.1 (*n* = 7)	0.8 ± 0.1 * (*n* = 9)	0.8 ± 0.1 * (*n* = 8)			
LDL/VLDL cholesterol (mg/dL)	472.5 ± 45.0 (*n* = 4)	543.0 ± 6.8 * (*n* = 4)	528.3 ± 11.6 (*n* = 4)	**		
hs-CRP (pg/mL)	102.7 ± 20.9 (*n* = 4)	307.4 ± 121.5 * (*n* = 4)	248.2 ± 40.1 * (*n* = 4)			
Aortic plaque lesion coverage (%)	5.2 ± 3.2 (*n* = 3)	20.6 ± 5.9 * (*n* = 5)	12.7 ± 6.1 (*n* = 4)			
Aortic root Inflammatory cell invasion area (%)	2.7 ± 1.0 (*n* = 4)	15.4 ± 1.2 * (*n* = 4)	4.8 ± 1.2 * (*n* = 4)	**		
Aortic root Macrophage cell invasion area (%)	0.3 ± 0.1 (*n* = 4)	10.2 ± 2.2 * (*n* = 4)	3.5 ± 0.4 * (*n* = 4)	**		
Macrophage M1 cell area (%)	0.6 ± 0.3 (*n* = 4)	11.4 ± 2.5 * (*n* = 4)	1.9 ± 0.4 * (*n* = 4)	**		
Macrophage M2 cell area (%)	0.2 ± 0.2 (*n* = 4)	1.1 ± 0.5 * (*n* = 4)	2.1 ± 0.3 * (*n* = 4)	**		
TGFβ receptor II area (%)	1.1 ± 0.7 (*n* = 4)	25.7 ± 3.9 * (*n* = 4)	5.7 ± 1.4 * (*n* = 4)	**		
Abdominal aorta Inflammatory cell invasion area (%)	3.3 ± 1.2 (*n* = 4)	9.5 ± 0.9 * (*n* = 4)	4.2 ± 0.4 (*n* = 4)	**		
Abdominal aorta Macrophage cell invasion area (%)	0.2 ± 0.1 (*n* = 4)	2.0 ± 1.6 * (*n* = 4)	1.1 ± 0.2 * (*n* = 4)			
10 weeks after CAWS administration	Control	CAWS	CAWS+statin		Late-statin	
Body weight (g)	33.1 ± 4.2 (*n* = 8)	28.0 ± 2.3 * (*n* = 9)	26.2 ± 1.6* (*n* = 9)		26.9 ± 1.5 * (*n* = 9)	
Spleen weight (g)	0.14 ± 0.02 (*n* = 8)	0.27 ± 0.04 * (*n* = 9)	0.25 ± 0.04 * (*n* = 9)		0.24 ± 0.06 * (*n* = 9)	**
Spleen/body weight ratio (%)	0.4 ± 0.1 (*n* = 8)	1.0 ± 0.1 * (*n* = 9)	0.9 ± 0.1 * (*n* = 9)		0.9 ± 0.2 * (*n* = 9)	**
LDL/VLDL cholesterol (mg/dL)	513.3 ± 49.8 (*n* = 4)	532.9 ± 13.9 (*n* = 4)	532.0 ± 8.6 ** (*n* = 4)		446.5 ± 52.7 ** (*n* = 4)	
hs-CRP (pg/mL)	79.4 ± 17.8 (*n* = 4)	396.3 ± 197.5 * (*n* = 4)	314.6 ± 108.9 * (*n* = 4)		141.0 ± 86.0 ** (*n* = 4)	
Aortic plaque lesion coverage (%)	6.5 ± 2.4 (*n* = 5)	33.0 ± 7.6 * (*n* = 5)	17.5 ± 3.9 * (*n* = 5)	**	21.0 ± 4.0 * (*n* = 5)	**
Aortic root Inflammatory cell invasion area (%)	2.0 ± 0.5 (*n* = 3)	16.9 ± 1.3 (*n* = 4)	6.0 ± 2.0 ** (*n* = 4)		6.9 ± 1.1 ** (*n* = 4)	

Table 1. *Cont.*

Aortic root Macrophage cell invasion area (%)	0.7 ± 0.4 (*n* = 3)	6.5 ± 1.1 (*n* = 4)	2.5 ± 0.7 * (*n* = 4)	**	3.0 ± 0.8 (*n* = 4)	**
Macrophage M1 cell area (%)	1.0 ± 0.3 (*n* = 3)	6.3 ± 1.3 * (*n* = 4)	1.9 ± 0.3 * (*n* = 4)	**	2.1 ± 0.7 * (*n* = 4)	**
Macrophage M2 cell area (%)	0.2 ± 0.1 (*n* = 3)	1.0 ± 0.4 (*n* = 4)	1.8 ± 0.3 * (*n* = 4)		0.9 ± 0.2 (*n* = 4)	
TGFβ receptor II area (%)	0.4 ± 0.3 (*n* = 3)	17.5 ± 2.5 * (*n* = 4)	6.2 ± 1.8 (*n* = 4)	**	7.6 ± 1.4 * (*n* = 4)	**
Abdominal aorta Inflammatory cell invasion area (%)	3.3 ± 0.7 (*n* = 3)	11.1 ± 1.9 * (*n* = 4)	4.6 ± 0.9 (*n* = 4)	**	5.8 ± 0.6 (*n* = 4)	
Abdominal aorta Macrophage cell invasion area (%)	0.2 ± 0.2 (*n* = 3)	6.3 ± 3.4 * (*n* = 4)	2.1 ± 1.9 * (*n* = 4)		2.0 ± 2.3 (*n* = 4)	

* $p < 0.05$ compaird to Control. ** $p < 0.05$ compaird to CAWS.

2) **Evaluation of Aortic root samples**

2a) **Inflammatory cells were evaluated by HE staining (Figure 3)**

HE staining showed severe inflammatory cell invasion at the aortic root and pericoronary arteries in all CAWS-administered individuals. The inflammatory cell invasion area ratio was significantly higher in the CAWS group than in the control group 6 weeks after CAWS administration and remained unchanged 10 weeks after CAWS (6 weeks: 2.7 ± 1.0% vs. 15.5 ± 1.2% [$p = 0.030$]; 10 weeks: 2.0 ± 0.5% vs. 16.9 ± 1.3% [$p = 0.052$]; Table 1 and Figure 3a,b). Compared to the CAWS group, the CAWS+statin group showed a significant decrease at both 6 and 10 weeks after CAWS (6 weeks: 15.5 ± 1.2% vs. 4.8 ± 1.2% [$p < 0.001$]; 10 weeks: 16.9 ± 1.3% vs. 6.0 ± 2.0% [$p = 0.030$]; Table 1 and Figure 3a,b). The inflammatory cell invasion area ratio in the late-statin group significantly decreased to 6.9 ± 1.1% compared to the CAWS group ($p = 0.030$) but did not show a significant difference compared to values in the CAWS+statin group ($p = 0.471$).

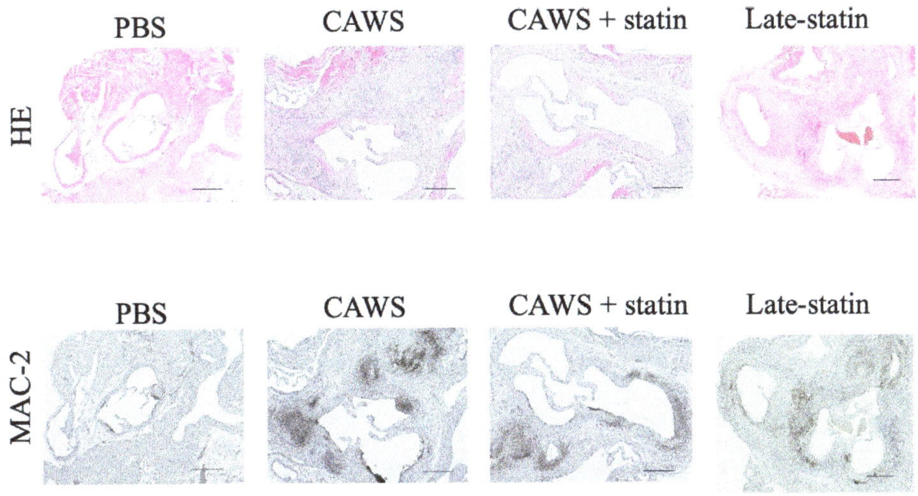

(a)

Figure 3. *Cont.*

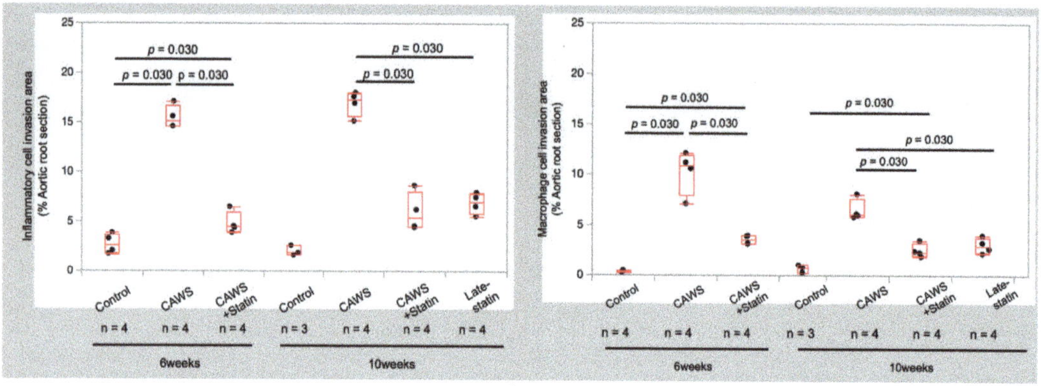

(b)

Figure 3. Histological findings at the aortic root. (**a**) Histological findings at the aortic root after 10 weeks of CAWS administration. Histological evaluation of inflammatory or macrophage cell invasion into the sinus area of the aortic root of Apo E-/- mice. Statin treatment suppressed the infiltration of inflammatory cells and macrophage cells at the aortic root; the anti-inflammatory effect of statins was also observed in the late-statin group. (**b**) Inflammatory cell invasion area and macrophage cell invasion area ratios. Inflammatory cell and macrophage cell invasion area ratios were calculated and compared between groups. Area ratios were expressed as percentages. CAWS promoted inflammatory and macrophage cell infiltration at both 6 and 10 weeks after CAWS administration. Statins also significantly inhibited the infiltration of inflammatory cells and macrophage cells at both 6 and 10 weeks after CAWS administration. In the late statin group, statins suppressed the inflammatory cell invasion area ratio. Means and SD are shown.Upper, H&E staining; lower, anti-MAC-2 antibody staining. (Bar = 500 μm).

2b) Immunostaining of macrophage cells with anti-Galectin 3 (MAC-2) antibody (Figure 3)

A large number of macrophage cells remained at the aortic root and pericoronary arteries of all individuals treated with CAWS, even after 10 weeks of CAWS administration. The macrophage cell invasion area ratio significantly increased in the CAWS group compared to the control group at 6 weeks after CAWS administration and slightly decreased at 10 weeks after CAWS administration (6 weeks: 0.3 ± 0.1% vs. 10.2 ± 2.2% [p = 0.030]; 10 weeks: 0. 7 ± 0.4% vs. 6.5 ± 1.1% [p = 0.052]; Table 1 and Figure 3a,b).

The macrophage cell invasion area ratio of the CAWS+statin group significantly decreased at both 6 and 10 weeks after CAWS compared to values in the CAWS group (6 weeks: 10.2 ± 2.2% vs. 3.5 ± 0.4% [p = 0.030]; 10 weeks: 6.5 ± 1.1% vs. 2.5 ± 0.7% [p = 0.030]; Table 1 and Figure 3a,b).

The late-statin group had a macrophage cell invasion area ratio of 3.0 ± 0.8%, significantly lower than that of the CAWS group (p = 0.030). There was no significant difference between the CAWS+statin group (p = 0.471).

2c) Immunostaining of macrophage M1 cells with anti-CD80 antibody

The macrophage M1 cell invasion area ratio significantly increased in the CAWS group compared to the control group at both 6 and 10 weeks after CAWS administration (6 weeks: 0.6 ± 0.3% vs. 11.4 ± 2.5% [p = 0.030]; 10 weeks: 1.0 ± 0.3% vs. 6.3 ± 1.3% [p = 0.030]; Table 1). The macrophage M1 cell invasion area ratio was significantly lower in the CAWS+statin group than in the CAWS group at both 6 and 10 weeks (6 weeks: 11.4 ± 2.5% vs. 1.9 ± 0.4% [p = 0.030]; 10 weeks: 6.3 ± 1.3% vs. 1.9 ± 0.3% [p = 0.030]; Table 1). The macrophage M1 cell invasion area ratio for the late-statin group was 2.1 ± 0.7%, which was significantly lower than that in the CAWS group (p = 0.030) but not statistically different compared to values in the CAWS + statin group (p = 0.883).

2d) Immunostaining of macrophage M2 cells with the anti-CD206 antibody

The macrophage M2 cell invasion area ratio was significantly higher in the CAWS group than in the control group at both 6 and 10 weeks after CAWS administration (6 weeks: 0.2 ± 0.2% vs. 1.1 ± 0.5% [p = 0.030]; 10 weeks: 0.3 ± 0.1% vs. 1.0 ± 0.4% [p = 0.052]; Table 1).

The macrophage M2 cell invasion area ratio was significantly higher in the CAWS+statin group than in the CAWS group 6 weeks after CAWS administration. This value peaked at 10 weeks after CAWS administration (6 weeks: 1.1 ± 0.5% vs. 2.1 ± 0.3% [p = 0.030]; 10 weeks: 1.0 ± 0.4% vs. 1.8 ± 0.3% [p = 0.066]; Table 1).

The macrophage M2 cell invasion area ratio in the late-statin group was 0.9 ± 0.2%, which was not significantly different from either the CAWS or CAWS+statin group (p = 0.885 and p = 0.066, respectively).

2e) M2/M1 macrophage cell ratio

The M2/M1 macrophage cell ratio in the same sample was calculated; compared to the control group, the CAWS group showed a significant decrease at 6 weeks after CAWS administration and no significant difference at 10 weeks (6 weeks: 0.4 ± 0.1 vs. 0.1 ± 0.1 [p = 0.030]; 10 weeks: 0.3 ± 0.1 vs. 0.2 ± 0.1 [p = 0.112]; Table 1). The M2/M1 macrophage cell ratio was significantly higher in the CAWS+statin group than in the CAWS group both at 6 and 10 weeks after CAWS administration (6 weeks: 0.1 ± 0.1 vs. 1.1 ± 0.3 [p = 0.030]; 10 weeks: 0.2 ± 0.1 vs. 1.0 ± 0.2 [p = 0.030]; Table 1). The M2/M1 ratio for the late-statin group was 0.5 ± 0.2, which was significantly higher than that in the CAWS (p = 0.030) and significantly lower than that of the CAWS+statin group (p = 0.030).

2f) Immunostaining with anti-TGFβ receptor II antibody

The TGFβ receptor II expression area ratio was significantly higher in the CAWS group than in the control group both after 6 and 10 weeks of CAWS administration (6 weeks: 1.1 ± 0.7% vs. 25.7 ± 3.9% [p = 0.030]; 10 weeks: 0.4 ± 0.3% vs. 17.5 ± 2.5% [p = 0.030]; Table 1). The CAWS+statin group showed a significant decrease in both 6 and 10 weeks after CAWS administration compared to the CAWS group (6 weeks: 25.7 ± 3.9% vs. 5.7 ± 1.4% [p = 0.030]; 10 weeks: 17.5 ± 2.5% vs. 6.2 ± 1.8% [p = 0.020]; Table 1). The TGFβ receptor II expression area ratio in the late-statin group was 7.6 ± 1.4%, which was significantly lower than that in the CAWS group (p = 0.030), but not significantly different from that in the CAWS+statin group (p = 0.540).

3) Evaluation of Infrarenal abdominal aorta samples

3a) Inflammatory cells evaluated by HE staining

The inflammatory cell invasion area ratio was significantly increased in the CAWS group than in the control group both at 6 and 10 weeks after CAWS administration (6 weeks: 3.3 ± 1.2% vs. 9.5 ± 1.0% [p = 0.030]; 10 weeks: 3.3 ± 0.7% vs. 11.1 ± 1.9% [p = 0.030]; Table 1). The inflammatory cell invasion area was significantly reduced in the CAWS+statin group than in the CAWS group at both 6 and 10 weeks (6 weeks: 9.5 ± 0.9% vs. 4.2 ± 0.4% [p = 0.030]; 10 weeks: 11.1 ± 1.9% vs. 4.6 ± 0.9% [p = 0.030]; Table 1). The inflammatory cell invasion area ratio in the late-statin group was 5.8 ± 0.6% compared to the CAWS group, showing a decreasing trend (p = 0.052).

3b) Immunostaining of macrophage cells with anti-galectin 3 (MAC-2) antibody

Macrophage cells were detected in aneurysm walls of specimens that formed abdominal aortic aneurysms. The macrophage cell invasion area ratio was significantly higher in the CAWS group than in the control group at both 6 and 10 weeks after CAWS administration (6 weeks: 0.2 ± 0.1% vs. 2.0 ± 1.6% [p = 0.030]; 10 weeks: 0.2 ± 0.2% vs. 6.3 ± 3.4% [p = 0.030]; Table 1). The macrophage cell invasion area ratio was reduced in the CAWS+statin group than in the CAWS group at both 6 and 10 weeks (6 weeks: 2.0 ± 1.6% vs. 1.1 ± 0.2% [p = 1.000]; 10 weeks: 6.3 ± 3.4% vs. 2.1 ± 1.9% [p = 0.112]; Table 1). The macrophage cell invasion area ratio of the late-statin group was 2.0 ± 2.3%, showing a decreasing trend compared to the CAWS group ($p = 0.112$) and no significant difference compared to the values in the CAWS+statin group ($p = 1.000$).

4) Serological examination

4a) Serum LDL/VLDL cholesterol levels in each group

LDL/VLDL cholesterol levels were significantly higher in the CAWS group than in the control group only at 6 weeks after CAWS administration, yet results were similar between groups at 10 weeks of administration (6 weeks: 472.6 ± 45.0 mg/dL vs. 543.0 ± 6.8 mg/dL [$p = 0.005$]; 10 weeks: 513.3 ± 49.8 mg/dL vs. 532.9 ± 13.9 mg/dL [$p = 0.379$]; Table 1). LDL/VLDL cholesterol levels were significantly lower in the CAWS+statin group than in the CAWS group only 6 weeks after CAWS administration (6 weeks: 543.0 ± 6.8 mg/dL vs. 528.3 ± 11.6 mg/dL [$p = 0.031$]; 10 weeks: 532.9 ± 13.9 mg/dL vs. 532.0 ± 8.6 mg/dL [$p = 0.810$]; Table 1). LDL/VLDL was significantly lower in the late-statin group than in the control group at 10 weeks after CAWS administration ($p = 0.005$).

4b) Serum hs-CRP levels in each group

hs-CRP was significantly higher in the CAWS group than in the control group both at 6 weeks and 10 weeks after CAWS administration (6 weeks: 102.7 ± 20.9 pg/mL vs. 307.4 ± 121.5 pg/mL [$p < 0.001$]; 10 weeks: 79.4 ± 17.8 pg/mL vs. 396.3 ± 197.5 pg/mL [$p < 0.001$]; Table 1). hs-CRP was not significantly different between the CAWS and CAWS+statin groups at either 6 or 10 weeks after CAWS administration (6 weeks: 307.4 ± 121.5 pg/mL vs. 248.2 ± 121.5 pg/mL [$p = 0.318$]; 10 weeks: 396.3 ± 197.5 pg/mL vs. 314.6 ± 108.9 pg/mL [$p = 0.471$]; Table 1).

4c) Correlations between aortic plaque area, inflammatory cell invasion area, and macrophage invasion area

Correlations were examined between aortic plaque coverage ratio, inflammatory cell area ratio at the aortic root, and macrophage cell area ratio. A significant correlation was found between the aortic plaque coverage ratio and aortic root inflammatory cell invasion area ratio ($\rho = 0.786$; $p < 0.0001$) and between the aortic plaque coverage ratio and aortic root macrophage area ratio ($\rho = 0.641$; $p = 0.0004$) (Figure 4).

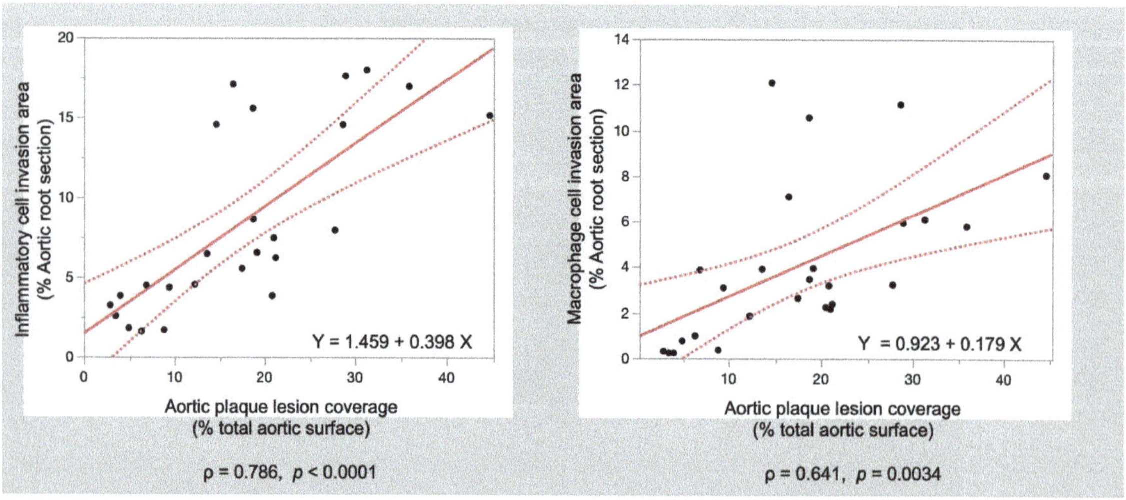

Figure 4. Correlation coefficient scatter diagrams. The aortic plaque lesion coverage ratio was significantly associated with the inflammatory cell invasion area ratio and macrophage cell invasion area ratio at the aortic root.

CRP, C-reactive protein; LDL/VLDL, low-density lipoprotein/very low-density lipoprotein; and TGFβ receptor II, Transforming growth factor beta receptor II.

4. Discussion

Endothelial cell dysfunction has been reported to persist in patients with sequelae of KD vasculitis [10,22–24]. However, there is no evidence of a direct relationship between KD vasculitis and early progression to atherosclerosis. In this study, we formed a KD-like vasculitis model of atherosclerosis with CAWS in Apo E-/- mice. Through this animal model of atherosclerosis, we confirmed that CAWS vasculitis promotes atherosclerosis. Subsequently, we showed in animal models that statins have anti-inflammatory and anti-atherosclerotic effects against atherosclerosis after CAWS vasculitis. Recently, Miyabe et al. reported that macrophage cell activation in response to CAWS deposits in the adventitia of the aortic root caused a large and persistent inflammatory cell infiltration around the coronary arteries [25]. In this study, a long-term infiltration of inflammatory cells, including macrophage cells, was also observed at the aortic root and in the abdominal aorta. TGFβ receptor II was strongly expressed in the vessel wall of the aortic root, supporting the presence of CAWS-induced vasculitis [21]. Statins significantly suppressed the infiltration of inflammatory cells, including macrophage cells, TGFβ receptor II expression, and atherosclerosis in the aorta. Atherosclerosis is a chronic inflammatory disease of the arterial wall, and macrophages have an important role as the starting point of plaque formation in patients with atherosclerosis [26,27].

Macrophages are broadly classified into the M1 phenotype, which produces inflammatory cytokines and has tissue-destructive properties, and the M2 phenotype, which produces anti-inflammatory cytokines and promotes angiogenesis and tissue repair [28] and whose phenotype and function depend on the microenvironmental stimuli [29]. We have previously reported that the phenotype of macrophages in coronary aneurysms of human KD patients differs from that of normal atherosclerotic lesions, with a predominance of the M1 phenotype. We have also reported that statins have an anti-inflammatory effect on the coronary atrial wall of patients treated with statins for more than 3 years [30]. In this study, most macrophages infiltrating the aortic root were the M1 phenotype. Statins might suppress the tissue infiltration of M1 macrophages or inhibit their differentiation into the M1 phenotype. Zhang et al. reported that statins induced monocyte differentiation into M2 macrophages [31]. This analysis showed that the M2 macrophage cell invasion area ratio was lower, and although statin administration increased the count of M2 macrophage cells, the difference between groups was insignificant. However, statins significantly suppressed the overall macrophage cell area, and when the M2/M1 macrophage cell area ratio was calculated, the M2/M1 ratio was significantly increased by statin administration. After 10 weeks of CAWS administration, the effect of statins on increasing M2 macrophages was also observed in the late-statin group, even though it was inferior to the group that had received statins from the start. The results suggested that statins might suppress inflammation and promote tissue repair by increasing M2 macrophages relatively.

In contrast to the 10 to 150 mg/dL serum LDL/VLDL cholesterol levels of non-Apo E-/- mice fed a high-fat diet [32–34], the serum LDL/VLDL levels of Apo E-/- mice fed a high-fat diet are known to be significantly higher than those of normal mice [27,33–37]. Interestingly, although the lipid-lowering effect of statins was confirmed in this analysis, the degree of reduction was mild. There were no animals whose LDL/VLDL cholesterol levels were normalized with statins. Statins have not only LDL-lowering but also pleiotropic effects, including anti-inflammatory effects, improvements in endothelial cell function, effects on macrophage fractions, and antioxidant effects [38]. It was considered that the therapeutic effect of statins observed in this analysis was more likely due to pleiotropic effects rather than lipid-lowering effects. The late-statin group in this analysis was designed to determine whether statins have a therapeutic effect in adult patients with KD who have started statins in young adulthood. Statins are not approved for infants, and it is difficult for patients with KD to take statins continuously from infancy. Remarkably, the late-statin group showed almost the same anti-inflammatory and anti-atherosclerotic effects as the group that was given statins after 2 weeks of CAWS administration (CAWS+statin group) (Figures 2a and 3a). These results suggest that statins have a sufficient therapeutic effect, even if statins were started as patients with KD were adults.

In this study, it was confirmed that statins have an inhibitory effect on vascular inflammation and atherosclerosis in CAWS vasculitis-infected mice. However, this is an animal study using mice only. Moreover, CAWS vasculitis is not exactly the same as KD vasculitis, although CAWS vasculitis is considered an experimental model that has been generalized as KD-like vasculitis. Newly designed statin clinical trials should be conducted to establish the efficacy of statin against KD. Second, we have not been shown that CAWS vasculitis and statins affect vascular endothelial cell function. We have previously reported that KD vasculitis may impair endothelial cell function [10], and several centers have reported that statins may improve endothelial cell function and suppress vasculitis [12,13,39,40]. It is considered important to research the relationship between inflammatory cell invasion, including macrophages, in the remote stage and the statin-induced improvement of vascular endothelial cell function to elucidate the usefulness of statins in patients with KD.

5. Conclusions

It was found that the administration of CAWS to an atherosclerosis model mice induced KD-like vasculitis and that vasculitis promoted atherosclerosis in the remote stage of KD. Moreover, statin therapy inhibited the development of KD-like vasculitis, which induces atherosclerosis, and had an anti-inflammatory effect on the vasculitis itself. The results suggest that statin therapy is effective for patients with KD accompanied by cardiovascular sequelae. The indication for statin therapy in such patients may be extended in the future.

Author Contributions: Conceptualization, R.F.; methodology and validation, Y.M. (Yusuke Motoji), R.M., N.N.-M. and R.F.; formal analysis, Y.M. (Yusuke Motoji) and R.F.; investigation, Y.M. (Yusuke Motoji); resources, Y.M. (Yasuo Miyagi) and R.F.; data curation, Y.M. (Yasuo Miyagi) and R.F.; writing—original draft preparation, Y.M. (Yusuke Motoji); writing—review and editing, R.F. and Y.I. (Yasuhiko Itoh); visualization, Y.M. (Yusuke Motoji); supervision, Y.I. (Yosuke Ishii); project administration, R.F.; and funding acquisition, R.F. All authors have read and agreed to the published version of the manuscript.

Funding: This work was supported by a Grant-inAid for Scientific Research Japan (JP12K34567) from the Ministry of Education, Culture, Sports, Science, and Technology.

Institutional Review Board Statement: The study was conducted in accordance with the Declaration of Helsinki and approved by the Ethics Committee of the Nippon Medical School (approval number: 2020-029).

Informed Consent Statement: Not applicable.

Data Availability Statement: Not applicable.

Conflicts of Interest: The authors declare no conflict of interest.

References

1. Kawasaki, T.; Kosaki, F.; Okawa, S.; Shigematsu, I.; Yanagawa, H. A new infantile acute febrile mucocutaneous lymph node syndrome (MLNS) prevailing in Japan. *Pediatrics* **1974**, *54*, 271–276. [CrossRef] [PubMed]
2. Fukazawa, R.; Kobayashi, J.; Ayusawa, M.; Hamada, H.; Miura, M.; Mitani, Y.; Tsuda, E.; Nakajima, H.; Matsuura, H.; Ikeda, K.; et al. JCS/JSCS 2020 guideline on diagnosis and management of cardiovascular sequelae in Kawasaki disease. *Circ. J.* **2020**, *84*, 1348–1407. [CrossRef] [PubMed]
3. Report of the 24th Nation-Wide Survey for Kawasaki Disease. Available online: http://www.jichi.ac.jp/dph/kawasakibyou/20170928/mcls24report.pdf (accessed on 31 January 2021). (In Japanese).
4. Fukazawa, R. Long-Term prognosis of Kawasaki disease: Increased cardiovascular risk? *Curr. Opin. Pediatr.* **2010**, *22*, 587–592. [CrossRef] [PubMed]

5. Ueno, K.; Ninomiya, Y.; Hazeki, D.; Masuda, K.; Nomura, Y.; Kawano, Y. Disruption of endothelial cell homeostasis plays a key role in the early pathogenesis of coronary artery abnormalities in Kawasaki disease. *Sci. Rep.* **2017**, *7*, 43719. [CrossRef]
6. Zeng, Y.-Y.; Zhang, M.; Ko, S.; Chen, F. An update on cardiovascular risk factors after Kawasaki disease. *Front. Cardiovasc. Med.* **2021**, *8*, 671198. [CrossRef]
7. Mitani, Y.; Ohashi, H.; Sawada, H.; Ikeyama, Y.; Hayakawa, H.; Takabayashi, S.; Maruyama, K.; Shimpo, H.; Komada, Y. In vivo plaque composition and morphology in coronary artery lesions in adolescents and young adults long after Kawasaki disease: A virtual histology—Intravascular ultrasound study. *Circulation* **2009**, *119*, 2829–2836. [CrossRef]
8. Matsubara, T.; Ichiyama, T.; Furukawa, S. Immunological profile of peripheral blood lymphocytes and monocytes/macrophages in Kawasaki disease. *Clin. Exp. Immunol.* **2005**, *141*, 381–387. [CrossRef]
9. Burns, J.C.; Glodé, M.P. Kawasaki syndrome. *Lancet* **2004**, *364*, 533–544. [CrossRef]
10. Fukazawa, R.; Ikegam, E.; Watanabe, M.; Hajikano, M.; Kamisago, M.; Katsube, Y.; Yamauchi, H.; Ochi, M.; Ogawa, S. Coronary artery aneurysm induced by Kawasaki disease in children show features typical senescence. *Circ. J.* **2007**, *71*, 709–715. [CrossRef]
11. Chen, S.; Lee, Y.; Crother, T.R.; Fishbein, M.; Zhang, W.; Yilmaz, A.; Shimada, K.; Schulte, D.J.; Lehman, T.J.A.; Shah, P.K.; et al. Marked acceleration of atherosclerosis after *Lactobacillus casei*–induced coronary arteritis in a mouse model of Kawasaki disease. *Arterioscler. Thromb. Vasc. Biol.* **2012**, *32*, e60–e71. [CrossRef]
12. Huang, S.-M.; Weng, K.-P.; Chang, J.-S.; Lee, W.-Y.; Huang, S.-H.; Hsieh, K.-S. Effects of statin therapy in children complicated with coronary arterial abnormality late after Kawasaki disease. *Circ. J.* **2008**, *72*, 1583–1587. [CrossRef]
13. Duan, C.; Du, Z.-D.; Wang, Y.; Jia, L.-Q. Effect of pravastatin on endothelial dysfunction in children with medium to giant coronary aneurysms due to Kawasaki disease. *World J. Pediatr.* **2014**, *10*, 232–237. [CrossRef] [PubMed]
14. McCrindle, B.W.; Rowley, A.H.; Newburger, J.W.; Burns, J.C.; Bolger, A.F.; Gewitz, M.; Baker, A.L.; Jackson, M.A.; Takahashi, M.; Shah, P.B.; et al. Diagnosis, treatment, and long-term management of Kawasaki disease: A scientific statement for health professionals from the American Heart Association. *Circulation* **2017**, *135*, e927–e999. [CrossRef] [PubMed]
15. Hashimoto, Y.; Fukazawa, R.; Nagi-Miura, N.; Ohno, N.; Suzuki, N.; Katsube, Y.; Kamisago, M.; Akao, M.; Watanabe, M.; Hashimoto, K.; et al. Interleukin-1beta inhibition attenuates vasculitis in a mouse model of Kawasaki disease. *J. Nippon Med. Sch.* **2019**, *86*, 108–116. [CrossRef] [PubMed]
16. Ohno, N. Chemistry and biology of angiitis inducer, *Candida albicans* water-soluble mannoprotein-β-glucan complex (CAWS). *Microbiol. Immunol.* **2003**, *47*, 479–490. [CrossRef]
17. Nagi-Miura, N.; Harada, T.; Shinohara, H.; Kurihara, K.; Adachi, Y.; Ishida-Okawara, A.; Oharaseki, T.; Takahashi, K.; Naoe, S.; Suzuki, K.; et al. Lethal and severe coronary arteritis in DBA/2 mice induced by fungal pathogen, CAWS, *Candida albicans* water-soluble fraction. *Atherosclerosis* **2006**, *186*, 310–320. [CrossRef] [PubMed]
18. Science Council of Japan Guidelines for Proper Conduct of Animal Experiments. Available online: https://www.scj.go.jp/ja/info/kohyo/pdf/kohyo-20-k16-2e.pdf (accessed on 20 July 2022).
19. Uchiyama, M.; Ohno, N.; Miura, N.N.; Adachi, Y.; Aizawa, M.W.; Tamura, H.; Tanaka, S.; Yadomae, T. Chemical and immunochemical characterization of limulus factor G-activating substance of *Candida* spp. *FEMS Immunol. Med. Microbiol.* **1999**, *24*, 411–420. [CrossRef] [PubMed]
20. Centa, M.; Ketelhuth, D.F.J.; Malin, S.; Gisterå, A. Quantification of atherosclerosis in mice. *J. Vis. Exp.* **2019**, e59828.
21. Lee, A.M.; Shimizu, C.; Oharaseki, T.; Takahashi, K.; Daniels, L.B.; Kahn, A.; Adamson, R.; Dembitsky, W.; Gordon, J.B.; Burns, J.C. Role of TGF-β Signaling in remodeling of noncoronary artery aneurysms in Kawasaki disease. *Pediatr. Dev. Pathol.* **2015**, *18*, 310–317. [CrossRef]
22. Sugimura, T.; Kato, H.; Inoue, O.; Fukuda, T.; Sato, N.; Ishii, M.; Takagi, J.; Akagi, T.; Maeno, Y.; Kawano, T. Intravascular ultrasound of coronary arteries in children. assessment of the wall morphology and the lumen after Kawasaki disease. *Circulation* **1994**, *89*, 258–265. [CrossRef]
23. Yamakawa, R.; Ishii, M.; Sugimura, T.; Akagi, T.; Eto, G.; Iemura, M.; Tsutsumi, T.; Kato, H. Coronary endothelial dysfunction after Kawasaki disease: Evaluation by intracoronary injection of acetylcholine. *J. Am. Coll. Cardiol.* **1998**, *31*, 1074–1080. [CrossRef]
24. Grundy, S.M.; Stone, N.J.; Bailey, A.L.; Beam, C.; Birtcher, K.K.; Blumenthal, R.S.; Braun, L.T.; de Ferranti, S.; Faiella-Tommasino, J.; Forman, D.E.; et al. 2018 AHA/ACC/AACVPR/AAPA/ABC/ACPM/ADA/AGS/APhA/ASPC/NLA/PCNA Guideline on the management of blood cholesterol: A report of the American College of Cardiology/American Heart Association Task Force on clinical practice guidelines. *J. Am. Coll. Cardiol.* **2019**, *73*, e285–e350. [CrossRef] [PubMed]
25. Miyabe, C.; Miyabe, Y.; Bricio-Moreno, L.; Lian, J.; Rahimi, R.A.; Miura, N.N.; Ohno, N.; Iwakura, Y.; Kawakami, T.; Luster, A.D. Dectin-2–induced CCL2 production in tissue-resident macrophages ignites cardiac arteritis. *J. Clin. Investig.* **2019**, *129*, 3610–3624. [CrossRef] [PubMed]
26. Momi, S.; Monopoli, A.; Alberti, P.F.; Falcinelli, E.; Corazzi, T.; Conti, V.; Miglietta, D.; Ongini, E.; Minuz, P.; Gresele, P. Nitric oxide enhances the anti-inflammatory and anti-atherogenic activity of atorvastatin in a mouse model of accelerated atherosclerosis. *Cardiovasc. Res.* **2012**, *94*, 428–438. [CrossRef] [PubMed]
27. Nie, P.; Li, D.; Hu, L.; Jin, S.; Yu, Y.; Cai, Z.; Shao, Q.; Shen, J.; Yi, J.; Xiao, H.; et al. Atorvastatin improves plaque stability in ApoE-knockout mice by regulating chemokines and chemokine receptors. *PLoS ONE* **2014**, *9*, e97009. [CrossRef] [PubMed]
28. Mantovani, A.; Biswas, S.K.; Galdiero, M.R.; Sica, A.; Locati, M. Macrophage plasticity and polarization in tissue repair and remodelling. *J. Pathol.* **2013**, *229*, 176–185. [CrossRef]

29. Da Silva, R.F.; Lappalainen, J.; Lee-Rueckert, M.; Kovanen, P.T. Conversion of human M-CSF macrophages into foam cells reduces their proinflammatory responses to classical M1-polarizing activation. *Atherosclerosis* **2016**, *248*, 170–178. [CrossRef]
30. Ohashi, R.; Fukazawa, R.; Shimizu, A.; Ogawa, S.; Ochi, M.; Nitta, T.; Itoh, Y. M1 macrophage is the predominant phenotype in coronary artery lesions following Kawasaki disease. *Vasc. Med.* **2019**, *24*, 484–492. [CrossRef]
31. Zhang, T.; Shao, B.; Liu, G.-A. Rosuvastatin promotes the differentiation of peripheral blood monocytes into M2 macrophages in patients with atherosclerosis by activating PPAR-γ. *Eur. Rev. Med. Pharmacol. Sci.* **2017**, *21*, 4464–4471.
32. Wang, Y.; Krishna, S.M.; Moxon, J.; Dinh, T.N.; Jose, R.J.; Yu, H.; Golledge, J. Influence of apolipoprotein E, age and aortic site on calcium phosphate induced abdominal aortic aneurysm in mice. *Atherosclerosis* **2014**, *235*, 204–212. [CrossRef]
33. Widjaja, A.A.; Singh, B.K.; Adami, E.; Viswanathan, S.; Dong, J.; D'Agostino, G.A.; Ng, B.; Lim, W.W.; Tan, J.; Paleja, B.S.; et al. Inhibiting interleukin 11 signaling reduces hepatocyte death and liver fibrosis, inflammation, and steatosis in mouse models of nonalcoholic steatohepatitis. *Gastroenterology* **2019**, *157*, 777–792.e14. [CrossRef] [PubMed]
34. Dai, B.; Li, X.; Xu, J.; Zhu, Y.; Huang, L.; Tong, W.; Yao, H.; Chow, D.H.-K.; Qin, L. Synergistic effects of magnesium ions and simvastatin on attenuation of high-fat diet-induced bone loss. *Bioact. Mater.* **2021**, *6*, 2511–2522. [CrossRef] [PubMed]
35. Plump, A.S.; Smith, J.D.; Hayek, T.; Aalto-Setälä, K.; Walsh, A.; Verstuyft, J.G.; Rubin, E.M.; Breslow, J.L. Severe hypercholesterolemia and atherosclerosis in apolipoprotein E-deficient mice created by homologous recombination in ES cells. *Cell* **1992**, *71*, 343–353. [CrossRef]
36. Nachtigal, P.; Pospisilova, N.; Jamborova, G.; Pospechova, K.; Solichova, D.; Andrys, C.; Zdansky, P.; Micuda, S.; Semecky, V. Atorvastatin has hypolipidemic and anti-inflammatory effects in ApoE/LDL receptor-double-knockout mice. *Life Sci.* **2008**, *82*, 708–717. [CrossRef] [PubMed]
37. Fukuda, D.; Enomoto, S.; Nagai, R.; Sata, M. Inhibition of renin–angiotensin system attenuates periadventitial inflammation and reduces atherosclerotic lesion formation. *Biomed. Pharmacother.* **2009**, *63*, 754–761. [CrossRef]
38. Oesterle, A.; Laufs, U.; Liao, J.K. Pleiotropic effects of statins on the cardiovascular system. *Circ. Res.* **2017**, *120*, 229–243. [CrossRef] [PubMed]
39. Hamaoka, A.; Hamaoka, K.; Yahata, T.; Fujii, M.; Ozawa, S.; Toiyama, K.; Nishida, M.; Itoi, T. Effects of HMG-CoA reductase inhibitors on continuous post-inflammatory vascular remodeling late after Kawasaki disease. *J. Cardiol.* **2010**, *56*, 245–253. [CrossRef]
40. Niedra, E.; Chahal, N.; Manlhiot, C.; Yeung, R.S.M.; McCrindle, B.W. Atorvastatin safety in Kawasaki disease patients with coronary artery aneurysms. *Pediatr. Cardiol.* **2014**, *35*, 89–92. [CrossRef]

biomedicines

Brief Report

The Effect of Nutritional Ketosis on Aquaporin Expression in Apolipoprotein E-Deficient Mice: Potential Implications for Energy Homeostasis

Inês V. da Silva [1,2], Sean Gullette [3], Cristina Florindo [2], Neil K. Huang [4,5], Thomas Neuberger [3,6], A. Catharine Ross [4], Graça Soveral [1,2,*] and Rita Castro [2,4,*]

1 Research Institute for Medicines (iMed.ULisboa), Faculty of Pharmacy, Universidade de Lisboa, 1649-003 Lisbon, Portugal; imsilva1@campus.ul.pt
2 Department of Pharmaceutical Sciences and Medicines, Faculty of Pharmacy, Universidade de Lisboa, 1649-003 Lisbon, Portugal; cristinaflorindo@ff.ulisboa.pt
3 Huck Institutes of the Life Sciences, The Pennsylvania State University, State College, PA 16802, USA; sqg5746@psu.edu (S.G.); tun3@psu.edu (T.N.)
4 Department of Nutritional Sciences, The Pennsylvania State University, State College, PA 16802, USA; neil.huang@tufts.edu (N.K.H.); acr6@psu.edu (A.C.R.)
5 Cardiovascular Nutrition Laboratory, Jean Mayer USDA Human Nutrition Research Center on Aging, Tufts University, Boston, MA 02111, USA
6 Department of Biomedical Engineering, The Pennsylvania State University, University Park, State College, PA 16802, USA
* Correspondence: gsoveral@ff.ulisboa.pt (G.S.); mum689@psu.edu (R.C.)

Citation: da Silva, I.V.; Gullette, S.; Florindo, C.; Huang, N.K.; Neuberger, T.; Ross, A.C.; Soveral, G.; Castro, R. The Effect of Nutritional Ketosis on Aquaporin Expression in Apolipoprotein E-Deficient Mice: Potential Implications for Energy Homeostasis. *Biomedicines* **2022**, *10*, 1159. https://doi.org/10.3390/biomedicines10051159

Academic Editors: Tânia Martins-Marques, Gonçalo F. Coutinho and Attila Kiss

Received: 12 April 2022
Accepted: 13 May 2022
Published: 18 May 2022

Publisher's Note: MDPI stays neutral with regard to jurisdictional claims in published maps and institutional affiliations.

Copyright: © 2022 by the authors. Licensee MDPI, Basel, Switzerland. This article is an open access article distributed under the terms and conditions of the Creative Commons Attribution (CC BY) license (https://creativecommons.org/licenses/by/4.0/).

Abstract: Ketogenic diets (KDs) are very low-carbohydrate, very high-fat diets which promote nutritional ketosis and impact energetic metabolism. Aquaporins (AQPs) are transmembrane channels that facilitate water and glycerol transport across cell membranes and are critical players in energy homeostasis. Altered AQP expression or function impacts fat accumulation and related comorbidities, such as the metabolic syndrome. Here, we sought to determine whether nutritional ketosis impacts AQPs expression in the context of an atherogenic model. To do this, we fed $ApoE^{-/-}$ (apolipoprotein E-deficient) mice, a model of human atherosclerosis, a KD (Kcal%: 1/81/18, carbohydrate/fat/protein) or a control diet (Kcal%: 70/11/18, carbohydrate/fat/protein) for 12 weeks. Plasma was collected for biochemical analysis. Upon euthanasia, livers, white adipose tissue (WAT), and brown adipose tissue (BAT) were used for gene expression studies. Mice fed the KD and control diets exhibited similar body weights, despite the profoundly different fat contents in the two diets. Moreover, KD-fed mice developed nutritional ketosis and showed increased expression of thermogenic genes in BAT. Additionally, these mice presented an increase in Aqp9 transcripts in BAT, but not in WAT, which suggests the participation of Aqp9 in the influx of excess plasma glycerol to fuel thermogenesis, while the up-regulation of *Aqp7* in the liver suggests the involvement of this aquaporin in glycerol influx into hepatocytes. The relationship between nutritional ketosis, energy homeostasis, and the AQP network demands further investigation.

Keywords: aquaporins; very low-carbohydrate diet; ketogenic diet; obesity

1. Introduction

Despite significant advances in atherosclerotic cardiovascular disease (CVD) treatment, CVD remains the leading cause of mortality among adults [1]. The mechanistic underpinnings of dietary pattern-related CVD risk are still not well understood. An apolipoprotein E-deficient ($ApoE^{-/-}$) mouse model is a pre-clinical model that is widely used to study the pathophysiology of vascular plaque formation and atherosclerosis [2,3]. The global obesity epidemic is one of the main risk factors for atherosclerosis and CVD [4]. Recent scientific evidence highlights the ketogenic diet (KD)—a very low-carbohydrate, high-fat diet—as a promising strategy to treat obesity [5]. Interestingly, this diet was initially used to treat refractory epilepsy in children [6] and is now being tested as a dietary approach to treat other disorders, such as a common renal pathology (autosomal dominant polycystic kidney disease) [7–9]. The fundamental principle of the KD is a severe restriction of dietary carbohydrate consumption with a concurrent increase in dietary fat consumption to compensate for the energy deficit, resulting in a metabolic state of nutritional ketosis. The impact of a KD on cardiometabolic and vascular health is still a subject under intense debate [10,11]. In fact, in being used to achieve ketosis, KDs are typically inconsistent with nutritional recommendations for CVD prevention; cardio-protective foods are severely restricted (e.g., fruits, legumes) and foods associated with increased CVD risk are promoted (e.g., meats rich in saturated fat) [12,13].

The unique macronutrient profile of the KD results in the promotion of lipid oxidation to produce ketones as an energy source [12]. In fact, this metabolic adaptation is based on coordinated responses of the liver and an altered energetic metabolism at the cellular level. When there is an adequate supply of carbohydrates, cells primarily rely on glucose metabolism, whereas under carbohydrate-depletion conditions, cells use ketone bodies as their primary energy source [14,15]. Ketone bodies are produced in the liver via mitochondrial β-oxidation and are then released into the blood for uptake and utilization by peripheral tissues. Moreover, a critical metabolic change in response to a KD involves the mobilization of lipids stored in adipose tissue, with triacylglycerols (TAG) being hydrolyzed to yield glycerol and fatty acids to be taken up by the liver to feed hepatic ketogenesis [16].

Several proteins have been identified as key players in energy balance and vascular homeostasis. Aquaporins (AQPs) are transmembrane proteins that function as channels, allowing the permeation of water, glycerol, and other small non-charged molecules across biological membranes driven by osmotic or solute gradients [17]. Thirteen AQP isoforms have been identified in humans (AQP0–AQP12) and they are widely distributed in tissue-specific manners [18]. AQPs have been categorized into three subgroups, according to their transport selectivity and primary structure: classical aquaporins (AQP0, AQP1, AQP2, AQP4, AQP5, AQP6, and AQP8), which are primarily selective to water; aquaglyceroporins (AQP3, AQP7, AQP9, and AQP10), which also facilitate the permeation of other small solutes (glycerol, urea); and S-aquaporins (AQP11 and AQP12), comprising intracellular isoforms that have impacts on organelle homeostasis [19,20]. Recently, a few isoforms have also been reported to transport hydrogen peroxide and have been termed peroxiporins (AQP1, AQP3, AQP5, AQP8, AQP9, and AQP11) [21–23], opening new perspectives in understanding the potential roles of AQPs in physiological redox balance as well as oxidative stress [24,25].

The involvement of aquaglyceroporins in adipose tissue homeostasis has possible implications for metabolic disorders [26]. In fasting conditions, glycerol from TAG lipolysis in adipose tissue is released via AQP7 into the bloodstream and is taken up via AQP9 by the liver as a substrate for gluconeogenesis [27]. The adipose AQP7–hepatic AQP9 axis has been extensively characterized, and the synchronization of these AQPs ensures glycerol metabolism for gluconeogenesis [28]. In addition, the involvement of AQPs in obesity-induced inflammation [29], as well as their ability to trigger inflammatory processes involved in metabolic disorders [30,31], suggests their importance in the maintenance of vascular homeostasis.

In this study, we have investigated how nutritional ketosis induced by a KD diet in the context of an atherogenic model impacts the transcript levels of AQPs involved in energy homeostasis. We fed $ApoE^{-/-}$ mice a KD or control diet for 12 weeks and assessed metabolic disturbances. After confirming the presence of nutritional ketosis in the KD-fed mice versus the control group, we performed gene expression studies for key metabolic tissues.

2. Materials and Methods

2.1. Animals

Male $ApoE^{-/-}$ mice were purchased from Jackson Laboratory (Bar Harbor, ME, USA) and housed in a temperature- and humidity-controlled room. Only male mice were included to control for the known effect of sex on atherosclerosis in this strain [32]. Wild-type mice, known to be resistant to atherosclerosis, were not included because we wanted to specifically address the consequences of a ketogenic diet in the context of an atherosclerosis-prone model [2]. At the age of 7 weeks, mice were weighed and divided into two groups (7–8 animals/group). All procedures were performed in compliance with the Institutional Animal Care and Use Committee of Pennsylvania State University (PRAMS#201747911 to R.C.), which specifically approved this study.

2.2. Diets and Feeding

Mice were fed, for 12 weeks, one of the following diets based on AIN 93G [33] (Research Diets, New Brunswick, NJ, USA) with modifications (Kcal%; fat/carbohydrate/protein): a very-low-carbohydrate KD diet (81/1/18) with 0.15% cholesterol or a control diet (12/70/18). Details about diet composition are shown in Table 1. Diets were replaced once a week, at which time the animals and the remaining food were weighed to determine food consumption and body weight progression.

Table 1. Macronutrients in the experimental diets.

Macronutrient (g/Kg of Diet)	Control Diet	Ketogenic Diet
Casein	180	280
Corn Starch	430	0
Maltodextrin 10	155	0
Sucrose	100	0
Cocoa Butter	0	155
Corn Oil	25	40
Primex (Non-Trans-Fat)	25	365

2.3. Blood Sampling and Measuring of Biochemical/Metabolic Markers

Every four weeks, approximately 200 µL of blood was collected from the retroorbital cavity into heparinized tubes and immediately placed on ice. Plasma was isolated by centrifugation at 4 °C and immediately stored at −80 °C prior to biochemical analyses.

2.3.1. β-Hydroxybutyrate, Triacylglycerols, Total Cholesterol, and Glucose

Plasma collected at 8 weeks was tested for ß-hydroxybutyrate (BHB), TAG, and total cholesterol contents using colorimetric assays kits (Randox, Ann Arbor, MI, USA) following the manufacturer's instructions. Blood glucose was measured using a glucometer (Contour, Bayer, Tarrytown, NY, USA) following the manufacturer's instructions.

2.3.2. Glutathione Amino Acids Precursors

In plasma collected at 12 weeks, the circulating levels of glutamate and glycine were determined by gas chromatography–flame ionization detector (GC-FID) using the Phenomenex EZ:faastTM kit for physiological amino acid analysis (Phenomenex, Torrance, CA, USA) as previously described [34]. Moreover, plasma cysteine concentrations were quantified by high-performance liquid chromatography (HPLC) analysis with fluorometric detection, as previously described [35].

2.4. Tissue Collection

After 12 weeks, mice were euthanized by carbon dioxide inhalation. Aortas were obtained and subjected to Oil Red O staining, as previously described in detail [36]. Livers were removed and immediately snap-frozen in liquid nitrogen and stored at −80 °C. Interscapular brown adipose tissue (BAT) and epididymal white adipose tissue (WAT) were dissected following the protocol previously described by Bagchi and MacDougald [37], immediately snap-frozen in liquid nitrogen, and stored at −80 °C.

2.5. Quantification of Inflammatory Cytokines and Aortic Atheroma

The effects of the experimental diets at 4 and 12 weeks on the plasma concentrations of tumor necrosis factor-alpha (TNF-α) and interleukin 6 (IL-6) were evaluated using ELISA assays (Meso Scale Diagnostics, Rockville, MD, USA), following the manufacturer's instructions.

An advanced imaging technique (14 T magnetic resonance imaging (MRI)) was used to quantify the volume of vascular lesions in the mouse aortas, as previously described in detail [36,38,39].

2.6. RNA Extraction

Total RNAs were extracted from WAT, BAT, and liver tissue using the Qiagen RNeasy lipid tissue mini kit and Qiagen RNeasy mini kit, respectively (Qiagen, Germantown, MD, USA), followed by a DNAase treatment. All procedures were conducted following the manufacturer's protocols. RNA concentrations were determined using the NanoDrop 2000c (Thermo Fisher Scientific, Waltham, MA, USA). Only samples with 260/280 nm ratios between 1.8 and 2.2 were used for cDNA synthesis. Additionally, agarose bleach gels were used to confirm RNA integrity, as previously described [40].

2.7. Quantitative PCR Analysis

Reverse transcription of 1µg RNA was performed using M-MLV reverse transcriptase and oligo dT primers (Promega, Madison, WI, USA). Amplification by quantitative PCR was executed using a CFX96 Real-Time System C1000 (BioRad, Hercules, CA, USA) after cDNA was mixed with TaqMan Universal PCR Master Mix and the following specific TaqMan gene expression assays (Applied Biosystems, Foster City, CA, USA) were performed, as described in the manufacturer's protocol and previous publications of our group [41]. The following probes and primers were used in this study: *Aqp1* (#Mm00431834_m1), *Aqp3* (#Mm01208559_m1), *Aqp5* (#Mm00437578_m1), *Aqp7* (#Mm00431839_m10), *Aqp9* (#Mm00508094_m1), *Ucp1* (#Mm01244861_m1), and *Eef2* (#Mm00833287_g1). Relative gene expression was calculated using a variation of the Livak method with a houskeeping gene *Eef2* normalization step [42,43].

2.8. Statistical Analyses

Analyses were performed in GraphPad Prism 7 (GraphPad Software, La Jolla, CA, USA), with statistical significance set to $p < 0.05$. For comparison of two groups, an unpaired Student's t-test was used.

3. Results

3.1. KD-Fed Mice Weight Gain Was Similar to Controls

KD-fed mice consumed significantly less food than mice fed the control diet (Figure 1A). Nevertheless, due to the higher energy density of this diet (6.2 kcal/g KD versus 3.9 kcal/g control diet), KD-fed mice consumed more calories than the controls (Figure 1B). Notably, despite the higher number of calories consumed by the KD group than the controls, both groups of mice presented similar body weights (Figure 1C).

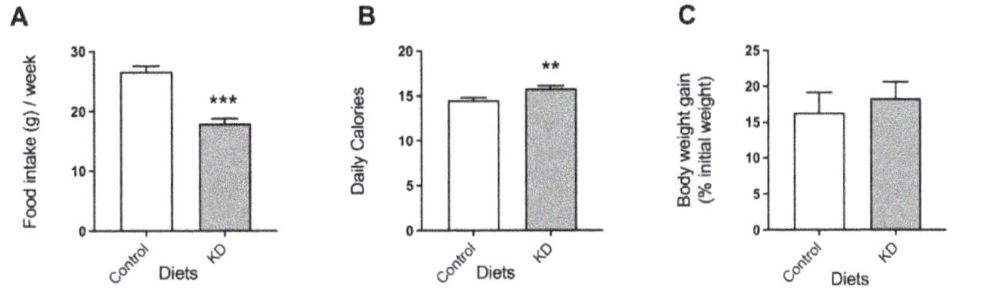

Figure 1. The effect of the experimental diets on (**A**) food consumption, (**B**) calories consumed, and (**C**) body weight. Data shown are the mean ± SEM, n = 7–8/group. **, $p < 0.01$; ***, $p < 0.001$, KD versus control.

3.2. KD Promoted Nutritional Ketosis and a Distinct Metabolic Profile

The plasma concentration of the major ketone body, ß-hydroxybutyrate (BHB), was significantly elevated in KD mice, confirming the presence of nutritional ketosis under this dietary condition (Figure 2A). Specifically, BHB concentrations (μM, mean ± SEM) were 2427 ± 347 for KD and 460 ± 85 for control mice. On the other hand, plasma TAG (Figure 2C) and total cholesterol levels (Figure 2D) were significantly increased by the KD (Figure 2D). Blood glucose levels showed opposite results, with KD-fed mice presenting significantly lower blood glucose concentrations than the controls (Figure 2B). We also assessed the effect of the KD on the sum of plasma glutamate, glycine, and cysteine, which determine glutathione availability [44]. No significant differences were detected between the two groups of animals (Figure 2E). Glutathione is a tripeptide that constitutes the most abundant intra-cellular antioxidant system and it is a major determinant of redox balance. Thus, this observation suggests that glutathione availability was preserved in KD mice.

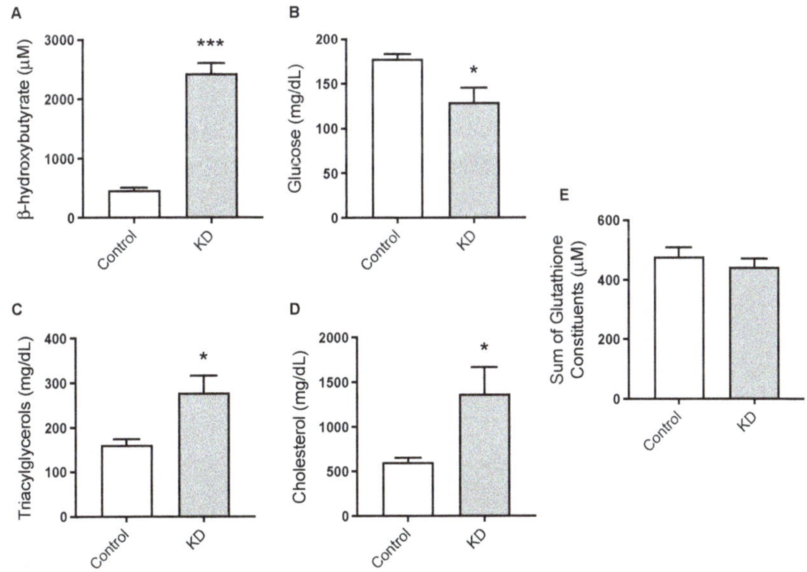

Figure 2. The effect of the experimental diets on the circulating levels of (**A**) ß-hydroxybutyrate, (**B**) glucose, (**C**) triacylglycerols, (**D**) total cholesterol, and (**E**) the sum of glutamate, glycine, and cysteine. Data shown are the mean ± SEM, n = 7–8/group. *, $p < 0.05$; ***, $p < 0.001$, KD versus control.

3.3. KD Augmented Plasma TNF-α and IL-6 and Vascular Lesions

We next examined the effect of nutritional ketosis on the circulating levels of the pro-inflammatory cytokines TNF-α and IL-6 and on the volume of atheroma. The results suggested a sustained pro-inflammatory effect of the KD when compared to the control diet. Specifically, feeding the experimental diets for as little as 4 weeks resulted in significantly higher plasma concentrations of TNF-α and IL-6 in the KD mice than in the controls, which were maintained at 12 weeks (Figure 3A,B). Moreover, levels of aortic atherosclerotic plaque in KD-fed mice, as assessed by MRI, were significantly elevated compared to control levels (0.23 ± 0.03 versus 0.03 ± 0.01 mm^3, mean ± SEM, n = 7–8) (Figure 3C).

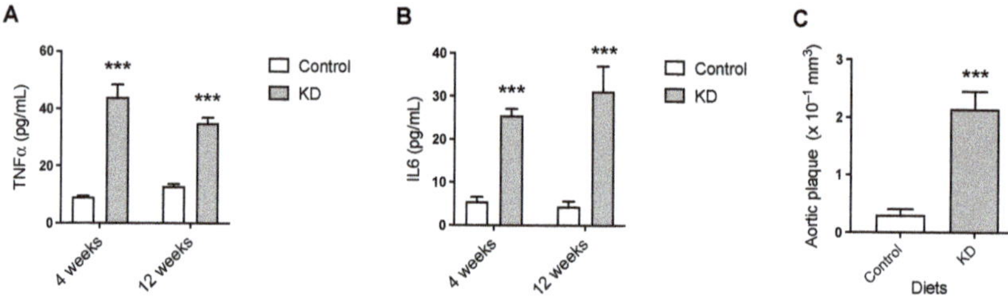

Figure 3. The effect of 12 weeks of KD on systemic inflammation and atherosclerotic plaque burden. Plasma concentrations of (**A**) interleukin 6 (IL-6) and (**B**) tumor necrosis factor α (TNF-α). (**C**) Ex vivo 14T-MRI volumetric assessment of total atheroma. Data shown are the mean ± SEM, n = 4–8/group. ***, $p < 0.001$, KD versus control.

3.4. KD Up-Regulated the Expression of Thermogenic Genes

The transcript level of Ucp1 in BAT of KD-fed mice was four times higher compared to control mice (Figure 4A), revealing a thermogenic effect of the KD. In contrast, the transcript level for Ucp1 in WAT was significantly decreased compared to the controls (around a 0.15-fold change; Figure 4B).

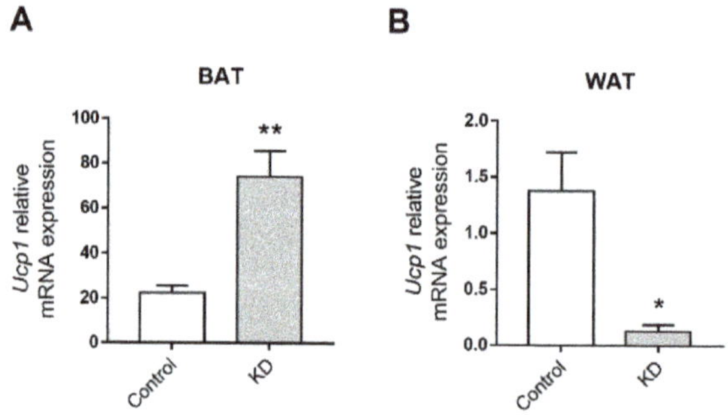

Figure 4. Effect of the ketogenic diet (KD) on the transcription levels of Uncoupling protein 1 (Ucp1) in (**A**) BAT and (**B**) WAT of ApoE$^{-/-}$ mice fed the experimental diets for 12 weeks. Data shown are the mean ± SEM (n = 6–8/group). *, $p < 0.05$; **, $p < 0.01$, KD versus control.

3.5. KD Altered AQPs Expression in Adipose Tissues and Liver

The gene expression of aquaporins implicated in endothelial homeostasis (Aqp1 and Aqp5) [45] and energetic metabolism (Aqp3, Aqp7, and Aqp9) [26] were evaluated in BAT, WAT, and the liver, which are tissues mainly involved in energy homeostasis. All the AQP isoforms were detected in all the investigated tissues, which is in accordance with previous reports [38]. However, each AQP showed a tissue-specific profile (Figure 5). For example, in BAT, high Aqp7 and Aqp1 transcripts levels were detected, while Aqp3, Aqp5, and Aqp9 were present in lower amounts (Figure 5A,B). As with BAT, WAT contained abundant levels of Aqp7 and Aqp1 transcripts, followed by lower expression of Aqp3, Aqp9, and Aqp5 (Figure 5C,D). In the WAT of mice fed the KD, Aqp7 mRNA levels were low, but the expression of the other isoforms was not altered by the diets. The BAT of these animals, however, presented *Aqp9* up-regulation. Further, this same isoform was the most representative aquaglyceroporin in the liver, as previously reported [46], but expression of Aqp1, Aqp3, Aqp7, and Aqp5 was also detected (Figure 5E,F). In KD-fed mice, an intense 20-fold hepatic induction of Aqp7 transcript was also observed.

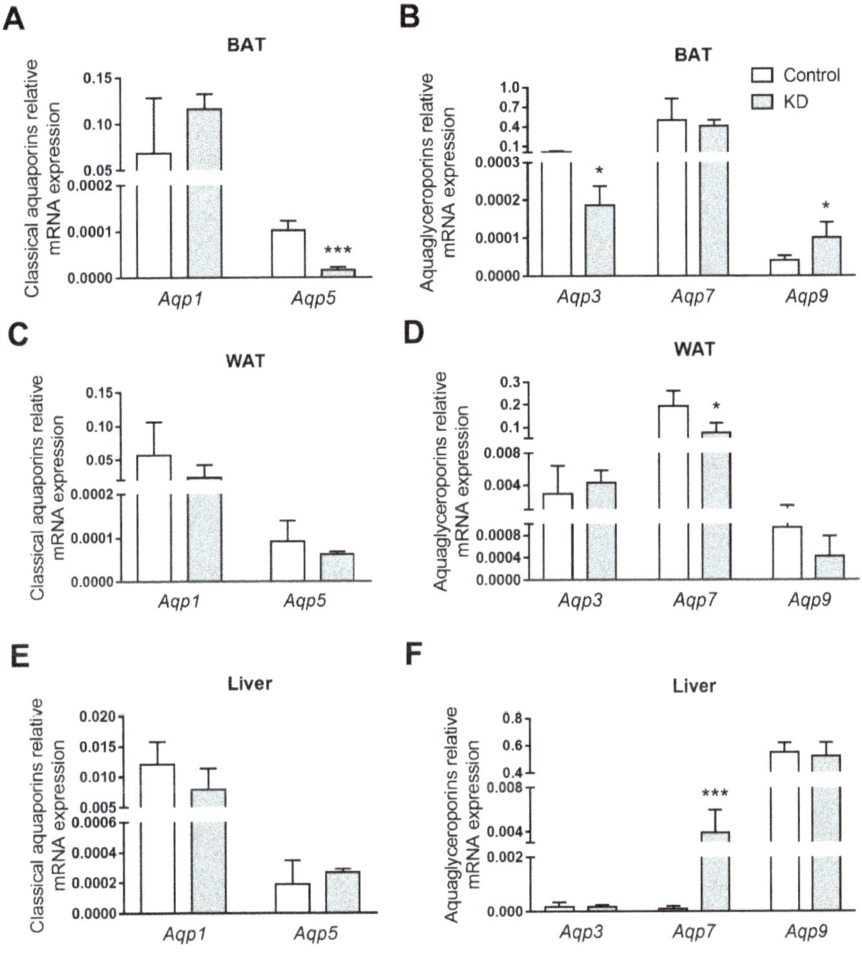

Figure 5. Aquaporin gene expression in (**A**,**B**) white adipose tissue (WAT), (**C**,**D**) brown adipose tissue (BAT), and (**E**,**F**) the livers of *ApoE*$^{-/-}$ mice fed control or a ketogenic diet (KD). Relative gene expression of (**A**,**C**,**E**) classical aquaporins and (**B**,**D**,**F**) aquaglyceroporins. Data shown are the mean ± SEM (n = 6–8/group). *, $p < 0.05$; ***, $p < 0.001$, KD versus control.

4. Discussion

In recent years, KDs have been considered a promising strategy to treat obesity—a major CVD risk factor. KDs contain 70–80% of kcal from fat but very little carbohydrate, which stimulates endogenous ketogenesis, thus yielding high levels of BHB. Previous studies with mice have reported an anti-obesogenic effect for the KD, despite its extremely high fat content [47–50]. Accordingly, in this study, notwithstanding a caloric intake that was markedly higher in KD-fed mice than in controls and 12 weeks of ad libitum feeding, no significant differences were observed in body weight between KD- and control-fed *ApoE*$^{-/-}$ mice. As we wanted to focus on atherogenesis, we did not include wild-type mice, which are not atherosclerosis-prone. However, we acknowledge that additional studies in wild-type mice could be of interest regarding other health outcomes.

The absence of dietary carbohydrates in KD stimulates the hepatic production of ketone bodies that are further transported to extrahepatic tissues for terminal oxidation as the primary energy source [51]. In this study, the plasma concentration of the major ketone body, BHB, was significantly elevated in KD-fed mice and within the range reported by others [49,51]. This finding confirms the presence of nutritional ketosis in KD-fed mice and indicates the reliability of this dietary mouse model in understanding the effect of nutritional ketosis on aquaporin expression. On the other hand, and as previously reported, KD promoted an accumulation of circulating TAG and cholesterol levels together with significantly lowered blood glucose levels compared to controls [52–54].

The ratio of the concentration of the universal methyl donor S-adenosylmethionine (AdoMet) to S-adenosylhomocysteine (AdoHcy) (AdoMet:AdoHcy) is an index of cellular transmethylation reactions [55]. In previous cell studies, dysregulation of the transmethylation capacity impacted *AQP1* expression and promoted an atherogenic phenotype [45]. Moreover, an increased dietary fat content in the presence of carbohydrates was associated with a diminished systemic methylation index in wild-type mice [56] but not in *ApoE*$^{-/-}$ mice [38]. The effect of the diets used in this study on AdoMet:AdoHcy ratio has recently been published [39]. A significantly decreased plasma AdoMet:AdoHcy ratio was detected in KD-fed *ApoE*$^{-/-}$ mice versus controls, revealing systemic hypomethylation under nutritional ketosis. This observation led us to focus in the present study on the glutathione system. In fact, a link between hypomethylation and oxidative stress due to the dysregulation of the antioxidant glutathione system has been previously reported [57]. Moreover, disturbances in redox balance contribute to susceptibility and pathology in human diseases, including atherosclerosis. Interestingly, cumulative evidence points to the role of AQPs as facilitators of ROS membrane permeation [58]. These intriguing observations led us to assess the effect of the KD on the sum of glutamate, glycine, and cysteine concentrations. Nevertheless, no significant differences were detected between the two groups of animals, suggesting that glutathione availability was preserved with the KD.

Inflammation plays an important role in the progression of atherosclerosis [59]. Furthermore, compelling evidence strongly suggests that AQPs are key regulators of inflammation [31]. The pro-inflammatory cytokines TNF-α and IL-6 play crucial roles in inflammation and exacerbate atherosclerosis in murine species [60–62]. The results suggested a sustained pro-inflammatory effect of the KD when compared to the control diet. As previously mentioned, the KD contains approximately 80% of Kcal from fat. Thus, although other cytokines could have been quantified in plasma to unequivocally address and compare the levels of systemic inflammation between the KD and control groups, our data are consistent with the well-established positive correlation of dietary fat and systemic inflammation [63]. KD-fed mice developed more extensive atherosclerosis than control mice, which may have been contributed to, in part, by elevated systemic inflammation in the KD mice. Previous studies in which high-fat diets with very low carbohydrate contents were fed to murine models of atherosclerosis also reported atherogenic effects of the diets [53,64]. In the present study, our goal was to test whether a KD, in the context of an atherogenic model, impacted AQPs expression. Thus, we used a diet that was very low in carbohydrates, very high in fat, and contained 0.15% of cholesterol, this being typical of atherogenic diets described in the literature. We acknowledge that cholesterol, as a well-established dietary atherogenic component in *ApoE*$^{-/-}$ mice, may have contributed to the observed KD-induced exacerbated systemic inflammation and vascular toxicity [65,66]. Thus, future studies conducted without dietary cholesterol are needed to unequivocally address the vascular effect of nutritional ketosis in this mouse model. In fact, increasing evidence suggests a positive impact of nutritional ketosis on vascular physiology and homeostasis in humans [67–69].

The finding of similar body weight in all mice, despite the higher caloric intake in the KD-fed group, was intriguing. BAT plays a vital role in energy homeostasis and heat production [70]. The thermogenic function of BAT is dependent on Ucp1, a protein expressed in the inner mitochondrial membrane of brown adipocytes [71]. Ucp1 disrupts the electrochemical gradient across the inner mitochondrial membrane, causing energy derived from the metabolism of food to be released as heat [71]. The exceptional function of BAT to increase energy expenditure is shown by its anti-obesogenic role. In fact, mice deficient in Ucp1 are susceptible to weight gain [72], whereas an excess of Ucp1 protects against diet-induced obesity [73]. In the present study, *Ucp1* transcription in BAT of KD-fed mice was observed to be up-regulated as compared to the controls. Thus, it is tempting to speculate that this KD-induced up-regulation of *Ucp1* may have contributed to the similar body weight of KD-fed mice and controls despite the first group having a higher caloric intake. Additional studies assessing basal metabolism are warranted to explore this exciting possibility.

Lastly, it was observed that the transcriptional levels of several AQPs were different among the tissues involved in energy homeostasis, namely, the liver, BAT, and WAT. Further, and interestingly, this tissue-specific profile was altered when mice were subjected to KD (Figure 5). In BAT, KD induced a two-fold increase in Aqp9 expression, suggesting that this isoform provides a route for the influx of excess plasma glycerol to be used as fuel in thermogenesis. Thermogenesis is a very intensive process in terms of energy; thus, BAT has a high metabolic demand. Recent studies have identified the need for a variety of metabolic substrates to initiate and maintain thermogenesis, such as intracellular triglycerides and glucose, in addition to the well-established free fatty acids [74]. The high expression of BAT glycerol kinase, the enzyme that converts glycerol to glycerol-3-phosphate for triacylglycerol synthesis in both rodents and humans [75], suggests an important role for glycerol in maintaining intracellular levels of TAG during thermogenesis and highlights the role of AQPs in the process. In addition, KD impaired Aqp3 and Aqp5 expression, both peroxiporins acting as hydrogen peroxide facilitators, suggesting that they might be involved in an unbalanced redox potential induced by KD. Interestingly, the same AQP response pattern was observed in our previous study with mice fed a high-fat diet with a higher carbohydrate content [38], suggesting that high-calorie diets affect BAT similarly independently of their macronutrient compositions. In the WAT of mice fed the KD, Aqp7 mRNA levels were low, suggesting that, due to increased circulating TAG, adipocyte lipid droplets were not used as fuel in KD-fed mice. AQP7 is the main gateway for glycerol efflux from white adipocytes following the breakdown of TAG. Finally, in KD-fed mice, an intense up-regulation of hepatic *Aqp7* was observed, which may indicate an additional glycerol uptake route as a compensatory strategy favoring gluconeogenesis.

5. Conclusions

The current study contributes to a better characterization of an atherogenic-susceptible *ApoE*$^{-/-}$ mouse model by reporting that the pattern of AQPs expression in these mice is disturbed in a tissue-specific manner by nutritional ketosis, which, in turn, was found to be associated with the up-regulation of thermogenic genes in BAT. Our data represent the first experimental approach ever reported on the modulation of AQPs expression in *ApoE*$^{-/-}$ mice under this metabolic condition and warrant further investigation to establish and validate the relationship between nutritional ketosis, the AQP network, and energy homeostasis.

Author Contributions: Methodology, I.V.d.S., S.G. and C.F.; investigation, I.V.d.S., S.G. and C.F.; formal analysis, I.V.d.S., N.K.H. and T.N.; writing—original draft preparation, I.V.d.S., G.S. and R.C.; writing—reviewing and editing, I.V.d.S., S.G., N.K.H., T.N., A.C.R., G.S. and R.C.; conceptualization, G.S. and R.C.; resources, T.N., A.C.R., G.S. and R.C.; supervision, G.S. and R.C. All authors have read and agreed to the published version of the manuscript.

Funding: This work was supported by the following funding sources at Pennsylvania State University: the Department of Nutritional Sciences, the High-Field Magnetic Resonance Imaging Facility, and the Huck Institutes of the Life Sciences; as well as by the Fundação para a Ciência e Tecnologia (FCT), Portugal (grant PTDC/BTM-SAL/28977/2017) and strategic projects UIDB/04378/2020 and UIDP/04138/2020 (iMed.ULisboa).

Institutional Review Board Statement: The study was conducted according to the guidelines of the Declaration of Helsinki and approved by the Institutional Animal Care and Use Committee of Pennsylvania State University, which specifically approved this study (PRAMS#201747911 to R.C.) in November 2018.

Informed Consent Statement: Not applicable.

Data Availability Statement: Not applicable.

Acknowledgments: The authors wish to thank Courtney A. Whalen and Floyd J. Mattie (Penn State University, State College, PA, USA) for expert technical assistance and Isabel Tavares de Almeida (University of Lisbon, Lisbon, Portugal) for her valuable support.

Conflicts of Interest: The authors declared no potential conflict of interest with respect to the research, authorship, and/or publication of this article.

References

1. Powell-Wiley, T.M.; Poirier, P.; Burke, L.E.; Despres, J.P.; Gordon-Larsen, P.; Lavie, C.J.; Lear, S.A.; Ndumele, C.E.; Neeland, I.J.; Sanders, P.; et al. Obesity and Cardiovascular Disease: A Scientific Statement From the American Heart Association. *Circulation* **2021**, *143*, e984–e1010. [CrossRef] [PubMed]
2. Getz, G.S.; Reardon, C.A. ApoE knockout and knockin mice: The history of their contribution to the understanding of atherogenesis. *J. Lipid Res.* **2016**, *57*, 758–766. [CrossRef] [PubMed]
3. Getz, G.S.; Reardon, C.A. Do the Apoe$^{-/-}$ and Ldlr$^{-/-}$ Mice Yield the Same Insight on Atherogenesis? *Arter. Thromb. Vasc. Biol.* **2016**, *36*, 1734–1741. [CrossRef] [PubMed]
4. Rochlani, Y.; Pothineni, N.V.; Kovelamudi, S.; Mehta, J.L. Metabolic syndrome: Pathophysiology, management, and modulation by natural compounds. *Adv. Cardiovasc. Dis.* **2017**, *11*, 215–225. [CrossRef]
5. Harvey, K.L.; Holcomb, L.E.; Kolwicz, S.C., Jr. Ketogenic Diets and Exercise Performance. *Nutrients* **2019**, *11*, 2296. [CrossRef]
6. Giordano, C.; Marchio, M.; Timofeeva, E.; Biagini, G. Neuroactive peptides as putative mediators of antiepileptic ketogenic diets. *Front. Neurol.* **2014**, *5*, 63. [CrossRef]
7. Testa, F.; Marchiò, M.; Belli, M.; Giovanella, S.; Ligabue, G.; Cappelli, G.; Biagini, G.; Magistroni, R. A pilot study to evaluate tolerability and safety of a modified Atkins diet in ADPKD patients. *PharmaNutrition* **2019**, *9*, 100154. [CrossRef]
8. Testa, F.; Marchiò, M.; D'Amico, R.; Giovanella, S.; Ligabue, G.; Fontana, F.; Alfano, G.; Cappelli, G.; Biagini, G.; Magistroni, R. Grease II. A phase II randomized, 12-month, parallel-group, superiority study to evaluate the efficacy of a Modified Atkins Diet in Autosomal Dominant Polycystic Kidney Disease patients. *PharmaNutrition* **2020**, *13*, 100206. [CrossRef]
9. Magistroni, R.; Biagini, G. Response letter to the Editorial: "Ketogenic Diet in ADPKD Patients". *PharmaNutrition* **2021**, *16*, 100268. [CrossRef]
10. Ludwig, D.S. The Ketogenic Diet: Evidence for Optimism but High-Quality Research Needed. *J. Nutr.* **2020**, *150*, 1354–1359. [CrossRef]
11. Fechner, E.; Smeets, E.; Schrauwen, P.; Mensink, R.P. The Effects of Different Degrees of Carbohydrate Restriction and Carbohydrate Replacement on Cardiometabolic Risk Markers in Humans—A Systematic Review and Meta-Analysis. *Nutrients* **2020**, *12*, 991. [CrossRef] [PubMed]
12. Ludwig, D.S.; Willett, W.C.; Volek, J.S.; Neuhouser, M.L. Dietary fat: From foe to friend? *Science* **2018**, *362*, 764–770. [CrossRef] [PubMed]
13. Kirkpatrick, C.F.; Bolick, J.P.; Kris-Etherton, P.M.; Sikand, G.; Aspry, K.E.; Soffer, D.E.; Willard, K.E.; Maki, K.C. Review of current evidence and clinical recommendations on the effects of low-carbohydrate and very-low-carbohydrate (including ketogenic) diets for the management of body weight and other cardiometabolic risk factors: A scientific statement from the National Lipid Association Nutrition and Lifestyle Task Force. *J. Clin. Lipidol.* **2019**, *13*, 689–711. [CrossRef] [PubMed]
14. Golonka, R.M.; Xiao, X.; Abokor, A.A.; Joe, B.; Vijay-Kumar, M. Altered nutrient status reprograms host inflammation and metabolic health via gut microbiota. *J. Nutr. Biochem.* **2020**, *80*, 108360. [CrossRef] [PubMed]
15. Dabek, A.; Wojtala, M.; Pirola, L.; Balcerczyk, A. Modulation of Cellular Biochemistry, Epigenetics and Metabolomics by Ketone Bodies. Implications of the Ketogenic Diet in the Physiology of the Organism and Pathological States. *Nutrients* **2020**, *12*, 788. [CrossRef]
16. Grabacka, M.; Pierzchalska, M.; Dean, M.; Reiss, K. Regulation of Ketone Body Metabolism and the Role of PPARα. *Int. J. Mol. Sci.* **2016**, *17*, 2093. [CrossRef]
17. Carbrey, J.M.; Agre, P. Discovery of the aquaporins and development of the field. *Aquaporins* **2009**, *190*, 3–28. [CrossRef]

18. King, L.S.; Kozono, D.; Agre, P. From structure to disease: The evolving tale of aquaporin biology. *Nat. Rev. Mol. Cell Biol.* **2004**, *5*, 687–698. [CrossRef]
19. Ishibashi, K.; Tanaka, Y.; Morishita, Y. The role of mammalian superaquaporins inside the cell. *Biochim. Biophys. Acta* **2014**, *1840*, 1507–1512. [CrossRef]
20. Madeira, A.; Fernandez-Veledo, S.; Camps, M.; Zorzano, A.; Moura, T.F.; Ceperuelo-Mallafre, V.; Vendrell, J.; Soveral, G. Human aquaporin-11 is a water and glycerol channel and localizes in the vicinity of lipid droplets in human adipocytes. *Obesity* **2014**, *22*, 2010–2017. [CrossRef]
21. Bestetti, S.; Galli, M.; Sorrentino, I.; Pinton, P.; Rimessi, A.; Sitia, R.; Medrano-Fernandez, I. Human aquaporin-11 guarantees efficient transport of H_2O_2 across the endoplasmic reticulum membrane. *Redox Biol.* **2020**, *28*, 101326. [CrossRef] [PubMed]
22. Bienert, G.P.; Moller, A.L.; Kristiansen, K.A.; Schulz, A.; Moller, I.M.; Schjoerring, J.K.; Jahn, T.P. Specific aquaporins facilitate the diffusion of hydrogen peroxide across membranes. *J. Biol. Chem.* **2007**, *282*, 1183–1192. [CrossRef]
23. Rodrigues, C.; Pimpao, C.; Mosca, A.F.; Coxixo, A.S.; Lopes, D.; da Silva, I.V.; Pedersen, P.A.; Antunes, F.; Soveral, G. Human Aquaporin-5 Facilitates Hydrogen Peroxide Permeation Affecting Adaption to Oxidative Stress and Cancer Cell Migration. *Cancers* **2019**, *11*, 932. [CrossRef] [PubMed]
24. Prata, C.; Hrelia, S.; Fiorentini, D. Peroxiporins in Cancer. *Int. J. Mol. Sci.* **2019**, *20*, 1371. [CrossRef] [PubMed]
25. Čipak Gašparović, A.; Milković, L.; Rodrigues, C.; Mlinarić, M.; Soveral, G. Peroxiporins Are Induced upon Oxidative Stress Insult and Are Associated with Oxidative Stress Resistance in Colon Cancer Cell Lines. *Antioxidants* **2021**, *10*, 1856. [CrossRef]
26. da Silva, I.V.; Rodrigues, J.S.; Rebelo, I.; Miranda, J.P.G.; Soveral, G. Revisiting the metabolic syndrome: The emerging role of aquaglyceroporins. *Cell. Mol. Life Sci.* **2018**, *75*, 1973–1988. [CrossRef]
27. Rodriguez, A.; Catalan, V.; Gomez-Ambrosi, J.; Fruhbeck, G. Aquaglyceroporins serve as metabolic gateways in adiposity and insulin resistance control. *Cell Cycle* **2011**, *10*, 1548–1556. [CrossRef]
28. Maeda, N.; Funahashi, T.; Shimomura, I. Metabolic impact of adipose and hepatic glycerol channels aquaporin 7 and aquaporin 9. *Nat. Clin. Pract. Endocrinol. Metab.* **2008**, *4*, 627–634. [CrossRef]
29. da Silva, I.V.; Cardoso, C.; Mendez-Gimenez, L.; Camoes, S.P.; Fruhbeck, G.; Rodriguez, A.; Miranda, J.P.; Soveral, G. Aquaporin-7 and aquaporin-12 modulate the inflammatory phenotype of endocrine pancreatic beta-cells. *Arch. Biochem. Biophys.* **2020**, *691*, 108481. [CrossRef]
30. da Silva, I.V.; Cardoso, C.; Martinez-Banaclocha, H.; Casini, A.; Pelegrin, P.; Soveral, G. Aquaporin-3 is involved in NLRP3-inflammasome activation contributing to the setting of inflammatory response. *Cell. Mol. Life Sci.* **2021**, *78*, 3073–3085. [CrossRef]
31. da Silva, I.V.; Soveral, G. Aquaporins in Immune Cells and Inflammation: New Targets for Drug Development. *Int. J. Mol. Sci.* **2021**, *22*, 1845. [CrossRef] [PubMed]
32. Caligiuri, G.; Nicoletti, A.; Zhou, X.; Tornberg, I.; Hansson, G.K. Effects of sex and age on atherosclerosis and autoimmunity in apoE-deficient mice. *Atherosclerosis* **1999**, *145*, 301–308. [CrossRef]
33. Reeves, P.G.; Nielsen, F.H.; Fahey, G.C., Jr. AIN-93 purified diets for laboratory rodents: Final report of the American Institute of Nutrition ad hoc writing committee on the reformulation of the AIN-76A rodent diet. *J. Nutr.* **1993**, *123*, 1939–1951. [CrossRef]
34. Badawy, A.A.; Morgan, C.J.; Turner, J.A. Application of the Phenomenex EZ: Faasttrade mark amino acid analysis kit for rapid gas-chromatographic determination of concentrations of plasma tryptophan and its brain uptake competitors. *Amino Acids* **2008**, *34*, 587–596. [CrossRef] [PubMed]
35. Barroso, M.; Rocha, M.S.; Esse, R.; Goncalves, I., Jr.; Gomes, A.Q.; Teerlink, T.; Jakobs, C.; Blom, H.J.; Loscalzo, J.; Rivera, I.; et al. Cellular hypomethylation is associated with impaired nitric oxide production by cultured human endothelial cells. *Amino Acids* **2012**, *42*, 1903–1911. [CrossRef] [PubMed]
36. Whalen, C.A.; Mattie, F.J.; Florindo, C.; van Zelst, B.; Huang, N.K.; Tavares de Almeida, I.; Heil, S.G.; Neuberger, T.; Ross, A.C.; Castro, R. No Effect of Diet-Induced Mild Hyperhomocysteinemia on Vascular Methylating Capacity, Atherosclerosis Progression, and Specific Histone Methylation. *Nutrients* **2020**, *12*, 2182. [CrossRef]
37. Bagchi, D.P.; MacDougald, O.A. Identification and Dissection of Diverse Mouse Adipose Depots. *J. Vis. Exp.* **2019**, *149*, e59499. [CrossRef]
38. da Silva, I.V.; Whalen, C.A.; Mattie, F.J.; Florindo, C.; Huang, N.K.; Heil, S.G.; Neuberger, T.; Ross, A.C.; Soveral, G.; Castro, R. An Atherogenic Diet Disturbs Aquaporin 5 Expression in Liver and Adipocyte Tissues of Apolipoprotein E-Deficient Mice: New Insights into an Old Model of Experimental Atherosclerosis. *Biomedicines* **2021**, *9*, 150. [CrossRef]
39. Castro, R.; Whalen, C.A.; Gullette, S.; Mattie, F.J.; Florindo, C.; Heil, S.G.; Huang, N.K.; Neuberger, T.; Ross, A.C. A Hypomethylating Ketogenic Diet in Apolipoprotein E-Deficient Mice: A Pilot Study on Vascular Effects and Specific Epigenetic Changes. *Nutrients* **2021**, *13*, 3576. [CrossRef]
40. Aranda, P.S.; LaJoie, D.M.; Jorcyk, C.L. Bleach gel: A simple agarose gel for analyzing RNA quality. *Electrophoresis* **2012**, *33*, 366–369. [CrossRef]
41. da Silva, I.V.; Diaz-Saez, F.; Zorzano, A.; Guma, A.; Camps, M.; Soveral, G. Aquaglyceroporins Are Differentially Expressed in Beige and White Adipocytes. *Int. J. Mol. Sci.* **2020**, *21*, 610. [CrossRef] [PubMed]
42. Fleige, S.; Pfaffl, M.W. RNA integrity and the effect on the real-time qRT-PCR performance. *Mol. Asp. Med.* **2006**, *27*, 126–139. [CrossRef] [PubMed]
43. Livak, K.J.; Schmittgen, T.D. Analysis of relative gene expression data using real-time quantitative PCR and the $2(-\Delta\Delta C(T))$ Method. *Methods* **2001**, *25*, 402–408. [CrossRef] [PubMed]

44. Ballatori, N.; Krance, S.M.; Notenboom, S.; Shi, S.; Tieu, K.; Hammond, C.L. Glutathione dysregulation and the etiology and progression of human diseases. *Biol. Chem.* **2009**, *390*, 191–214. [CrossRef] [PubMed]
45. da Silva, I.V.; Barroso, M.; Moura, T.; Castro, R.; Soveral, G. Endothelial Aquaporins and Hypomethylation: Potential Implications for Atherosclerosis and Cardiovascular Disease. *Int. J. Mol. Sci.* **2018**, *19*, 130. [CrossRef]
46. Calamita, G.; Gena, P.; Ferri, D.; Rosito, A.; Rojek, A.; Nielsen, S.; Marinelli, R.A.; Fruhbeck, G.; Svelto, M. Biophysical assessment of aquaporin-9 as principal facilitative pathway in mouse liver import of glucogenetic glycerol. *Biol. Cell.* **2012**, *104*, 342–351. [CrossRef]
47. Kostogrys, R.B.; Johann, C.; Czyzynska, I.; Franczyk-Zarow, M.; Drahun, A.; Maslak, E.; Jasztal, A.; Gajda, M.; Mateuszuk, L.; Wrobel, T.P.; et al. Characterisation of Atherogenic Effects of Low Carbohydrate, High Protein Diet (LCHP) in ApoE/LDLR$^{-/-}$ Mice. *J. Nutr. Health Aging* **2015**, *19*, 710–718. [CrossRef]
48. Kennedy, A.R.; Pissios, P.; Otu, H.; Roberson, R.; Xue, B.; Asakura, K.; Furukawa, N.; Marino, F.E.; Liu, F.F.; Kahn, B.B.; et al. A high-fat, ketogenic diet induces a unique metabolic state in mice. *Am. J. Physiol. Endocrinol. Metab.* **2007**, *292*, E1724–E1739. [CrossRef]
49. Ma, D.; Wang, A.C.; Parikh, I.; Green, S.J.; Hoffman, J.D.; Chlipala, G.; Murphy, M.P.; Sokola, B.S.; Bauer, B.; Hartz, A.M.S.; et al. Ketogenic diet enhances neurovascular function with altered gut microbiome in young healthy mice. *Sci. Rep.* **2018**, *8*, 6670. [CrossRef]
50. Merra, G.; Miranda, R.; Barrucco, S.; Gualtieri, P.; Mazza, M.; Moriconi, E.; Marchetti, M.; Chang, T.F.; De Lorenzo, A.; Di Renzo, L. Very-low-calorie ketogenic diet with aminoacid supplement versus very low restricted-calorie diet for preserving muscle mass during weight loss: A pilot double-blind study. *Eur. Rev. Med. Pharm. Sci.* **2016**, *20*, 2613–2621.
51. Roberts, M.N.; Wallace, M.A.; Tomilov, A.A.; Zhou, Z.; Marcotte, G.R.; Tran, D.; Perez, G.; Gutierrez-Casado, E.; Koike, S.; Knotts, T.A.; et al. A Ketogenic Diet Extends Longevity and Healthspan in Adult Mice. *Cell Metab.* **2017**, *26*, 539–546. [CrossRef] [PubMed]
52. Badman, M.K.; Kennedy, A.R.; Adams, A.C.; Pissios, P.; Maratos-Flier, E. A very low carbohydrate ketogenic diet improves glucose tolerance in *ob/ob* mice independently of weight loss. *Am. J. Physiol. Endocrinol. Metab.* **2009**, *297*, E1197–E1204. [CrossRef] [PubMed]
53. Foo, S.Y.; Heller, E.R.; Wykrzykowska, J.; Sullivan, C.J.; Manning-Tobin, J.J.; Moore, K.J.; Gerszten, R.E.; Rosenzweig, A. Vascular effects of a low-carbohydrate high-protein diet. *Proc. Natl. Acad. Sci. USA* **2009**, *106*, 15418–15423. [CrossRef]
54. Shimazu, T.; Hirschey, M.D.; Newman, J.; He, W.; Shirakawa, K.; Le Moan, N.; Grueter, C.A.; Lim, H.; Saunders, L.R.; Stevens, R.D.; et al. Suppression of oxidative stress by beta-hydroxybutyrate, an endogenous histone deacetylase inhibitor. *Science* **2013**, *339*, 211–214. [CrossRef]
55. Perla-Kajan, J.; Jakubowski, H. Dysregulation of Epigenetic Mechanisms of Gene Expression in the Pathologies of Hyperhomocysteinemia. *Int. J. Mol. Sci.* **2019**, *20*, 3140. [CrossRef] [PubMed]
56. Yun, K.U.; Ryu, C.S.; Oh, J.M.; Kim, C.H.; Lee, K.S.; Lee, C.H.; Lee, H.S.; Kim, B.H.; Kim, S.K. Plasma homocysteine level and hepatic sulfur amino acid metabolism in mice fed a high-fat diet. *Eur. J. Nutr.* **2013**, *52*, 127–134. [CrossRef]
57. Barroso, M.; Florindo, C.; Kalwa, H.; Silva, Z.; Turanov, A.A.; Carlson, B.A.; de Almeida, I.T.; Blom, H.J.; Gladyshev, V.N.; Hatfield, D.L.; et al. Inhibition of cellular methyltransferases promotes endothelial cell activation by suppressing glutathione peroxidase 1 protein expression. *J. Biol. Chem.* **2014**, *289*, 15350–15362. [CrossRef]
58. Tamma, G.; Valenti, G.; Grossini, E.; Donnini, S.; Marino, A.; Marinelli, R.A.; Calamita, G. Aquaporin Membrane Channels in Oxidative Stress, Cell Signaling, and Aging: Recent Advances and Research Trends. *Oxid. Med. Cell. Longev.* **2018**, *2018*, 1501847. [CrossRef]
59. Esse, R.; Barroso, M.; Tavares de Almeida, I.; Castro, R. The Contribution of Homocysteine Metabolism Disruption to Endothelial Dysfunction: State-of-the-Art. *Int. J. Mol. Sci.* **2019**, *20*, 867. [CrossRef]
60. Kleemann, R.; Zadelaar, S.; Kooistra, T. Cytokines and atherosclerosis: A comprehensive review of studies in mice. *Cardiovasc. Res.* **2008**, *79*, 360–376. [CrossRef]
61. Sukovich, D.A.; Kauser, K.; Shirley, F.D.; DelVecchio, V.; Halks-Miller, M.; Rubanyi, G.M. Expression of interleukin-6 in atherosclerotic lesions of male ApoE-knockout mice: Inhibition by 17beta-estradiol. *Arter. Thromb. Vasc. Biol.* **1998**, *18*, 1498–1505. [CrossRef]
62. Branen, L.; Hovgaard, L.; Nitulescu, M.; Bengtsson, E.; Nilsson, J.; Jovinge, S. Inhibition of tumor necrosis factor-α reduces atherosclerosis in apolipoprotein E knockout mice. *Arter. Thromb. Vasc. Biol.* **2004**, *24*, 2137–2142. [CrossRef] [PubMed]
63. Duan, Y.; Zeng, L.; Zheng, C.; Song, B.; Li, F.; Kong, X.; Xu, K. Inflammatory Links Between High Fat Diets and Diseases. *Front. Immunol.* **2018**, *9*, 2649. [CrossRef] [PubMed]
64. Kostogrys, R.B.; Franczyk-Zarow, M.; Maslak, E.; Gajda, M.; Mateuszuk, L.; Jackson, C.L.; Chlopicki, S. Low carbohydrate, high protein diet promotes atherosclerosis in apolipoprotein E/low-density lipoprotein receptor double knockout mice (apoE/LDLR($^{-/-}$)). *Atherosclerosis* **2012**, *223*, 327–331. [CrossRef] [PubMed]
65. Maeda, N. Development of apolipoprotein E-deficient mice. *Arter. Thromb. Vasc. Biol.* **2011**, *31*, 1957–1962. [CrossRef]
66. Plump, A.S.; Smith, J.D.; Hayek, T.; Aalto-Setala, K.; Walsh, A.; Verstuyft, J.G.; Rubin, E.M.; Breslow, J.L. Severe hypercholesterolemia and atherosclerosis in apolipoprotein E-deficient mice created by homologous recombination in ES cells. *Cell* **1992**, *71*, 343–353. [CrossRef]

67. Nasser, S.; Vialichka, V.; Biesiekierska, M.; Balcerczyk, A.; Pirola, L. Effects of ketogenic diet and ketone bodies on the cardiovascular system: Concentration matters. *World J. Diabetes* **2020**, *11*, 584–595. [CrossRef]
68. Hu, T.; Bazzano, L.A. The low-carbohydrate diet and cardiovascular risk factors: Evidence from epidemiologic studies. *Nutr. Metab. Cardiovasc. Dis.* **2014**, *24*, 337–343. [CrossRef]
69. Ebbeling, C.B.; Knapp, A.; Johnson, A.; Wong, J.M.W.; Greco, K.F.; Ma, C.; Mora, S.; Ludwig, D.S. Effects of a low-carbohydrate diet on insulin-resistant dyslipoproteinemia—A randomized controlled feeding trial. *Am. J. Clin. Nutr.* **2022**, *115*, 154–162. [CrossRef]
70. Schneider, K.; Valdez, J.; Nguyen, J.; Vawter, M.; Galke, B.; Kurtz, T.W.; Chan, J.Y. Increased Energy Expenditure, Ucp1 Expression, and Resistance to Diet-induced Obesity in Mice Lacking Nuclear Factor-Erythroid-2-related Transcription Factor-2 (Nrf2). *J. Biol. Chem.* **2016**, *291*, 7754–7766. [CrossRef]
71. Mills, E.L.; Harmon, C.; Jedrychowski, M.P.; Xiao, H.; Garrity, R.; Tran, N.V.; Bradshaw, G.A.; Fu, A.; Szpyt, J.; Reddy, A.; et al. UCP1 governs liver extracellular succinate and inflammatory pathogenesis. *Nat. Metab.* **2021**, *3*, 604–617. [CrossRef] [PubMed]
72. Kopecky, J.; Clarke, G.; Enerback, S.; Spiegelman, B.; Kozak, L.P. Expression of the mitochondrial uncoupling protein gene from the aP2 gene promoter prevents genetic obesity. *J. Clin. Investig.* **1995**, *96*, 2914–2923. [CrossRef] [PubMed]
73. Li, B.; Nolte, L.A.; Ju, J.S.; Han, D.H.; Coleman, T.; Holloszy, J.O.; Semenkovich, C.F. Skeletal muscle respiratory uncoupling prevents diet-induced obesity and insulin resistance in mice. *Nat. Med.* **2000**, *6*, 1115–1120. [CrossRef]
74. McNeill, B.T.; Morton, N.M.; Stimson, R.H. Substrate Utilization by Brown Adipose Tissue: What's Hot and What's Not? *Front. Endocrinol.* **2020**, *11*, 571659. [CrossRef]
75. Skowronski, M.T.; Lebeck, J.; Rojek, A.; Praetorius, J.; Fuchtbauer, E.M.; Frokiaer, J.; Nielsen, S. AQP7 is localized in capillaries of adipose tissue, cardiac and striated muscle: Implications in glycerol metabolism. *Am. J. Physiol. Ren. Physiol.* **2007**, *292*, F956–F965. [CrossRef] [PubMed]

biomedicines MDPI

Article

Gasdermin D Deficiency Limits the Transition of Atherosclerotic Plaques to an Inflammatory Phenotype in *ApoE* Knock-Out Mice

Pauline Puylaert, Melissa Van Praet, Frederik Vaes, Cédric H. G. Neutel, Lynn Roth, Pieter-Jan Guns, Guido R. Y. De Meyer and Wim Martinet *

Laboratory of Physiopharmacology, University of Antwerp, 2610 Antwerp, Belgium; pauline.puylaert@uantwerpen.be (P.P.); melissa.vanpraet@uantwerpen.be (M.V.P.); frederik.vaes@student.uantwerpen.be (F.V.); cedric.neutel@uantwerpen.be (C.H.G.N.); lynn.roth@uantwerpen.be (L.R.); pieter-jan.guns@uantwerpen.be (P.-J.G.); guido.demeyer@uantwerpen.be (G.R.Y.D.M.)
* Correspondence: wim.martinet@uantwerpen.be

Abstract: Gasdermin D (GSDMD) is the key executor of pyroptotic cell death. Recent studies suggest that GSDMD-mediated pyroptosis is involved in atherosclerotic plaque destabilization. We report that cleaved GSDMD is expressed in macrophage- and smooth muscle cell-rich areas of human plaques. To determine the effects of GSDMD deficiency on atherogenesis, *ApoE*$^{-/-}$ *Gsdmd*$^{-/-}$ (n = 16) and *ApoE*$^{-/-}$ *Gsdmd*$^{+/+}$ (n = 18) mice were fed a western-type diet for 16 weeks. Plaque initiation and formation of stable proximal aortic plaques were not altered. However, plaques in the brachiocephalic artery (representing more advanced lesions compared to aortic plaques) of *ApoE*$^{-/-}$ *Gsdmd*$^{-/-}$ mice were significantly smaller (115 ± 18 vs. $186 \pm 16 \times 10^3$ μm^2, p = 0.006) and showed features of increased stability, such as decreased necrotic core area (19 ± 4 vs. $37 \pm 7 \times 10^3$ μm^2, p = 0.03) and increased αSMA/MAC3 ratio (1.6 ± 0.3 vs. 0.7 ± 0.1, p = 0.01), which was also observed in proximal aortic plaques. Interestingly, a significant increase in TUNEL positive cells was observed in brachiocephalic artery plaques from *ApoE*$^{-/-}$ *Gsdmd*$^{-/-}$ mice (141 ± 25 vs. 62 ± 8 cells/mm^2, p = 0.005), indicating a switch to apoptosis. This switch from pyroptosis to apoptosis was also observed in vitro in *Gsdmd*$^{-/-}$ macrophages. In conclusion, targeting GSDMD appears to be a promising approach for limiting the transition to an inflammatory, vulnerable plaque phenotype.

Keywords: gasdermin D; pyroptosis; inflammation; atherosclerosis

Citation: Puylaert, P.; Van Praet, M.; Vaes, F.; Neutel, C.H.G.; Roth, L.; Guns, P.-J.; De Meyer, G.R.Y.; Martinet, W. Gasdermin D Deficiency Limits the Transition of Atherosclerotic Plaques to an Inflammatory Phenotype in *ApoE* Knock-Out Mice. *Biomedicines* **2022**, *10*, 1171. https://doi.org/10.3390/biomedicines10051171

Academic Editors: Tânia Martins-Marques, Gonçalo F. Coutinho and Attila Kiss

Received: 17 April 2022
Accepted: 17 May 2022
Published: 19 May 2022

Publisher's Note: MDPI stays neutral with regard to jurisdictional claims in published maps and institutional affiliations.

Copyright: © 2022 by the authors. Licensee MDPI, Basel, Switzerland. This article is an open access article distributed under the terms and conditions of the Creative Commons Attribution (CC BY) license (https://creativecommons.org/licenses/by/4.0/).

1. Introduction

Vulnerable atherosclerotic plaques are characterized by a large necrotic core formed by excessive necrotic cell death and inflammation [1]. Plaque cells can undergo different types of regulated necrosis although their significance in atherosclerosis is not always clear-cut. One of the best-defined forms of regulated necrosis is pyroptosis, a pro-inflammatory form of regulated cell death that is characterized by the formation of plasma membrane pores via members of the gasdermin (GSDM) protein family [2,3]. Six members of this family have been identified, including gasdermin D (GSDMD). GSDMD is N-terminally (NT) cleaved and is activated by caspase 1 and caspase 4/5 (homologous to caspase 11 in mouse) [3–5]. Subsequently, NT-GSDMDs oligomerize, translocate to the cell membrane and induce pore formation, which allows the release of cellular content and pro-inflammatory cytokines such as IL-1β and IL-18, and finally results in membrane disruption and cell lysis [6,7].

In canonical pyroptosis induction, cleavage and activation of caspase 1 is mediated by inflammasomes, which are large supramolecular complexes that sense intracellular danger signals through nucleotide-binding oligomerization (NOD)-like receptor sensor molecules [8]. NLRP3 (NOD-like, leucine-rich repeat (LRR) and pyrin domain containing receptor 3) inflammasome-mediated pyroptosis is currently the best characterized pathway described in atherosclerotic plaques. Activation of the NLRP3 inflammasome can be initiated by oxidized LDL, cholesterol and calcium phosphate crystals inducing lysosomal lysis, cathepsin release, and potassium efflux [9–12]. In human atherosclerotic plaques, mRNA and protein levels of caspase 1, NLRP3, IL-1β, and IL-18 are increased compared to normal arteries [13–16]. Immunohistochemical analyses of human carotid plaques have revealed that NLRP3 immunoreactivity colocalizes with CD68-positive macrophages and occasionally with smooth muscle cells [15,16], and that plaque rupture is associated with strong immunoreactivity for caspase 1 [17]. In mice, several studies have demonstrated beneficial effects on atherogenesis and plaque stability when components of the NLRP3 inflammasome, caspase 1 activity, IL-1 or IL-18 are lacking [18–26]. Together, these studies suggest that modulation of NLRP3- and caspase 1-mediated pyroptosis offers significant health benefits for patients with advanced atherosclerosis [27]. This assumption is supported by the CANTOS trial showing beneficial effects of IL-1β antibody canakinumab in statin-treated patients at risk [28]. In contrast to canonical pyroptosis, caspase 4/5 and caspase 11 can be directly activated without the need for an inflammasome, resulting in non-canonical pyroptosis induction [29,30]. Recently, caspase 11 deficiency was reported to reduce plaque burden and macrophage infiltration in $ApoE^{-/-}$ mice [30], demonstrating that also non-canonical pyroptosis plays a role in atherogenesis.

GSDMD is the common executor of both canonical pyroptosis (mediated by NLRP3 and other inflammasomes) and non-canonical pyroptosis. Moreover, GSDMD is required for IL-1β release, not only during pyroptosis but also in viable macrophages [7,31]. Interestingly, *Gsdmd* mRNA is upregulated in peripheral blood monocytes from patients with coronary artery disease and expression of GSDMD and NT-GSDMD is increased in $ApoE^{-/-}$ mice fed a high fat diet as compared to chow-fed controls [30]. These experiments indicate that GSDMD is actively involved in pyroptosis during atherogenesis in both humans and mice, and thus represents a promising target in plaques for inhibiting pyroptosis and inflammation. Genetic deletion of *Gsdmd* or pharmacological inhibition with necrosulfonamide reduces infarct size and heart failure in a mouse model of acute myocardial infarction [32], underlining the involvement of GSDMD in cardiovascular disease and the possibility for using it as a pharmacological target in atherosclerosis. Therefore, we aimed to evaluate the impact of *Gsdmd* gene deletion in atherogenesis. First, the effects of *Gsdmd* gene deletion were evaluated in macrophages and smooth muscle cells in vitro. We also analyzed the expression of cleaved GSDMD in human carotid lesions. Next, the effect of *Gsdmd* deletion on advanced atherogenesis was evaluated in $ApoE^{-/-}$ mice.

2. Materials and Methods

2.1. Human Atherosclerotic Plaques

Human carotid endarterectomy specimens were obtained from patients (71 ± 3 years, 70% men) with a carotid stenosis of >70% [33]. Specimens were fixed in 4% formaldehyde (pH 7.4) within 2 min after surgical removal and paraffin embedded. To identify specific cell types expressing cleaved GSDMD, double immunohistochemical staining was performed using anti-cleaved GSDMD (Cell signaling, 36425, Danvers, MA, USA) combined with anti-CD68 (clone PG-M1, ab9498, Abcam, Cambridge, UK), anti-α-smooth muscle actin (αSMA; clone 1A4, A2547, Sigma-Aldrich, St. Louis, MO, USA), or anti-CD31 (clone JC/70A, ab783, Abcam, Cambridge, UK). Images were acquired with an Olympus BX43 microscope and quantified using Image J software (National Institutes of Health, Bethesda, MD, USA).

2.2. Mice

Standard $ApoE^{-/-}$ mice (Jackson Laboratory, 002052, Bar Harbor, ME, USA) were crossbred with $Gsdmd^{-/-}$ mice (Genentech, South San Francisco, CA, USA), carrying a 1 bp insertion in the *Gsdmd* coding sequence (GAGTGATGTTGTCAGGCATGGGA becomes GAGTGATGTTtGTCAGGCATGGGA) created with CRISPR/Cas9. Litters were screened for the $Gsdmd^{-/-}$ genotype by PCR analysis using Gsdmd-specific primers (forward primer:

GTTTCTTGTCGATGGGAACATTCAG, reverse primer: TGAGTCACACGCAGTATA) followed by Sanger sequencing using the reverse primer. Genotyping of the *ApoE* alleles was performed by PCR according to the instructions from the Jackson Laboratory. Thereafter, *ApoE*$^{-/-}$ *Gsdmd*$^{-/-}$ mice and *ApoE*$^{-/-}$ *Gsdmd*$^{+/+}$ controls (all females, 6–8 weeks old) were fed a western-type diet (WD; TD.88137 supplemented with 21% fat and 0.2% cholesterol, Envigo, Indianapolis, IN, USA) to induce plaque formation. The animals were housed in a temperature-controlled room with a 12 h light/dark cycle and had free access to water and food. After 16 weeks WD, an overdose of sodium pentobarbital (250 mg/kg, i.p.) was administered and blood samples were collected via the retro-orbital plexus. Plasma levels of total cholesterol were measured using a commercially available kit (Randox laboratories, Crumlin, UK). Blood leukocyte subsets were analyzed on a BD accuri C6 flow cytometer as previously described [34]. All experiments were approved by the Ethical Committee of the University of Antwerp (Code 2019-24) and carried out in accordance with European Directive 2010/63/EEC.

2.3. Histological Analyses

The thoracic aorta was stained en face with Oil Red O to determine lipid burden. The heart, brachiocephalic artery, and proximal ascending aorta of *ApoE*$^{-/-}$ *Gsdmd*$^{-/-}$ and *ApoE*$^{-/-}$ *Gsdmd*$^{+/+}$ mice were fixed in 4% formaldehyde (pH 7.4) for 24 h, dehydrated overnight in 60% isopropanol, and subsequently embedded in paraffin. The proximal ascending aorta was marked on the distal arch end and the brachiocephalic artery on the distal carotid end to ensure that they were always cut on the proximal side. Serial cross-sections (5 μm) of the proximal parts of the brachiocephalic artery, proximal aorta and aortic root were prepared in random for histological analyses. Atherosclerotic plaque size and necrotic core area (defined as acellular areas with a threshold of 3000 μm^2) were analyzed on hematoxylin-eosin (H&E) stained sections. Total collagen content was measured on Sirius red stained sections. Apoptosis was analyzed using an ApopTag Plus Peroxidase In Situ Apoptosis kit (Millipore, S7101, Burlington, VT, USA). For immunohistochemistry, the following antibodies were used: anti-MAC3 (BD Pharmingen, 550292, San Diego, CA, USA) and anti-α-smooth muscle actin (αSMA, 12547, Sigma-Aldrich, St. Louis, MO, USA). Images were acquired with an Olympus BX43 microscope, which was calibrated for each magnification. Plaque size was measured based on pixels per μm, which was determined during the calibration of the microscope. Per mouse, one section was analyzed. Plaques were manually delineated in Image J software (National Institutes of Health) to establish the region of interest (ROI). Further analyses within the ROIs were performed using color thresholding or manual counting (apoptotic cells).

2.4. Cell Culture

Bone marrow-derived macrophages (BMDMs) were harvested by flushing bone marrow of the femur with a 25 Gauge needle and heparinized (10 IU/mL) RPMI 1640 medium (Gibco Life Technology, Merelbeke, Belgium). After washing and filtration, cells were cultured in RPMI 1640 medium supplemented with Glutamax (Gibco Life Technology, Merelbeke, Belgium) and 15% L929-cell conditioned medium (LCCM) containing monocyte colony stimulating factor (M-CSF) for 7 days in 95% air/5% CO_2 until 80–90% confluency was reached. To induce pyroptosis, BMDMs were primed with 100 ng/mL lipopolysaccharide (LPS, Sigma-Aldrich, St. Louis, MO, USA) for 4 h followed by treatment with 2.5–20 μM nigericin (Enzo Life Sciences, BML-CA421-0005, Brussels, Belgium) or 5 mM ATP (Calbiochem, 1191, Sigma-Aldrich, St. Louis, MO, USA). Vascular smooth muscle cells (VSMCs) were isolated as previously described [35,36]. Briefly, aortas were incubated in an enzyme solution containing 1 mg/mL collagenase type II (Worthington, Lakewood, NJ, USA), 1 mg/mL soybean trypsin inhibitor (Worthington, Lakewood, NJ, USA), and 0.744 units/mL elastase (Worthington) for 15 min at 37 °C to remove the adventitia. Subsequently, aortas were placed in a fresh enzyme solution for 75 min at 37 °C. Isolated cells were collected, washed, and resuspended in DMEM/F12 medium (Gibco Life Technology) supplemented with 20% heat-inactivated FBS (Gibco Life Technology). Cells were used from passage 4 till 10 and cultured in DMEM/F12 medium supplemented with 10% heat-inactivated FBS. To induce pyroptosis, VSMCs were primed with 50 ng/mL TNFα (Abcam, ab9740, Cambridge, UK) for 2 h followed by treatment with 5 mM ATP (Calbiochem, 1191) for 2 h. Alternatively, non-primed cells were treated with 300 µg/mL oxidized LDL (oxLDL, L34357, Life Technologies, Carlsbad, CA, USA) for 48 h. Necrosis was evaluated by labeling BMDMs or VSMCs with 1 µg/mL propidium iodide (PI, Molecular Probes, Eugene, OR, USA) and 10 µg/mL Hoechst (Life Technologies, Carlsbad, CA, USA), followed by visualization of PI/Hoechst-labeled cells using a Celena S digital Imaging System (Logos Biosystems, Dongan-gu, Anyang-si, Korea). To measure the release of IL-1β and IL-18, primed BMDMs were treated with 10 µM nigericin. After 2 h, cell supernatant was collected and cellular debris was removed by centrifugation. IL-1β and IL-18 secretions were quantified in the supernatant using a mouse IL-1β Quantikine ELISA kit (R&D Systems, MLB00C, Minneapolis, MN, USA) and IL-18 mouse ELISA kit (Invitrogen, BMS618-3, Waltham, MA, USA), respectively.

2.5. Flow Cytometry

Apoptosis was quantified with TdT-mediated dUTP-X nick end labeling (TUNEL). Briefly, BMDMs were detached with 0.25% trypsin-EDTA (Thermo Fischer Scientific, 25200072, Waltham, MA, USA). Detached BMDMs were fixed in 4% paraformaldehyde for 1 h at room temperature. After washing with PBS, BMDMs were permeabilized with 0.1% Triton X-100 in 0.1% sodium citrate solution for 2 min on ice. BMDMs were washed again and incubated with TUNEL reaction mixture using an in situ cell death detection kit (fluorescein, 11684795910, Roche, Switzerland) for 1 h at 37 °C. The samples were washed twice with FACS buffer (PBS with 0.1% bovine serum albumin and 0.05% sodium azide) and measured in the FL-1 channel on a BD Accuri C6 flow cytometer. At least 10,000 cells were measured. Debris was always gated out based on FSC/SSC scatter. Positive controls consisted of BMDMs treated with TNFα combined with cycloheximide. Negative controls consisted of untreated BMDMs. Unstained controls were included to exclude background signal.

2.6. Western Blotting

Tissues were homogenized in RIPA buffer containing protease and phosphatase inhibitors. Protein concentrations were determined using the BCA method. Samples were then 1:1 diluted in Laemmli sample buffer (Bio-Rad, Hercules, CA, USA) containing 5% β-mercaptoethanol (Sigma-Aldrich) and heat-denatured for 5 min at 100 °C. Samples were loaded on Bolt 4–12% Bis-Tris gels (Invitrogen) and after electrophoresis transferred to Immobilon-FL PVDF membranes (Millipore) according to standard procedures. Subsequently, membranes were blocked for one hour in Odyssey Li-COR blocking buffer. After blocking, membranes were probed with primary antibodies diluted in Odyssey Li-COR blocking buffer followed by 1 h incubation with IRDye-labeled secondary antibodies at room temperature. Membranes were visualized with an Odyssey SA infrared imaging system (Li-COR Biosciences, Lincoln, NE, USA).

The following primary antibodies were used: rabbit anti-GSDMD (Abcam, ab219800, Cambridge, UK), rabbit anti-cleaved GSDMD (Cell signaling, 10137, Danvers, MA, USA), rabbit anti-NLRP3 (Abcam, ab263899), rabbit anti-caspase 1 (Abcam, ab179515), and mouse anti-β-actin (Abcam, ab8226).

2.7. Statistical Analyses

Statistical analyses were performed using Graphpad Prism 9. All data were expressed as mean ± SEM, except in boxplots where medians are shown. Dots represent the number of samples from independent experiments or individual mice. Statistical tests are specified in the text and figure legends. Differences were considered significant when $p < 0.05$.

3. Results

3.1. Pyroptosis Is Inhibited in $Gsdmd^{-/-}$ BMDMs but a Switch to Apoptosis Is Induced

Because pyroptosis is mainly characterized in myeloid cells [37,38], BMDMs were isolated from $Gsdmd^{-/-}$ and $Gsdmd^{+/+}$ mice. First, GSDMD deficiency was confirmed in $Gsdmd^{-/-}$ BMDMs via Western blotting (Figure 1A). To induce pyroptosis, BMDMs were primed with LPS followed by treatment with nigericin, a canonical pyroptosis inducer. Significant induction of cell death was observed in $Gsdmd^{+/+}$ cells while $Gsdmd^{-/-}$ cells were clearly more resistant to pyroptosis induction (Figure 1B). The release of IL-1β and IL-18 was quantified in BMDM supernatant after LPS priming and treatment with nigericin. Both IL-1β and IL-18 were released by $Gsdmd^{+/+}$ BMDMs and this was significantly decreased in $Gsdmd^{-/-}$ BMDMs (Figure 1C). IL-1β and IL-18 were not detected in supernatant of untreated BMDMs (data not shown). LPS priming induced the expression of NLRP3 and procaspase 1, both in $Gsdmd^{+/+}$ and $Gsdmd^{-/-}$ BMDMs (Figure 1D). Accordingly, Western blot analysis further showed that LPS-primed $Gsdmd^{+/+}$ BMDMs treated with 2.5 to 20 μM nigericin expressed GSDMD and cleaved GSDMD, while $Gsdmd^{-/-}$ BMDMs did not (Figure 1E). Moreover, both $Gsdmd^{+/+}$ and $Gsdmd^{-/-}$ BMDMs expressed procaspase 1 and, upon treatment with nigericin, the active caspase 1 p10 subunit was expressed (Figure 1E). Altogether, these results confirm that GSDMD-pore formation and pyroptosis induction are inhibited in $Gsdmd^{-/-}$ BMDMs while upstream NLRP3 assembly and procaspase 1 recruitment (and cleavage) are not affected.

Interestingly, the expression of caspase 1 p10 was significantly higher in $Gsdmd^{-/-}$ BMDMs as compared to $Gsdmd^{+/+}$ controls (Figure 1E). Because active caspase 1 can also act in a pro-apoptotic fashion [39,40], a TUNEL assay was performed on LPS-primed BMDMs after treatment with nigericin for 1 h (Figure 1F). A significant increase in TUNEL positivity was observed after nigericin treatment in $Gsdmd^{-/-}$ BMDMs, while PI positivity was not increased, indicating that the plasma membranes of $Gsdmd^{-/-}$ cells were intact but that DNA fragmentation typical of apoptosis occurred. In contrast, TUNEL positivity did not change in $Gsdmd^{+/+}$ BMDMs after nigericin treatment while PI positivity did increase significantly (Figure 1F). Similar findings were observed after treatment with ATP, another classical caspase 1-dependent pyroptosis inducer (Figure 1G).

3.2. $Gsdmd^{-/-}$ VSMCs Are Less Sensitive to Pyroptosis Inducers

There is increasing evidence that pyroptosis is not limited to inflammatory cells [22,41–43]. Therefore, VSMCs were isolated from $Gsdmd^{+/+}$ and $Gsdmd^{-/-}$ mice. Absence of GSDMD in $Gsdmd^{-/-}$ VSMCs was confirmed via Western blotting (Figure 2A). To induce pyroptosis, VSMCs were primed with TNFα followed by ATP treatment. Cell death increased significantly in $Gsdmd^{+/+}$ VSMCs but not in $Gsdmd^{-/-}$ VSMCs (Figure 2B). Similar to TNFα/ATP treatment, oxLDL, which is also reported to induce pyroptosis [12,44], increased cell death significantly in $Gsdmd^{+/+}$ VSMCs but not in $Gsdmd^{-/-}$ cells (Figure 2C).

3.3. Cleaved GSDMD Is Expressed in Human Carotid Lesions

To confirm that GSDMD is not only active in mice, cell-specific expression of cleaved GSDMD was evaluated in human carotid lesions (Figure 3). In general, cleaved GSDMD was expressed in the plaque, especially in the shoulder regions, and media (Figure 3B–D). Double immunohistochemical staining of cleaved GSDMD and CD68 showed that GSDMD is cleaved in plaque areas rich in macrophages, albeit not all CD68-positive cells contained cleaved GSDMD (Figure 3B). On sections stained separately for cleaved GSDMD and αSMA (Figure 3C), the positivity pattern appeared very similar. Indeed, double immunohistochemical staining of cleaved GSDMD and αSMA showed a clear overlapping of red (cl-GSDMD) and purple (αSMA) signal, confirming colocalization of cleaved GSDMD and smooth muscle cells. The CD31-positive intima delineated the vessel lumen (Figure 3D). However, no cleaved GSDMD positivity was observed in CD31-positive endothelial cells.

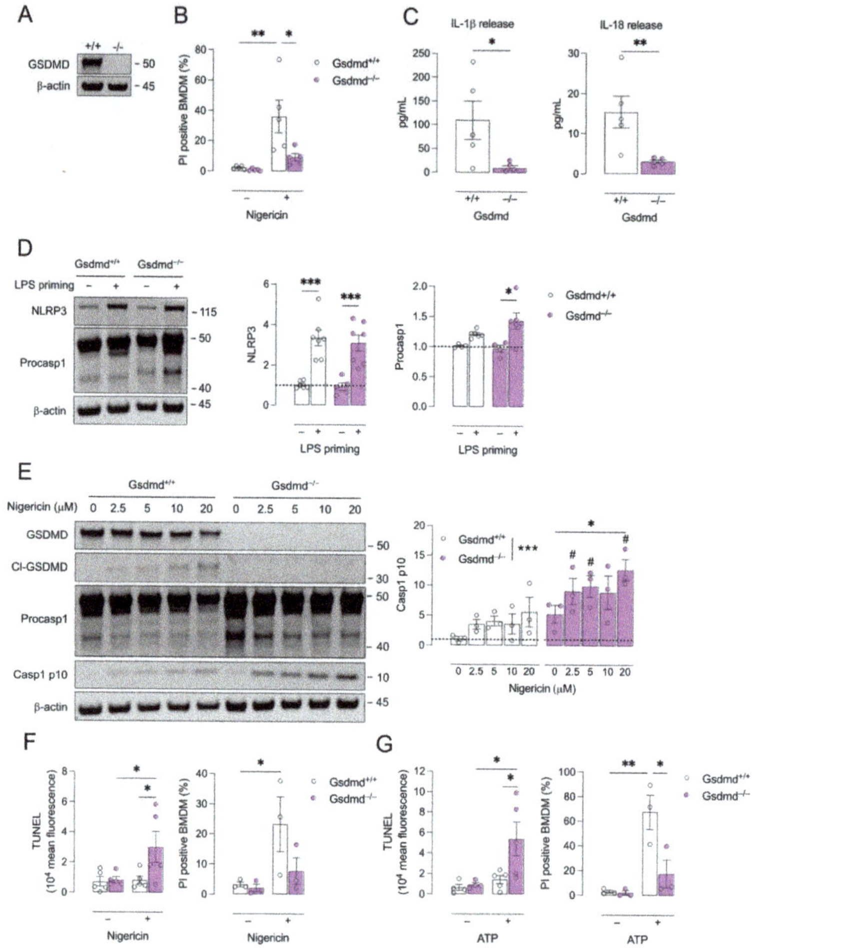

Figure 1. Pyroptosis is inhibited in *Gsdmd*$^{-/-}$ BMDMs and a switch to apoptosis occurs. Bone marrow-derived macrophages (BMDMs) were isolated from *Gsdmd*$^{+/+}$ and *Gsdmd*$^{-/-}$ mice. (**A**) Deficiency of GSDMD in *Gsdmd*$^{-/-}$ BMDMs was confirmed via Western blotting. (**B–D**) *Gsdmd*$^{-/-}$ and *Gsdmd*$^{+/+}$ BMDMs were primed with 100 ng/mL LPS for 4 h followed by treatment with 10 μM nigericin for 2 h. (**B**) Cell death was measured using propidium iodide (PI) labelling (two-way ANOVA followed by Sidak's multiple comparison, $n = 5$ independent experiments) and (**C**) the release of IL-1β and IL-18 was quantified in the cell supernatant (Mann–Whitney test, $n = 5$ independent experiments). (**D**) Western blot analyses of NLRP3 and procaspase (procasp) 1 on lysates of LPS-primed BMDMs (two-way ANOVA followed by Sidak's multiple comparison, $n = 4$–7 independent experiments, data are expressed as fold change of target/β-actin ratio). (**E**) *Gsdmd*$^{-/-}$ and *Gsdmd*$^{+/+}$ BMDMs were primed with 100 ng/mL LPS for 4 h followed by treatment with 2.5–20 μM nigericin for 2 h. Western blot analyses of GSDMD, cleaved (cl)-GSDMD, procaspase 1, and caspase (casp) 1 p10 (two-way ANOVA followed by Dunnett's multiple comparison between concentrations per genotype, * $p < 0.05$, *** $p < 0.001$; two-way ANOVA followed by Sidak's multiple comparison between genotypes per concentration, # $p < 0.05$; $n = 3$ independent experiments, data are expressed as fold change of target/β-actin ratio). (**F**,**G**) LPS-primed BMDMs were treated for 1 h with (**F**) 20 μM nigericin or (**G**) 5 mM ATP. TUNEL labeling was performed and analyzed via flow cytometry and cell death was measured using PI labelling (two-way ANOVA followed by Sidak's multiple comparison, $n = 3$–5 independent experiments). * $p < 0.05$, ** $p < 0.01$, *** $p < 0.001$.

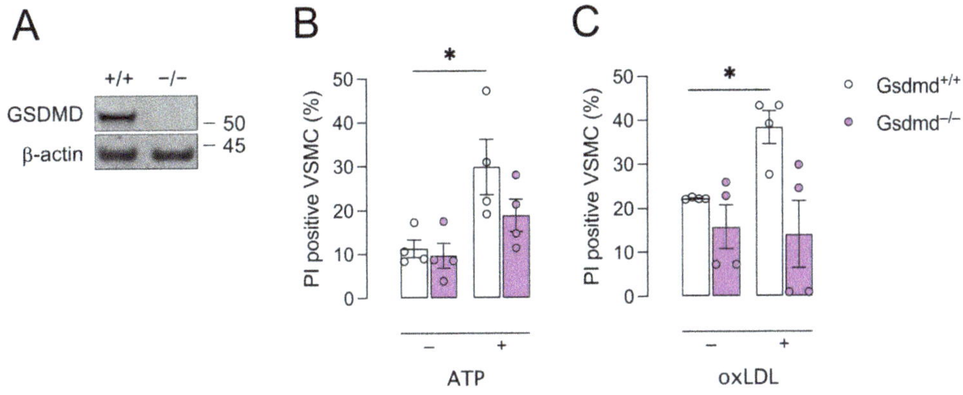

Figure 2. *Gsdmd*$^{-/-}$ VSMCs are less sensitive to pyroptosis inducers compared to controls. Vascular smooth muscle cells (VSMCs) were isolated from *Gsdmd*$^{+/+}$ and *Gsdmd*$^{-/-}$ mice. (**A**) Deficiency of GSDMD in *Gsdmd*$^{-/-}$ VSMCs was confirmed via Western blotting. (**B**) VSMCs were primed with 50 ng/mL TNFα for 2 h followed by treatment with 5mM ATP for 1 h. (**C**) VSMCs were treated with 300 µg/mL oxLDL for 48 h. Cell death was measured using PI labelling. * $p < 0.05$ (two-way ANOVA followed by Sidak's multiple comparison, $n = 4$ independent experiments).

3.4. Atherogenesis Is Delayed in ApoE$^{-/-}$ *Gsdmd*$^{-/-}$ *Mice but a Switch to Apoptosis Occurs in Plaques of the Brachiocephalic Artery*

To evaluate the effect of *Gsdmd* gene deletion on atherogenesis, *ApoE*$^{-/-}$ *Gsdmd*$^{-/-}$ and *ApoE*$^{-/-}$ *Gsdmd*$^{+/+}$ mice were fed a western-type diet (WD) for 16 weeks. First, deletion of *Gsdmd* was confirmed via Western blotting in the aortic arch, descending thoracic aorta (desc TA), lung, liver, heart, spleen, and kidney of *ApoE*$^{-/-}$ *Gsdmd*$^{-/-}$ mice (Supplementary Figure S1). After 16 weeks WD, body weight (27.2 ± 0.7 vs. 29.3 ± 1.0 g, 1.5 ± 0.03 vs. 1.5 ± 0.05 fold change compared to starting weight; independent samples *t*-test, $p > 0.05$), plasma cholesterol (685.8 ± 18.2 vs. 652.1 ± 20.0 mg/dL; independent samples *t*-test, $p > 0.05$), and circulating leukocyte subsets (Supplementary Figure S2) were not different between *ApoE*$^{-/-}$ *Gsdmd*$^{+/+}$ and *ApoE*$^{-/-}$ *Gsdmd*$^{-/-}$ mice. En face Oil Red O staining of the thoracic aorta showed no differences in lipid burden between *ApoE*$^{-/-}$ *Gsdmd*$^{+/+}$ and *ApoE*$^{-/-}$ *Gsdmd*$^{-/-}$ mice in the whole thoracic aorta, as well as in the aortic arch and descending aorta separately (Figure 4).

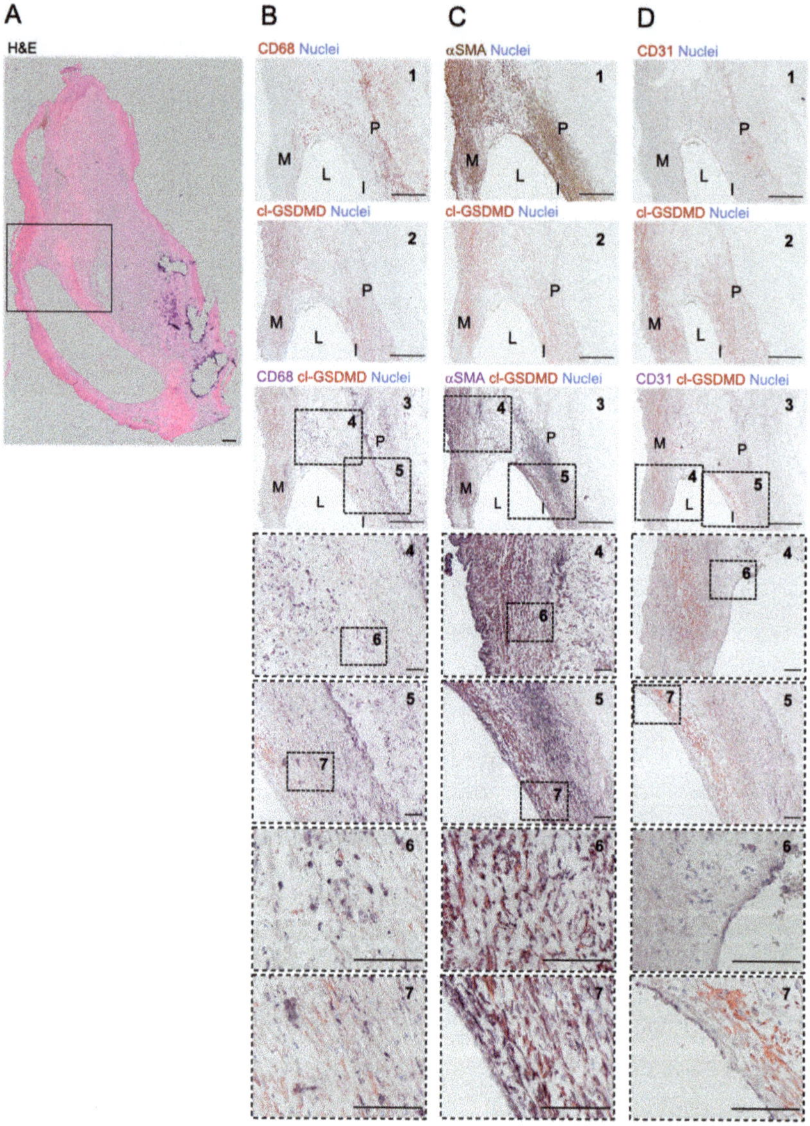

Figure 3. Cleaved GSDMD is expressed in human carotid plaques. (**A**) Overview image of a section from a human carotid artery lesion stained with hematoxylin/eosin. The boxed area corresponds with the region shown in (**B–D**). (**B–D**) Immunohistochemical staining of cleaved (cl)-GSDMD combined with (**B**) CD68, (**C**) α-smooth muscle actin (αSMA) or (**D**) CD31. From top to bottom: 1. Image of section stained for (B) CD68 (red), (**C**) αSMA (brown), or (**D**) CD31 (red). 2. Image of section stained for cl-GSDMD (red). 3. Image of double stained section for cl-GSDMD (red) combined with (**B**) CD68, (**C**) αSMA, or (**D**) CD31 (purple). 4–7. Magnifications of dotted frames in images 3–5. Scale bar = 500 μm (**A**,**B–D**: 1–3), 100 μm (**B–D**: 4–7). Representative images are shown. M = media, L = lumen, I = intima, P = plaque.

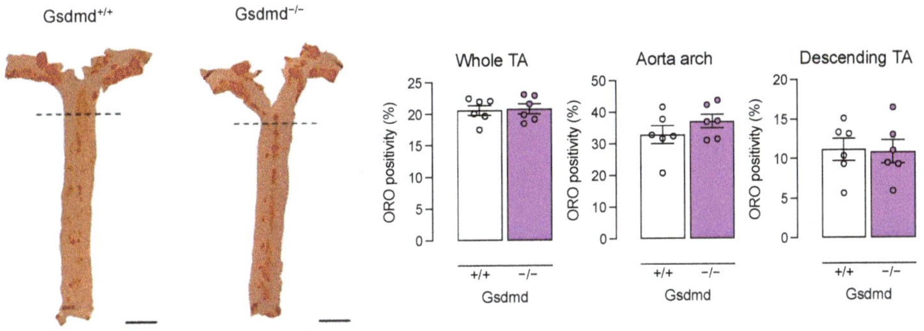

Figure 4. Lipid burden in the thoracic aorta is not different between *ApoE*$^{-/-}$ *Gsdmd*$^{-/-}$ and *ApoE*$^{-/-}$ *Gsdmd*$^{+/+}$ mice. *ApoE*$^{-/-}$ *Gsdmd*$^{-/-}$ and *ApoE*$^{-/-}$ *Gsdmd*$^{+/+}$ mice were fed a WD for 16 weeks. The aortic arch and descending thoracic aorta (TA) were stained en face with Oil Red O to evaluate the plaque burden (Mann–Whitney test, $n = 6$ mice per group). The dotted line seperates the aortic arch (top) from the descending TA (bottom). Scale bar = 5 mm.

In contrast, plaque size was significantly decreased in the brachiocephalic artery of *ApoE*$^{-/-}$ *Gsdmd*$^{-/-}$ mice as compared to *ApoE*$^{-/-}$ *Gsdmd*$^{+/+}$ controls (115 ± 18 vs. 186 ± 16 × 10^3 μm^2; independent samples *t*-test, $p = 0.006$, Figure 5A). This suggests that plaque progression is delayed in *ApoE*$^{-/-}$ *Gsdmd*$^{-/-}$ mice. Moreover, the necrotic core area was significantly decreased in *ApoE*$^{-/-}$ *Gsdmd*$^{-/-}$ plaques of the brachiocephalic artery as compared to *ApoE*$^{-/-}$ *Gsdmd*$^{+/+}$ controls (19 ± 4 vs. 37 ± 7 × 10^3 μm^2; independent samples *t*-test, $p = 0.03$), although the relative necrotic core area was not altered (Figure 5A). Similarly, the number of cells infiltrated in plaques of the brachiocephalic artery of *ApoE*$^{-/-}$ *Gsdmd*$^{-/-}$ mice was significantly lower as compared to *ApoE*$^{-/-}$ *Gsdmd*$^{+/+}$ mice (Figure 5A). The total collagen content did not differ between *ApoE*$^{-/-}$ *Gsdmd*$^{-/-}$ and *ApoE*$^{-/-}$ *Gsdmd*$^{+/+}$ plaques in the brachiocephalic artery (Figure 5B). Moreover, αSMA and MAC3 immunoreactivity did not change significantly, however, the ratio of αSMA to MAC3 immunoreactivity increased significantly in *ApoE*$^{-/-}$ *Gsdmd*$^{-/-}$ plaques of the brachiocephalic artery as compared to controls (Figure 5C), indicating a shift in plaque composition and inflammatory state. Because a switch in cell death modality was observed in vitro in *Gsdmd*$^{-/-}$ BMDMs, TUNEL staining was performed on plaques *of ApoE*$^{-/-}$ *Gsdmd*$^{+/+}$ and *ApoE*$^{-/-}$ *Gsdmd*$^{-/-}$ mice. Similarly to what was observed in vitro, TUNEL positivity in vivo was significantly increased in plaques of the brachiocephalic artery from *ApoE*$^{-/-}$ *Gsdmd*$^{-/-}$ mice compared to *ApoE*$^{-/-}$ *Gsdmd*$^{+/+}$ controls (Figure 5D).

In plaques of the proximal aorta (Figure 6) and aortic root (Supplementary Figure S3), no differences in plaque size, necrotic core area, cell infiltration, and total collagen content were observed between *ApoE*$^{-/-}$ *Gsdmd*$^{-/-}$ and *ApoE*$^{-/-}$ *Gsdmd*$^{+/+}$ mice. Nevertheless, MAC3 immunoreactivity was significantly decreased while the αSMA to MAC3 immunoreactivity ratio was significantly increased in plaques of the proximal aorta from *ApoE*$^{-/-}$ *Gsdmd*$^{-/-}$ mice, again indicating a shift in plaque composition and inflammatory state (Figure 6C). In contrast to what was observed in vitro and in the brachiocephalic artery from *ApoE*$^{-/-}$ *Gsdmd*$^{-/-}$ mice, TUNEL positivity was not increased in plaques in the proximal aorta (Figure 6D).

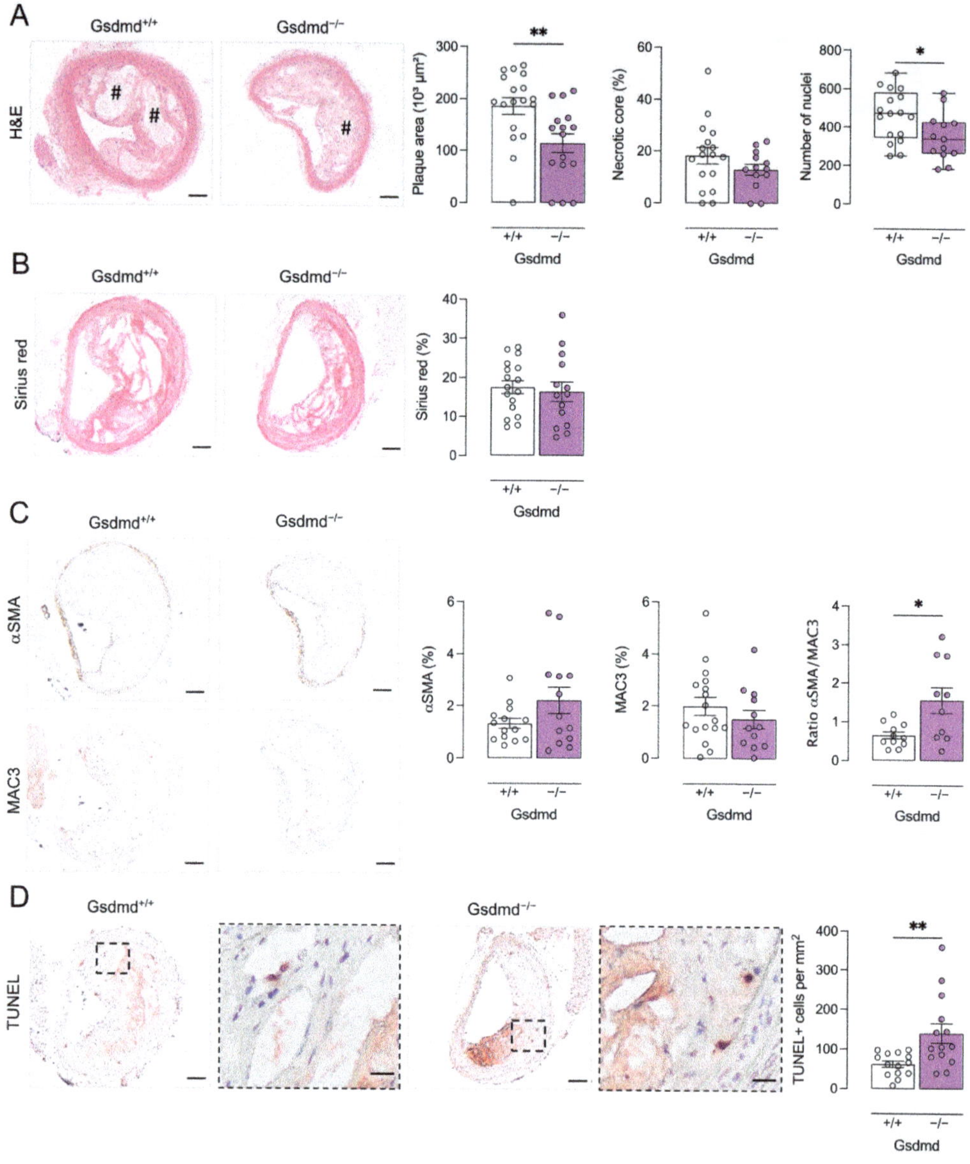

Figure 5. Plaques in the brachiocephalic artery from $ApoE^{-/-}$ $Gsdmd^{-/-}$ mice are smaller but show increased apoptosis. $ApoE^{-/-}$ $Gsdmd^{-/-}$ and $ApoE^{-/-}$ $Gsdmd^{+/+}$ mice were fed a WD for 16 weeks. Sections of the brachiocephalic artery were stained with (**A**) hematoxylin/eosin to quantify plaque size, necrotic cores (# hash signs), and cell infiltration; (**B**) Sirius red to measure total collagen content; (**C**) anti-MAC3 and anti-α-smooth muscle actin (αSMA) to determine macrophage and smooth muscle cell content, respectively, and to calculate the ratio of αSMA/MAC3 immunoreactivity; (**D**) TUNEL to count apoptotic cells (dotted boxes are magnified, scale bar = 20 µm). * $p < 0.05$, ** $p < 0.01$ (independent samples t-test, boxplot: Mann–Whitney test, n = 10–18 mice per group). Scale bar = 100 µm. Representative images are shown.

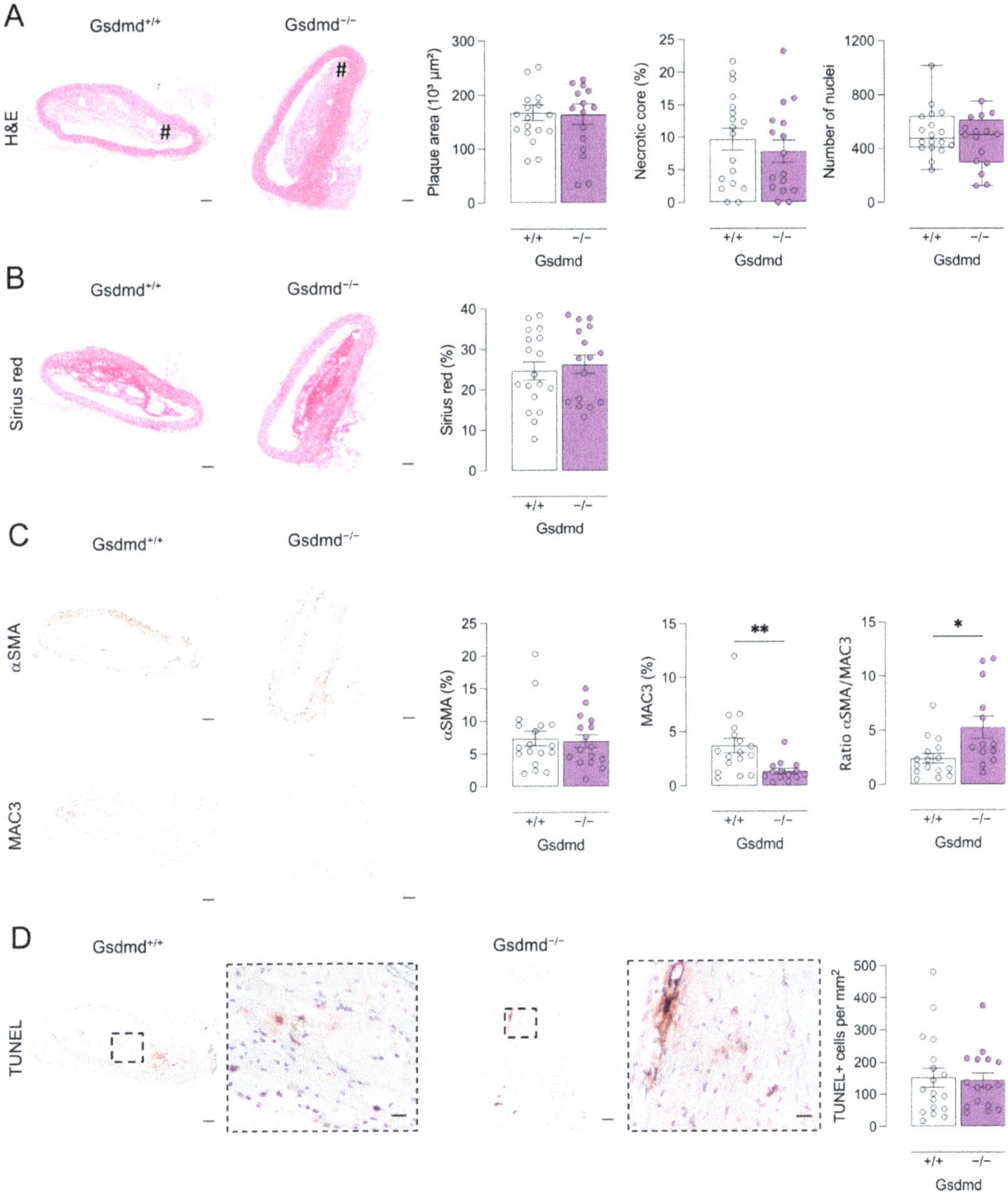

Figure 6. Plaques in the proximal aorta from *ApoE*$^{-/-}$ *Gsdmd*$^{-/-}$ mice show decreased macrophage infiltration as compared to *ApoE*$^{-/-}$ *Gsdmd*$^{+/+}$ controls. *ApoE*$^{-/-}$ *Gsdmd*$^{-/-}$ and *ApoE*$^{-/-}$ *Gsdmd*$^{+/+}$ mice were fed a WD for 16 weeks. Sections of the proximal aorta were stained with (**A**) hematoxylin/eosin to quantify plaque size, necrotic cores (# hash signs), and cell infiltration; (**B**) Sirius red to measure total collagen content; (**C**) anti-MAC3 and anti-α-smooth muscle actin (αSMA) to determine macrophage and smooth muscle cell content, respectively, and to calculate the ratio of αSMA/MAC3 immunoreactivity; (**D**) TUNEL to count apoptotic cells (dotted boxes are magnified, scale bar = 20 µm). * $p < 0.05$, ** $p < 0.01$ (independent samples *t*-test, boxplot: Mann–Whitney test, n = 13–18 mice per group). Scale bar = 100 µm. Representative images are shown.

4. Discussion

GSDMD is the common executor of canonical and non-canonical pyroptosis. Increasing evidence suggests that pyroptosis occurs in atherosclerotic plaques in both humans and mice [13–26]. In the present study, we isolated BMDMs and VSMCs from *Gsdmd*$^{+/+}$ and *Gsdmd*$^{-/-}$ mice and confirmed that canonical pyroptosis is GSDMD-dependent in BMDMs and VSMCs. Indeed, treatment of LPS-primed BMDMs with classical caspase 1 activators such as nigericin and ATP induced significant pyroptosis in *Gsdmd*$^{+/+}$ controls, but not in *Gsdmd*$^{-/-}$ cells. Concurrently, *Gsdmd*$^{-/-}$ BMDMs showed increased DNA fragmentation, measured with TUNEL staining, which was absent in *Gsdmd*$^{+/+}$ BMDMs. Thus, *Gsdmd*$^{-/-}$ BMDMs are resistant to canonical pyroptosis induction, albeit they switch to another type of cell death that is characterized by DNA fragmentation, strongly suggestive of apoptosis. A switch from pyroptosis to apoptosis is further supported by the observation that caspase 1 p10 is upregulated in LPS/nigericin-treated *Gsdmd*$^{-/-}$ BMDMs, and not in *Gsdmd*$^{+/+}$ controls, and may act in a pro-apoptotic manner in the absence of GSDMD [39,40,45].

Previous studies have demonstrated that GSDMD, NT-GSDMD, and *Gsdmd* mRNA are upregulated in aortas of hyperlipidemic mice and in LPS/oxLDL-treated mouse peritoneal macrophages [30]. Moreover, the expression of NT-GSDMD is increased in plaques of LDLr$^{-/-}$ mice [45]. *Gsdmd* mRNA is also upregulated in peripheral blood monocytes of atherosclerotic patients [30]. However, to the best of our knowledge, the expression of cleaved NT-GSDMD has not yet been evaluated in human atherosclerotic plaques. Therefore, we performed immunostaining on human carotid lesions. We demonstrated that NT-GSDMD was expressed in human plaques, both in macrophage- and VSMC-rich regions. However, it should be noted that this finding was based on analysis of a limited number of plaques and comparison with plaque-free arteries was not made. Other groups have shown that components of the canonical pyroptosis pathway colocalize with plaque macrophages and contribute to macrophage pyroptosis and atherogenesis [12,15,24,27]. Interestingly, in the present study, colocalization of NT-GSDMD with αSMA-positive smooth muscle cells was even more pronounced than with CD68-positive macrophages. This is in line with a recent study reporting that caspase 1-dependent pyroptosis occurs in VSMCs and contributes to the progression of atherosclerosis [22]. Importantly, pyroptosis has also been described in plaque endothelial cells [46–48]. However, we did not observe any NT-GSDMD immunoreactivity in the luminal endothelial layer of human carotid lesions. Similarly, Rajamäki and colleagues also did not observe any NLRP3 immunoreactivity in the endothelium of human coronary plaques [15].

Disulfiram (used to treat alcohol use disorder) has recently been identified as a potent GSDMD inhibitor [49]. Furthermore, dimethyl fumarate (used as an immunomodulator in multiple sclerosis) also covalently binds GSDMD thereby inhibiting pyroptosis in vitro and in vivo in a mouse model of LPS-induced shock [50]. The availability of safe, EMA- and FDA-approved drugs that inhibit GSDMD makes it an interesting therapeutic target to address pyroptosis in atherosclerosis. Therefore, we crossbred atherosclerotic *ApoE*$^{-/-}$ mice with *Gsdmd*$^{-/-}$ mice to evaluate the effect of GSDMD deficiency on atherogenesis. After 16 weeks WD, Oil red O staining was performed on the thoracic aorta but no change in overall lipid burden was observed. In line with this finding, plaque size was not altered in the proximal aorta. However, the plaque size in the brachiocephalic artery of *ApoE*$^{-/-}$ *Gsdmd*$^{-/-}$ mice was significantly decreased as compared to *ApoE*$^{-/-}$ *Gsdmd*$^{+/+}$ mice. Although both the ascending proximal aorta and the brachiocephalic artery of *ApoE*$^{-/-}$ mice are atherogenesis-prone sites [51], plaques in the brachiocephalic artery enter an advanced, human-like stage more rapidly [52,53]. Indeed, plaques in the brachiocephalic artery of *ApoE*$^{-/-}$ mice easily reveal a vulnerable plaque phenotype and thinning of the fibrous cap may even lead to plaque rupture, similarly to human plaques [53–55]. In contrast, aortic plaques of *ApoE*$^{-/-}$ mice resemble stable lesions in humans, based on fibrous cap stress analysis [53]. Altogether, these observations suggest that GSDMD deficiency does not affect plaque initiation (as lipid burden in the thoracic aorta and proximal aortic plaque size were not decreased in *ApoE*$^{-/-}$ *Gsdmd*$^{-/-}$ mice) but rather the transition to and growth of a vulnerable plaque. This is supported by the increased αSMA/MAC3-immunoreactivity ratio observed in plaques of both the brachiocephalic artery and the proximal aorta of *ApoE*$^{-/-}$ *Gsdmd*$^{-/-}$ mice as compared to controls. This finding suggests a lower degree of plaque inflammation and vulnerability, and was also observed in plaques of the proximal aorta despite the unchanged plaque size, demonstrating a general decrease in plaque inflammation. Moreover, macrophage content was significantly decreased in plaques in the proximal aorta of *ApoE*$^{-/-}$ *Gsdmd*$^{-/-}$ mice as compared to controls. This is in line with a recent study in which GSDMD expression was suppressed in *ApoE*$^{-/-}$ mice using an adeno-associated virus-5 (AAV-5) delivery system [30]. The authors reported a decrease in F4/80-positive macrophage content in the aorta and aortic valve of AAV-5-GSDMD-treated *ApoE*$^{-/-}$ mice as compared to AAV-5-control-treated *ApoE*$^{-/-}$ mice. Of note, the authors also reported a decrease in lipid burden and plaque size in the aorta and aortic valve when GSDMD expression was suppressed, however, only a limited number of mice was included and preliminary data were reported [30]. Another study reported a decreased lesion area in the aortic root of LDLr antisense oligonucleotide-treated *Gsdmd*$^{-/-}$ mice compared to *Gsdmd*$^{+/+}$ controls, which was attributed to decreased IL-1β release resulting in decreased foam cell formation and ATP release in plasma [45]. Importantly, experimental differences such as a longer period of feeding WD in the present study, evaluation of different vascular sites, and the use of different genetic models make comparison with these studies difficult.

In plaques of the brachiocephalic artery, the number of TUNEL positive cells per mm^2 was significantly increased in *ApoE*$^{-/-}$ *Gsdmd*$^{-/-}$ mice compared to controls. Increased TUNEL positivity is not specific and DNA fragmentation is reported to occur during apoptotic death as well as during different forms of necrotic death [56,57]. However, as necrosis is decreased in plaques of *ApoE*$^{-/-}$ *Gsdmd*$^{-/-}$ mice, it is plausible to conclude that a switch to apoptosis occurs when pyroptosis is defective. Accordingly, in vitro TUNEL positivity and levels of pro-apoptotic caspase 1 p10 were increased in *Gsdmd*$^{-/-}$ BMDMs while PI positivity did not increase. Similarly, Opoku and colleagues recently reported increased apoptosis, characterized by phosphatidyl serine exposure on the cell surface, when pyroptosis was defective in *Gsdmd*$^{-/-}$ macrophages [45]. Importantly, a switch to apoptosis, which is non-lytic and non-inflammatory (in contrast to pyroptosis), will limit plaque progression, inflammation, and destabilization and thus, is regarded to be atheroprotective at this stage of atherosclerosis in *ApoE*$^{-/-}$ mice. However, as previously reported by our lab, accumulation of apoptotic bodies can induce secondary necrosis, eventually resulting in expansion of the necrotic core and plaque size during later stages of atherosclerosis [58]. Accordingly, large plaques with an inflammatory phenotype were observed in the aortic root from both *ApoE*$^{-/-}$ *Gsdmd*$^{-/-}$ and *ApoE*$^{-/-}$ *Gsdmd*$^{+/+}$ mice, possibly because after 16 weeks WD plaques in the aortic root are in a more advanced stage compared to plaques in the brachiocephalic artery. Thus, the effects in the longer term of targeting GSDMD and the concomitant switch in cell death modality remain to be elucidated.

5. Conclusions

We report that cleaved GSDMD is present in human atherosclerotic plaques and is required for inflammatory pyroptosis in murine macrophages and smooth muscle cells in vitro. GSDMD deficiency in *ApoE*$^{-/-}$ mice does not inhibit plaque initiation and the formation of stable aortic plaques, but plays a role in the formation of inflammatory plaques in the brachiocephalic artery. Indeed, a shift toward a less inflammatory plaque composition and delayed plaque progression were observed in plaques of the brachiocephalic artery from *ApoE*$^{-/-}$ *Gsdmd*$^{-/-}$ mice as compared to plaques with a more vulnerable phenotype observed in *ApoE*$^{-/-}$ *Gsdmd*$^{+/+}$ controls. This is accompanied by less plaque necrosis and a switch to apoptosis when GSDMD is deficient, which is also observed in vitro in BMDMs. Therefore, targeting GSDMD appears to be a promising approach for limiting the transition to an inflammatory and vulnerable plaque phenotype, and subsequently plaque destabilization. However, the pharmacological translation and therapeutic value in atherosclerosis patients should be further investigated.

Supplementary Materials: The following supporting information can be downloaded at: https://www.mdpi.com/article/10.3390/biomedicines10051171/s1, Figure S1: Validation of GSDMD deficiency in tissues from *ApoE*$^{-/-}$ *Gsdmd*$^{-/-}$ mice; Figure S2: Analysis of leukocyte subsets in plasma from *ApoE*$^{-/-}$ *Gsdmd*$^{-/-}$ and *ApoE*$^{-/-}$ *Gsdmd*$^{+/+}$ mice. Figure S3: Plaque analysis in the aortic root from *ApoE*$^{-/-}$ *Gsdmd*$^{-/-}$ and *ApoE*$^{-/-}$ *Gsdmd*$^{+/+}$ mice.

Author Contributions: Conceptualization, P.P., W.M. and G.R.Y.D.M.; methodology, P.P., M.V.P., C.H.G.N., W.M. and G.R.Y.D.M.; validation, P.P., M.V.P. and F.V.; formal analysis, P.P.; investigation, P.P., M.V.P., F.V.; resources, W.M., G.R.Y.D.M., L.R. and P.-J.G.; data curation, P.P.; writing—original draft preparation, P.P.; writing—review and editing, P.P., W.M., G.R.Y.D.M., M.V.P., C.H.G.N., L.R. and P.-J.G.; visualization, P.P.; supervision, W.M. and G.R.Y.D.M.; project administration, P.P. and W.M.; funding acquisition, W.M., L.R., P.-J.G. and G.R.Y.D.M. All authors have read and agreed to the published version of the manuscript.

Funding: This work was supported by the Fund for Scientific Research (FWO)-Flanders (EOS 30826052) and the Hercules Foundation (grant number AUHA/13/03). Pauline Puylaert and Cédric H.G. Neutel are fellows of the FWO-Flanders.

Institutional Review Board Statement: The animal study protocol was approved by the Ethics Committee of the University of Antwerp (Code: 2019-24, date of approval: 11 December 2019).

Informed Consent Statement: Patient samples were collected anonymously and untraceably approximately 25 years ago.

Data Availability Statement: The data that support the findings of this study are available from the corresponding author upon reasonable request.

Acknowledgments: The authors are grateful to Mandy Vermont and Hermine Fret for excellent technical support. Bronwen Martin is acknowledged for critical reading of the manuscript. Gsdmd$^{-/-}$ mice were a kind gift from Genentech and delivered thanks to the help of Andy Wullaert (University of Ghent).

Conflicts of Interest: The authors declare no conflict of interest.

References

1. Virmani, R.; Burke, A.P.; Farb, A.; Kolodgie, F.D. Pathology of the vulnerable plaque. *J. Am. Coll. Cardiol.* **2006**, *47*, C13–C18. [CrossRef] [PubMed]
2. Galluzzi, L.; Vitale, I.; Aaronson, S.A.; Abrams, J.M.; Adam, D.; Agostinis, P.; Alnemri, E.S.; Altucci, L.; Amelio, I.; Andrews, D.W.; et al. Molecular mechanisms of cell death: Recommendations of the Nomenclature Committee on Cell Death 2018. *Cell Death Differ.* **2018**, *25*, 486–541. [CrossRef] [PubMed]
3. Liu, X.; Zhang, Z.; Ruan, J.; Pan, Y.; Magupalli, V.G.; Wu, H.; Lieberman, J. Inflammasome-activated gasdermin D causes pyroptosis by forming membrane pores. *Nature* **2016**, *535*, 153–158. [CrossRef] [PubMed]
4. Wu, D.; Chen, Y.; Sun, Y.; Gao, Q.; Yu, B.; Jiang, X.; Guo, M. Gasdermin family: A promising therapeutic target for cancers and inflammation-driven diseases. *J. Cell Commun. Signal.* **2020**, *14*, 293–301. [CrossRef] [PubMed]
5. Shi, J.; Zhao, Y.; Wang, K.; Shi, X.; Wang, Y.; Huang, H.; Zhuang, Y.; Cai, T.; Wang, F.; Shao, F. Cleavage of GSDMD by inflammatory caspases determines pyroptotic cell death. *Nature* **2015**, *526*, 660–665. [CrossRef]
6. Shi, J.; Gao, W.; Shao, F. Pyroptosis: Gasdermin-Mediated Programmed Necrotic Cell Death. *Trends Biochem. Sci.* **2017**, *42*, 245–254. [CrossRef]
7. He, W.-T.; Wan, H.; Hu, L.; Chen, P.; Wang, X.; Huang, Z.; Yang, Z.-H.; Zhong, C.-Q.; Han, J. Gasdermin D is an executor of pyroptosis and required for interleukin-1β secretion. *Cell Res.* **2015**, *25*, 1285–1298. [CrossRef]
8. Latz, E.; Xiao, T.S.; Stutz, A. Activation and regulation of the inflammasomes. *Nat. Rev. Immunol.* **2013**, *13*, 397–411. [CrossRef]
9. Rajamäki, K.; Lappalainen, J.; Oörni, K.; Välimäki, E.; Matikainen, S.; Kovanen, P.T.; Eklund, K.K. Cholesterol crystals activate the NLRP3 inflammasome in human macrophages: A novel link between cholesterol metabolism and inflammation. *PLoS ONE* **2010**, *5*, e11765. [CrossRef]
10. Chevriaux, A.; Pilot, T.; Derangère, V.; Simonin, H.; Martine, P.; Chalmin, F.; Ghiringhelli, F.; Rébé, C. Cathepsin B Is Required for NLRP3 Inflammasome Activation in Macrophages, Through NLRP3 Interaction. *Front. Cell Dev. Biol.* **2020**, *8*, 167. [CrossRef]
11. Karasawa, T.; Takahashi, M. The crystal-induced activation of NLRP3 inflammasomes in atherosclerosis. *Inflamm. Regen.* **2017**, *37*, 18. [CrossRef] [PubMed]

12. Lin, J.; Shou, X.; Mao, X.; Dong, J.; Mohabeer, N.; Kushwaha, K.K.; Wang, L.; Su, Y.; Fang, H.; Li, D. Oxidized low density lipoprotein induced caspase-1 mediated pyroptotic cell death in macrophages: Implication in lesion instability? *PLoS ONE* **2013**, *8*, e62148. [CrossRef]
13. Paramel Varghese, G.; Folkersen, L.; Strawbridge, R.J.; Halvorsen, B.; Yndestad, A.; Ranheim, T.; Krohg-Sorensen, K.; Skjelland, M.; Espevik, T.; Aukrust, P.; et al. NLRP3 Inflammasome Expression and Activation in Human Atherosclerosis. *J. Am. Heart Assoc.* **2016**, *5*, e003031. [CrossRef] [PubMed]
14. Mallat, Z.; Corbaz, A.; Scoazec, A.; Besnard, S.; Lesèche, G.; Chvatchko, Y.; Tedgui, A. Expression of interleukin-18 in human atherosclerotic plaques and relation to plaque instability. *Circulation* **2001**, *104*, 1598–1603. [CrossRef] [PubMed]
15. Rajamaki, K.; Mayranpaa, M.I.; Risco, A.; Tuimala, J.; Nurmi, K.; Cuenda, A.; Eklund, K.K.; Oorni, K.; Kovanen, P.T. p38delta MAPK: A Novel Regulator of NLRP3 Inflammasome Activation With Increased Expression in Coronary Atherogenesis. *Arterioscler. Thromb. Vasc. Biol.* **2016**, *36*, 1937–1946. [CrossRef] [PubMed]
16. Shi, X.; Xie, W.L.; Kong, W.W.; Chen, D.; Qu, P. Expression of the NLRP3 Inflammasome in Carotid Atherosclerosis. *J. Stroke Cerebrovasc. Dis.* **2015**, *24*, 2455–2466. [CrossRef]
17. Kolodgie, F.D.; Narula, J.; Burke, A.P.; Haider, N.; Farb, A.; Hui-Liang, Y.; Smialek, J.; Virmani, R. Localization of Apoptotic Macrophages at the Site of Plaque Rupture in Sudden Coronary Death. *Am. J. Pathol.* **2000**, *157*, 1259–1268. [CrossRef]
18. Duewell, P.; Kono, H.; Rayner, K.J.; Sirois, C.M.; Vladimer, G.; Bauernfeind, F.G.; Abela, G.S.; Franchi, L.; Nunez, G.; Schnurr, M.; et al. NLRP3 inflammasomes are required for atherogenesis and activated by cholesterol crystals. *Nature* **2010**, *464*, 1357–1361. [CrossRef]
19. Elhage, R.; Jawien, J.; Rudling, M.; Ljunggren, H.G.; Takeda, K.; Akira, S.; Bayard, F.; Hansson, G.K. Reduced atherosclerosis in interleukin-18 deficient apolipoprotein E-knockout mice. *Cardiovasc. Res.* **2003**, *59*, 234–240. [CrossRef]
20. Gage, J.; Hasu, M.; Thabet, M.; Whitman, S.C. Caspase-1 deficiency decreases atherosclerosis in apolipoprotein E-null mice. *Can. J. Cardiol.* **2012**, *28*, 222–229. [CrossRef]
21. Kirii, H.; Niwa, T.; Yamada, Y.; Wada, H.; Saito, K.; Iwakura, Y.; Asano, M.; Moriwaki, H.; Seishima, M. Lack of interleukin-1beta decreases the severity of atherosclerosis in ApoE-deficient mice. *Arterioscler. Thromb. Vasc. Biol.* **2003**, *23*, 656–660. [CrossRef] [PubMed]
22. Li, Y.; Niu, X.; Xu, H.; Li, Q.; Meng, L.; He, M.; Zhang, J.; Zhang, Z.; Zhang, Z. VX-765 attenuates atherosclerosis in ApoE deficient mice by modulating VSMCs pyroptosis. *Exp. Cell Res.* **2020**, *389*, 111847. [CrossRef] [PubMed]
23. Van der Heijden, T.; Kritikou, E.; Venema, W.; van Duijn, J.; van Santbrink, P.J.; Slutter, B.; Foks, A.C.; Bot, I.; Kuiper, J. NLRP3 Inflammasome Inhibition by MCC950 Reduces Atherosclerotic Lesion Development in Apolipoprotein E-Deficient Mice-Brief Report. *Arterioscler. Thromb. Vasc. Biol.* **2017**, *37*, 1457–1461. [CrossRef] [PubMed]
24. Zeng, W.; Wu, D.; Sun, Y.; Suo, Y.; Yu, Q.; Zeng, M.; Gao, Q.; Yu, B.; Jiang, X.; Wang, Y. The selective NLRP3 inhibitor MCC950 hinders atherosclerosis development by attenuating inflammation and pyroptosis in macrophages. *Sci. Rep.* **2021**, *11*, 19305. [CrossRef]
25. Zheng, F.; Xing, S.; Gong, Z.; Mu, W.; Xing, Q. Silence of NLRP3 suppresses atherosclerosis and stabilizes plaques in apolipoprotein E-deficient mice. *Mediat. Inflamm.* **2014**, *2014*, 507208. [CrossRef]
26. Wang, Y.; Ji, N.; Gong, X.; Ni, S.; Xu, L.; Zhang, H. Thioredoxin-1 attenuates atherosclerosis development through inhibiting NLRP3 inflammasome. *Endocrine* **2020**, *70*, 65–70. [CrossRef]
27. Martinet, W.; Coornaert, I.; Puylaert, P.; De Meyer, G.R.Y. Macrophage Death as a Pharmacological Target in Atherosclerosis. *Front. Pharmacol.* **2019**, *10*, 306. [CrossRef]
28. Ridker, P.M.; Everett, B.M.; Thuren, T.; MacFadyen, J.G.; Chang, W.H.; Ballantyne, C.; Fonseca, F.; Nicolau, J.; Koenig, W.; Anker, S.D.; et al. Antiinflammatory Therapy with Canakinumab for Atherosclerotic Disease. *N. Engl. J. Med.* **2017**, *377*, 1119–1131. [CrossRef]
29. Huang, X.; Feng, Y.; Xiong, G.; Whyte, S.; Duan, J.; Yang, Y.; Wang, K.; Yang, S.; Geng, Y.; Ou, Y.; et al. Caspase-11, a specific sensor for intracellular lipopolysaccharide recognition, mediates the non-canonical inflammatory pathway of pyroptosis. *Cell Biosci.* **2019**, *9*, 31. [CrossRef]
30. Jiang, M.; Sun, X.; Liu, S.; Tang, Y.; Shi, Y.; Bai, Y.; Wang, Y.; Yang, Q.; Yang, Q.; Jiang, W.; et al. Caspase-11-Gasdermin D-Mediated Pyroptosis Is Involved in the Pathogenesis of Atherosclerosis. *Front. Pharmacol.* **2021**, *12*, 657486. [CrossRef]
31. Evavold, C.L.; Ruan, J.; Tan, Y.; Xia, S.; Wu, H.; Kagan, J.C. The Pore-Forming Protein Gasdermin D Regulates Interleukin-1 Secretion from Living Macrophages. *Immunity* **2018**, *48*, 35–44.e36. [CrossRef] [PubMed]
32. Jiang, K.; Tu, Z.; Chen, K.; Xu, Y.; Chen, F.; Xu, S.; Shi, T.; Qian, J.; Shen, L.; Hwa, J.; et al. Gasdermin D inhibition confers antineutrophil-mediated cardioprotection in acute myocardial infarction. *J. Clin. Investig.* **2022**, *132*, e151268. [CrossRef] [PubMed]
33. Kockx, M.M.; De Meyer, G.R.; Muhring, J.; Jacob, W.; Bult, H.; Herman, A.G. Apoptosis and related proteins in different stages of human atherosclerotic plaques. *Circulation* **1998**, *97*, 2307–2315. [CrossRef] [PubMed]
34. Kurdi, A.; Roth, L.; Van der Veken, B.; Van Dam, D.; De Deyn, P.P.; De Doncker, M.; Neels, H.; De Meyer, G.R.Y.; Martinet, W. Everolimus depletes plaque macrophages, abolishes intraplaque neovascularization and improves survival in mice with advanced atherosclerosis. *Vasc. Pharmacol.* **2019**, *113*, 70–76. [CrossRef] [PubMed]
35. Owens, G.K.; Loeb, A.; Gordon, D.; Thompson, M.M. Expression of smooth muscle-specific alpha-isoactin in cultured vascular smooth muscle cells: Relationship between growth and cytodifferentiation. *J. Cell Biol.* **1986**, *102*, 343–352. [CrossRef]

36. Geisterfer, A.A.; Peach, M.J.; Owens, G.K. Angiotensin II induces hypertrophy, not hyperplasia, of cultured rat aortic smooth muscle cells. *Circ. Res.* **1988**, *62*, 749–756. [CrossRef]
37. Zychlinsky, A.; Prevost, M.C.; Sansonetti, P.J. Shigella flexneri induces apoptosis in infected macrophages. *Nature* **1992**, *358*, 167–169. [CrossRef]
38. Yu, P.; Zhang, X.; Liu, N.; Tang, L.; Peng, C.; Chen, X. Pyroptosis: Mechanisms and diseases. *Signal Transduct. Target. Ther.* **2021**, *6*, 128. [CrossRef]
39. Syed, F.M.; Hahn, H.S.; Odley, A.; Guo, Y.; Vallejo, J.G.; Lynch, R.A.; Mann, D.L.; Bolli, R.; Dorn, G.W., 2nd. Proapoptotic effects of caspase-1/interleukin-converting enzyme dominate in myocardial ischemia. *Circ. Res.* **2005**, *96*, 1103–1109. [CrossRef]
40. Tsuchiya, K.; Nakajima, S.; Hosojima, S.; Nguyen, D.T.; Hattori, T.; Le, T.M.; Hori, O.; Mahib, M.R.; Yamaguchi, Y.; Miura, M.; et al. Caspase-1 initiates apoptosis in the absence of gasdermin D. *Nat. Commun.* **2019**, *10*, 2091. [CrossRef]
41. Hakimi, M.; Peters, A.; Becker, A.; Böckler, D.; Dihlmann, S. Inflammation-related induction of absent in melanoma 2 (AIM2) in vascular cells and atherosclerotic lesions suggests a role in vascular pathogenesis. *J. Vasc. Surg.* **2014**, *59*, 794–803. [CrossRef] [PubMed]
42. Wu, X.; Zhang, H.; Qi, W.; Zhang, Y.; Li, J.; Li, Z.; Lin, Y.; Bai, X.; Liu, X.; Chen, X.; et al. Nicotine promotes atherosclerosis via ROS-NLRP3-mediated endothelial cell pyroptosis. *Cell Death Dis.* **2018**, *9*, 171. [CrossRef]
43. Xu, Y.J.; Zheng, L.; Hu, Y.W.; Wang, Q. Pyroptosis and its relationship to atherosclerosis. *Clin. Chim. Acta* **2018**, *476*, 28–37. [CrossRef] [PubMed]
44. Zhaolin, Z.; Jiaojiao, C.; Peng, W.; Yami, L.; Tingting, Z.; Jun, T.; Shiyuan, W.; Jinyan, X.; Dangheng, W.; Zhisheng, J.; et al. OxLDL induces vascular endothelial cell pyroptosis through miR-125a-5p/TET2 pathway. *J. Cell Physiol.* **2019**, *234*, 7475–7491. [CrossRef] [PubMed]
45. Opoku, E.; Traughber, C.A.; Zhang, D.; Iacano, A.J.; Khan, M.; Han, J.; Smith, J.D.; Gulshan, K. Gasdermin D mediates inflammation-induced defects in reverse cholesterol transport and promotes atherosclerosis. *Front. Cell Dev. Biol.* **2021**, *9*, 715211. [CrossRef]
46. Lopez-Pastrana, J.; Ferrer, L.M.; Li, Y.F.; Xiong, X.; Xi, H.; Cueto, R.; Nelson, J.; Sha, X.; Li, X.; Cannella, A.L.; et al. Inhibition of Caspase-1 Activation in Endothelial Cells Improves Angiogenesis: A Novel Therapeutic Potential For Ischemia. *J. Biol. Chem.* **2015**, *290*, 17485–17494. [CrossRef]
47. Bai, B.; Yang, Y.; Ji, S.; Wang, S.; Peng, X.; Tian, C.; Sun, R.C.; Yu, T.; Chu, X.M. MicroRNA-302c-3p inhibits endothelial cell pyroptosis via directly targeting NOD-, LRR- and pyrin domain-containing protein 3 in atherosclerosis. *J. Cell Mol. Med.* **2021**, *25*, 4373–4386. [CrossRef]
48. Xing, S.S.; Yang, J.; Li, W.J.; Li, J.; Chen, L.; Yang, Y.T.; Lei, X.; Li, J.; Wang, K.; Liu, X. Salidroside Decreases Atherosclerosis Plaque Formation via Inhibiting Endothelial Cell Pyroptosis. *Inflammation* **2020**, *43*, 433–440. [CrossRef]
49. Hu, J.J.; Liu, X.; Xia, S.; Zhang, Z.; Zhang, Y.; Zhao, J.; Ruan, J.; Luo, X.; Lou, X.; Bai, Y.; et al. FDA-approved disulfiram inhibits pyroptosis by blocking gasdermin D pore formation. *Nat. Immunol.* **2020**, *21*, 736–745. [CrossRef]
50. Humphries, F.; Shmuel-Galia, L.; Ketelut-Carneiro, N.; Li, S.; Wang, B.; Nemmara, V.V.; Wilson, R.; Jiang, Z.; Khalighinejad, F.; Muneeruddin, K.; et al. Succination inactivates gasdermin D and blocks pyroptosis. *Science* **2020**, *369*, 1633–1637. [CrossRef]
51. Nakashima, Y.; Plump, A.S.; Raines, E.W.; Breslow, J.L.; Ross, R. ApoE-deficient mice develop lesions of all phases of atherosclerosis throughout the arterial tree. *Arterioscler. Thromb.* **1994**, *14*, 133–140. [CrossRef] [PubMed]
52. Williams, H.; Johnson, J.L.; Carson, K.G.; Jackson, C.L. Characteristics of intact and ruptured atherosclerotic plaques in brachiocephalic arteries of apolipoprotein E knockout mice. *Arterioscler. Thromb. Vasc. Biol.* **2002**, *22*, 788–792. [CrossRef] [PubMed]
53. Vengrenyuk, Y.; Kaplan, T.J.; Cardoso, L.; Randolph, G.J.; Weinbaum, S. Computational stress analysis of atherosclerotic plaques in ApoE knockout mice. *Ann. Biomed. Eng.* **2010**, *38*, 738–747. [CrossRef] [PubMed]
54. Rosenfeld, M.E.; Polinsky, P.; Virmani, R.; Kauser, K.; Rubanyi, G.; Schwartz, S.M. Advanced atherosclerotic lesions in the innominate artery of the ApoE knockout mouse. *Arterioscler. Thromb. Vasc. Biol.* **2000**, *20*, 2587–2592. [CrossRef]
55. Meir, K.S.; Leitersdorf, E. Atherosclerosis in the apolipoprotein-E-deficient mouse: A decade of progress. *Arterioscler. Thromb. Vasc. Biol.* **2004**, *24*, 1006–1014. [CrossRef]
56. Charriaut-Marlangue, C.; Ben-Ari, Y. A cautionary note on the use of the TUNEL stain to determine apoptosis. *Neuroreport* **1995**, *7*, 61–64. [CrossRef]
57. Grasl-Kraupp, B.; Ruttkay-Nedecky, B.; Koudelka, H.; Bukowska, K.; Bursch, W.; Schulte-Hermann, R. In situ detection of fragmented DNA (TUNEL assay) fails to discriminate among apoptosis, necrosis, and autolytic cell death: A cautionary note. *Hepatology* **1995**, *21*, 1465–1468. [CrossRef]
58. Coornaert, I.; Puylaert, P.; Marcasolli, G.; Grootaert, M.O.J.; Vandenabeele, P.; De Meyer, G.R.Y.; Martinet, W. Impact of myeloid RIPK1 gene deletion on atherogenesis in ApoE-deficient mice. *Atherosclerosis* **2021**, *322*, 51–60. [CrossRef]

biomedicines

MDPI

Systematic Review

Worsening Thrombotic Complication of Atherosclerotic Plaques Due to Neutrophils Extracellular Traps: A Systematic Review

Francesco Nappi [1,*], Francesca Bellomo [2] and Sanjeet Singh Avtaar Singh [3]

1 Department of Cardiac Surgery, Centre Cardiologique du Nord of Saint-Denis, 93200 Saint-Denis, France
2 Department of Clinical and Experimental Medicine, University of Messina, 98122 Messina, Italy
3 Department of Cardiothoracic Surgery, Royal Infirmary of Edinburgh, Edinburgh EH16 4SA, UK
* Correspondence: francesconappi2@gmail.com; Tel.: +33-(14)-9334104; Fax: +33-149334119

Abstract: Neutrophil extracellular traps (NETs) recently emerged as a newly recognized contributor to venous and arterial thrombosis. These strands of DNA, extruded by activated or dying neutrophils, decorated with various protein mediators, become solid-state reactors that can localize at the critical interface of blood with the intimal surface of diseased arteries alongside propagating and amplifying the regional injury. NETs thus furnish a previously unsuspected link between inflammation, innate immunity, thrombosis, oxidative stress, and cardiovascular diseases. In response to disease-relevant stimuli, neutrophils undergo a specialized series of reactions that culminate in NET formation. DNA derived from either nuclei or mitochondria can contribute to NET formation. The DNA liberated from neutrophils forms a reticular mesh that resembles morphologically a net, rendering the acronym NETs particularly appropriate. The DNA backbone of NETs not only presents intrinsic neutrophil proteins (e.g., MPO (myeloperoxidase) and various proteinases) but can congregate other proteins found in blood (e.g., tissue factor procoagulant). This systematic review discusses the current hypothesis of neutrophil biology, focusing on the triggers and mechanisms of NET formation. Furthermore, the contribution of NETs to atherosclerosis and thrombosis is extensively addressed. Again, the use of NET markers in clinical trials was considered. Ultimately, given the vast body of the published literature, we aim to integrate the experimental evidence with the growing body of clinical information relating to NET critically.

Keywords: neutrophil extracellular trap; atherosclerosis; atherosclerotic plaques; thrombosis

Citation: Nappi, F.; Bellomo, F.; Avtaar Singh, S.S. Worsening Thrombotic Complication of Atherosclerotic Plaques Due to Neutrophils Extracellular Traps: A Systematic Review. *Biomedicines* **2023**, *11*, 113. https://doi.org/10.3390/biomedicines11010113

Academic Editors: Tânia Martins-Marques, Gonçalo F. Coutinho and Attila Kiss

Received: 4 December 2022
Revised: 19 December 2022
Accepted: 22 December 2022
Published: 2 January 2023

Copyright: © 2023 by the authors. Licensee MDPI, Basel, Switzerland. This article is an open access article distributed under the terms and conditions of the Creative Commons Attribution (CC BY) license (https://creativecommons.org/licenses/by/4.0/).

1. Introduction

Fucks et al. [1] further emphasized the role of neutrophils as the master cells of the innate immune system, anticipating one of the mechanisms of action of neutrophils which is the generation of neutrophil extracellular traps (NETs). Since the publication of this landmark study, research consideration for the neutrophil extracellular trap (NET) has steadily increased. Recently, it affirmed a preponderant presence in the panorama of cardiovascular biology, recognizing a crucial role of the NET in the pathogenetic mechanisms that support venous and arterial thrombosis [1–6]. The NET is composed of DNA strands squeezed out from activated or dying neutrophils, decorated with various protein mediators such as neutrophil elastase or azurocidine. The latter belongs to the serprocidin family and has the function of promoting the adhesion and transmigration of monocytes, as well as influencing the proinflammatory inclination that macrophages have among their functions. These inherent features of the NET offer further insight into a formerly unlooked-for connection between inflammation, innate immunity, thrombosis, and cardiovascular disease described by Lim et al. [5] Neutrophils have the ability to respond to an expanding range of stimuli by initiating a specialized series of reactions that culminate in the formation of the NET. DNA derived from nuclei or mitochondria can contribute to the generation of the NET. The fundamental step in the formation of the NET is linked to the release from the ionic bond constraints that hold DNA together with histones. In this way, the neutrophil spreads out the linear deoxyribonucleic acid (DNA) in the extracellular space thus constituting a reticular mesh that morphologically looks like a network which confers the acronym NET extremely suitable [7–9]. It was observed that the existence of mitochondrial oxidative stress can be considered as an important triggering factor specifically in human aging and more particularly in cardiovascular diseases. It was also suggested that a causal link exists between the mitochondrial oxidative stress of endogenous neutrophils and the subsequent NETosis during aging with a tendency towards a greater formation of atherosclerotic lesions [1,6–8].

Although we have learned that DNA once assembled to the proteins present within the neutrophil itself constitutes the backbone of the NET, however, this DNA can collect other circulating proteins in the blood. NET-associated neutrophil-specific proteins include MPO (myeloperoxidase), the serine proteinases of neutrophils with specific functions such as cathepsin G, neutrophil elastase, and proteinase and again the proinflammatory IL (interleukin)-1a molecule. Regarding the circulating proteins of non-neutrophilic derivation aggregated to the NET, the procoagulant tissue factor is of recognized importance because it plays a crucial role in thrombogenesis [1–9].

Several studies [10–17] described the presence of neutrophil-generated NETs in atherosclerotic lesions in mice and humans [17]. In atherosclerotic plaques studied in humans, neutrophils and NETs were isolated near atherosclerotic lesion segments rich in apoptotic smooth muscle and endothelial cells (SMCs). This typical localization implies that NETs contribute not only to plaque rupture but are differentially distributed depending on the classification of the plaque in its degree of complication [12]. Human coronary plaques recovered from subjects with CAD including 44 complicated plaques, characterized by intraplaque hemorrhages, erosions, and ruptures, as well as 20 intact plaques, were evaluated by immunohistochemistry. The presence of neutrophils with myeloperoxidase, neutrophil elastase and CD177 and of NETs with citrullinated histone-3 and PAD4 were closely evaluated highlighting some specific differences between the complicated lesions compared to the intact ones. Concerning the former, neutrophils and NETs were recorded in all recovered samples, without significant differences in their extension or in the presence of ruptures, erosions, and intraplate hemorrhages. In contrast, intact plaques did not reveal the presence of NETs. Furthermore, evidence of the fundamental importance for the evolution of atheromatous plaque in patients with CAD involved the adjacent perivascular tissue. In complicated plaques, it also contained a high number of neutrophils and NETs which was not detected in intact plaques [16].

A vast body of literature is now available to corroborate that the presence of NETs in human lesions supports their substantial involvement in promoting the process of atherothrombosis. It is true that while the strong experimental evidence has been well established, evidence for the causality of NETs in human disease remains elusive. Furthermore, it remains to be clarified whether their formation precedes or follows the initial rupture of the atherothrombotic plaque. Here, we present a systematic review that offers an update on this field that is experiencing rapid and extensive scientific speculation today.

We discuss the fundamental principles of neutrophil biology, the triggers, and the mechanisms of NET formation. Furthermore, we focus our analysis on the contribution that NET provides to atherosclerosis and thrombosis. Recent evidence produced in the literature focused attention on the use of NETs' markers in clinical trials which were considered for the completeness of the review. Our effort was aimed at providing a synthesis between the evidence suggested in the experimental literature and the results emerging from the large body of clinical information relating to NET. We believe the evidence dispensed here could supply a basis for further discernment on the knowledge of the NET and could help the physician–patient discussion of the risks, benefits, and expectations about the role of the NET.

2. Material and Methods

The authors rule that all supporting data are ready for use within the article and its online supplementary files. This systematic review complies with the Preferred Reporting Items for Systematic Reviews statement.

2.1. Data Sources and Systematic Literature Review

Searches were run on 1 September 2022, to date examining Ovid's version of MEDLINE and EMBASE in order to ensure the update of the manuscripts of interest. Inclusion criteria were English language publications, research articles, adjusted or matched observational studies or RCTs evaluating neutrophil function (15,428 to date), NETs (1628 to date), NETs' function (746 to date), and NETosis (132 to date) and coupled with the neutrophil NET (306 to date), neutrophil NETosis (117 to date), neutrophil atherosclerosis (211 to date), neutrophil extracellular traps' atherosclerosis (19 to date). In addition, we searched recent meta-analyses and reviews of this topic for potential additional studies. A total of 18,587 studies were retrieved, and after deduplication, 3 reviewers (S.S.A.S., F.B., and F.N.) independently screened a total of 6240 citations. All citations were reviewed by two investigators independently (S.S.A.S., F.B.), and any dissents were resolved by a third author (F.N). In case of overlapping studies, the greater series were included. A total of 465 citations were evaluated of which 40 studies met inclusion criteria and were included in the final systematic review as reported in Figure 1 including the PRISMA flow diagram. In the supplementary material, the PRISMA 2020 checklist is illustrated. The systematic review was registered in the OSF platform https://osf.io/hs2yj/, accessed on 12 December 2022.

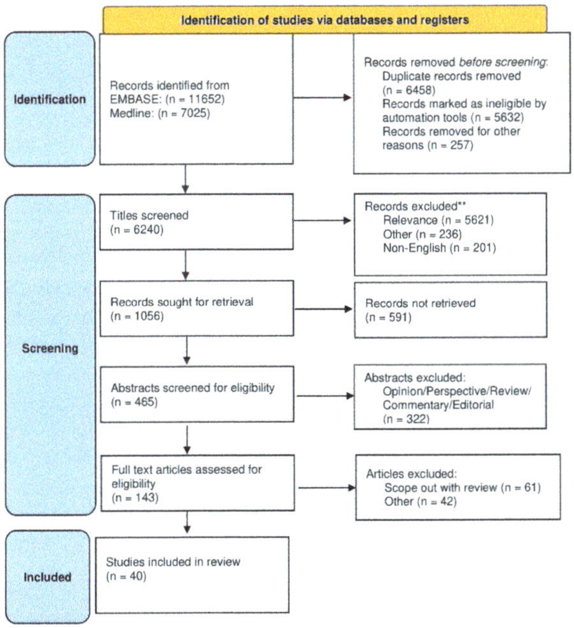

Figure 1. PRISMA 2021 Flow diagram for new systematic review which includes searches of database and registers only. Consider, if feasible to do so, reporting the number of records identified from each database or register searched (rather than the total number across all databases/registers). ** If automation tools were used, indicate how many records were excluded by a human and how many were excluded by automation tools. From: McKenzie et al. [18] (Supplementary Table S1).

2.2. Data Extraction, Quality Assessment, and Aims

Data extraction was performed independently by 2 investigators (S.S.A.S., F.B.). The following variables were included: study demographics (sample size, number of centers, publication year, study period), type of study (animal, human), aims, and major findings of the data. Only animal models, RCTs, and observational studies of high quality were included in the final analysis. The primary objective of the systematic review was, in light of recent discoveries, the evaluation of the mechanisms that determine an alteration of inflammatory responses with the release of NETs in arterial thrombosis. The secondary objective was the evaluation of the potential clinical cardiovascular implication of biomarkers of NETs and the therapeutic implication of NETs in cardiovascular conditions. The selected studies are reported in Tables 1–3.

3. Results

3.1. Recent Advances in the Biology of Neutrophils

Neutrophils are terminally differentiated cell lines with short-lived phagocyte function but with an action that is expressed in a quite unrestrained manner. To perform their function adequately, the circulating number must amount to a production of 1011 neutrophils per day in the adult human body. In medullary hematopoietic tissues, neutrophils develop and mature from progenitor cells (HSPC), and after a cascade of proliferation and differentiation phases, they are stored as a rapidly mobilizable pool. The production of neutrophils adapts to the conditions of stress when the request for their intervention supports accelerated cell production to satisfy further requests. Substantial evidence shows that hypercholesterolemia and hyperglycemia, which are considered two main actors for the stratification of cardiovascular disease risk, determine an alteration of inflammatory responses through a reprogramming of the HSPC function. The result lies in the enhancement of the subsequent myelopoiesis including the triggering of a number of granulocytes necessary to cope with the new role. Again, the proliferation towards the differentiation and functional maturation of neutrophils is evoked by the accumulation of cholesterol in the cell membrane of HSPCs. The latter mobilize even if misrepresenting their maturation towards a myeloid bias [2]. Likewise, Negardy et al. [3] reported an altered process of granulopoiesis sustained by hyperglycemia and mediated by the release of S100 calcium-binding protein A9 (S100A8/A9) from circulating neutrophils. It is important to underline that the metabolic risk factors intervene on neutrophils on two levels, both favoring an accelerated production and inducing a state of greater reactivity of the cells. Indeed, Wong et al. [4] observed that hyperglycemia is a strong conditioning factor for the production of NETs in humans and mice, through a process that requires the generation of reactive oxygen species (ROS). Therefore, the overproduction of ROS in patients with diabetes mellitus is closely related to the release of extracellular DNA from neutrophils. Similarly, the evidence reported by Tall et al. [6] suggested a substantial role of NETS in accelerating the formation of atherosclerotic lesions. This process is aided by the disgregated efflux of cholesterol in neutrophils that can also increase the activation of the inflammasome and encourage arterial infiltration of neutrophils and NETs' release [8]. Cellular aging of HSPCs leads to somatic mutations in subjects even in the absence of overt hematological neoplasms. Substantial evidence was reported which demonstrated how HSPCs with inherent specific somatic mutations can spread out to form clones in peripheral blood according to the phenomenon known as clonal hematopoiesis. Clonal hematopoiesis primarily affects cells of the myeloid lineage, and this phenomenon is notably prone to lead to an increased risk of adverse cardiovascular outcomes [9].

A look at specific mutations that intervene in clonal hematopoiesis is crucial for implication in the process of the infiltration of atheromatous plaque by neutrophils. As for the genes that are most frequently mutated, they include DNA (cytosine-5)-methyltransferase 3A (DNMT3a), Ten-eleven-translocation 2 (TET2), Putative Polycomb group protein (ASXL1), and JAK2. In particular, a specific somatic signaling gain-of-function mutation in JAK2 (Janus kinase 2, V617F) activates STAT (signal transducer and activator) and deserves to be underlined. Studies on hyperlipidemia in mice by Wang et al. [10] demonstrated that *Jak2*^{V617F} in myeloid cells induces marked erythrophagocytosis associated with neutrophil infiltration. The events that emerge lead to a speeding up of the atherogenesis process associated with an increase in the characteristics of the propensity to rupture. Furthermore, two independent studies revealed that the JakV617F mutation is also coupled to spontaneous strengthening in NET release [10,11] as well as thrombus generation [11].

Once the acute inflammatory process has been established, the neutrophils are suddenly recalled through a well-defined recruitment cascade divided into several steps. There is no single pattern in which this process is articulated with peculiar differences that occur between tissues and with marked structural disorders that occur in the great arteries [12,13]. Several studies clarified the mechanisms that regulate the inflammatory process of atherosclerotic plaque. Drechsler et al. [14] highlighted the role of arterial neutrophil adhesion favored by platelet-producing chemokines by suggesting that this mechanism plays a substantially minor role in microcirculation. In particular, the role offered by platelet-derived Chemokine (C-C motif) ligand 5 (CCL5) and chemokine (C-X-C motif) ligand 4 (CXCL4) is crucial in activating neutrophils, especially as regards the release of the NET, thus providing a plausible explanation of the finding of the NET on the luminal side of the developing atherosclerotic lesion [15–17,19,20]. A relevant discovery was reported by relating the behavior of neutrophils to circadian rhythms in mice that are surprisingly interconnected. Winter et al. [21] suggested that not only an oscillation in the neutrophil count is detectable in peripheral blood during the course of the day but that the diurnal variation in the adhesion of neutrophils to the large arteries shows a 12 h phase shift compared to the same phenomenon in the microcirculation. Furthermore, recent evidence proved the existence of an intrinsic clock within neutrophils that regulates the activity of these cells including the ability to release the NET [19,22,23]. Of note, it is observed that neutrophils can invade non-inflamed tissues such as the liver and intestines through the microvasculature with a mechanism similar to that carried out by non-classical patrolling monocytes [24–27]. The point that deserves to be clarified concerns the possibility that this phenomenon also occurs in the great arteries. The anti-inflammatory offensive that can be carried out by neutrophils includes a wide range of preconstituted "weapons" which are therefore distributed in the inflammatory site. This line of attack includes the production and release of ROS as well as bioactive lipid mediators. A second offensive line is warranted by neutrophil storage of preformed granular proteins including various alarmins (e.g., cathelicidins, defensins) or serine protease and MMP-8 (matrix metalloproteinase-8). Likewise, neutrophils store high levels of MPO in the granules which can generate the avidly oxidant and chlorinating species hypochlorous acid (HOCl) that is generated in situ from sodium hypochlorite [28]. In this way, the anti-inflammatory effect is synergistically guaranteed by NADPH (nicotinamide adenine dinucleotide phosphate), associated with the surface, whose role of oxidase favors the production of superoxide anion (O2−), while myeloperoxidase, once released from the granules, induces the production of ROS with a highly oxidizing and chlorinating function such as extracellular HOCl [26–28]. Again, the specific function of granular proteins can be used in the inflammatory site after further degranulation and the rapid release of anti-inflammatory substances that act outside the neutrophil cells. Microbiology studies taught us a recognized and crucial antimicrobial role played by neutrophil granules. Recent findings suggested that granular proteins have an effect as important stimulant molecules on immune cells [29]. Researchers today are very actively engaged in explaining the interaction between enzymes contained in granules stored by neutrophils and NETs. In fact, when this bond is exerted, it is capable of favoring the local activity of the NETs [30,31]. The most stimulating studies are aimed at understanding how the interaction of granular proteins with the chromatin NET alters their function [2,4,32–34].

Mechanisms for the Release of NETs and Composition of NETs

The mechanism that supports the genesis of NETs is dependent on both NADPH oxidase-dependent and NADPH-independent activity. We recognize a wide range of stimuli that induce a NETosis favored by oxidase-dependent NADPH. An extrinsic stimulus involves the role of bacteria [35–37], as opposed to intrinsic stimuli induced by hydrogen peroxide [38,39]. Again, other stimuli that induce NADPH oxidase-dependent NETosis involve the role of concanavalin A [40] or phorbol myristate acetate [36]. Studies by Amulic et al. [40] and Hakkim et al. [41] revealed that the stimuli described above trigger the activation of downstream signaling molecules such as extracellular signal-regulated kinases (ERKs), capable of activating NADPH oxidase. It was observed that the levels of O2, CO2, bicarbonate, and pH have a modulation function on NETosis. In this regard, in the presence of phorbol myristate acetate, normoxia is not required to generate NET from a Staphylococcus aureus infection [42]. Moreover, Dölling et al. found a wide infiltration of the hypoxic heart muscle after myocardial infarction. Most of these neutrophils had viable morphology, and only a few showed signs of nuclear decondensation, a hallmark of early NET formation [43]. The stimuli work to activate NADPH oxidase which transforms molecular oxygen into superoxide. Therefore, as it was suggested, the mechanisms supporting a pharmacological inhibition of NADPH oxidase or ROS scavengers lock the genesis of the NET [41]. The superoxide is broken down into hydrogen peroxide which provides the substrate for the MPO-catalyzed production of hypochlorous acid which is a ROS with a highly oxidizing and chlorinating function. It was reported that inhibitors of the function of MPO have a substantial role in stopping the process of NETosis [44–46]. Two other substances of bacterial origin can induce NET formation independent of NADPH oxidase or MPO activity. These inducers are the calcium ionophore A23187 processed by Streptomyces chartreusensis and the potassium nigericin ionophore created by Streptomyces hygroscopicus [47,48].

Among the other inducers of NETosis, we recognize the nuclear peptidylarginine deiminase 4 (PAD4) which supports the conversion of positively charged arginyl residues and which are abundantly found in histones, in citrulline. The latter has the peculiar characteristic of an uncharged amino acid. The dense network of DNA and histones are tightly bound together. The intervention of PAD4 has the substantial function of breaking the ionic bonds that favor the close interconnection between the negatively charged DNA with the histones in the nucleosomes. The intervention of PAD4 leads to the release of DNA that can take place to form strands which, once extruded, generate the NETs [49,50]. Li et al. [51] suggested the key role involving PAD4 in NET formation and the killing of bacteria. The mouse animal model also revealed a strong susceptibility to bacterial infections. Evidence proven by the authors stated that NET formation depends on PAD4-mediated histone hypercitrullination. However, the role of PAD4 must be contextualized. This aspect emerges in a paper by Martinod et al. [52] who did not report increased susceptibility to bacterial infections induced by cecal ligation and puncture in PAD4-deficient mice. The formation of NETs can evade the role of PAD4 and be due to independent mechanisms which therefore do not give importance to the detection of a PAD4 deficiency as reported in a model of mice with anticitrullinated histone antibodies [53]. Although the role of PAD4 in NETosis has been downsized, however, the reduced production of NETs in PAD4-deficient mice caused by an anti-histone H3 antibody remains important evidence [54–56]. A landmark paper suggesting the role of circadian rhythms in the neutrophil discharge mechanism of the NET was established by Adrover et al. [57]. The authors observed that neutrophils release the NET at a certain time of day. Furthermore, they demonstrated that there is a neutrophil-intrinsic trigger, represented by the CXCR2-dependent timer that controls this rhythmicity. Together with the observed circadian oscillation of the neutrophil proteome, it was possible to demonstrate that the NET composition and therefore the functionality differ with the time of day [57]. Likewise, a marked neutrophilic degranulation activity was noted in hemodialysis patients who disclose a higher rate of mortality from bacterial infections in hemodialysis that is estimated to be 100–1000 times as compared to the healthy population. Talal et al. [58] recently described a massive neutrophil degranulation with a considerable reduction in ROS production that was noted in hemodialysis patients whose protein levels and transcriptome of neutrophils were found highly expressed. These patients experienced defective oxidative cellular signaling with a compromised function of neutrophils that exhibit a severely impaired ability to generate NETosis due to both NADPH oxidase-dependent and independent pathways, thus reflecting their loss of capacity to kill extracellular bacteria. Evidence suggested that severe and chronic impairment of NET formation led to substantial clinical vulnerability to bacteremia that most likely results from the metabolic and environmental context representative to hemodialysis patients and not by the common human genetic shortage. Importantly, aberrant gene expression and differential exocytosis of well-defined granule populations could ponder the chronic flaw in neutrophil functionality and their diminished ability to induce NETosis. These findings support the conclusion that targeting NETosis in hemodialysis patients may reduce infections, minimize their severity, and decrease the mortality rate from infections in this cohort of patients [58]. In Table 1 are reported the characteristics of more recent studies discussed.

Table 1. Characteristics of the included studies of biology of neutrophils, NET composition, and release.

First Author/Year Ref	Type of Study	Cohort	Aims	Finding
Gu et al. (2019) Science [2]	Multicenter Center (USA/China)	Animal Model Zebrafish lines and Ldlr−/− (B6) mice	Whether AIBP orchestrates HSPC emergence from the hemogenic endothelium	AIBP-regulated Srebp2-dependent paradigm for HPSC expansion in atherosclerotic cardiovascular disease.
Lim et al. (2018) Front Immunol [5]	Prospective Multicenter Center (China/Sweden)	Blood human donors	To evaluate effects of human protease thrombin and plasmin during injury and wounding on NETome	Exogenous proteases present during wounding and inflammation influence the NETome.
Keitelman (2022) Front Immunol [7]	Prospective Multicenter Center (Argentina)	Blood human donors. Healthy vs. GOF NLRP3-mutation	To study the role of caspase-1 upon inflammasome activation to release Interleukin-1 beta (IL-1β)	Caspase-1 regulates human neutrophil IL-1β secretion.
Jaiswal et al. (2017) NEJM [9]	Prospective Multicenter Center (USA/UK/Spain)	Blood human donors/mice model Prospective CAD (4726 pt.) Retrospective Control (3529 pt.). Hypercholesterolemia-prone mice BM homozygous or heterozygous Tet2 knockout mice BM control	Association between CHIP and atherosclerotic cardiovascular disease using whole-exome sequencing	Double risk of CAD in humans with CHIP. Higher risk of atherosclerosis in mice. With higher macrophages infiltrate in Tet2 knockout mice. Higher expression of chemokine and cytokine genes.
Wang et al. (2018) Circ Res. [10]	Multicenter Center (USA/China)	Animal Model Ldlr−/− vs. BM wild-type or Mice Jak2 VF	To evaluate atherosclerosis and mechanisms in hypercholesterolemic mice with hematopoietic Jak2 VF expression	Lesion formation with increased complexity in advanced atherosclerosis hematopoietic Jak2 VF expression led. In early lesion neutrophil and macrophages infiltration.

Table 1. *Cont.*

First Author/Year Ref	Type of Study	Cohort	Aims	Finding
Wolach et al. (2018) Sci Transl Med [11]	Multicenter Center (USA/Israel/UK/ Netherlands)	Human/Animal Model Human 10,000 without MPNs Mice knock-in of *Jak2*^{V617F}	Whether neutrophils from patients with MPNs are triggered for NET formation	*JAK2*^{V617F} expression is linked to NET formation and thrombosis. JAK2 inhibition may reduce thrombosis in MPNs through cell-intrinsic effects on neutrophil function. PAD4 is required for NET formation increasing in *Jak2*^{V617}.
Li et al. (2022) Int J Biochem Cell Biol [13]	Single Center (China)	Animal model BAP31 knockdown mice	Whether BAP31 regulates CD11b/CD18 neutrophils	BAP31 depletion exerted a protective effect on ALI. Decreased neutrophil adhesion and infiltration by blocking the expression of adhesion molecules CD11b/CD18 and PSGL-1.
Winter (2018) Cell Metab [21]	Multicenter Center (Germany/Netherlands/ Spain/Sweden)	Animal model Mice Cx3cr1GFP/WTApoe−/− monocyte BM control	To evaluate that diurnal invasion of the arterial wall could sustain atherogenic growth by CCL2 and CCR2	In activity phase, chronic inflammation of large vessels nourishes on cadenced myeloid cell recruitment. Inhibition of atherosclerosis by means of pharmacological CCR2 neutralization.
Adrover (2019) Immunity [22]	Multicenter Center (Spain/USA/Germany/ France/Singapore)	Animal model. Mice engineered for constitutive neutrophil aging	To identify a neutrophil-intrinsic program that works as efficient anti-microbial defense while preserving vascular health	In mice, engineered diurnal compartmentalization of neutrophils coordinates immune defense and vascular protection becoming resistant to infection. Nevertheless, higher evidence of thrombo-inflammation and death
Yan (2021) Front Immunol [23]	Single center (USA)	Animal model C57Bl/6 mouse neutrophils containing a genomic knock-in BM control	Mutation disabling RGS protein-Gαi2 interactions (G184S) lead to worse chemoattractant receptor signaling; compromised response to inflammatory insults	Neutrophil Gαi2/RGS protein interactions limit and facilitate Gαi2 signaling. Promoting normal neutrophil trafficking, aging, and clearance.

Table 1. *Cont.*

First Author/Year Ref	Type of Study	Cohort	Aims	Finding
Casanova-Acebes (2018) J Exp Med [24]	Multicenter (Spain/Singapore/ Canada/Germany/ Netherlands/Japan)	Animal model	To investigate neutrophils' capacity to infiltrate multiple tissues in the steady-state leading to process that follows tissue-specific dynamics	Homeostatic infiltration of tissues unveils a facet of neutrophil biology that sustains organ function inducing pathological states.
Zheng (2022) Proc Natl Acad Sci U S A [25]	Multicenter Center (USA)	Animal model * MISTRGGR mice	To evaluate the role of humanized mouse model MISTRGGR The mouse G-CSF was replaced with human G-CSF, and the mouse G-CSF receptor gene was deleted in existing MISTRG mice	MISTRGGR mice represent a unique mouse model that permits the study of human neutrophils in health and disease.
Amulic (2017) Dev Cell. [40]	Multicenter Center (Germany, USA)	Animal model	To investigate NETs' formation induced by mitogens; role of phosphorylation	In neutrophils, CDK6 is required for clearance of the fungal pathogen Candida albicans. CDK4/6 is implicated in immunity.
Dölling (2022) Front Immunol. [43]	Multicenter Center (Germany)	Human autopsy samples	To investigate the neutrophil infiltration in cardiac tissue of patients with AMI	Nuclear HIF-1α is associated with prolonged neutrophil survival and enhanced oxidative stress in hypoxic areas of AMI.
Silvestre-Roig (2019) Nature [54]	Multicenter (Germany/Netherlands/ USA/France/ Spain/Sweden)	Animal model Mouse models of atherosclerosis	To investigate chronic inflammation and its cellular and molecular mediators	Histone H4 binds to and lyses SMCs, leading to the destabilization of plaques. The neutralization of histone H4 prevents cell death of SMCs and stabilizes atherosclerotic lesions.
Talal et al. (2022) BMC Med [58]	Multicenter (Israel)	Human HD patients	To investigate mortality from bacterial infections in HD patients and NET role	Targeting NETosis in HD patients may reduce infections, minimize their severity, and decrease the mortality rate from infections in this patient population.

Abbreviations; AIPB, accelerated irrigation benefit program; ALI, Acute lung injury; AMI, acute myocardial infarction; BAP31, B-cell receptor associated protein 31; BM, bone marrow; CAD, coronary artery disease; CCL2, chemokine ligand 2; CCR2, C-C chemokine receptor type 2; CDK4, cyclin-dependent kinases 4; CHIP, Clonal hematopoiesis of indeterminate potential; Gαi2, g protein 2; GOF, gain-of-function; G-CSF, granulocyte colony-stimulating factor; HD, hemodialysis; HSPCs, hematopoietic stem and progenitor cells; JAK, Janus kinase; Ldlr$-/-$, low-density lipoprotein receptor knockout; MPNs, myeloproliferative neoplasms; NETome, neutrophil extracellular trap proteome; NLRP3, NOD-like receptor family, pyrin domain containing 3; PSGL-1, P-selectin glycoprotein ligand-1; RGS, Regulator of G protein signaling; SMC, smooth muscle cell; SREBP2, Sterol Regulatory Element-binding Protein-2; TET2, ten-eleven-translocation 2. * MISTRG acronym for the 7 modified genes in the genome of these mice: M-CSFh/h IL-3/GM-CSFh/h SIRPah/h TPOh/h RAG2$-/-$ IL2Rg$-/-$.

3.2. The NETs Induce a Worsening of the Thrombotic Process Complicating the Atherosclerotic Plaque

The atherosclerotic plaques are responsible for causing fatal events and are characterized by the presence of large lipid nuclei, significant infiltration of macrophages, and depletion of SMC which underlines the presence of thin fibrous caps as well as collagen exhaustion. The occurrence of fibrous cap thinning due to an imbalance between collagen formation and increased collagen degradation, caused by SMC death, can frequently be complicated by plaque rupture and the evolution to myocardial infarction with a fatal outcome [59–61]. In advanced lesions, an anatomopathological examination of the atheromatous plaque highlights the greater presence of infiltrates of macrophages and subgroups of T cells [62,63] compared to neutrophils which are found with a much lower frequency. However, this can be interpreted as misleading as the low counts observed at any given time may simply reflect the typical characteristic of neutrophils being present in tissue for a typically short duration. Although evidence suggests that neutrophil infiltration is poor in atheromatous injuries characterized by older lesions, clinical data support that they may play a key role in plaque complications. Although evidence suggests that neutrophil infiltration is poor in atheromas characterized by more advanced lesions, clinical data support that they may play a key role in plaque complications. Peripheral blood neutrophil counts or neutrophil/lymphocyte ratios were shown to be closely related to the risk of developing cardiovascular events and outcomes [64–68]. Similarly, Zhang et al. [69] demonstrated that plasma MPO levels are positively correlated with the risk of developing coronary heart disease. Plasma nucleosome levels were also observed to be associated with an increased risk of coronary stenosis, and the presence of MPO-DNA complexes supports a close link with the occurrence of major adverse cardiac events. This evidence proves that NET-derived markers can predict atherosclerotic disease charges and events [70] as reported in Figure 2.

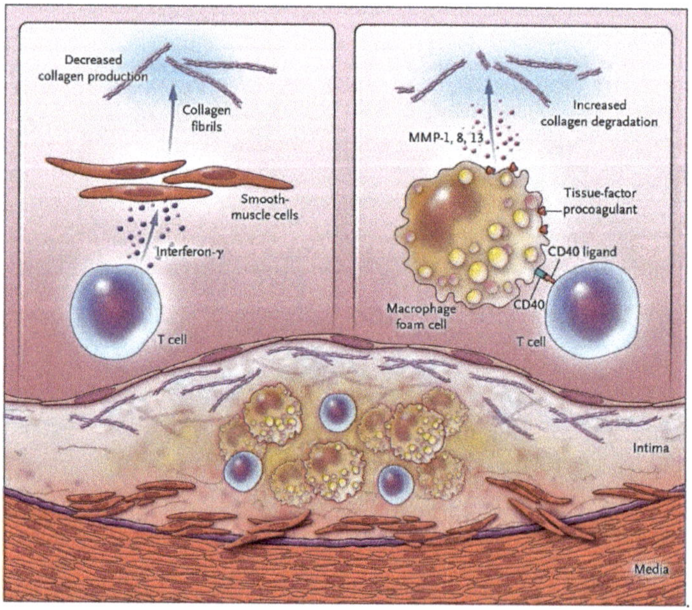

Figure 2. The inflammatory process intervenes in the mechanisms of rupture and thrombosis of atheromatous plaques. Mechanisms that drive plaque inflammation may accelerate and precipitate thrombotic complications of atherosclerosis by promoting symptom onset. The central lipid core of

the plaque is infiltrated by foam cells of macrophages (yellow) and T cells (blue) as evidenced in the cross-section of an atheromatous plaque (bottom of figure). The cellular component of the intima and media is characterized by the presence of arterial- smooth muscle cells (red), which are producing arterial collagen, which is represented by the triple helix spiral structures. The activated T lymphocytes present in the lesion are of the type 1 helper T lymphocyte subtype with the specific function of secreting the cytokine interferon-γ, which inhibits the production of the new interstitial collagen necessary to repair and maintain the protective fibrous cap of the plaque (top left). Another T cell function leads to the activation of macrophages in the intimal lesion by expressing the CD40 ligand (CD154), which engages its cognate receptor (CD40) on the phagocyte. This inflammatory signaling results in an overproduction of interstitial collagenases consisting of matrix metalloproteinases (MMPs) that catalyze the initial rate-limiting step in collagen breakdown (top right). The ligand effect of CD40 also induces macrophages to produce more of the procoagulant than tissue factor. Thus, the inflammatory response generated in the plaque exposes the collagen in the fibrous cap to a double risk. It decreases synthesis and increases breakage, thus making the cap more susceptible to breakage. The inflammatory activation also induces an increased production of tissue factors, which trigger thrombus formation in the damaged plaque [71–75].

Role of Nets in Plaque Destabilization: The New Challenge

An important contribution to the knowledge of the pathophysiological process that favors the progression of the atheromatous plaque and its complications was provided by the study on the infiltration of neutrophils and consequent destabilization of the atheromatous lesion. Ionita et al. [76] studied the morpho-functional alterations of the atheromatous plaque emphasizing the peculiar characteristic of the cellular infiltrate. The authors described lesions characterized by marked infiltration of neutrophils substantially present in plaques with a large lipid core. Again, they recorded an abundant concentration of macrophages and a marked modification of the vascular architecture, which was testified by a reduction in the amount of collagen fiber and SMC expression. Taken together, these alterations suggested the crucial role of neutrophils in plaque destabilization processes, with two key elements to underline, necrotic nucleus growth and thinning of the fibrous cap [76]. Given the importance of the lesions sustained by an abnormal growth of the necrotic nucleus and the thinning of the fibrous cap, Silvestre-Roig et al. [54] observed that these two anatomopathological conditions were the indication of an excessive concentration of neutrophils-triggering inflammation and in particular of the generation of NETs in promoting the destabilization of the plaque [77].

Substantial evidence demonstrated a mechanical interaction between the SMCs populating the fibrous cap and neutrophils with the effect of producing their activation leading to ROS production and NET release. This sequence of events is mediated by the action of CCL7 conveyed by the SMC which supports the generation and release of NETs. Inhibition of the NET release can be achieved after treatment with a PAD inhibitor in mice with pre-existing lesions or those lacking PAD4, revealing the presence of lesions with characteristics of reduced vulnerability. The latter appears to be an expression of smaller necrotic nuclei in association with higher levels of SMC, compared with a control cohort of mice. One fact emerges with fundamental relevance in the mouse model studied and concerns the cytotoxic action exerted by NETs. This is triggered in the circuit involving the close interaction between neutrophils and NETs on the one hand and dying SMCs, dimensions of the necrotic nucleus, and structural changes typical of the thinning of the fibrous caps. Of note, the killer action of NETs on SMCs was reported in ex vivo studies which suggested that this effect was not exerted by granule-derived neutrophil proteins or the cytoplasm, but rather by a specific intervention afforded by nuclear histones. Among these, histone H4 works with well-understood cytotoxic activity, in the context of the specific action offered by histones transmitted by the NET [2,78–80]. Histone H4 deserves attention for its function as a strongly cationic protein, which can interact with negatively charged SMC surfaces. This bond interferes with membrane activity leading to membrane flexure and ultimately pore formation, prodromal of the lytic effect, resulting in cell death.

Furthermore, the depletion of the SMC reservoir can lead to a reduction in the thickness of the fibrous cap, a phenomenon supported by marked collagen removal, which is crucial for the preservation of a suitable interstitial architecture. The demonstrated centrality of histone H4 and its involvement in the mechanisms leading to plaque rupture were suggested in further studies using a neutralizing histone H4 antibody and cyclic HIPE, exhibiting the function of histone interference peptides that are involved in targeting the N-terminus of histone H4. The two intervention strategies reported above involve the pathoanatomic characteristics of the developing atheromatous plaque and have the effect of reinforcing its stability. First, it is important to underline that in mice with advanced atherosclerosis, activation of the dsDNA sensing AIM2 inflammasome induced a marked release of the proatherogenic cytokines IL-1β and IL-18. Second, the lack or neutralization of AIM2 generated plaques that disclosed a poorer propensity to rupture; in particular, they expressed a smaller necrotic nucleus in association with a thicker fibrous cap [81–83] However, the exciting promises that emerged about AIM2's role in promoting greater human plaque stability warrant further study as highlighted in Figure 3 [71–75].

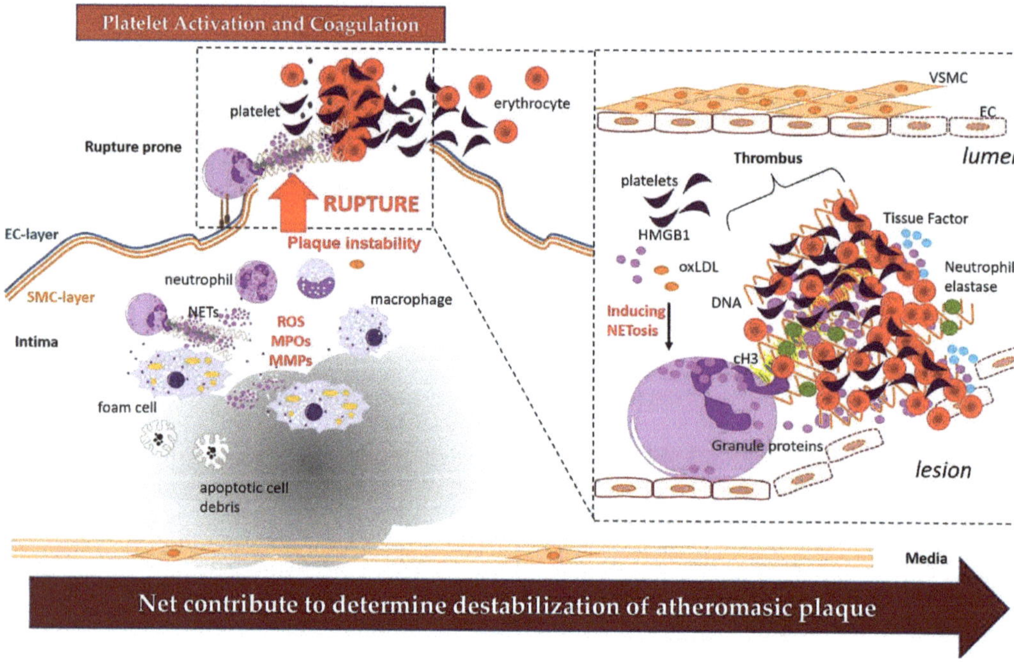

Figure 3. NETs' direct arterial occlusion. During the process of atherothrombosis, neutrophil-generated NETs can intervene by supporting a series of biochemical phenomena that can ultimately trigger the activation of the coagulation cascade. Through the destabilization of the atheromatous plaque up to its rupture, NETs help to increase the stability of the thrombus. Abbreviations; EC, endothelial cell; MMPs, Matrix metalloproteinases; MPO, myeloperoxidase; ROS, reactive oxygen species; SMC, smooth muscle cell. Other abbreviations in text and tables [18,71–75].

4. NETs' Actors of Arterial Thrombosis

Given the sizeable evidence recently offered, new insights into the implication of neutrophils and the role of neutrophil-derived inflammatory mediators in thrombosis have emerged [84]. The role of platelets in NET generation was discussed by Carestia et al. who highlighted the mediators, stimuli, and molecular mechanisms involved in the process of NET generation that was platelet-induced, both in human and murine models [85]. Gould et al. [86] studied the procoagulant potential of intact NETs released from activated neutrophils. The authors highlighted the relative contribution of cell-free DNA (cfDNA) and histones to thrombin generation in plasmas from patients with sepsis. A total of 1000 patients were enrolled in the DYNAMICS STUDY (DNA as a prognostic marker in ICU patients) from 2010 to 2012. The cohort included a prospectively independent population of severe sepsis patients (n = 400) and a broad cohort of non-septic ICU patients (n = 600). The first confirmation provided by the study was that NETs released by phorbyl myristate-activated neutrophils increased thrombin generation in platelet-poor plasma. This effect was closely related to DNA, after confirmation obtained from DNase treatment, and occurred through the intrinsic pathway of coagulation. In fact, in the plasma, there was a depletion of factor XII and factor XI of coagulation. In the platelet-rich plasma treated with a corn trypsin inhibitor, the addition of neutrophils activated by phorbol myristate showed an increased production of thrombin which was associated with a decreased lag time due to a mechanism dependent on toll-like receptor-2- and toll-like receptor-4-dependent. After the addition of DNase, a further increase in thrombin genesis was observed, suggesting that dismantling of the NET scaffold increases histone-mediated and platelet-dependent thrombin generation. Another key piece of evidence that emerged showed that in platelet-poor plasma samples from patients with sepsis, a positive correlation was found between endogenous cfDNA and thrombin generation, and the increase in thrombin generation was attenuated by DNAse. The results of the DYNAMICS STUDY highlighted the procoagulant activities of cfDNA and histones in the context of NETs. The study suggested the key role played by the intrinsic coagulation pathway in the pathogenesis of sepsis [86].

Pivotal studies by Fuchs et al. [87] and von Brul et al. [88] reported the presence of NETs associated with prothrombotic molecules such as the tissue factor, factor XII, histones H3 and H4, and fibrinogen, which promote thrombosis. Again, the NETs and the fibrin network generated by the same traps can also constitute a scaffold that entraps platelets and red blood cells. However, Noubouossie et al. [89] reported substantial evidence showing that the in vitro activation of coagulation was caused by human neutrophil DNA and histone proteins but not by neutrophil extracellular traps. Firstly, human neutrophils were incited to generate NETs in platelet-free plasma (PFP) or buffer using phorbol myristate acetate or calcium ionophore. DNA and histone proteins were also separately purified from normal human neutrophils and used to reconstitute chromatin using a salt-gradient dialysis method. Neutrophil stimulation resulted in a robust NET release. The second key finding of the analysis disclosed that in recalcified PFP, purified DNA triggered contact-dependent thrombin generation (TG) and amplified TG initiated by low concentrations of the tissue factor. Likewise, in a buffer milieu, DNA started the contact pathway and magnified thrombin-dependent factor XI activation. Recombinant human histones H3 and H4 triggered TG in recalcified human plasma in a platelet-dependent manner. Neither intact NETs, reconstituted chromatin, individual nucleosome particles, nor octameric core histones repeated any of these procoagulant effects. This study suggested that unlike DNA or individual histone proteins, human intact NETs do not directly launch or intensify coagulation in vitro. This difference is likely unfolded by the complex histone-histone and histone-DNA interactions within the nucleosome unit and higher-order supercoiled chromatin that determined the neutralization of the negative charges on polyanionic DNA leading to modification of the binding properties of individual histone proteins [89].

Accumulating evidence revealed that NETs and their components are located in human coronaries [90–94] as they disperse in the thrombus related to ischemic strokes [95]. This process occurs regardless of plaque type, but NETs were observed numerically to dominate more recent thrombi rather than older, more organized thrombi as demonstrated by Fucks et al. [87]. In this way, of crucial relevance are investigations of Riegger et al. [90] and de Boer et al. [92].

The study of Riegger et al. [90] evaluated 253 thrombus specimens from a large-scale multicenter study in patients with ST across Europe (Prevention of Late Stent Thrombosis by an Interdisciplinary Global European Effort (PRESTIGE) Investigators). Of the population studied, 79 (31.2%) patients had thrombosis with an early ST while 174 (68.8%) had a late ST.

A total of 79 patients (31.2%) were treated with PCI with bare metal stenting, as opposed to 166 (65.6%) patients in whom a drug-eluting stent was used. For 8 (3.2%) PCI recipients, the thrombus examined was from a stent of an unknown type. Regarding the morphology of the thrombus samples examined, heterogeneity was observed. However, an increased abundance of platelet-rich thrombi and fibrin/fibrinogen fragments was revealed; again, 57% of the overall thrombus surface was covered with platelets. An important finding was the amount of thrombus-infiltrating leukocytes that was distinctive for both early and late ST (early: 2260 $\pm$ 1550 per mm (2) vs. late: 2485 $\pm$ 1778 per mm (2); $p = 0.44$). Of the leukocytes examined, the subset of neutrophils constituting the most important cellular component was preponderant (early: 1364 $\pm$ 923 per mm (2) vs. late: 1428 $\pm$ 1023 per mm (2); $p = 0.81$). Leukocyte counts were markedly higher than in a control group of patients with thrombus aspiration in spontaneous myocardial infarction. Extracellular neutrophil traps were found in 23% of these samples. Eosinophils were expressed as a line in all stent types, albeit the largest number was recorded in patients with late ST normalization in sirolimus and everolimus-eluting stents [90].

De Boer et al. [92] studied samples of different blood clots from patients who experienced acute myocardial infarction in various developmental stages. The neutrophilic cell component, NET, and immunoreactive IL-17A were preferentially localized in fresh rather than organized thrombi, thus suggesting a substantial contribution of these elements to thrombus stabilization and growth. A detailed histological analysis performed on 64 human coronary artery plaque segments included 44 complicated plaques and 20 intact plaques. In the 44 complicated plaques, where intraplaque hemorrhages, erosions, and ruptures occurred, and in the 20 intact plaques, neutrophils and NETs were most frequently observed in more recent unorganized thrombi and acute developmental intraplaque processes with ongoing hemorrhages versus older and more organized thrombi. This evidence supports the presence of NETs in distinct types of atherothrombosis, in particular, with all evidence, in fresher arterial thrombi [16,96]. The investigation of 111 different thrombi obtained from patients with acute coronary elevation syndrome reported a concentration of higher NETs, and it was correlated to a less favorable evolution of the ST elevation resolution associated with a greater extension of the infarct size. These suggest two key roles for NETs. In the former, they offer negative support to infarct lesions through the propagation of thrombosis. Instead, the second role exerted by NETs leads to the inflammation that develops distally in the infarcted myocardium and to the death of myocytes during athero-embolism. The use of DNAse, an enzyme that acts to eliminate NETs by digesting DNA strands, in regions of the infarcted myocardium, revealed a smaller extent of lesions that were related to both a reduced size of the infarction but also to a positive resolution of ST-segment elevation acute myocardial infarction (STEMI). Mangold et al. suggested that the lysis of these thrombi was accelerated by an ex vivo addition of DNAse [93]. It is important to note that observational studies performed on human samples generate intriguing hypotheses but do not allow definitive conclusions on causality to be drawn.

The role of the procoagulant tissue factor as one of the key factors for the triggering of coronary syndromes and the characteristic that they can decorate NETs was recently extensively investigated. Stakos et al. [94] studied the in vivo relevance of NETs during atherothrombosis in humans by means of selective sampling of thrombotic material. They retrieved surrounding blood from the infarct-related coronary artery (IRA) and the non-IRA during primary percutaneous revascularization in 18 patients with STEMI. Authors suggested that thrombi detached from the IRA encompassed PMNs and NETs decorated with the tissue factor (TF). The relevant findings suggested that although TF was expressed intracellularly in circulating PMNs of patients in which a STEMI occurred, however, an active TF was specifically exposed by NETs obtained from a complicated atheromatous plaque with exhibited rupture. Authors again generated NETs and subsequent TF exposure after a second necessary step consisting of the interaction between PMN and thrombin-activated platelets. The conclusions reached by Stakos et al. established that the interaction between neutrophils and platelets at the sites of plaque rupture in patients with STEMI promoted local NETs' generation associated with the delivery of an active TF and the stimulation of prothrombotic events. The evidence highlighting this new role of NETs offers adjunctive findings to explain the mechanism by which PMNs release thrombogenic signals during atherothrombosis and may offer novel therapeutic targets [94].

Maugeri et al. [97] evaluated the presence of NETs in 26 retrieved thrombi from patients with acute myocardial infarction using immunohistochemistry and immunofluorescence and markers of NETs assessed in the plasma. In addition, in vitro NET generation was studied both in static and physiological flow conditions. Investigators revealed that coronary thrombi mainly consist of activated platelets, neutrophils, and NETs in close proximity to platelets. Activated platelets engaged neutrophils to NET generation. The event decreased in the presence of competitive antagonists of the high mobility group box 1 (HMGB1) protein. Hmgb1(−/−) platelets were lacking to elicit NETs, whereas the HMGB1 alone engaged neutrophils to NET generation. The integrity of the HMGB1 receptor, Receptor for Advanced Glycation End-products (RAGE), was required for NET generation, as evaluated using pharmacologic and genetic tools. Exposure to HMGB1 anticipated depletion of the mitochondrial potential that prompted autophagosome formation and extended neutrophil survival. These metabolic effects were determined by the activation of autophagy. Of note, stoppage of the autophagic flux regressed platelet HMGB1-elicited NET generation. These findings proved that activated platelets presented HMGB1 to neutrophils and engaged them in autophagy and NET generation. This chain of events may be responsible for some types of thrombo-inflammatory injuries thus designating novel tracks for molecular intervention [97].

We learned that optimal medical therapy for arterial thrombosis is mainly directed against platelets and coagulation factors. However, the treatment can lead to bleeding complications in an environment where the cells of the immune response and the atheromatous plaque interfere [98–100]. To circumvent concerns related to bleeding after treatment with antiplatelet drugs, new antithrombotic therapies, currently being studied in animals, targeted immune cells and neutrophil extracellular traps (NETs). Several studies addressed whether the immune cell composition of arterial thrombi induced in mouse models of thrombosis resembles that of human acute myocardial infarction (AMI) patients [101–103]. A previous report by Novotny et al. [104] studied the immune component of human thrombi with that of a mouse model. The authors evaluated 81 human arterial thrombi obtained during a percutaneous coronary procedure in patients suffering from acute myocardial infarction that were compared to arterial thrombi retrieved from mice in which experimental thrombosis was induced using ferric chloride (FeCl3) or carotid artery wire injury. The results evidenced by a detailed histological analysis demonstrated that the FeCl3-induced murine arterial thrombi and those of human coronary thrombi consisted of similar immune cells. The component of human thrombi and those withdrawn from the mouse experimental model consisted of neutrophils combined with plenty of NETs and coagulation factors. The authors found that the addition of the pharmacological administration of mice with the protein arginine deiminase (PAD)-inhibitor Cl-amidine abolished NET formation, and was related to a reduction in arterial thrombosis associated with a restriction of injury in a model of myocardial infarction. Neutrophils are a hallmark of arterial thrombi in patients suffering from acute myocardial infarction and in mouse models of arterial thrombosis. Inhibition of PAD could represent an interesting strategy for the treatment of arterial thrombosis to reduce neutrophil-associated tissue damage and improve functional outcomes. The evidence reported in Novotny's study supports the presence of extensive neutrophil infiltration. Again, the results observed with PAD inhibition may guide the study and application of new strategies for the treatment of arterial thrombosis. The ultimate goal is aimed at obtaining a reduction in tissue damage associated with the concentration of neutrophils at the infarct site and at improving the functional outcome [104].

A convergence of findings emerges in several independent reports revealing a body of evidence supporting the substantial role that NETs play in acute ischemic stroke (AIS) and acute myocardial infarction thrombi [100]. Laridan et al. [95] investigated sixty-eight thrombi retrieved from patients who experienced an ischemic stroke and received an endovascular procedure. The analysis of thrombi was performed using immunostaining neutrophil markers such us CD66b and neutrophil elastase as well as specified NET markers such as citrullinated histone H3 (H3Cit) and extracellular DNA. For completion of the histopathology evaluation, neutrophils and NETs were quantified, and the addition of DNase 1 to the tissue plasminogen activator was used to facilitate ex vivo lysis of patient-derived thrombi. The evidence showed the marked existence of neutrophils throughout all thrombi. Again, H3Cit, a hallmark of NETs, was disclosed in nearly all thrombi. Importantly, investigators suggested that the colocalization of H3Cit with extracellular DNA released from neutrophils could be related to the peculiar occurrence of NETs in the damage site. Moreover, H3Cit was highlighted abundantly in patients with thrombi generated from a cardioembolic origin compared to other causes. Likewise, high levels of neutrophils and H3Cit were recorded in older thrombi as compared to fresh thrombi. Interestingly, using DNAse 1 added to standard t-PA, the ex vivo lysis of the patient thrombi was achieved with greater success. These findings suggest that neutrophils and NETs constitute significant components of cerebral thrombi and may offer a novel diagnostic/therapeutic pathway for targeting NETs with DNase that could potentially be used as a thrombolytic in the treatment of acute ischemic stroke [95].

Recently, Novotny et al. [100] performed a detailed histological analysis of arterial thrombi in which the structure of the thrombus but especially the abundance of leukocyte subsets differed between patients with acute ischemic stroke (AIS) and acute myocardial infarction (AMI). Investigators revealed that although amounts of leukocytes ($p = 0.133$) and neutrophils ($p = 0.56$) were alike between AIS and AMI thrombi, however, monocytes ($p = 0.0052$), eosinophils ($p < 0.0001$), B cells ($p < 0.0001$), and T cells ($p < 0.0001$) were found to be more abundant in patients with a stroke as to patients with AMI thrombi. In addition, the quantity and appearance of NETs revealed an inhomogeneous pattern of NETs that were experienced in 100% of patients with AIS as compared to only 20.8% of patients with AMI. The wealth of NETs in thrombi was associated with poor outcome scores in patients with AIS while patients with AMI recorded a decreased ejection fraction. This difference outlined in patients' outcomes after AIS and AMI supports a critical impact of NETs on thrombus stability in both conditions [100]. Ducrox et al. [105] reported histological evidence in a randomized controlled trial enrolling 108 acute ischemic stroke patients from whom thrombi were recovered. The objective of the RCT was to assess the presence of NETs in thrombi retrieved during endovascular therapy in patients with AIS and their impact on tPA-induced thrombolysis. The authors demonstrated the presence of conglomerates of NETs in all thrombi with a higher concentration of networks in the outer layers of the thrombus. They concluded that the thrombus NET content was responsible for reperfusion resistance, and included both mechanistic and pharmacological approaches with intravenous tPA, regardless of their etiology. Thus, the efficacy of a strategy involving the administration of DNAse 1 in addition to tPA could be considered a new way to be explored in the context of AIS. Studies by Novotny et al. and Ducrox et al. offered an explanation of the importance of NETs in thrombosis and the clinical potential they represent. Above all, the substantial result is represented by the fact that ex vivo, recombinant DNAse 1 accelerated thrombolysis induced by the tissue plasminogen activator. However, DNAse 1 alone did not have the same effect [105].

These findings should be correlated with those suggested by Farkas et al. [91] which highlighted a difference between the content of NETs, the relative concentration of platelets, and the fibrin structure of thrombi as well as their distribution in thrombi recovered from patients with acute ischemic stroke, myocardial infarction, or peripheral artery disease. The authors reported the presence of DNA in the thrombi of acute ischemic stroke patients during the comparison, perhaps explaining why DNase 1 alone did not effectively destroy the thrombi. Furthermore, the DNA/fibrin ratio was significantly lower in thrombi recovered from patients who had experienced an acute ischemic stroke than in patients in whom thrombi were recovered from peripheral arteries. Again, the thrombi from peripheral artery disease contained fewer platelets. Bucking, thrombi from the three sites studied expressed comparable histone citrullinated levels-3 [91].

From these data, it emerges that the levels of DNA substantially condition the effect of the therapy. Thus, the combination of the tissue plasminogen activator and DNAse I demonstrated greater efficacy in thrombus lysis, just as it was suggested that the addition of DNAse I to DNA-poor thrombi, such as those obtained from patients with ischemic stroke, is not sufficient to lyse the thrombus. Furthermore, the presence and involvement of NETs in thrombosis may vary according to the type of artery involved. This aspect is crucial because it can direct both the type of intervention to be performed and the time necessary to obtain a benefit from the treatment. Therefore, potential therapeutic targeting should take due consideration of the age of the thrombus and the aspiration procedure which play a key role in the success of the procedure. The latter will require adaptation depending on the site and the timing of the treatment.

Studies performed in mice revealed that neutrophil-derived externalized nucleosomes associated with NETs are implicated in arterial thrombosis induced by FeCl3 application. Massemberg et al. [106] in a wild-type FeCl3-mouse model disclosed that administration of a histone-neutralizing antibody resulted in a prolonged time to occlusion associated with decreased thrombus stability in the carotid arteries. This result was not obtained after infusion of antibodies in mice with selective neutrophil elastase/cathepsin G deficiency in which vascular damage was induced. This evidence supports the substantial role played by externalized nucleosomes which allow the assembly of the complex constituted by neutrophil elastase and its substrate-tissue factor pathway inhibitor located on the surface of activated neutrophils. This mechanism may sustain the series of events leading to thrombosis. The evidence emerging in the study by Massemberg et al. supports substantial data that, in sterile inflammation that develops in large vessels, neutrophil-derived serine proteases and nucleosomes may favor the initiation of a thrombotic process responsible for myocardial infarction and heart attack stroke [106]. However, it is important to underline the independent role of neutrophils with respect to NETs which, through the release of single components, can exert stronger procoagulant effects than those induced by NETs by completing the action of the latter. This aspect emerges during in vitro activation of coagulation by human neutrophil DNA and histone proteins but not by neutrophil extracellular traps [89]. Knight et al. [102] studying an in vivo model of New Zealand mixed mice, which have the characteristic of developing a lupus-like systemic pathological condition driven by IFN type I, reported increased NETs' formation with manifested accelerated vascular dysfunction and increased prothrombotic risk. The authors administered chloramidine to New Zealand mixed mice and recorded a substantial reduction in the generation of NETs associated with both an improvement in endothelium-dependent vasodilation and a marked reduction in the time to arterial thrombosis [102]. In a more recent study, Sorvillo et al. [107] supported the role of PAD4 in accelerating thrombus formation and stability. This process was mediated by a reduction in the clearance of von Willebrand factor (vWf) -platelet thread structures in the circulation. The proposed role of PAD4 lies in the inhibition of the vWf degrading proteinase ADAMTS-13 which acts as a disintegrin and metalloprotease with a thrombospondin-1-like domain 13 [107].

NETs are not always generated. Cherpokova et al. [108] evaluated neutrophil function in a mouse model using the specialized resolution mediator resolvin D4. The authors recorded a reduced susceptibility to NETosis induced by ionomycin administration, explaining the role of modulators exerted by these resolution mediators after administration. Their modulatory action in directing the severity of thrombo-inflammatory diseases in vivo is exerted through interference with the generation of NETs [108]. Franck et al. [109] identified a crucial role played by NETs in exacerbating the aspects of thrombosis supported by superficial erosion. Evaluations were performed on mice in which an arterial lesion was previously induced followed by a periarterial cuff-induced flow disturbance. A model thus generated allowed reproduction of the set of characteristics of human lesions complicated by superficial erosion, therefore to study the plaque complication that can previously be associated with NETs. The number of NETs generated and the initiation of thrombosis were strictly dependent on a PAD4 deficiency in myeloid cells or DNAse treatment. Either a PAD4 deficiency in bone-marrow-derived cells or administration of DNAse1 to disrupt NETs reduced the extent of arterial intimal injury in mice with arterial lesions tailored to summarize environments reproduced in human atheroma complicated by erosion. Results derived from Franck et al. suggested that PAD4 from bone-marrow-derived cells and NETs did not influence the chronic processes on the basis of induced experimental atherogenesis. In contrast, they fortuitously shared in the acute thrombotic complications of tunica intima lesions that restate features of superficial erosion [109]. The main studies identified in the systematic review are reported in Table 2.

Table 2. Characteristics of the included studies investigating the role of NET in arterial thrombosis.

First Author/Year Ref	Type of Study	Cohort	Aims	Finding
Fernandez et al. (2019) Nat Med [62]	Multicenter Center (USA/Sweden)	Human carotid artery plaques	To investigate the role of T cells and macrophages in carotid artery plaques of patients with recent stroke or transient ischemic attack compared to no recent stroke	In plaques from asymptomatic patients, T cells and macrophages were activated and displayed evidence of IL-1β signaling.
Zhao et al. (2022) J Hum Hypertens [68]	Multicenter Center (China/Sweden)	Human 4667 pts > 40 yrs	To study the relationship between NLR and the risk of CVD	When NLR was categorized into tertiles, participants in the top tertile had a significantly higher risk of CVDs (HR 1.61, 95% CI: 1.06, 2.44) and MI (HR 1.88, 95% CI: 1.09, 3.27) relative to those in the bottom tertile.
Locke et al. (2020) Sci Rep. [78]	Multicenter (UK)	Human blood donors	Whether formation of fibrinogen/fibrin-histone aggregates prevented cell death	Fibrinogen did not bind to or protect neutrophils stimulated with PMA. The role of fibrinogen in NETosis.
Noubouossie et al. (2017) Blood [89]	Multicenter (USA/Taiwan)	Human blood patients' neutrophils	To investigate the mechanism leading to thrombosis promoted by intact NETs	Recombinant human histones H3 and H4 triggered TG in recalcified human plasma in a platelet-dependent manner. However, human intact NETs do not directly initiate or amplify coagulation in vitro.
Riegger et al. (2016) Eur Heart J [90]	Multicenter (Germany /USA/UK/ Netherlands/Belgium/ France/Spain/Polland)	Human 253 thrombus specimens	To evaluate thrombus specimens in patients with ST and presence of NET	Patients with ST revealed thrombus with leukocytes, particularly neutrophils in ST group. The presence of NETs supports their pathophysiological relevance.
Farkas et al. (2019) Thromb Res [91]	Multicenter (Hungary/Australia)	Human CAD 66 pts PAD 64 pts	To investigate the NET-related structural features of thrombi retrieved from different arterial localizations	NET content of thrombi was correlated with systemic inflammatory markers and with patients' age. Evidence of NET-related variations in thrombus structure.

Table 2. *Cont.*

First Author/Year Ref	Type of Study	Cohort	Aims	Finding
Laridan et al. (2017) Ann Neurol [95]	Multicenter (Belgium/Sweden)	Human 68 thrombus specimens	To investigate the presence of neutrophils and NETs in ischemic stroke thrombi	H3Ci higher in cardioembolic thrombi. Older thrombi contained significantly more neutrophils and H3Cit compared to fresh thrombi.
Shimony et al. (2010) Acute Card Care [110]	Single Center (Israel)	Human 16 randomly acute STEMI pts vs. 47 healthy individuals	To evaluate detection of CFD in patients with ST STEMI; to study correlation with established markers of necrosis and ventricular function	CFD levels were linked with the levels of established markers of myocardial necrosis but not with EF.
Langseth et al. (2020) PLoS One [111]	Multicenter (Norway)	Human 61 randomly STEMI pts	To investigate NETs' associate to MF and IL 8 in STEMI patients with symptomatic acute HF	NETs' components higher in acute heart failure and associated to myocardial function and interleukin 8 levels.
Langseth et al. (2020) Sci Rep [112]	Multicenter (Norway)	Human 956 cohort STEMI pts	To investigate association between circulating NETs-related components, clinical outcome, and hypercoagulability in STEMI	dsDNA levels were associated with increased all-cause mortality and with hypercoagulability in STEMI patients.
Helseth et al. (2016) Mediators Inflamm [113]	Single center (Norway)	Human 20 pts with STEMI vs. 10 pts with AP	To study infarct size and NETs' markers in STEMI and AP	Higher levels of NETs in STEMI
Valles et al. (2017) Thromb Haemost [114]	Single center (Spain)	Human 43 pts with AIS	To evaluate NETs in the plasma of patients with acute ischemic stroke	At one-year follow-up, NETs were associated with severity and mortality. Relevant specific marker of NETs citH3 was higher and independently associated with AF and all-cause mortality. Significant role of NETs in the pathophysiology of stroke.
Hirose et al. (2014) PLoS One [115]	Multicenter (Japan)	Human 263 pts in ICU	To evaluate whether NETs and Cit-H3 were correlated with clinical and biological parameters	Crucial role of NETs in the biological defense against the dissemination of pathogens from the respiratory tract to the bloodstream in potentially infected patients.
Helseth (2019) Mediators Inflamm [116]	Multicenter Center (Germany)	Human 259 pts with STEMI	To explore circulating NET markers and associations to myocardial injury	dsDNA levels after STEMI were associated with myocardial infarct size, adverse left ventricular remodeling, and clinical outcome.
Langseth (2018) Eur J Prev Cardiol [117]	Multicenter (Norway)	Human 1001 pts with AP	To investigate the role of NETs' markers, dsDNA, and myeloperoxidase-DNA in clinical outcome and hypercoagulability	Double-stranded DNA levels were significantly related to adverse clinical outcome after 2 years, but only weakly associated with hypercoagulability.

Table 2. *Cont.*

First Author/Year Ref	Type of Study	Cohort	Aims	Finding
Martinod et al. (2016) Thromb Haemost [118]	Multicenter (USA)	Animal model WT vs. †NE (−/−) vs. SB1 vs. † NE (−/−) SB1 (−/−) mice.	Whether neutrophils from NE (−/−) mice have a defect in NETosis, similar to PAD4 (−/−)	Neutrophil elastase is not required for NET formation. NE (−/−) mice, which form pathological venous thrombi containing NETs, do not phenocopy PAD4 (−/−) mice in in vitro NETosis assays or experimental venous thrombosis.
Kim et al. (2017) J Immunol Res [119]	Multicenter (Korea)	60 pts in MHD	Whether NET formation was responsible for ESRD leading to higher incidence of CAD	Uremia-associated-increased NET generation may be a sign of increased burden of atherosclerosis.
Bang et al. (2019) Stroke [120]	Multicenter (Korea)	Human 138 randomly 38 pts cancer-related stroke 36 pts healthy-controls 27 pts cancer-controls (active cancer but no stroke) 40 pts stroke-controls (acute ischemic stroke but no cancer)	Whether NETs were increased in cancer-related stroke; whether higher NETs levels were associated with coagulopathy	Increased circulating DNA levels were associated with cancer-related stroke. NETosis was one of the molecular mechanisms of cancer-related stroke

Abbreviations; AF, atrial fibrillation; AIS, acute ischemic stroke; AP, stable angina pectoris; CAD, coronary artery disease; CFD, circulating cell free DNA; CVD, cardiovascular diseases; dsDNA, double-stranded DNA; end-stage renal disease (ESRD Il, interleukine; EF, ejection fraction; H, histone; H3Cit, citrullinated histone H3; HF, heart failure; MI, myocardial infarct; MF, myocardial function; MHD, maintenance hemodialysis; NE, neutrophil elastase, NET, neutrophil extracellular trap; NLR, neutrophils-to-lymphocyte ratio; PAD, peripheral artery disease; PAD4 (−/−), peptidylarginine deiminase 4; PMA, phorbol 12-myristate 13 acetate; ST, stent thrombosis; SB1, SerpinB1; STEMI, ST elevation myocardial infarction; TG, thrombin generation; WT, wilt type. † (−/−) Knock minu.

5. The Cardiovascular Clinical Potential Derived from Biomarker Applications of NETs

The literature has a large number of studies in which a relationship between potential biomarkers of NETs and clinical variables emerged, including some that are discussed in the text in Table 3 [93,94,97,104,105,111]. To date, there are no solid data to support a definitive validation of these biomarkers in clinical practice, whose full consideration, therefore, remains incomplete and still open to discussion. However, the biomarkers, which several investigations adopted as reliable parameters to establish the effective involvement of NETs in the destabilization process of atheromatous plaque, are double-stranded DNA, MPO-linked DNA, citrullinated histones, neutrophil elastase, among other NET constituents in the blood or aspirated thrombi. In clinical practice, the challenge that will allow to develop and validate more specific biomarkers of NETs becomes of utmost importance. Such studies can be directed to explore different hypotheses relating to NETs' involvement in the progression of plaque which explains this phenomenon in purely physical or deterministic terms. Furthermore, these studies can contribute to providing useful information for prognosis, and possibly could be used to designate specific therapies based on the degree of NET involvement, in the utility of personalized or precision medicine [121].

One study correlated NETs' markers with the manifestation of more severe degrees of coronary atherosclerosis in CAD associated with stable angina as assessed by various imaging modalities [69]. Several studies suggest the presence of elevated markers of NETs in patients exhibiting acute coronary syndromes. Patients who develop CAD with ST-elevation in the course of IMA reveal an increased level of cell-free DNA and other NET constituents, compared with the control group [15,110–113,122,123]. Likewise, patients with an acute ischemic stroke disclose increased blood NET markers [114,115,124]. NET markers were observed in aspirated thrombi from patients with acute coronary syndromes, stent thrombosis, ischemic stroke, and acute peripheral arterial disease [90–97,105]. In patients with documented atherosclerosis, several clinical conditions were highlighted which may be associated with the different concentrations of NET markers. For example, NET load is closely related to infarct size as observed with magnetic resonance imaging [93,116]; it was also associated with worse outcomes at a 2-year follow-up in patients with stable coronary artery disease [117,125]. We have a growing experimental literature, which agrees with the data above, reporting that NET-derived products can be detected in systemic chronic diseases complicated by cardiovascular disease. This condition is typical of patients who associate atherosclerosis with rheumatoid arthritis [118,126]. A total of 60 maintenance hemodialysis patients were studied in a cross-sectional analysis. Of these, 30 matched for age and sex were healthy individuals and represented the negative control while the other 30 patients manifested an acute infection and represented the positive control. In the latter, a correlation was reported between the increased production of NETs associated with uremia and with a worsening of atherosclerosis disease [119]. Of no less importance is the evidence linking the formation of NETs to the vexing clinical problem of thrombotic complications associated with cancer [127]. Another example of an increase in the generation of NETs was reported in a study that enrolled 138 patients with stroke-related cancer [120]. The main studies identified in the systematic review are reported in Table 3.

Table 3. Characteristics of the included studies investigating the role of NET in cardiovascular clinical biomarker applications.

First Author/Year Ref	Type of Study	Cohort	Aims	Finding
Mangold et al. (2015) Circ Res [93]	Single Center (Germany)	Human 111 pts with STEMI	To investigate relationships between CLS -NETs, bacterial components as triggers of NETosis, activity of endogenous deoxyribonuclease, ST-segment resolution, and infarct size	PMNs were highly activated in STEMI with NETosis at the CLS and coronary NET burden. Deoxyribonuclease activity was predictor of ST-segment resolution and myocardial infarct size.
Stakos et al. (2015) Eur Heart J [94]	Multicenter Center (Greece/Germany)	Human 18 pts with STEMI	To assess the in vivo importance of NETs during atherothrombosis	NETs by mean PMNs mediated thrombogenic signals during atherothrombosis.
Maugeri et al. (2020) Sci Rep [97]	Single center (IT)	Human	To assess the mechanism of platelets induced NETs	Activated platelets were related to HMGB1 in neutrophils. HMGB1 led to autophagy and NET generation.
Novotny et al. (2018) PLoS One [104]	Multicenter (Germany)	Human/animal model 81 human arterial thrombi retrieved mice with injury of carotid artery	To evaluate composition of arterial thrombi in mice compared to those of human patients with AMI	Inhibition of PAD was useful for the treatment of arterial thrombosis and to reduce NET generation.
Ducrox et al. (2018) Stroke [105]	Multicenter (France)	Human 108 pts with AIS	To determine the occurrence of NETs in thrombi retrieved during endovascular therapy. Impact on tPA-induced thrombolysis	The efficacy of a strategy involving an administration of DNAse 1 in addition to tPA was effective in the setting of AIS.
Cui et al. (2013) Cardiology [91]	Single center (China)	Human 137 pts with ACS 60 healthy individuals 13 pts with stable angina (SA)	To investigate cf-DNA concentrations in ACS and their relationship with clinical features	cf-DNA as a new valuable marker for diagnosing and predicting the severity of coronary artery lesions and risk stratification in ACS.

Abbreviations; ACS, acute coronary syndrome; AIS, acute ischemic stroke; AMI, acute myocardial infarction; cf-DNA, circulating cell free DNA; dsDNA, double-stranded DNA; HMGB1, high mobility group box 1 protein; NET, neutrophil extracellular trap; PMNs; polymorphonuclear neutrophils; SA, stable angina; STEMI, ST elevation myocardial infarction; tPA, tissue-type plasminogen activator.

6. Therapeutic Implications of NETs in Cardiovascular Conditions

The afforded scientific literature states that NETs contribute to thrombosis and are considered effectors in the process of the amplification of inflammation. The development of vascular damage offers additional potential for the generation of the NET and NETs as therapeutic objectives. In the course of this reported analysis, we cited numerous experimental studies that used chloramidine as an inhibitor of PAD4. While on the one hand, the use of chloramidine can be useful to demonstrate the mechanisms of interference for the formation of NET, chloramidine works to inhibit different isoforms of PAD, and for this reason, it is characterized by a lack of specificity. This constitutes an important aspect of the clinical application limiting its use. Increasing research aimed at studying other PAD4 inhibitors could provide new avenues toward a direct approach in which the goal is to limit the formation of NETs. If, on the one hand, limiting the generation of NETs can be considered a valid approach, also the acceleration of the disintegration of these reticular structures can guarantee an effective result considering another therapeutic route. It was described how DNAse-1 can dissolve NETs by breaking their DNA strands. The use of DNAse was successful when it was applied to counteract the inflammatory response of the bronchial mucosa in patients with cystic fibrosis resulting in neutrophil-rich bronchial mucosa and NET generation. Another open front for the application of DNAse in clinical therapy, supported by clinical and experimental studies, is directed towards the cure of myocardial infarction. Improvements in this area were reported in IMA injuries due to ischemia-reperfusion after DNase infusion [128–131]. As far as myeloperoxidase inhibitors are concerned, they could be effective in reducing the production of ROS associated with the generation of NETs. Furthermore, the formation of NETs can be reduced, thus limiting the harmful consequences, with the administration of drugs for preventive purposes which include colchicine, inhibitors of complement, or phosphodiesterase 4 [132–134].

Two randomized trials were designed for the administration of colchicine in patients with recent myocardial infarction. In the CALCOT (Colchicine Cardiovascular Outcomes Trial) study, [135] 2366 patients treated with low-dose colchicine versus 2379 placebo patients revealed a significantly lower risk of myocardial infarction [0.91 (95% CI, 0.68 to 1.21)]. Again, the hazard ratios were 0.84 (95% CI, 0.46 to 1.52) for death from cardiovascular causes, and 0.83 (95% CI, 0.25 to 2.73) for resuscitated cardiac arrest in the cohort colchicine vs. placebo. The administration of colchicine at a dose of 0.5 mg daily was associated with a significantly lower risk of ischemic cardiovascular events than the placebo in patients with recent myocardial infarction.

Likewise in the other CONVINCE RCT (Colchicine for the prevention of vascular inflammation in non-cardioembolic patients Stroke) currently underway [136], 154 patients were enrolled from 17 countries in a multicenter international Prospective, Randomized Open-label, Blinded-Endpoint assessment (PROBE) controlled Phase 3 clinical trial. The study was designed to establish the safety and efficacy of low-dose colchicine anti-inflammatory therapy plus usual care with the goal of reducing recurrent vascular events in patients with non-severe, non-cardioembolic stroke and TIA compared with care alone. Colchicine proved safety and effectiveness as a novel therapeutic agent to induce inhibition of the inflammation cascade in patients with extra- or intracranial atherosclerosis or arteriolosclerosis, resulting in decreased vascular events [137,138]. The action exerted by colchicine could be directed towards the inhibition of inflammasome activity, [139–141] with the associated effect of reducing the burden of NETs generated in recent myocardial infarction or stroke tissue and could contribute strongly to support the outcome reported in the COLCOT study. Instead, the supposed action of reducing the accumulation of NETs in acute coronary syndrome suggested by the administration of anticoagulants and antiplatelet therapies provides sufficient evidence to demonstrate efficacy [142–144]. It is important to underline the contradiction that emerged from experimental evidence which proved how NETs can induce the polarization of macrophages in the infarct zones, promoting the healing of the affected myocardium through a push towards connectivity [145,146]. Therefore, caution in the use of NET inhibitors is necessary because their inhibition could be a sword of Damocles [144,146,147], in circumstances in which the inflammatory response can produce beneficial effects [145]. It is appropriate to affirm how the need to promote other clinical and experimental studies on the role of NETs can feed our knowledge and lead to an adequate use of drugs that interfere with the generation of NETs.

7. Limitations

The initial description of NETs is relatively recent and dates back to 2004, albeit it sprouted quickly. For this reason, the current review is limited to the relative novelty of NETs even if links between data emerging in the experimental literature and those present in the clinic focus on the role of NETs in numerous cardiovascular conditions. This evidence opens new windows for the mechanistic exploration of pathophysiology. The current availability of data is also limited regarding the use of NET biomarkers that can provide a step towards personalized precision medicine capable of identifying groups of subjects particularly susceptible to certain therapies. Importantly, the limitation of the data presented could be related to the lack of manuscripts in languages other than English that were not considered for this review, which could have reduced the number of available studies. While data are abundant, exemplified by the number of hits on primary search criteria, a large proportion of these came from secondary sources (reviews and editorials) that were not included as part of the systematic review process but discussed later. Since the recognition of the role of NETs in cardiovascular diseases is of fundamental importance, even more because it identifies a new series of therapeutic targets currently under intense consideration, the analysis of data coming from randomized clinical trials registry studies including a large number of patients becomes crucial. The provenance and certainty of the data are of substantial importance to the role played by the NET in cardiovascular disease which was the subject of some criticism, reinforcing the need for rigor in basic and clinical research in this field [53].

8. Conclusions

Our current knowledge of NETs' evolution and its influence on the propagation of cardiovascular insults such as acute ischemic stroke, peripheral artery disease or acute myocardial infarction, and subsequent development of heart failure are still in their infancy. However, therapeutic options look to be on the horizon. The immediate therapeutic options have the caveat of negating the beneficial effects of inflammation which play an important role in homeostasis. This should be the next avenue to the cardiovascular disease spectrum and its potential sequelae.

Supplementary Materials: The following supporting information can be downloaded at: https://www.mdpi.com/article/10.3390/biomedicines11010113/s1, Supplementary Table S1: Prisma checklist 2020. References [18,71–75] are cited in supplementary materials.

Author Contributions: Conceptualization, F.N.; methodology, F.N., F.B. and S.S.A.S.; software, F.B. and S.S.A.S.; validation, F.N. and S.S.A.S.; formal analysis, F.N., F.B. and S.S.A.S.; investigation, F.N.; data curation, F.N., F.B. and S.S.A.S.; writing—original draft preparation, F.N.; writing—review and editing, F.N. and S.S.A.S.; visualization, F.N. and S.S.A.S.; and supervision, F.N. and S.S.A.S. All authors have read and agreed to the published version of the manuscript.

Funding: This research received no external funding.

Institutional Review Board Statement: Not applicable.

Informed Consent Statement: Not applicable.

Data Availability Statement: Not applicable.

Conflicts of Interest: The authors declare no conflict of interest.

References

1. Fuchs, T.A.; Abed, U.; Goosmann, C.; Hurwitz, R.; Schulze, I.; Wahn, V.; Weinrauch, Y.; Brinkmann, V.; Zychlinsky, A. Novel cell death program leads to neutrophil extracellular traps. *J. Cell Biol.* **2007**, *176*, 231–241. [CrossRef]
2. Gu, Q.; Yang, X.; Lv, J.; Zhang, J.; Xia, B.; Kim, J.D.; Wang, R.; Xiong, F.; Meng, S.; Clements, T.P.; et al. AIBP-mediated cholesterol efflux instructs hematopoietic stem and progenitor cell fate. *Science* **2019**, *363*, 1085–1088. [CrossRef]
3. Nagareddy, P.R.; Murphy, A.J.; Stirzaker, R.A.; Hu, Y.; Yu, S.; Miller, R.G.; Ramkhelawon, B.; Distel, E.; Westerterp, M.; Huang, L.S.; et al. Hyperglycemia promotes myelopoiesis and impairs the resolution of atherosclerosis. *Cell Metab.* **2013**, *17*, 695–708. [CrossRef]
4. Wong, S.L.; Demers, M.; Martinod, K.; Gallant, M.; Wang, Y.; Goldfine, A.B.; Kahn, C.R.; Wagner, D.D. Diabetes primes neutrophils to undergo NETosis, which impairs wound healing. *Nat. Med.* **2015**, *21*, 815–819. [CrossRef]
5. Lim, C.H.; Adav, S.S.; Sze, S.K.; Choong, Y.K.; Saravanan, R.; Schmidtchen, A. Thrombin and plasmin alter the proteome of neutrophil extracellular traps. *Front. Immunol.* **2018**, *9*, 1554. [CrossRef]
6. Tall, A.R.; Westerterp, M. Inflammasomes, neutrophil extracellular traps, and cholesterol. *J. Lipid Res.* **2019**, *60*, 721–727. [CrossRef]
7. Keitelman, I.A.; Shiromizu, C.M.; Zgajnar, N.R.; Danielián, S.; Jancic, C.C.; Martí, M.A.; Fuentes, F.; Yancoski, J.; Vera Aguilar, D.; Rosso, D.A.; et al. The interplay between serine proteases and caspase-1 regulates the autophagy-mediated secretion of Interleukin-1 beta in human neutrophils. *Front. Immunol.* **2022**, *13*, 832306. [CrossRef]
8. Moschonas, I.C.; Tselepis, A.D. The pathway of neutrophil extracellular traps towards atherosclerosis and thrombosis. *Atherosclerosis* **2019**, *288*, 9–16. [CrossRef]
9. Jaiswal, S.; Natarajan, P.; Silver, A.J.; Gibson, C.J.; Bick, A.G.; Shvartz, E.; McConkey, M.; Gupta, N.; Gabriel, S.; Ardissino, D.; et al. Clonal hematopoiesis and risk of atherosclerotic cardiovascular disease. *N. Engl. J. Med.* **2017**, *377*, 111–121. [CrossRef]
10. Wang, W.; Liu, W.; Fidler, T.; Wang, Y.; Tang, Y.; Woods, B.; Welch, C.; Cai, B.; Silvestre-Roig, C.; Ai, D.; et al. Macrophage inflammation, erythrophagocytosis, and accelerated atherosclerosis in $Jak2^{V617F}$ mice. *Circ. Res.* **2018**, *123*, e35–e47. [CrossRef]
11. Wolach, O.; Sellar, R.S.; Martinod, K.; Cherpokova, D.; McConkey, M.; Chappell, R.J.; Silver, A.J.; Adams, D.; Castellano, C.A.; Schneider, R.K.; et al. Increased neutrophil extracellular trap formation promotes thrombosis in myeloproliferative neoplasms. *Sci. Transl. Med.* **2018**, *10*, eaan8292. [CrossRef] [PubMed]
12. Quillard, T.; Araújo, H.A.; Franck, G.; Shvartz, E.; Sukhova, G.; Libby, P. TLR2 and neutrophils potentiate endothelial stress, apoptosis and detachment: Implications for superficial erosion. *Eur. Heart J.* **2015**, *36*, 1394–1404. [CrossRef]
13. Li, G.X.; Jiang, X.H.; Zang, J.N.; Zhu, B.Z.; Jia, C.C.; Niu, K.W.; Liu, X.; Jiang, R.; Wang, B. B-cell receptor associated protein 31 deficiency decreases the expression of adhesion molecule CD11b/CD18 and PSGL-1 in neutrophils to ameliorate acute lung injury. *Int. J. Biochem. Cell Biol.* **2022**, *152*, 106299. [CrossRef]
14. Drechsler, M.; Megens, R.T.; van Zandvoort, M.; Weber, C.; Soehnlein, O. Hyperlipidemia-triggered neutrophilia promotes early atherosclerosis. *Circulation* **2010**, *122*, 1837–1845. [CrossRef]
15. Xu, X.; Wu, Y.; Xu, S.; Yin, Y.; Ageno, W.; De Stefano, V.; Zhao, Q.; Qi, X. Clinical significance of neutrophil extracellular traps biomarkers in thrombosis. *Thromb. J.* **2022**, *20*, 63. [CrossRef]
16. Pertiwi, K.R.; van der Wal, A.C.; Pabittei, D.R.; Mackaaij, C.; van Leeuwen, M.B.; Li, X.; de Boer, O.J. Neutrophil extracellular traps participate in all different types of thrombotic and haemorrhagic complications of coronary atherosclerosis. *Thromb. Haemost.* **2018**, *118*, 1078–1087. [CrossRef]
17. Megens, R.T.; Vijayan, S.; Lievens, D.; Döring, Y.; van Zandvoort, M.A.; Grommes, J.; Weber, C.; Soehnlein, O. Presence of luminal neutrophil extracellular traps in atherosclerosis. *Thromb. Haemost.* **2012**, *107*, 597–598. [CrossRef]
18. Page, M.J.; McKenzie, J.E.; Bossuyt, P.M.; Boutron, I.; Hoffmann, T.C.; Mulrow, C.D.; Shamseer, L.; Tetzlaff, J.M.; Akl, E.A.; Brennan, S.E.; et al. The PRISMA 2020 statement: An updated guideline for reporting systematic reviews. *BMJ* **2021**, *372*, n71. [CrossRef]
19. Herre, M.; Cedervall, J.; Mackman, N.; Olsson, A.K. Neutrophil extracellular traps in the pathology of cancer and other inflammatory diseases. *Physiol. Rev.* **2023**, *103*, 277–312. [CrossRef]
20. Rossaint, J.; Herter, J.M.; Van Aken, H.; Napirei, M.; Döring, Y.; Weber, C.; Soehnlein, O.; Zarbock, A. Synchronized integrin engagement and chemokine activation is crucial in neutrophil extracellular trap-mediated sterile inflammation. *Blood J. Am. Soc. Hematol.* **2014**, *123*, 2573–2584. [CrossRef]
21. Winter, C.; Silvestre-Roig, C.; Ortega-Gomez, A.; Lemnitzer, P.; Poelman, H.; Schumski, A.; Winter, J.; Drechsler, M.; de Jong, R.; Immler, R.; et al. Chrono-pharmacological targeting of the CCL2-CCR2 axis ameliorates atherosclerosis. *Cell Metab.* **2018**, *28*, 175–182.e5. [CrossRef] [PubMed]
22. Adrover, J.M.; Del Fresno, C.; Crainiciuc, G.; Cuartero, M.I.; Casanova-Acebes, M.; Weiss, L.A.; Huerga-Encabo, H.; Silvestre-Roig, C.; Rossaint, J.; Cossío, I.; et al. A neutrophil timer coordinates immune defense and vascular protection. *Immunity* **2019**, *51*, 966–967. [CrossRef] [PubMed]
23. Yan, S.L.; Hwang, I.Y.; Kamenyeva, O.; Kabat, J.; Kim, J.S.; Park, C.; Kehrl, J.H. Unrestrained Gαi2 Signaling Disrupts Neutrophil Trafficking, Aging, and Clearance. *Front. Immunol.* **2021**, *12*, 679856. [CrossRef] [PubMed]
24. Casanova-Acebes, M.; Nicolás-Ávila, J.A.; Li, J.L.; García-Silva, S.; Balachande, A.; Rubio-Ponce, A.; Weiss, L.A.; Adrover, J.M.; Burrows, K.; A-González, N.; et al. Neutrophils instruct homeostatic and pathological states in naive tissues. *J. Exp. Med.* **2018**, *215*, 2778–2795. [CrossRef]

25. Zheng, Y.; Sefik, E.; Astle, J.; Karatepe, K.; Öz, H.H.; Solis, A.G.; Jackson, R.; Luo, H.R.; Bruscia, E.M.; Halene, S.; et al. Human neutrophil development and functionality are enabled in a humanized mouse model. *Proc. Natl. Acad. Sci. USA* **2022**, *119*, e2121077119. [CrossRef]
26. Wigerblad, G.; Kaplan, M.J. Neutrophil extracellular traps in systemic autoimmune and autoinflammatory diseases. *Nat. Rev. Immunol.* 2022; 1–15, *online ahead of print*. [CrossRef]
27. Lee, M.; Lee, S.Y.; Bae, Y.S. Emerging roles of neutrophils in immune homeostasis. *BMB Rep.* **2022**, *55*, 473–480. [CrossRef]
28. Palmer, L.J.; Cooper, P.R.; Ling, M.R.; Wright, H.J.; Huissoon, A.; Chapple, I.L. Hypochlorous acid regulates neutrophil extracellular trap release in humans. *Clin. Exp. Immunol.* **2011**, *167*, 261–268. [CrossRef]
29. Cassatella, M.A.; Östberg, N.K.; Tamassia, N.; Soehnlein, O. Biological roles of neutrophil-derived granule proteins and cytokines. *Trends Immunol.* **2019**, *40*, 648–664. [CrossRef]
30. Urban, C.F.; Ermert, D.; Schmid, M.; Abu-Abed, U.; Goosmann, C.; Nacken, W.; Brinkmann, V.; Jungblut, P.R.; Zychlinsky, A. Neutrophil extracellular traps contain calprotectin, a cytosolic protein complex involved in host defense against Candida albicans. *PLoS Pathog.* **2009**, *5*, e1000639. [CrossRef]
31. Zhong, H.; Lu, R.Y.; Wang, Y. Neutrophil extracellular traps in fungal infections: A seesaw battle in hosts. *Front. Immunol.* **2022**, *13*, 977493. [CrossRef]
32. Subramaniam, S.; Kothari, H.; Bosmann, M. Tissue factor in COVID-19-associated coagulopathy. *Thromb. Res.* **2022**, *220*, 35–47. [CrossRef]
33. Wei, Z.; Cheng, Q.; Xu, N.; Zhao, C.; Xu, J.; Kang, L.; Lou, X.; Yu, L.; Feng, W. Investigation of CRS-associated cytokines in CAR-T therapy with meta-GNN and pathway crosstalk. *BMC Bioinform.* **2022**, *23*, 373. [CrossRef]
34. Wang, H.; Shao, J.; Lu, X.; Jiang, M.; Li, X.; Liu, Z.; Zhao, Y.; Zhou, J.; Lin, L.; Wang, L.; et al. Potential of immune-related genes as promising biomarkers for premature coronary heart disease through high throughput sequencing and integrated bioinformatics analysis. *Front. Cardiovasc. Med.* **2022**, *9*, 893502. [CrossRef]
35. Arazna, M.; Pruchniak, M.P.; Zycinska, K.; Demkow, U. Neutrophil extracellular trap in human diseases. *Adv. Exp. Med. Biol.* **2013**, *756*, 1–8. [CrossRef]
36. Brinkmann, V.; Reichard, U.; Goosmann, C.; Fauler, B.; Uhlemann, Y.; Weiss, D.S.; Weinrauch, Y.; Zychlinsky, A. Neutrophil extracellular traps kill bacteria. *Science* **2004**, *303*, 1532–1535. [CrossRef]
37. Brinkmann, V.; Laube, B.; Abu Abed, U.; Goosmann, C.; Zychlinsky, A. Neutrophil extracellular traps: How to generate and visualize them. *J. Vis. Exp.* **2010**, *36*, e1724. [CrossRef]
38. Leung, H.H.L.; Perdomo, J.; Ahmadi, Z.; Zheng, S.S.; Rashid, F.N.; Enjeti, A.; Ting, S.B.; Chong, J.J.H.; Chong, B.H. NETosis and thrombosis in vaccine-induced immune thrombotic thrombocytopenia. *Nat. Commun.* **2022**, *13*, 5206. [CrossRef]
39. Remijsen, Q.; Vanden Berghe, T.; Wirawan, E.; Asselbergh, B.; Parthoens, E.; De Rycke, R.; Noppen, S.; Delforge, M.; Willems, J.; Vandenabeele, P. Neutrophil extracellular trap cell death requires both autophagy and superoxide generation. *Cell Res.* **2011**, *21*, 290–304. [CrossRef]
40. Amulic, B.; Knackstedt, S.L.; Abu Abed, U.; Deigendesch, N.; Harbort, C.J.; Caffrey, B.E.; Brinkmann, V.; Heppner, F.L.; Hinds, P.W.; Zychlinsky, A. Cell-Cycle proteins control production of neutrophil extracellular traps. *Dev. Cell* **2017**, *43*, 449–462.e5. [CrossRef]
41. Hakkim, A.; Fuchs, T.A.; Martinez, N.E.; Hess, S.; Prinz, H.; Zychlinsky, A.; Waldmann, H. Activation of the Raf-MEK-ERK pathway is required for neutrophil extracellular trap formation. *Nat. Chem. Biol.* **2011**, *7*, 75–77. [CrossRef]
42. Branitzki-Heinemann, K.; Möllerherm, H.; Völlger, L.; Husein, D.M.; de Buhr, N.; Blodkamp, S.; Reuner, F.; Brogden, G.; Naim, H.Y.; von Köckritz-Blickwede, M. Formation of neutrophil extracellular traps under low oxygen level. *Front. Immunol.* **2016**, *7*, 518. [CrossRef]
43. Dölling, M.; Eckstein, M.; Singh, J.; Schauer, C.; Schoen, J.; Shan, X.; Bozec, A.; Knopf, J.; Schett, G.; Muñoz, L.E.; et al. Hypoxia Promotes Neutrophil Survival After Acute Myocardial Infarction. *Front. Immunol.* **2022**, *13*, 726153. [CrossRef]
44. Metzler, K.D.; Fuchs, T.A.; Nauseef, W.M.; Reumaux, D.; Roesler, J.; Schulze, I.; Wahn, V.; Papayannopoulos, V.; Zychlinsky, A. Myeloperoxidase is required for neutrophil extracellular trap formation: Implications for innate immunity. *Blood* **2011**, *117*, 953–959. [CrossRef]
45. Björnsdottir, H.; Welin, A.; Michaëlsson, E.; Osla, V.; Berg, S.; Christenson, K.; Sundqvist, M.; Dahlgren, C.; Karlsson, A. Bylund Neutrophil NET formation is regulated from the inside by myeloperoxidase-processed reactive oxygen species. *Free Radic. Biol. Med.* **2015**, *89*, 1024–1035. [CrossRef]
46. Lockhart, J.S.; Sumagin, R. Non-Canonical Functions of Myeloperoxidase in Immune Regulation, Tissue Inflammation and Cancer. *Int. J. Mol. Sci.* **2022**, *23*, 12250. [CrossRef]
47. Kenny, E.F.; Herzig, A.; Kruger, R.; Muth, A.; Mondal, S.; Thompson, P.R.; Brinkmann, V.; Bernuth, H.V.; Zychlinsky, A. Diverse stimuli engage different neutrophil extracellular trap pathways. *eLife* **2017**, *6*, e24437. [CrossRef]
48. Neeli, I.; Radic, M. Opposition between PKC isoforms regulates histone deimination and neutrophil extracellular chromatin release. *Front. Immunol.* **2013**, *4*, 38. [CrossRef]
49. Wang, Y.; Li, M.; Stadler, S.; Correll, S.; Li, P.; Wang, D.; Hayama, R.; Leonelli, L.; Han, H.; Grigoryev, S.A.; et al. Histone hypercitrullination mediates chromatin decondensation and neutrophil extracellular trap formation. *J. Cell Biol.* **2009**, *184*, 205–213. [CrossRef]

50. Leshner, M.; Wang, S.; Lewis, C.; Zheng, H.; Chen, X.A.; Santy, L.; Wang, Y. PAD4 mediated histone hypercitrullination induces heterochromatin decondensation and chromatin unfolding to form neutrophil extracellular trap-like structures. *Front. Immunol.* **2012**, *3*, 307. [CrossRef]
51. Li, P.; Li, M.; Lindberg, M.R.; Kennett, M.J.; Xiong, N.; Wang, Y. PAD4 is essential for antibacterial innate immunity mediated by neutrophil extracellular traps. *J. Exp. Med.* **2010**, *207*, 1853–1862. [CrossRef]
52. Martinod, K.; Fuchs, T.A.; Zitomersky, N.L.; Wong, S.L.; Demers, M.; Gallant, M.; Wang, Y.; Wagner, D.D. PAD4-deficiency does not affect bacteremia in polymicrobial sepsis and ameliorates endotoxemic shock. *Blood* **2015**, *125*, 1948–1956. [CrossRef]
53. Boeltz, S.; Amini, P.; Anders, H.J.; Andrade, F.; Bilyy, R.; Chatfield, S.; Cichon, I.; Clancy, D.M.; Desai, J.; Dumych, T.; et al. To NET or not to NET: Current opinions and state of the science regarding the formation of neutrophil extracellular traps. *Cell Death Differ.* **2019**, *26*, 395–408. [CrossRef]
54. Silvestre-Roig, C.; Braster, Q.; Wichapong, K.; Lee, E.Y.; Teulon, J.M.; Berrebeh, N.; Winter, J.; Adrover, J.M.; Santos, G.S.; Froese, A.; et al. Externalized histone H4 orchestrates chronic inflammation by inducing lytic cell death. *Nature* **2019**, *569*, 236–240. [CrossRef]
55. Zhang, Y.; Jian, W.; He, L.; Wu, J. Externalized histone H4: A novel target that orchestrates chronic inflammation by inducing lytic cell death. *Acta Biochim. Biophys. Sin.* **2020**, *52*, 336–338. [CrossRef]
56. Gorabi, A.M.; Penson, P.E.; Banach, M.; Motallebnezhad, M.; Jamialahmadi, T.; Sahebkar, A. Epigenetic control of atherosclerosis via DNA methylation: A new therapeutic target? *Life Sci.* **2020**, *253*, 117682. [CrossRef]
57. Adrover, J.M.; Aroca-Crevillén, A.; Crainiciuc, G.; Ostos, F.; Rojas-Vega, Y.; Rubio-Ponce, A.; Cilloniz, C.; Bonzón-Kulichenko, E.; Calvo, E.; Rico, D.; et al. Programmed 'disarming' of the neutrophil proteome reduces the magnitude of inflammation. *Nat. Immunol.* **2020**, *21*, 135–144. [CrossRef]
58. Talal, S.; Mona, K.; Karem, A.; Yaniv, L.; Reut, H.M.; Ariel, S.; Moran, A.K.; Harel, E.; Campisi-Pinto, S.; Mahmoud, A.A.; et al. Neutrophil degranulation and severely impaired extracellular trap formation at the basis of susceptibility to infections of hemodialysis patients. *BMC Med.* **2022**, *20*, 364. [CrossRef]
59. Kolte, D.; Libby, P.; Jang, I.K. New Insights into Plaque Erosion as a Mechanism of Acute Coronary Syndromes. *JAMA* **2021**, *325*, 1043–1044. [CrossRef]
60. Libby, P. Mechanisms of acute coronary syndromes and their implications for therapy. *N. Engl. J. Med.* **2013**, *368*, 2004–2013. [CrossRef]
61. Brezinski, M.; Willard, F.; Rupnick, M. Inadequate Intimal Angiogenesis as a Source of Coronary Plaque Instability: Implications for Healing. *Circulation* **2019**, *140*, 1857–1859. [CrossRef]
62. Fernandez, D.M.; Rahman, A.H.; Fernandez, N.F.; Chudnovskiy, A.; Amir, E.D.; Amadori, L.; Khan, N.S.; Wong, C.K.; Shamailova, R.; Hill, C.A.; et al. Single cell immune landscape of human atherosclerotic plaques. *Nat. Med.* **2019**, *25*, 1576–1588. [CrossRef] [PubMed]
63. Mangge, H.; Prüller, F.; Schnedl, W.; Renner, W.; Almer, G. Beyond Macrophages and T Cells: B Cells and Immunoglobulins Determine the Fate of the Atherosclerotic Plaque. *Int. J. Mol. Sci.* **2020**, *21*, 4082. [CrossRef] [PubMed]
64. Dentali, F.; Nigro, O.; Squizzato, A.; Gianni, M.; Zuretti, F.; Grandi, A.M.; Guasti, L. Impact of neutrophils to lymphocytes ratio on major clinical outcomes in patients with acute coronary syndromes: A systematic review and meta-analysis of the literature. *Int. J. Cardiol.* **2018**, *266*, 31–37. [CrossRef] [PubMed]
65. Guasti, L.; Dentali, F.; Castiglioni, L.; Maroni, L.; Marino, F.; Squizzato, A.; Ageno, W.; Gianni, M.; Gaudio, G.; Grandi, A.M.; et al. Neutrophils and clinical outcomes in patients with acute coronary syndromes and/or cardiac revascularization. A systematic review on more than 34,000 subjects. *Thromb. Haemost.* **2011**, *106*, 591–599. [CrossRef] [PubMed]
66. Dahdah, A.; Johnson, J.; Gopalkrishna, S.; Jaggers, R.M.; Webb, D.; Murphy, A.J.; Hanssen, N.M.J.; Hanaoka, B.Y.; Nagareddy, P.R. Neutrophil Migratory Patterns: Implications for Cardiovascular Disease. *Front. Cell Dev. Biol.* **2022**, *10*, 795784. [CrossRef] [PubMed]
67. Friedman, G.D.; Klatsky, A.L.; Siegelaub, A.B. The leukocyte count as a predictor of myocardial infarction. *N. Engl. J. Med.* **1974**, *290*, 1275–1278. [CrossRef]
68. Zhao, Y.; Zhang, S.; Yi, Y.; Qu, T.; Gao, S.; Lin, Y.; Zhu, H. Neutrophil-to-lymphocyte ratio as a predictor for cardiovascular diseases: A cohort study in Tianjin, China. *J. Hum. Hypertens.* 2022, *online ahead of print*. [CrossRef]
69. Zhang, R.; Brennan, M.L.; Fu, X.; Aviles, R.J.; Pearce, G.L.; Penn, M.S.; Topol, E.J.; Sprecher, D.L.; Hazen, S.L. Association between myeloperoxidase levels and risk of coronary artery disease. *JAMA* **2001**, *286*, 2136–2142. [CrossRef]
70. Borissoff, J.I.; Joosen, I.A.; Versteylen, M.O.; Brill, A.; Fuchs, T.A.; Savchenko, A.S.; Gallant, M.; Martinod, K.; Ten Cate, H.; Hofstra, L.; et al. Elevated levels of circulating DNA and chromatin are independently associated with severe coronary atherosclerosis and a prothrombotic state. *Arter. Thromb. Vasc. Biol.* **2013**, 33, 2032–2040. [CrossRef]
71. Libby, P. The molecular bases of the acute coronary syndromes. *Circulation* **1995**, *91*, 2844–2850. [CrossRef]
72. Narula, J.; Garg, P.; Achenbach, S.; Motoyama, S.; Virmani, R.; Strauss, H.W. Arithmetic of vulnerable plaques for noninvasive imaging. *Nat. Clin. Pract. Cardiovasc. Med.* **2008**, *5*, S2–S10. [CrossRef]
73. Libby, P.; Theroux, P. Pathophysiology of coronary artery disease. *Circulation* **2005**, *111*, 3481–3488. [CrossRef] [PubMed]
74. Falk, E.; Nakano, M.; Benton, J.F.; Finn, A.V.; Virmani, R. Update on acute coronary syndromes: The pathologists' view. *Eur. Heart J.* **2013**, *34*, 719–728. [CrossRef] [PubMed]
75. Crea, F.; Liuzzo, G. Pathogenesis of acute coronary syndromes. *J. Am. Coll. Cardiol.* **2013**, *61*, 1–11. [CrossRef] [PubMed]

76. Ionita, M.G.; van den Borne, P.; Catanzariti, L.M.; Moll, F.L.; de Vries, J.P.; Pasterkamp, G.; Vink, A.; de Kleijn, D.P. High neutrophil numbers in human carotid atherosclerotic plaques are associated with characteristics of rupture-prone lesions. *Arter. Thromb. Vasc. Biol.* **2010**, *30*, 1842–1848. [CrossRef]
77. Lee, M.W.; Luo, E.W.; Silvestre-Roig, C.; Srinivasan, Y.; Akabori, K.; Lemnitzer, P.; Schmidt, N.W.; Lai, G.H.; Santangelo, C.D.; Soehnlein, O.; et al. Apolipoprotein Mimetic Peptide Inhibits Neutrophil-Driven Inflammatory Damage via Membrane Remodeling and Suppression of Cell Lysis. *ACS Nano* **2021**, *26*, 15930–15939. [CrossRef] [PubMed]
78. Locke, M.; Francis, R.J.; Tsaousi, E.; Longstaff, C. Fibrinogen protects neutrophils from the cytotoxic effects of histones and delays neutrophil extracellular trap formation induced by ionomycin. *Sci. Rep.* **2020**, *10*, 11694. [CrossRef] [PubMed]
79. Saffarzadeh, M.; Juenemann, C.; Queisser, M.A.; Lochnit, G.; Barreto, G.; Galuska, S.P.; Lohmeyer, J.; Preissner, K.T. Neutrophil extracellular traps directly induce epithelial and endothelial cell death: A predominant role of histones. *PLoS ONE* **2012**, *7*, e32366. [CrossRef] [PubMed]
80. Wildhagen, K.C.; García de Frutos, P.; Reutelingsperger, C.P.; Schrijver, R.; Aresté, C.; Ortega-Gómez, A.; Deckers, N.M.; Hemker, H.C.; Soehnlein, O.; Nicolaes, G.A. Nonanticoagulant heparin prevents histone-mediated cytotoxicity in vitro and improves survival in sepsis. *Blood* **2014**, *123*, 1098–1101. [CrossRef] [PubMed]
81. Paulin, N.; Viola, J.R.; Maas, S.L.; de Jong, R.; Fernandes-Alnemri, T.; Weber, C.; Drechsler, M.; Döring, Y.; Soehnlein, O. Double-Strand DNA sensing aim2 inflammasome regulates atherosclerotic plaque vulnerability. *Circulation* **2018**, *138*, 321–323. [CrossRef]
82. Tall, A.R.; Fuster, J.J. Clonal hematopoiesis in cardiovascular disease and therapeutic implications. *Nat. Cardiovasc. Res.* **2022**, *1*, 116–124. [CrossRef]
83. Liao, Y.; Liu, K.; Zhu, L. Emerging Roles of Inflammasomes in Cardiovascular Diseases. *Front. Immunol.* **2022**, *13*, 834289. [CrossRef] [PubMed]
84. Thålin, C.; Hisada, Y.; Lundström, S.; Mackman, N.; Wallén, H. Neutrophil extracellular traps: Villains and targets in arterial, venous, and cancer-associated thrombosis. *Arterioscler. Thromb. Vasc. Biol.* **2019**, *39*, 1724–1738. [CrossRef] [PubMed]
85. Carestia, A.; Kaufman, T.; Schattner, M. Platelets: New bricks in the building of neutrophil extracellular traps. *Front. Immunol.* **2016**, *7*, 271. [CrossRef] [PubMed]
86. Gould, T.J.; Vu, T.T.; Swystun, L.L.; Dwivedi, D.J.; Mai, S.H.; Weitz, J.I.; Liaw, P.C. Neutrophil extracellular traps promote thrombin generation through platelet dependent and platelet-independent mechanisms. *Arter. Thromb. Vasc. Biol.* **2014**, *34*, 1977–1984. [CrossRef] [PubMed]
87. Fuchs, T.A.; Brill, A.; Duerschmied, D.; Schatzberg, D.; Monestier, M.; Myers DDJr Wrobleski, S.K.; Wakefield, T.W.; Hartwig, J.H.; Wagner, D.D. Extracellular DNA traps promote thrombosis. *Proc. Natl. Acad. Sci. USA* **2010**, *107*, 15880–15885. [CrossRef]
88. Von Brühl, M.L.; Stark, K.; Steinhart, A.; Chandraratne, S.; Konrad, I.; Lorenz, M.; Khandoga, A.; Tirniceriu, A.; Coletti, R.; Köllnberger, M.; et al. Monocytes, neutrophils, and platelets cooperate to initiate and propagate venous thrombosis in mice in vivo. *J. Exp. Med.* **2012**, *209*, 819–835. [CrossRef]
89. Noubouossie, D.F.; Whelihan, M.F.; Yu, Y.B.; Sparkenbaugh, E.; Pawlinski, R.; Monroe, D.M.; Key, N.S. In vitro activation of coagulation by human neutrophil DNA and histone proteins but not neutrophil extracellular traps. *Blood* **2017**, *129*, 1021–1029. [CrossRef]
90. Riegger, J.; Byrne, R.A.; Joner, M.; Chandraratne, S.; Gershlick, A.H.; Ten Berg, J.M.; Adriaenssens, T.; Guagliumi, G.; Godschalk, T.C.; Neumann, F.J.; et al. Prevention of Late Stent Thrombosis by an Interdisciplinary Global European Effort (PRESTIGE) Investigators. Histopathological evaluation of thrombus in patients presenting with stent thrombosis. A multicenter European study: A report of the prevention of late stent thrombosis by an interdisciplinary global European effort consortium. *Eur. Heart J.* **2016**, *37*, 1538–1549. [CrossRef]
91. Farkas, Á.Z.; Farkas, V.J.; Gubucz, I.; Szabó, L.; Bálint, K.; Tenekedjiev, K.; Nagy, A.I.; Sótonyi, P.; Hidi, L.; Nagy, Z.; et al. Neutrophil extracellular traps in thrombi retrieved during interventional treatment of ischemic arterial diseases. *Thromb. Res.* **2019**, *175*, 46–52. [CrossRef]
92. De Boer, O.J.; Li, X.; Teeling, P.; Mackaay, C.; Ploegmakers, H.J.; van der Loos, C.M.; Daemen, M.J.; de Winter, R.J.; van der Wal, A.C. Neutrophils, neutrophil extracellular traps and interleukin-17 associate with the organization of thrombi in acute myocardial infarction. *Thromb. Haemost.* **2013**, *109*, 290–297. [CrossRef]
93. Mangold, A.; Alias, S.; Scherz, T.; Hofbauer, T.; Jakowitsch, J.; Panzenböck, A.; Simon, D.; Laimer, D.; Bangert, C.; Kammerlander, A.; et al. Coronary neutrophil extracellular trap burden and deoxyribonuclease activity in ST-elevation acute coronary syndrome are predictors of ST-segment resolution and infarct size. *Circ Res.* **2015**, *116*, 1182–1192. [CrossRef] [PubMed]
94. Stakos, D.A.; Kambas, K.; Konstantinidis, T.; Mitroulis, I.; Apostolidou, E.; Arelaki, S.; Tsironidou, V.; Giatromanolaki, A.; Skendros, P.; Konstantinides, S.; et al. Expression of functional tissue factor by neutrophil extracellular traps in culprit artery of acute myocardial infarction. *Eur. Heart J.* **2015**, *36*, 1405–1414. [CrossRef] [PubMed]
95. Laridan, E.; Denorme, F.; Desender, L.; François, O.; Andersson, T.; Deckmyn, H.; Vanhoorelbeke, K.; De Meyer, S.F. Neutrophil extracellular traps in ischemic stroke thrombi. *Ann. Neurol.* **2017**, *82*, 223–232. [CrossRef] [PubMed]
96. Pertiwi, K.R.; de Boer, O.J.; Mackaaij, C.; Pabittei, D.R.; de Winter, R.J.; Li, X.; van der Wal, A.C. Extracellular traps derived from macrophages, mast cells, eosinophils and neutrophils are generated in a time-dependent manner during atherothrombosis. *J. Pathol.* **2019**, *247*, 505–512. [CrossRef]

97. Maugeri, N.; Campana, L.; Gavina, M.; Covino, C.; De Metrio, M.; Panciroli, C.; Maiuri, L.; Maseri, A.; D'Angelo, A.; Bianchi, M.E.; et al. Activated platelets present high mobility group box 1 to neutrophils, inducing autophagy and promoting the extrusion of neutrophil extracellular traps. *J. Thromb. Haemost.* **2014**, *12*, 2074–2088. [CrossRef]
98. Zaid, Y.; Merhi, Y. Implication of Platelets in Immuno-Thrombosis and Thrombo-Inflammation. *Front. Cardiovasc. Med.* **2022**, *9*, 863846. [CrossRef]
99. Theofilis, P.; Sagris, M.; Oikonomou, E.; Antonopoulos, A.S.; Tsioufis, K.; Tousoulis, D. Factors Associated with Platelet Activation-Recent Pharmaceutical Approaches. *Int. J. Mol. Sci.* **2022**, *23*, 3301. [CrossRef]
100. Novotny, J.; Oberdieck, P.; Titova, A.; Pelisek, J.; Chandraratne, S.; Nicol, P.; Hapfelmeier, A.; Joner, M.; Maegdefessel, L.; Poppert, H.; et al. Thrombus NET content is associated with clinical outcome in stroke and myocardial infarction. *Neurology* **2020**, *94*, e2346–e2360. [CrossRef]
101. Chilingaryan, Z.; Deshmukh, T.; Leung, H.H.L.; Perdomo, J.; Emerson, P.; Kurup, R.; Chong, B.H.; Chong, J.J.H. Erythrocyte interaction with neutrophil extracellular traps in coronary artery thrombosis following myocardial infarction. *Pathology* **2022**, *54*, 87–94. [CrossRef]
102. Knight, J.S.; Luo, W.; O'Dell, A.A.; Yalavarthi, S.; Zhao, W.; Subramanian, V.; Guo, C.; Grenn, R.C.; Thompson, P.R.; Eitzman, D.T.; et al. Peptidylarginine deiminase inhibition reduces vascular damage and modulates innate immune responses in murine models of atherosclerosis. *Circ. Res.* **2014**, *114*, 947–956. [CrossRef]
103. Bonnard, T.; Hagemeyer, C.E. Ferric Chloride-induced Thrombosis Mouse Model on Carotid Artery Mesentery Vessel. *J. Vis. Exp.* **2015**, e52838. [CrossRef] [PubMed]
104. Novotny, J.; Chandraratne, S.; Weinberger, T.; Philippi, V.; Stark, K.; Ehrlich, A.; Pircher, J.; Konrad, I.; Oberdieck, P.; Titova, A.; et al. Histological comparison of arterial thrombi in mice and men and the influence of Cl-amidine on thrombus formation. *PLoS ONE* **2018**, *13*, e0190728. [CrossRef] [PubMed]
105. Ducroux, C.; Di Meglio, L.; Loyau, S.; Delbosc, S.; Boisseau, W.; Deschildre, C.; Ben Maacha, M.; Blanc, R.; Redjem, H.; Ciccio, G.; et al. Thrombus neutrophil extracellular traps content impair tPA-induced thrombolysis in acute ischemic stroke. *Stroke* **2018**, *49*, 754–757. [CrossRef] [PubMed]
106. Massberg, S.; Grahl, L.; von Bruehl, M.L.; Manukyan, D.; Pfeiler, S.; Goosmann, C.; Brinkmann, V.; Lorenz, M.; Bidzhekov, K.; Khandagale, A.B.; et al. Reciprocal coupling of coagulation and innate immunity via neutrophil serine proteases. *Nat. Med.* **2010**, *16*, 887–896. [CrossRef] [PubMed]
107. Sorvillo, N.; Mizurini, D.M.; Coxon, C.; Martinod, K.; Tilvawala, R.; Cherpokova, D.; Salinger, A.J.; Seward, R.J.; Staudinger, C.; Weerapana, E.; et al. Plasma peptidylarginine deiminase IV promotes VWF-platelet string formation and accelerates thrombosis after vessel injury. *Circ. Res.* **2019**, *125*, 507–519. [CrossRef] [PubMed]
108. Cherpokova, D.; Jouvene, C.C.; Libreros, S.; DeRoo, E.P.; Chu, L.; de la Rosa, X.; Norris, P.C.; Wagner, D.D.; Serhan, C.N. Resolvin D4 attenuates the severity of pathological thrombosis in mice. *Blood* **2019**, *134*, 1458–1468. [CrossRef]
109. Franck, G.; Mawson, T.L.; Folco, E.J.; Molinaro, R.; Ruvkun, V.; Engelbertsen, D.; Liu, X.; Tesmenitsky, Y.; Shvartz, E.; Sukhova, G.K.; et al. Roles of PAD4 and NETosis in experimental atherosclerosis and arterial injury: Implications for superficial erosion. *Circ. Res.* **2018**, *123*, 33–42. [CrossRef] [PubMed]
110. Shimony, A.; Zahger, D.; Gilutz, H.; Goldstein, H.; Orlov, G.; Merkin, M.; Shalev, A.; Ilia, R.; Douvdevani, A. Cell free DNA detected by a novel method in acute ST-elevation myocardial infarction patients. *Acute Card. Care* **2010**, *12*, 109–111. [CrossRef]
111. Cui, M.; Fan, M.; Jing, R.; Wang, H.; Qin, J.; Sheng, H.; Wang, Y.; Wu, X.; Zhang, L.; Zhu, J.; et al. Cell-Free circulating DNA: A new biomarker for the acute coronary syndrome. *Cardiology* **2013**, *124*, 76–84. [CrossRef]
112. Langseth, M.S.; Helseth, R.; Ritschel, V.; Hansen, C.H.; Andersen, G.Ø.; Eritsland, J.; Halvorsen, S.; Fagerland, M.W.; Solheim, S.; Arnesen, H.; et al. Double-Stranded DNA and NETs Components in Relation to Clinical Outcome After ST-Elevation Myocardial Infarction. *Sci. Rep.* **2020**, *10*, 5007. [CrossRef]
113. Helseth, R.; Solheim, S.; Arnesen, H.; Seljeflot, I.; Opstad, T.B. The time course of markers of neutrophil extracellular traps in patients undergoing revascularization for acute myocardial infarction or stable angina pectoris. *Mediat. Inflamm.* **2016**, *2016*, 2182358. [CrossRef] [PubMed]
114. Vallés, J.; Lago, A.; Santos, M.T.; Latorre, A.M.; Tembl, J.I.; Salom, J.B.; Nieves, C.; Moscardó, A. Neutrophil extracellular traps are increased in patients with acute ischemic stroke: Prognostic significance. *Thromb. Haemost.* **2017**, *117*, 1919–1929. [CrossRef] [PubMed]
115. Hirose, T.; Hamaguchi, S.; Matsumoto, N.; Irisawa, T.; Seki, M.; Tasaki, O.; Hosotsubo, H.; Yamamoto, N.; Yamamoto, K.; Akeda, Y.; et al. Presence of neutrophil extracellular traps and citrullinated histone H3 in the bloodstream of critically ill patients. *PLoS Med.* **2007**, *4*, e297. [CrossRef] [PubMed]
116. Helseth, R.; Shetelig, C.; Andersen, G.Ø.; Langseth, M.S.; Limalanathan, S.; Opstad, T.B.; Arnesen, H.; Hoffmann, P.; Eritsland, J.; Seljeflot, I. Neutrophil extracellular trap components associate with infarct size, ventricular function, and clinical outcome in STEMI. *Mediat. Inflamm.* **2019**, *2019*, 7816491. [CrossRef] [PubMed]
117. Langseth, M.S.; Opstad, T.B.; Bratseth, V.; Solheim, S.; Arnesen, H.; Pettersen, A.Å.; Seljeflot, I.; Helseth, R. Markers of neutrophil extracellular traps are associated with adverse clinical outcome in stable coronary artery disease. *Eur. J. Prev. Cardiol.* **2018**, *25*, 762–769. [CrossRef]
118. Martinod, K.; Witsch, T.; Farley, K.; Gallant, M.; Remold-O'Donnell, E.; Wagner, D.D. Neutrophil elastase-deficient mice form neutrophil extracellular traps in an experimental model of deep vein thrombosis. *J. Thromb. Haemost.* **2016**, *14*, 551–558. [CrossRef]

119. Kim, J.K.; Hong, C.W.; Park, M.J.; Song, Y.R.; Kim, H.J.; Kim, S.G. Increased neutrophil extracellular trap formation in uremia is associated with chronic inflammation and prevalent coronary artery disease. *J. Immunol. Res.* **2017**, *2017*, 8415179. [CrossRef]
120. Bang, O.Y.; Chung, J.W.; Cho, Y.H.; Oh, M.J.; Seo, W.K.; Kim, G.M.; Ahn, M.J. Circulating DNAs, a marker of neutrophil extracellular traposis and cancer-related stroke: The OASIS-Cancer Study. *Stroke* **2019**, *50*, 2944–2947. [CrossRef]
121. Libby, P.; King, K. Biomarkers: A challenging conundrum in cardiovascular disease. *Arterioscler Thromb. Vasc. Biol.* **2015**, *35*, 2491–2495. [CrossRef]
122. Lee, K.H.; Cavanaugh, L.; Leung, H.; Yan, F.; Ahmadi, Z.; Chong, B.H.; Passam, F. Quantification of NETs-associated markers by flow cytometry and serum assays in patients with thrombosis and sepsis. *Int. J. Lab. Hematol.* **2018**, *40*, 392–399. [CrossRef]
123. Langseth, M.S.; Andersen, G.Ø.; Husebye, T.; Arnesen, H.; Zucknick, M.; Solheim, S.; Eritsland, J.; Seljeflot, I.; Opstad, T.B.; Helseth, R. Neutrophil extracellular trap components and myocardial recovery in post-ischemic acute heart failure. *PLoS ONE* **2020**, *15*, e0241333. [CrossRef]
124. Jiménez-Alcázar, M.; Kim, N.; Fuchs, T.A. Circulating Extracellular DNA: Cause or Consequence of Thrombosis? *Semin. Thromb. Hemost.* **2017**, *43*, 553–561. [CrossRef]
125. Kithcart, A.P.; Libby, P. Casting NETs to predict cardiovascular outcomes. *Eur. J. Prev. Cardiol.* **2018**, *25*, 759–761. [CrossRef] [PubMed]
126. De Arriba-Arnau, A.; Dalmau, A.; Soria, V.; Salvat-Pujol, N.; Ribes, C.; Sánchez-Allueva, A.; Menchón, J.M.; Urretavizcaya, M. Protocolized hyperventilation enhances electroconvulsive therapy. *J. Affect. Disord.* **2017**, *217*, 225–232. [CrossRef] [PubMed]
127. Demers, M.; Wagner, D.D. NETosis: A new factor in tumor progression and cancer-associated thrombosis. *Semin. Thromb. Hemost.* **2014**, *40*, 277–283. [CrossRef] [PubMed]
128. Arroyo, A.B.; de Los Reyes-García, A.M.; Rivera-Caravaca, J.M.; Valledor, P.; García-Barberá, N.; Roldán, V.; Vicente, V.; Martínez, C.; González-Conejero, R. MiR-146a Regulates Neutrophil Extracellular Trap Formation That Predicts Adverse Cardiovascular Events in Patients with Atrial Fibrillation. *Arter. Thromb. Vasc. Biol.* **2018**, *38*, 892–902. [CrossRef] [PubMed]
129. Ge, L.; Zhou, X.; Ji, W.J.; Lu, R.Y.; Zhang, Y.; Zhang, Y.D.; Ma, Y.Q.; Zhao, J.H.; Li, Y.M. Neutrophil extracellular traps in ischemia-reperfusion injury-induced myocardial no-reflow: Therapeutic potential of DNase-based reperfusion strategy. *Am. J. Physiol. Circ. Physiol.* **2015**, *308*, H500–H509. [CrossRef]
130. Gómez-Moreno, D.; Adrover, J.M.; Hidalgo, A. Neutrophils as effectors of vascular inflammation. *Eur. J. Clin. Investig.* **2018**, *48*. [CrossRef]
131. Gaul, D.S.; Stein, S.; Matter, C.M. Neutrophils in cardiovascular disease. *Eur. Heart J.* **2017**, *38*, 1702–1704. [CrossRef]
132. Jorch, S.K.; Kubes, P. An emerging role for neutrophil extracellular traps in noninfectious disease. *Nat. Med.* **2017**, *23*, 279–287. [CrossRef]
133. Van Avondt, K.; Maegdefessel, L.; Soehnlein, O. Therapeutic targeting of neutrophil extracellular traps in atherogenic inflammation. *Thromb. Haemost.* **2019**, *119*, 542–552. [CrossRef] [PubMed]
134. Zhai, M.; Gong, S.; Luan, P.; Shi, Y.; Kou, W.; Zeng, Y.; Shi, J.; Yu, G.; Hou, J.; Yu, Q.; et al. Extracellular traps from activated vascular smooth muscle cells drive the progression of atherosclerosis. *Nat. Commun.* **2022**, *13*, 7500. [CrossRef]
135. Tardif, J.C.; Kouz, S.; Waters, D.D.; Bertrand, O.F.; Diaz, R.; Maggioni, A.P.; Pinto, F.J.; Ibrahim, R.; Gamra, H.; Kiwan, G.S.; et al. Efficacy and safety of low-dose colchicine after myocardial infarction. *N. Engl. J. Med.* **2019**, *381*, 2497–2505. [CrossRef] [PubMed]
136. Kelly, P.; Weimar, C.; Lemmens, R.; Murphy, S.; Purroy, F.; Arsovska, A.; Bornstein, N.M.; Czlonkowska, A.; Fischer, U.; Fonseca, A.C.; et al. Colchicine for prevention of vascular inflammation in non-CardioEmbolic stroke (CONVINCE)—Study protocol for a randomised controlled trial. *Eur. Stroke J.* **2021**, *6*, 222–228. [CrossRef] [PubMed]
137. Tsivgoulis, G.; Katsanos, A.H.; Giannopoulos, G.; Panagopoulou, V.; Jatuzis, D.; Lemmens, R.; Deftereos, S.; Kelly, P.J. The role of colchicine in the prevention of cerebrovascular ischemia. *Curr. Pharm. Des.* **2018**, *24*, 668–674. [CrossRef] [PubMed]
138. Kelly, P.J.; Murphy, S.; Coveney, S.; Purroy, F.; Lemmens, R.; Tsivgoulis, G.; Price, C. Anti-inflammatory approaches to ischaemic stroke prevention. *J. Neurol. Neurosurg. Psychiatry* **2018**, *89*, 211–218. [CrossRef]
139. Martínez, G.J.; Celermajer, D.S.; Patel, S. The NLRP3 inflammasome and the emerging role of colchicine to inhibit atherosclerosis-associated inflammation. *Atherosclerosis* **2018**, *269*, 262–271. [CrossRef]
140. Otani, K.; Watanabe, T.; Shimada, S.; Takeda, S.; Itani, S.; Higashimori, A.; Nadatani, Y.; Nagami, Y.; Tanaka, F.; Kamata, N.; et al. Colchicine prevents NSAID-induced small intestinal injury by inhibiting activation of the NLRP3 inflammasome. *Sci. Rep.* **2016**, *6*, 32587. [CrossRef]
141. Hoss, F.; Latz, E. Inhibitory effects of colchicine on inflammasomes. *Atherosclerosis* **2018**, *273*, 153–154. [CrossRef]
142. Zhang, S.; Diao, J.; Qi, C.; Jin, J.; Li, L.; Gao, X.; Gong, L.; Wu, W. Predictive value of neutrophil to lymphocyte ratio in patients with acute ST segment elevation myocardial infarction after percutaneous coronary intervention: A meta-analysis. *BMC Cardiovasc. Disord.* **2018**, *18*, 75. [CrossRef]
143. Liu, J.; Yang, D.; Wang, X.; Zhu, Z.; Wang, T.; Ma, A.; Liu, P. Neutrophil extracellular traps and dsDNA predict outcomes among patients with STelevation myocardial infarction. *Sci. Rep.* **2019**, *9*, 11599. [CrossRef] [PubMed]
144. Ma, Y. Role of Neutrophils in Cardiac Injury and Repair Following Myocardial Infarction. *Cells* **2021**, *10*, 1676. [CrossRef] [PubMed]
145. Eghbalzadeh, K.; Georgi, L.; Louis, T.; Zhao, H.; Keser, U.; Weber, C.; Mollenhauer, M.; Conforti, A.; Wahlers, T.; Paunel-Görgülü, A. Compromised anti-inflammatory action of neutrophil extracellular traps in PAD4-deficient mice contributes to aggravated acute inflammation after myocardial infarction. *Front. Immunol.* **2019**, *10*, 2313. [CrossRef]

146. Du, M.; Yang, W.; Schmull, S.; Gu, J.; Xue, S. Inhibition of peptidyl arginine deiminase-4 protects against myocardial infarction induced cardiac dysfunction. *Int. Immunopharmacol.* **2020**, *78*, 106055. [CrossRef] [PubMed]
147. NLi, C.; Xing, Y.; Zhang, Y.; Hua, Y.; Hu, J.; Bai, Y. Neutrophil Extracellular Traps Exacerbate Ischemic Brain Damage. *Mol. Neurobiol.* **2022**, *59*, 643–656. [CrossRef]

Disclaimer/Publisher's Note: The statements, opinions and data contained in all publications are solely those of the individual author(s) and contributor(s) and not of MDPI and/or the editor(s). MDPI and/or the editor(s) disclaim responsibility for any injury to people or property resulting from any ideas, methods, instructions or products referred to in the content.

biomedicines

MDPI

Article

Purinergic Signaling in Pathologic Osteogenic Differentiation of Aortic Valve Interstitial Cells from Patients with Aortic Valve Calcification

Polina Klauzen [1], Daria Semenova [1], Daria Kostina [1], Vladimir Uspenskiy [2] and Anna Malashicheva [1,*]

1 Laboratory of Regenerative Biomedicine, Institute of Cytology Russian Academy of Science, Tikhoretskiy Avenue, 4, Saint Petersburg 194064, Russia
2 Almazov National Medical Research Centre, Akkuratova Street, 2, Saint-Petersburg 197341, Russia
* Correspondence: malashicheva@incras.ru

Abstract: Purinergic signaling is associated with a vast spectrum of physiological processes, including cardiovascular system function and, in particular, its pathological calcifications, such as aortic valve stenosis. Aortic valve stenosis (AS) is a degenerative disease for which there is no cure other than surgical replacement of the affected valve. Purinergic signaling is known to be involved in the pathologic osteogenic differentiation of valve interstitial cells (VIC) into osteoblast-like cells, which underlies the pathogenesis of AS. ATP, its metabolites and related nucleotides also act as signaling molecules in normal osteogenic differentiation, which is observed in pro-osteoblasts and leads to bone tissue development. We show that stenotic and non-stenotic valve interstitial cells significantly differ from each other, especially under osteogenic stimuli. In osteogenic conditions, the expression of the ecto-nucleotidases ENTPD1 and ENPP1, as well as ADORA2b, is increased in AS VICs compared to normal VICs. In addition, AS VICs after osteogenic stimulation look more similar to osteoblasts than non-stenotic VICs in terms of purinergic signaling, which suggests the stronger osteogenic differentiation potential of AS VICs. Thus, purinergic signaling is impaired in stenotic aortic valves and might be used as a potential target in the search for an anti-calcification therapy.

Keywords: aortic valve; calcification; purinergic signaling; osteogenic differentiation

Citation: Klauzen, P.; Semenova, D.; Kostina, D.; Uspenskiy, V.; Malashicheva, A. Purinergic Signaling in Pathologic Osteogenic Differentiation of Aortic Valve Interstitial Cells from Patients with Aortic Valve Calcification. *Biomedicines* **2023**, *11*, 307. https://doi.org/10.3390/biomedicines11020307

Academic Editors: Tânia Martins-Marques, Gonçalo F. Coutinho and Attila Kiss

Received: 19 December 2022
Revised: 14 January 2023
Accepted: 17 January 2023
Published: 21 January 2023

Copyright: © 2023 by the authors. Licensee MDPI, Basel, Switzerland. This article is an open access article distributed under the terms and conditions of the Creative Commons Attribution (CC BY) license (https://creativecommons.org/licenses/by/4.0/).

1. Introduction

Calcific aortic valve stenosis (AS) is a degenerative disease for which there is no cure other than surgical replacement of the affected valve. Pathological conditions associated with the disease include mechanical stress, endothelial dysfunction, lipid deposition, oxidative stress, bicuspid aortic valve, inflammation, extracellular matrix remodeling and biomineralization (comprehensively reviewed in [1]). Purinergic signaling is associated with a vast spectrum of physiological processes, including cardiovascular system function and also with its pathological calcification, such as calcific aortic valve stenosis.

It is known that the pathologic osteogenic differentiation of valve interstitial cells (VIC) into osteoblast-like cells underlies the pathogenesis of AS, and purinergic signaling is involved in this [2]. During normal osteogenic differentiation, which is observed in pro-osteoblasts and leads to bone tissue development, ATP, its metabolites and related nucleotides also act as signaling molecules.

Purinergic signaling involves ATP, its metabolites and related molecules, which may be implicated in multidirectional cellular effects. For example, while ATP and ADP act as damage-associated molecular patterns (DAMPs) and induce sterile inflammatory response, adenosine and inosine are known to be anti-inflammatory agents. Regarding vascular and valvular calcification, it has been shown that ATP and UTP decrease osteogenic the differentiation of VICs and vascular smooth muscles cells (VSMC), while uridine adenosine tetraphosphate (Up4A) stimulates it [3].

Purin metabolites are formed during their degradation by ecto-nucleotidases, which produce derivative products also exhibiting their effects via purinergic receptors. The first ectonucleotidase in the chain of extracellular ATP metabolites is ENTPD1 (CD39, E-NTPDase1), which converts ATP to AMP and phosphate (Pi). AMP is further converted to adenosine by the ectonucleotidase NT5E (CD73, Ecto5′NTase) with a release of inorganic phosphate (Pi); after that, adenosine is converted to inosine by the ectonucleotidase ADA. ENTPD1 and NT5E are significantly expressed in two main types of valve cells, namely, valve endothelial cells (VEC) and VICs. Cell cultivation with an addition of extracellular nucleotides showed a more rapid formation of NT5E products on VICs than on VECs, while the opposite pattern was found for ENTPD1 [4].

There is also an alternative mechanism of nucleotide degradation in which ATP and some other purines may be degraded to AMP by another ectonucleotidase—ENPP1 (CD203a)—with a release of pyrophosphate (PPi). While Pi promotes hydroxyapatite deposition, Ppi is known for its inhibiting influence on hydroxyapatite deposition via chemisorption of pyrophosphate on the surface of hydroxyapatite, which prevents further crystal growth. Nevertheless, PPi may be converted to Pi by alkaline phosphatase. ATP may decrease the calcification rate via PPi production. In addition, ATP decreases calcification via reduction in apoptosis [3], which is known to trigger vascular calcification by serving apoptotic bodies as nucleating structures for calcium crystal formation [5]. It has been shown that ENPP1 is highly upregulated in AS VICs and, despite the production of PPi, promotes the mineralization of cells [3,6]. This might be due to the fact that the overexpression of ENPP1 depletes the pool of extracellular ATP, which acts as a survival signal for VICs and prevents apoptosis [3]. In this way, apoptosis, which triggers calcium crystal formation, might be a stronger factor of calcification than Pi/PPi production.

Nucleotides act in cells via specific receptors. P2 purinergic receptors (P2X and P2Y subtypes) are mainly sensitive to ATP and UTP, while P1 (ADORA) receptors are activated by adenosine and inosine. The precise role of the purinergic receptors in cardiovascular calcification is controversial and not clear. Several works show an inhibitory effect of P2Y2 receptors, activated by ATP, on vascular calcification through PI3K/Akt [3,7]. In contrast, other works showed that the effects of ATP and UTP on VSMC calcification were not mediated via the P2Y2 receptor [8]. In addition, an activation of P2Y2/6 receptors by Up4A was described, which involved phosphorylation of the mitogen-activated kinases MEK and ERK1/2, followed by enhanced calcification of vascular cells [9,10].

Adenosine and its receptor ADORA2b may play a major role in cardiovascular calcification. Adenosine is a mediator of intercellular signaling in many tissues. In the cardiovascular system, adenosine reduces heart rate, inhibits inflammation, protects cells from hypoxia caused by this inflammation and increases resistance to ischemia [11]. Adenosine also impacts AS pathogenesis. It has been reported that stimulating receptors ADORA2A and ADORA2B produces strong pro-degenerative effects on VICs [12]. There are data showing that the expression of ADORA2a and ADORA2b is diminished in uncalcified fragments of stenotic valves compared to non-stenotic ones [13]. It has also been shown that the activity of adenosine deaminase, which degrades adenosine, is increased in stenotic aortic valves [13]. This leads to increased vascular adenosine degradation and hypertension. Due to turbulent blood flow on the aortic side of the valve, its stenotic part is prone to attachment and infiltration of immune cells with subsequent endothelial damage and even more severe calcification.

VICs play a crucial role in valvular calcification and are thought to differentiate into osteoblast-like cells. Nevertheless, the process is orchestrated by other cell types, such as VECs. Multiple effects of endothelial cells on the underlying interstitial cells have been shown. Endothelial cells biosynthesize Up4A via the activation of vascular endothelial growth factor receptor (VEGFR). Up4A synthesis performed by the endothelium is enhanced by stimulation with ATP, UTP and mechanical stress [14], which takes place in aortic stenosis. Up4A has a much longer half-life compared to ATP/UTP, contains both purine and pyrimidine parts [14] and, in contrast to ATP/UTP, enhances vascular calcification in rat VSMCs [9].

This study is devoted to the identification of the distinctive features of cells derived from stenotic aortic valves in the context of purinergic signaling, as well as searching for similarities and differences between the processes of pathogenic and normal osteogenic differentiation of mesenchymal cells. We show here that AS valvular cells react to extracellular ATP and adenosine in a distinct manner in comparison to non-stenotic cells; AS VICs also undergo pathogenic osteogenic differentiation more similar to osteoblast-like cells compared to the osteogenic differentiation of non-stenotic cells. Elevated expression levels of ATP-degrading nucleotidases and adenosine receptors indicate a general elevation of purinergic signaling in AS VICs influenced and orchestrated by the endothelial cells. Nevertheless, some components of the signaling turn out to be downregulated. Our data indicate a general imbalance in the purinergic signaling in AS VICs.

2. Materials and Methods

2.1. Isolation of Primary Cells

The study protocols were approved by the local Ethics Committee of Almazov Federal Medical Research Centre and Vreden Institute of Traumatology and Orthopedics and performed in accordance with the principles of the Declaration of Helsinki. All patients gave written informed consent. Human valve interstitial cells (VIC) and human valve endothelial cells (VEC) were isolated from tricuspid aortic valves explanted during aortic valve replacement due to calcific aortic valve disease (CAVD). Patients with known infective endocarditis and rheumatic disease were excluded from the study. Clinical data of the patients are represented in Table 1. VICs and VECs from normal aortic valves, which were used as a control group, were isolated from healthy tricuspid aortic valves obtained from explanted hearts from recipients of heart transplantation. Human umbilical vein endothelial cells (HUVEC) were obtained at Almazov National Medical Research Center from cords after healthy delivery. Femur bone epiphysis for osteoblast-like cell isolation was acquired during surgery at Vreden Institute of Traumatology and Orthopedics [15].

Table 1. Clinical data of the patients.

Characteristics (m, w)	CAVD Patients (n = 17) (m, w)
Age	63.0 (58.0; 67.0)
Gender	W—15 (46.8%) M—17 (53.2%)
Proximal ascending aortic diameter, mm	36 (32; 40)
Body mass index (BMI) kg/m^2	32.62 (26.79; 34.94)
General cholesterol, mM/L	4.91 (4.23; 6.11)
High density lipoproteins (HDL), mM/L	1.16 (1.04; 1.46)
Low density lipoproteins (LDL), mM/L	3.54 (2.77; 4.17)
C-reactive protein, mg/L	3.01 (1.06; 4.70)

VICs and VECs were isolated with a standard technique. Leaflets of the heart valves were cleared from aorta fragments. VECs were isolated by vortexing collagenase–treated leaflets (10 min with collagenase type IV) (Worthington Biochemical Corporation, Lakewood, NJ, USA). The resulting pellet after vortexing was suspended and seeded in a flask covered with 0.2% gelatin in an ECM (ScienCell, Logan, UT, USA). When a VEC confluence of 70–80% was attained, cells were magnetically sorted by CD31 antigen and passaged in a ratio of 1:3 in the ECM (ScienCell, Logan, UT, USA). To isolate VICs, leaflets were incubated again in collagenase type IV overnight. The resulting pellet after vortexing was suspended and seeded in a flask in DMEM supplemented with 15% FBS, 2 mM of L-glutamine and penicillin/streptomycin (100 mg/L) (all—Invitrogen, Grand Island, NJ, USA).

HUVEC were acquired from the umbilical vein by collagenase dissociation. The vein was filled with 0.1% collagenase solution (Collagenase, Type II) (Worthington Biochemical Corporation, USA) and incubated in PBS at 37 °C for 10 min. The suspension was centrifuged at 300× *g* for 5 min. The cell pellet was suspended and seeded on a Petri dish covered with 0.2% gelatin in the ECM (ScienCell, Logan, UT, USA).

Isolation of osteoblast-like cells from bone tissue was carried out using the enzymatic method. Bone tissue was crushed into fragments using carbide cutters. Cancellous bone fragments were selected, washed in PBS and incubated with 0.2% collagenase type II (Worthington Biochemical Corporation, USA) solution for 30 min at 37 °C. The homogenized mass was washed in PBS and transferred into 0.2% collagenase type IV solution (Worthington Biochemical Corporation, USA) for 16 h at 37 °C. The collagenase type IV was then inactivated with high-glucose DMEM (Gibco, Grand Island, NJ, USA) supplemented with 15% FBS (HyClone, New York, NY, USA) and 0.1% ascorbic acid solution (Sigma-Aldrich, St. Louis, MO, USA). The resulting suspension was transferred into a flask and cultivated for several weeks. Then, the resulting osteoblast-like cells were cultured in Petri dishes according to standard procedures [15].

2.2. Culture and Osteogenic Differentiation

For the experiments, VICs, VECs and HUVECs of 3–5 passages were used. The cells were seeded in 6-well plates with a density of 220 × 103 cells per well. Osteoblasts of 3–4 passages were plated in 12-well plates with a density of 90 × 103 cells per well covered with 0.2% gelatin. VICs and VICs with HUVECs were cultured under normal conditions in DMEM supplemented with 15% bovine serum (all—Gibco, USA), 2 mM glutamine, 50 IU/mL penicillin and 50 IU/mL streptomycin (Invitrogen, USA). VEC cultivation was carried out using an ECM medium (ScienCell, USA) with the addition of the appropriate supplements from the manufacturer. Osteoblasts were cultured in 4.5% glucose DMEM with the addition of 15% Hyclone bovine serum (all—Gibco, USA), 2 mM glutamine, 50 IU/mL penicillin and 50 IU/mL streptomycin (Sigma, USA). To induce osteogenic differentiation, cells were placed in DMEM medium supplemented with 10% Hyclone bovine serum (Gibco, USA), 2 mM glutamine, 50 IU/mL penicillin, 50 IU/mL streptomycin, 100 mM b-glycerophosphate, 50 mg/mL ascorbic acid and 0.1 μM dexamethasone (Sigma, USA).

All cells were cultured for 4 days, after which total RNA was isolated using the ExtractRNA reagent (Evrogen, Moscow, Russia) in accordance with the manufacturer's recommendations. Calcium deposition was detected by Alizarin Red staining (Sigma, USA) on day 18 of differentiation for the VIC and on day 15 for osteoblast cells.

2.3. Real-Time PCR

Real-time PCR was used to assess the expression levels of exo-nucleotidase and purinergic receptors. PCRs for the VEC and VIC populations (pure and in co-cultivation with HUVEC) were run without technical and biological replicates. PCRs were performed in technical and biological duplicates; all the resulting values were independently used in the analysis. Total RNA (1 μg) was reverse-transcribed with an MMLV RT kit (Eurogen, Moscow, Russia). Real-time PCR was performed with 1 μL of cDNA and TaqMan PCR Mastermix (Thermo Fisher Scientific, Grand Island, NJ, USA) in the Light Cycler system. The thermocycling conditions were as follows: 95 °C for 5 min, followed by 45 cycles at 95 °C for 15 s and 60 °C for 1 min. A final heating step of 65 °C to 95 °C was performed to obtain melting curves of the final PCR products. The corresponding gene expression level was normalized to GAPDH from the same samples. Changes in the target genes expression levels were calculated as fold differences using the comparative ΔΔCT method. ENTPD1, ENPP1, ADA, NT5E, ADORA2b, P2RX1 and P2RX7 primers were bought at Thermofisher Scientific (USA) as Single Tube TaqMan Gene Expression Assays (Hs01054040_m1, Hs00159686_m1, Hs00969559_m1, Hs01110945_m1, Hs00386497_m1, Hs00175721_m1, Hs00175686_m1, Hs00925146_m1, respectively).

Values were compared using Student's *t*-test and nonparametric Spearman correlation. A value of $p \leq 0.05$ was considered significant. Statistical analysis was performed using R software (version 2.12.0; R Foundation for Statistical Computing, Vienna, Austria).

3. Results

3.1. Purinergic Signaling in Osteogenic Differentiation of Healthy and Stenotic Aortic Valve Interstitial Cells Versus Osteoblasts

As pathologic osteogenic differentiation of VICs underlies the pathogenesis of AS, we induced the osteogenic differentiation of VICs to simulate this process. In all differentiated cell lineages, calcium depositions were positively stained (Figure 1).

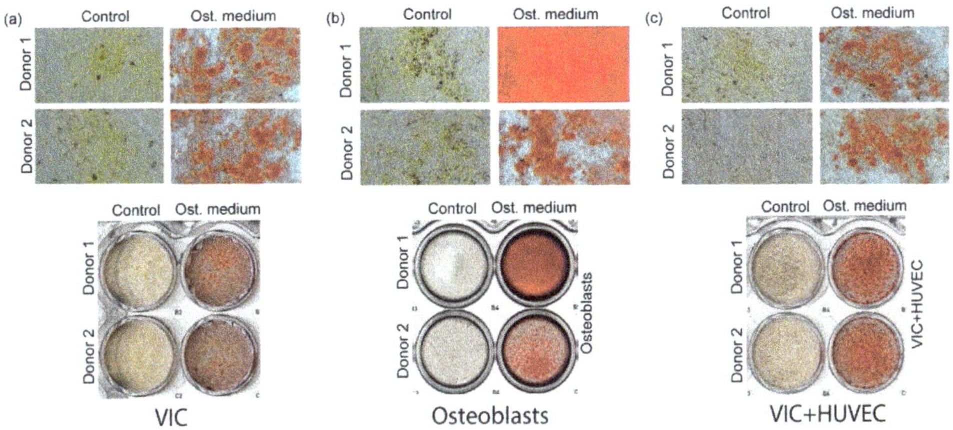

Figure 1. Microscopic images of alizarin-red staining of cells after osteogenic differentiation; calcium depositions are stained in red. Valve interstitial cells (VIC) (**a**), osteoblasts (**b**) and VIC co-cultivated with human umbilical cord endothelial cells (HUVEC) (**c**) before and after osteogenic differentiation.

Osteogenic conditions strongly increased the expression of the ecto-nucleotidases ENPP1 and ENTPD1 in both AS and non-stenotic VICs (Figure 2). Importantly, this elevation in ENPP1 and ENTPD1 expression in osteogenic media was higher in VICs derived from AS patients compared to the control group. Osteogenic differentiation also raised the expression of the ecto-nucleotidase ADA in non-stenotic cells and reduced the expression level of the ATP receptor P2RX7 and the adenosine receptor ADORA2b in AS VICs.

Osteoblast-like cells normally undergo osteogenic differentiation. We analyzed the expression profile of the components of purinergic signaling in human osteoblasts in osteogenic conditions (Figure 2). In differentiated osteoblasts, we observed an elevated expression level of the ecto-nucleotidases ENPP1 and ENTPD1 similarly to AS and non-stenotic VICs. The ADA level in osteoblast-like cells was not changed. The expression level of the ATP receptor P2RX7 declined in differentiated osteoblasts similarly to AS VICs. Concurrently, the expression level of adenosine receptor ADORA2b remained the same in osteogenic conditions and in non-stenotic cells, but not in AS VIC. The osteogenic differentiation of osteoblasts did not influence the expression of the ecto-nucleotidase NT5E, similarly to that in VICs.

To summarize, pro-osteoblast differentiation looks similar to VIC differentiation in terms of purinergic signaling component expression and in most cases corresponds to AS VICs but not to non-stenotic control cells. Nevertheless, the expression levels of ENPP1, ADA, NT5E and P2RX7 in osteoblast-like cells are reliably higher than in VICs, both before and after osteogenic differentiation. At the same time, the expression level of ADORA2b is not elevated in osteoblasts compared to VICs.

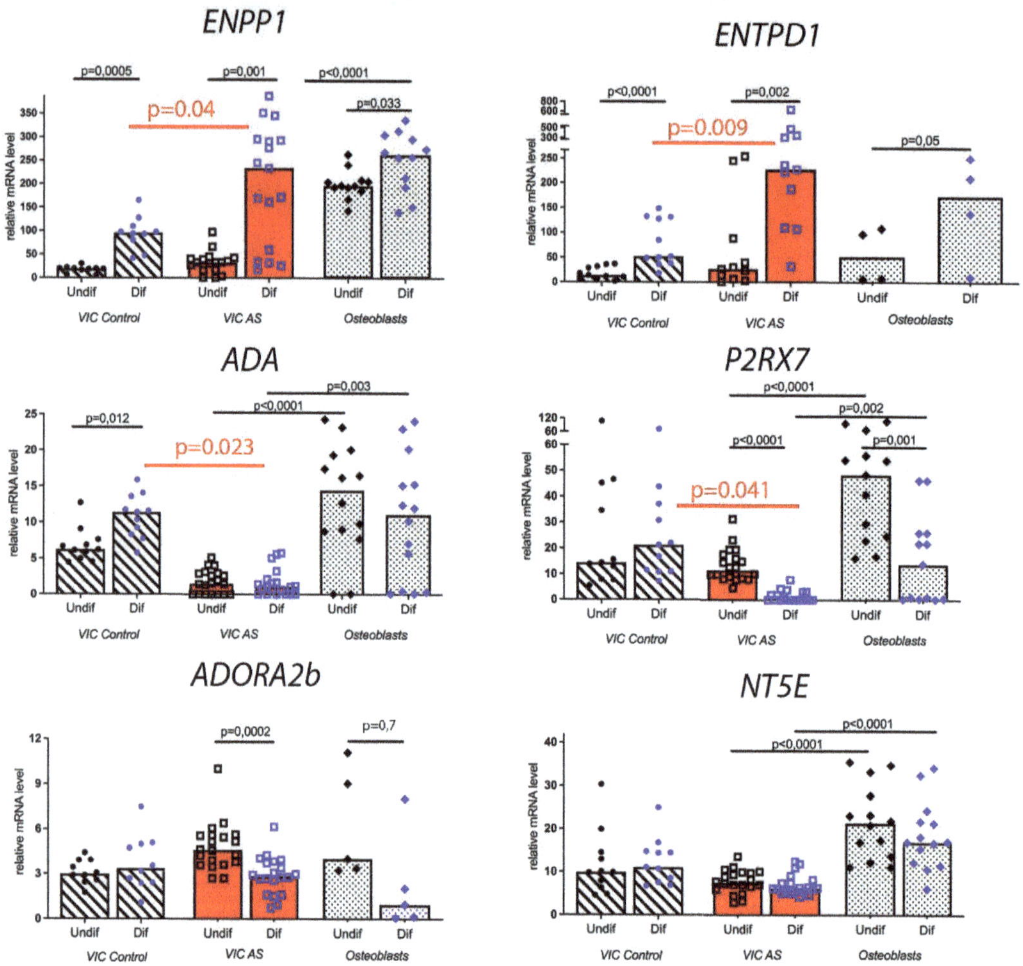

Figure 2. Expression levels of ENPP1, ENTPD1, ADA, P2RX7, ADORA2b and NT5E in valve interstitial cells (VIC) and osteoblast-like cells. Relative mRNA levels in non-stenotic VICs (VIC Control) are compared with stenotic VICs (AS VIC); relative mRNA levels in stenotic VICs are compared with osteoblast-like cells (Osteoblasts). Relative mRNA levels in undifferentiated cells (Undif) are compared with cells in which osteogenic differentiation (Dif) was induced.

Osteogenic conditions help to explain the difference between AS and non-stenotic cells regarding the expression levels of purinergic signaling components. In our experiments (Figure 2), the osteogenic medium triggered elevated expression of the ecto-nucleotidases ENTPD1 and ENPP1 in AS VICs compared to non-stenotic cells and a decreased expression of the ecto-nucleotidase ADA and ATP receptor P2RX7.

3.2. Purinergic Signaling in Valve Endothelial Cells

We also analyzed the expression of purinergic genes in valve endothelial cells (VEC) (Figure 3). In VEC, the expression level of ADORA2b also varied between AS and non-stenotic cells, but, in contrast to VIC, its expression decreased in AS cells compared to the control cells from healthy valves. In addition, AS VEC displayed an elevated level of the ATP receptor P2RX7 compared to non-stenotic VEC.

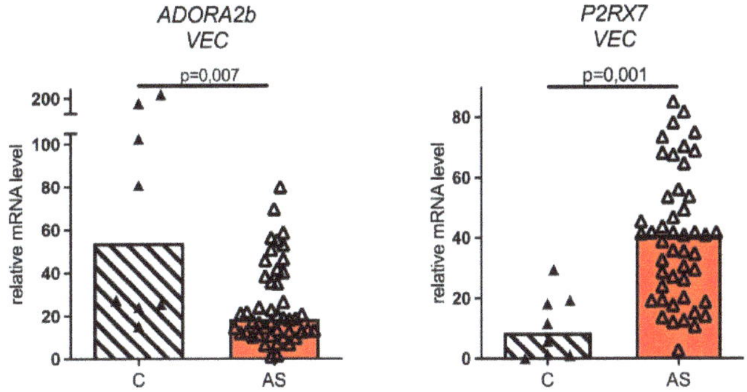

Figure 3. Expression levels of ADORA2b and P2RX7 in VEC. Relative mRNA levels in non-stenotic control cells (C) are compared with relative mRNA levels in stenotic cells (AS).

To see whether the presence of endothelial cells could influence purinergic signaling in VICs, we analyzed the expression of purinergic genes in a co-culture of VICs and endothelial cells. A co-culture of VICs with HUVECs revealed the difference between AS and non-stenotic VICs (Figure 4). The expression levels of the ecto-nucleotidase ENTPD1 and ATP-receptor P2RX1 in AS were lower compared to non-stenotic cells in the co-culture. Notably, this was not observed in a pure VIC culture (Figure 2). At the same time, the expression of ATP-receptor P2RX7 was higher in stenotic VICs co-cultured with endothelial cells in comparison to control cells from healthy valves (Figure 4).

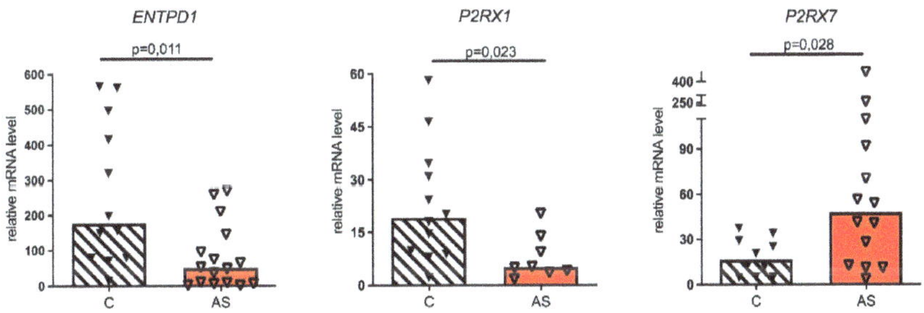

Figure 4. Expression levels of ENTPD1, P2RX1 and P2RX7 in VIC co-cultured with endothelial cells. Relative mRNA levels in non-stenotic control cells (C) are compared with relative mRNA levels in stenotic cells (AS).

The small molecule adenosine is a purine nucleoside widely used as a vasodilator and neuromodulator. Recently, a growing amount of attention has been focused on adenosine and adenosine receptors as factors implicated in bone regeneration. Exogenous adenosine supplementation has been reported as an effective method for successful osteogenic differentiation of mouse bone-marrow-derived MSC [16]. Therefore, we analyzed whether exogenous adenosine influenced osteogenic differentiation of valve interstitial cells. We induced the osteogenic differentiation of VICs in the presence of 10 uM of adenosine and analyzed whether that influenced alizarin staining efficiency. We did not detect any influence of adenosine on the effectivity of osteogenic differentiation of VICs. However, we observed a change in the expression profile of CD39/CD73 (ENTPD1/NT5E) using flow cytometric analysis (Supplementary Figure S1).

In summary, our data suggest an important role that changes in purinergic signaling could play in the sensitivity of cells from diseased valves undergoing pathologic osteogenic differentiation.

4. Discussion

We found that aortic valve cells derived from stenotic patients significantly differ in purinergic signaling from their counterparts derived from healthy donors. This fact is also proven by PCA plotting, which shows that different kinds of cells (AS and non-stenotic) create two clusters (Figure 5). Valve interstitial cells from stenotic and non-stenotic valves significantly differ in their expression of ENTPD1, ADORA2b and P2RX7 genes, while endothelial cells from stenotic and non-stenotic valves mainly differ in their expression of genes ADORA2b, P2RX7 and P2RX1.

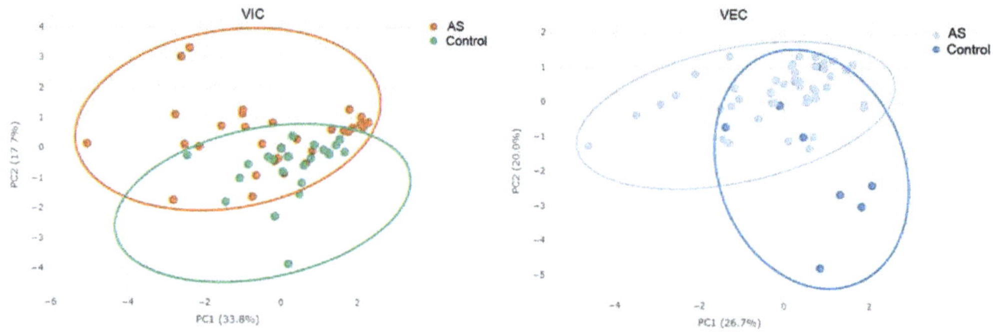

Figure 5. Principal component analysis (PCA) plot representing ENTPD1, ENPP1, ADA, NT5E, P2RX1, P2RX7 and ADORA2b gene expression in non-stenotic (Control) and stenotic (AS) VICs and VECs.

The endothelial influence on purinergic signaling component expression in VICs and their pathological differentiation potential is significant. It is known that the ecto-nucleotidase ENTPD1 is involved in the catabolism of the dinucleotide Up4A [17], which is produced by the endothelium and increases the intensity of the pathological calcification process in the cardiovascular system [9].

In the course of ATP degradation, the ectonucleotidases ENTPD1 and NT5E release Pi, while ENPP1 releases PPi—an inhibitor of hydroxyapatite formation and, hence, a mineralization deprivator. The relative amounts of Pi and PPi play an important role in the development of tissue calcification. In a co-culture of VICs with endothelial cells, the ENTPD1-pathway also predominates because endothelium increases ENTPD1 but does not affect ENPP1. It turns out that the endothelium changes the balance of two ATP-degradation pathways in VIC and hence reduces the relative amount of pyrophosphate and increases the relative amount of phosphate. Thus, the endothelium of the diseased valves could induce hydroxyapatite deposits.

It has been reported that the P2RX7 receptor participates in the differentiation of osteoclasts and osteoblasts and its downregulation inhibits osteogenic potential of pre-osteoblastic cells [18]. It also serves as a pattern recognition receptor for extracellular ATP-mediated apoptotic cell death [19]. Our results evidence a significant increase in P2RX7 expression in normal conditions in AS VECs and AS VICs co-cultured with endothelium (Figures 3 and 4). Nevertheless, pure AS VIC in normal conditions do not display this over-expression, nor do AS cells and osteoblasts in the osteogenic medium where we observed the downregulation of P2RX7. These data may indicate that in osteogenic conditions, when the work of ENTPD1 and ENPP1 is strongly intensified in AS VICs, the pool of extracellular ATP is largely exhausted and the signal does not reach the P2RX7 receptor. Here, our data suggest that in normal conditions AS VECs, unlike AS VICs intensify P2RX7 expression. Since the P2RX7 receptor takes part in ATP-mediated apoptotic cell death, AS VECs may intensify the apoptosis-mediated mineralization of VICs. These data further indicate the crucial orchestrating role of endothelium in the pathological calcification of VICs.

Adenosine, an important signaling molecule and a stress regulator, is produced by the ecto-nucleotidase NT5E from AMP. It is known that increasing adenosine levels produces strong pro-degenerative effects in VIC, and vice versa, lowering adenosine levels by the inhibition of NT5E leads to protective effects against VIC osteogenic degeneration [12].

According to our results, NT5E expression in VICs, VECs and pro-osteoblasts remained constant in all experiments, even in osteogenic conditions when AMP production was reinforced (elevated ENTPD1 and ENPP1 in osteo-differentiated cells, Figure 2). However, some evidence suggests that NT5E expression varies in non-stenotic and AS valves tissues when analyzed using immunohistochemical or immunofluorescent staining [13]. Changing NT5E expression might represent an effect of different cell type interactions in valve tissue and could be influenced by other tissue components, such as inflammatory cells, while isolated and co-cultured VICs and VECs exhibit stable NT5E expression. In addition, there are data showing that cells isolated from different locations of stenotic valves express different levels of NT5E [13], although in our work, we did not take the latter parameter into consideration.

We show that ecto-nucleotidase ADA, which is supposed to degrade over-accumulated adenosine, is not elevated in AS VICs in normal conditions, as might be expected. Moreover, its expression is even decreased in AS VICs under osteogenic stimulation (Figure 2). It is known that adenosine might be deaminated by adenosine deaminase, whose activity in AS cells is higher than in non-stenotic cells [20]. So, while normal VICs tend to reduce extra adenosine level via ADA, AS cells probably reduce extra adenosine with the help of adenosine deaminase.

Adenosine takes part in cardiovascular pathologies through the activation of adenosine receptors, including ADORA2b. It has been reported that stimulating adenosine receptors ADORA2a and ADORA2b results in strong pro-osteogenic effects on VICs [12]. We show that ADORA2b expression was elevated in AS VICs, both in normal and osteogenic conditions (Figure 2). The intensified ADORA2b pathway in AS VICs suggests their strong potential for osteogenic transition and also the elevated amount of adenosine in the stenotic valve, which is an important stress marker. Despite the stable NT5E expression observed in our experiments, adenosine level may be increased in stress caused by stenosis.

This study has limitations associated with the 2D in vitro culture of isolated valve cells and also with the fact that we isolated the cells from a patient with end-stage disease. Unfortunately, the animal models for calcific aortic valve stenosis are still very limited [21], and this lack of relevant models limits our understanding of the disease pathogenesis. Nevertheless, we suggest that our findings are relevant for finding potential targets for calcific aortic stenosis, as understanding the mechanisms of pathological osteogenic differentiation of valve cells could help to find potential anti-calcification therapies [22,23].

5. Conclusions

We show here that valve interstitial cells from stenotic valves react to extracellular ATP and adenosine in a distinct manner in comparison to non-stenotic valve cells. AS VICs also undergo pathologic osteogenic differentiation more similarly to osteoblast-like cells in comparison to cells from non-stenotic valves. Elevated expression levels of ATP-degrading nucleotidases and adenosine receptors suggest general elevation of purinergic signaling in AS VICs, influenced and orchestrated by the endothelial cells. However, some components of the signaling were found to be downregulated. Our data indicate a general imbalance in the purinergic signaling in aortic valve interstitial cells of the stenotic valves. Moreover, in osteogenic conditions, the diseased interstitial cells look more similar to osteoblasts than non-stenotic cells in terms of purinergic signaling, which suggest the stronger osteogenic differentiation potential of AS VICs. In summary, purinergic signaling is impaired in stenotic aortic valves and might be used as a potential target in the search for anti-calcification therapies.

Supplementary Materials: The following supporting information can be downloaded at: https://www.mdpi.com/article/10.3390/biomedicines11020307/s1. Figure S1: Analysis of the influence of adenosine on the expression of CD39 (ENTPD1) and CD73 (NT5E) in valve interstitial cells from the patients with aortic stenosis. The cells were induced to osteogenic differentiation in control conditions and in the presence of 10 uM of adenosine. CD39/CD73 expression was analyzed by flow cytometry using specific antibodies to CD39 and CD73 (Biolegend) and flowcytometer Cytoflex. * $p < 0.01$, ** $p < 0.05$.

Author Contributions: (1) The conception and design of the study, A.M.; (2) acquisition of data, or analysis and interpretation of data, P.K., D.K., D.S. and V.U.; (3) drafting the article or revising it critically for important intellectual content, P.K. and A.M.; (4) final approval of the version to be submitted, A.M. and P.K. All authors have read and agreed to the published version of the manuscript.

Funding: This work was supported by a grant from the Russian Science Foundation 18-14-00152.

Institutional Review Board Statement: The studies involving human participants were reviewed and approved by the Local Ethics Committee of the Almazov Federal Medical Research Center (ethical permit number: 12.26/2014).

Informed Consent Statement: Consent was obtained from all subjects involved in the study.

Data Availability Statement: All data related to this work can be made available upon request to the corresponding authors.

Acknowledgments: We cordially thank Vitaly Karelkin for the assistance in obtaining bone samples and Alexey Golovkin for his help in analyzing flow cytometry data.

Conflicts of Interest: The authors declare no conflict of interest.

References

1. Tanase, D.M.; Valasciuc, E.; Gosav, E.M.; Floria, M.; Costea, C.F.; Dima, N.; Tudorancea, I.; Maranduca, M.A.; Serban, I.L. Contribution of Oxidative Stress (OS) in Calcific Aortic Valve Disease (CAVD): From Pathophysiology to Therapeutic Targets. *Cells* **2022**, *11*, 2663. [CrossRef] [PubMed]
2. Weber, A.; Leuders, P.; Barth, M.; Selig, J.; Hesse, J.; Schrader, J.; Lichtenberg, A.; Akhyari, P. P5089The role of the purinergic signaling system in the degeneration process of aortic valves. *Eur. Heart J.* **2018**, *39*, ehy566.P5089. [CrossRef]
3. Côté, N.; El Husseini, D.; Pépin, A.; Guauque-Olarte, S.; Ducharme, V.; Bouchard-Cannon, P.; Audet, A.; Fournier, D.; Gaudreault, N.; Derbali, H. ATP acts as a survival signal and prevents the mineralization of aortic valve. *J. Mol. Cell. Cardiol.* **2012**, *52*, 1191–1202. [CrossRef] [PubMed]
4. Kudryavtsev, I.; Serebriakova, M.; Zhiduleva, E.; Murtazalieva, P.; Titov, V.; Malashicheva, A.; Shishkova, A.; Semenova, D.; Irtyuga, O.; Isakov, D.; et al. CD73 Rather Than CD39 Is Mainly Involved in Controlling Purinergic Signaling in Calcified Aortic Valve Disease. *Front. Genet.* **2019**, *10*, 604. [CrossRef] [PubMed]
5. Proudfoot, D.; Skepper, J.N.; Hegyi, L.; Bennett, M.R.; Shanahan, C.M.; Weissberg, P.L. Apoptosis regulates human vascular calcification in vitro: Evidence for initiation of vascular calcification by apoptotic bodies. *Circ. Res.* **2000**, *87*, 1055–1062. [CrossRef]
6. Johnson, K.; Hashimoto, S.; Lotz, M.; Pritzker, K.; Goding, J.; Terkeltaub, R. Up-regulated expression of the phosphodiesterase nucleotide pyrophosphatase family member PC-1 is a marker and pathogenic factor for knee meniscal cartilage matrix calcification. *Arthritis Rheum. Off. J. Am. Coll. Rheumatol.* **2001**, *44*, 1071–1081. [CrossRef]
7. Qian, S.; Regan, J.N.; Shelton, M.T.; Hoggatt, A.; Mohammad, K.S.; Herring, P.B.; Seye, C.I. The P2Y2 nucleotide receptor is an inhibitor of vascular calcification. *Atherosclerosis* **2017**, *257*, 38–46. [CrossRef]
8. Patel, J.J.; Zhu, D.; Opdebeeck, B.; d'Haese, P.; Millán, J.L.; Bourne, L.E.; Wheeler-Jones, C.P.; Arnett, T.R.; MacRae, V.E.; Orriss, I.R. Inhibition of arterial medial calcification and bone mineralization by extracellular nucleotides: The same functional effect mediated by different cellular mechanisms. *J. Cell. Physiol.* **2018**, *233*, 3230–3243. [CrossRef]
9. Schuchardt, M.; Tölle, M.; Prüfer, J.; Prüfer, N.; Huang, T.; Jankowski, V.; Jankowski, J.; Zidek, W.; Van Der Giet, M. Uridine adenosine tetraphosphate activation of the purinergic receptor P2Y enhances in vitro vascular calcification. *Kidney Int.* **2012**, *81*, 256–265. [CrossRef]
10. Zhou, Z.; Chrifi, I.; Xu, Y.; Pernow, J.; Duncker, D.J.; Merkus, D.; Cheng, C. Uridine adenosine tetraphosphate acts as a proangiogenic factor in vitro through purinergic P2Y receptors. *Am. J. Physiol. Heart Circ. Physiol.* **2016**, *311*, H299–H309. [CrossRef]
11. Eltzschig, H.K. Extracellular adenosine signaling in molecular medicine. *J. Mol. Med.* **2013**, *91*, 141–146. [CrossRef] [PubMed]
12. Weber, A.; Barth, M.; Selig, J.I.; Raschke, S.; Dakaras, K.; Hof, A.; Hesse, J.; Schrader, J.; Lichtenberg, A.; Akhyari, P. Enzymes of the purinergic signaling system exhibit diverse effects on the degeneration of valvular interstitial cells in a 3-D microenvironment. *FASEB J. Off. Publ. Fed. Am. Soc. Exp. Biol.* **2018**, *32*, 4356–4369. [CrossRef] [PubMed]

13. Kutryb-Zajac, B.; Jablonska, P.; Serocki, M.; Bulinska, A.; Mierzejewska, P.; Friebe, D.; Alter, C.; Jasztal, A.; Lango, R.; Rogowski, J.; et al. Nucleotide ecto-enzyme metabolic pattern and spatial distribution in calcific aortic valve disease; its relation to pathological changes and clinical presentation. *Clin. Res. Cardiol.* **2020**, *109*, 137–160. [CrossRef] [PubMed]
14. Jankowski, V.; Tölle, M.; Vanholder, R.; Schönfelder, G.; van der Giet, M.; Henning, L.; Schlüter, H.; Paul, M.; Zidek, W.; Jankowski, J. Uridine adenosine tetraphosphate: A novel endothelium-derived vasoconstrictive factor. *Nat. Med.* **2005**, *11*, 223–227. [CrossRef] [PubMed]
15. Kostina, D.; Lobov, A.; Klausen, P.; Karelkin, V.; Tikhilov, R.; Bozhkova, S.; Sereda, A.; Ryumina, N.; Enukashvily, N.; Malashicheva, A. Isolation of Human Osteoblast Cells Capable for Mineralization and Synthetizing Bone-Related Proteins In Vitro from Adult Bone. *Cells* **2022**, *11*, 3356. [CrossRef]
16. Ahmadi, A.; Mazloomnejad, R.; Kasravi, M.; Gholamine, B.; Bahrami, S.; Sarzaeem, M.M.; Niknejad, H. Recent advances on small molecules in osteogenic differentiation of stem cells and the underlying signaling pathways. *Stem Cell Res. Ther.* **2022**, *13*, 518. [CrossRef]
17. Burnstock, G. Purinergic signaling in the cardiovascular system. *Circ. Res.* **2017**, *120*, 207–228. [CrossRef] [PubMed]
18. Yang, J.; Ma, C.; Zhang, M. High glucose inhibits osteogenic differentiation and proliferation of MC3T3-E1 cells by regulating P2X7. *Mol. Med. Rep.* **2019**, *20*, 5084–5090. [CrossRef] [PubMed]
19. Kawano, A.; Tsukimoto, M.; Noguchi, T.; Hotta, N.; Harada, H.; Takenouchi, T.; Kitani, H.; Kojima, S. Involvement of P2X4 receptor in P2X7 receptor-dependent cell death of mouse macrophages. *Biochem. Biophys. Res. Commun.* **2012**, *419*, 374–380. [CrossRef]
20. Kutryb-Zajac, B.; Mierzejewska, P.; Slominska, E.M.; Smolenski, R.T. Therapeutic perspectives of adenosine deaminase inhibition in cardiovascular diseases. *Molecules* **2020**, *25*, 4652. [CrossRef] [PubMed]
21. Bogdanova, M.; Zabirnyk, A.; Malashicheva, A.; Semenova, D.; Kvitting, J.E.; Kaljusto, M.L.; Perez, M.D.M.; Kostareva, A.; Stensløkken, K.O.; Sullivan, G.J.; et al. Models and Techniques to Study Aortic Valve Calcification In Vitro, Ex Vivo and In Vivo. An Overview. *Front. Pharmacol.* **2022**, *13*, 835825. [CrossRef] [PubMed]
22. Lobov, A.; Malashicheva, A. Osteogenic differentiation: A universal cell program of heterogeneous mesenchymal cells or a similar extracellular matrix mineralizing phenotype? *Biol. Commun.* **2022**, *67*, 32–48. [CrossRef]
23. Lobov, A.A.; Boyarskaya, N.V.; Kachanova, O.S.; Gromova, E.S.; Shishkova, A.A.; Zainullina, B.R.; Pishchugin, A.S.; Filippov, A.A.; Uspensky, V.E.; Malashicheva, A.B. Crenigacestat (LY3039478) inhibits osteogenic differentiation of human valve interstitial cells from patients with aortic valve calcification in vitro. *Front. Cardiovasc. Med.* **2022**, *9*, 969096. [CrossRef] [PubMed]

Disclaimer/Publisher's Note: The statements, opinions and data contained in all publications are solely those of the individual author(s) and contributor(s) and not of MDPI and/or the editor(s). MDPI and/or the editor(s) disclaim responsibility for any injury to people or property resulting from any ideas, methods, instructions or products referred to in the content.

biomedicines

MDPI

Review

Suppression of Cardiogenic Edema with Sodium–Glucose Cotransporter-2 Inhibitors in Heart Failure with Reduced Ejection Fraction: Mechanisms and Insights from Pre-Clinical Studies

Ryan D. Sullivan †, Mariana E. McCune †, Michelle Hernandez, Guy L. Reed and Inna P. Gladysheva *

Department of Medicine, University of Arizona College of Medicine–Phoenix, Phoenix, AZ 85004, USA
* Correspondence: innagladysheva@arizona.edu; Tel.: +1-(602)-827-2919
† These authors contributed equally to this work.

Abstract: In heart failure with reduced ejection fraction (HFrEF), cardiogenic edema develops from impaired cardiac function, pathological remodeling, chronic inflammation, endothelial dysfunction, neurohormonal activation, and altered nitric oxide-related pathways. Pre-clinical HFrEF studies have shown that treatment with sodium–glucose cotransporter-2 inhibitors (SGLT-2i) stimulates natriuretic and osmotic/diuretic effects, improves overall cardiac function, attenuates maladaptive cardiac remodeling, and reduces chronic inflammation, oxidative stress, and endothelial dysfunction. Here, we review the mechanisms and effects of SGLT-2i therapy on cardiogenic edema in various models of HFrEF. Overall, the data presented suggest a high translational importance of these studies, and pre-clinical studies show that SGLT-2i therapy has a marked effect on suppressing the progression of HFrEF through multiple mechanisms, including those that affect the development of cardiogenic edema.

Keywords: edema; HFrEF; dilated cardiomyopathy; excessive extracellular fluid; fluid management; endothelial dysfunction; inflammation; cardiac remodeling

Citation: Sullivan, R.D.; McCune, M.E.; Hernandez, M.; Reed, G.L.; Gladysheva, I.P. Suppression of Cardiogenic Edema with Sodium–Glucose Cotransporter-2 Inhibitors in Heart Failure with Reduced Ejection Fraction: Mechanisms and Insights from Pre-Clinical Studies. *Biomedicines* **2022**, *10*, 2016. https://doi.org/10.3390/biomedicines10082016

Academic Editor: Tânia Martins-Marques

Received: 4 August 2022
Accepted: 17 August 2022
Published: 19 August 2022

Publisher's Note: MDPI stays neutral with regard to jurisdictional claims in published maps and institutional affiliations.

Copyright: © 2022 by the authors. Licensee MDPI, Basel, Switzerland. This article is an open access article distributed under the terms and conditions of the Creative Commons Attribution (CC BY) license (https://creativecommons.org/licenses/by/4.0/).

1. Introduction

Heart failure (HF) affects about 64.3 million people worldwide [1], including an estimated 6.2 million in the United States [2]. The prevalence of symptomatic HF is expected to increase 46% by 2030, in comparison to 2012 [3]. HF with reduced ejection fraction (HFrEF) is characterized by progressive heart enlargement and declining contraction that leads to clinical symptoms (breathlessness, fatigue, swelling, etc.) from pathological fluid and sodium retention (edema) [4,5]. Sodium retention and the dysregulation of neurohumoral systems, including the sympathetic nervous system, renin–angiotensin–aldosterone system (RAAS), and the natriuretic peptide (NP) system, lead to excessive extracellular fluid accumulation in the interstitial space or edema (pulmonary, pleural effusion, ascites and/or gross/systemic peripheral fluid retention), which are hallmarks of symptomatic HF [5–12]. Symptomatic HF adversely impacts quality of life, is the primary cause of patient hospitalization, and is associated with premature mortality [13–20]. The prognostic role of edema is confirmed by clinical trials and post hoc analysis [13,19,21–23]. Maintaining physiologically relative fluid homeostasis is one of the primary goals of HF management [17,24].

During the last decade, significant progress has been made in the management of HFrEF with clearly defined, guideline-directed therapies [5,25–31]. Pharmaco-therapeutic interventions for HFrEF include modulating the neurohumoral response by targeting RAAS with angiotensin-converting enzyme inhibitors and angiotensin II (Ang II) receptor blockers; the sympathetic nervous system with beta-blockers; the mineralocorticoid system with mineralocorticoid receptor blockers (MRBs); and both RAAS and NP systems with combined Ang II receptor–neprilysin inhibitors (ARNI). However, better pharmacological interventions for reversing or preventing cardiogenic edema are needed. Early treatment at the pre-clinical stages may prevent HF progression to symptomatic stages with or without decompensation and improve outcomes. Most recently, the US Food and Drug Administration approved a new class of drug for the management of HF with reduced, mid-range or preserved ejection fraction called sodium–glucose cotransporter-2 inhibitors (SGLT-2i, originally developed as glucose-lowering agents), including canagliflozin, dapagliflozin, empagliflozin, and ertugliflozin [32,33]. In patients with HFrEF, SGLT-2i-based therapies enhance natriuresis/diuresis, modulate neurohumoral activation, improve cardiac and renal functions, functional status, duration, and quality of life [34]. These agents reduce HF-related hospitalization rates that are due to the aggravation of HF signs and symptoms caused by edema, independent from the reduction in glycemic level and the co-administration of guideline-directed HF therapy [34].

In HF, the retention of sodium and water by the kidneys leads to an expansion of free fluid in the interstitial compartments or edema. The expansion of the interstitial fluid volume is aggravated by vascular leakage caused by inflammation and endothelial dysfunction [35,36] and is associated with impaired extra fluid removal from the interstitial to intravascular space, which is in part caused by pathological alteration in the capillary dynamics and lymphatic system [24,37–39]. Fluid accumulation in the lungs manifests as cardiogenic pulmonary edema, while peripheral interstitial fluid overload manifests as soft tissue or third space edema. In contrast to classical diuretics, which affect blood plasma volume, SGLT-2i efficiently lowers edema by reducing pathological HF-related sodium retention and intestinal fluid volume [34,40–43]. Treatment with SGLT-2i also reduces the plasma level of N-terminal pro B-type natriuretic peptide (NT-pro-BNP) as an indicator of HF-associated reduced edema [44]. SGLT-2i receptor SGLT-2 is not expressed by cardiac tissue [45], excluding direct action of this class of drug on the heart. Therefore, the molecular and pathophysiological mechanisms of action of SGLT-2i in HFrEF remain inconclusive, although many intriguing hypotheses and mechanisms of action have been proposed [46–49].

This review aims to summarize the current evidence from pre-clinical translational studies, providing mechanistic experimental support for SGLT-2i action targeting the normalization of sodium–water homeostasis and the attenuation or prevention of edema signs and symptoms in HFrEF.

2. Contribution of Neurohumoral Activation, Cardiorenal Dysfunction and Cardiac Remodeling to Edema in HFrEF

Regardless of the etiology causing ventricular dysfunction, the ensuing compensatory mechanisms that occur in response to decreased cardiac output will be the same. Decreased cardiac output leads to a reduction in the intra-arterial effective circulating volume [24]. Subsequent sympathetic nervous system activation leads to an increase in heart rate and peripheral vasoconstriction. Concurrent RAAS activation and decreased renal perfusion lead to a reduction in sodium and water excretion from the kidneys [26,50]. Increased adrenergic tone and (Ang II)-induced vasoconstriction cause cardiac pressure overload. Increased intra-cardiac filling pressure leads to a backup of fluid in the pulmonary vasculature and the development of pulmonary congestion symptoms [9]. Increased water and sodium retention leads to intravascular blood volume expansion as well as interstitial fluid accumulation (extracellular water). Increased blood volume leads to increased central filling pressures and contributes to the development of peripheral congestion symptoms [9,24]. Edema directly increases pre-load as a feedback mechanism, contributing to left ventricular wall stress, cardiac remodeling and overall further decline in cardiac function [17].

During the asymptomatic stage of ventricular dysfunction (at risk for HF or at pre-HF stage [5]), the effects of elevated intra-cardiac filling pressures are mitigated by structural ventricular remodeling and the compensatory interdependent activation of neurohumoral systems. Thus, the early activation of the sympathetic nervous system and RAAS initially may help to maintain cardiac output and systemic perfusion. This compensatory response prevents the development of clinically evident signs or symptoms associated with the dysregulation of sodium–water homeostasis. As systolic dysfunction progresses, increased volume within the ventricles leads to an alteration in ventricular architecture, including myocyte hypertrophy/dilation, myocyte apoptosis, myofibroblast proliferation, and interstitial fibrosis [51]. Decreased forward flow as a result of decreased contractility leads to pathological neurohumoral activation, and the increased production of pro-atrial (ANP) and pro-B-type (BNP) natriuretic peptides by atria and ventricles as a compensatory mechanism. However, as HFrEF progresses, the NP system becomes impaired [52–54] and unbalanced persistent neurohumoral activation eventually becomes maladaptive and contributes to extracellular fluid retention or edema via vascular exchange to the interstitial space [26,41,50,55–57].

3. Evidence Supporting Edema Attenuation by SGLT-2i in Pre-Clinical HF Models

SGLT-2i inhibits renal sodium reabsorption and increases the urinary excretion of sodium, which may attenuate symptomatic HFrEF by reducing pulmonary and systemic edema and/or preventing edema formation [34]. Pre-clinical studies in animal models of HFrEF have shown that, in addition to natriuretic and osmotic/diuretic effects, treatment with SGLT-2i positively modulates edema-related plasma biomarkers and physiological outcomes contributing to edema modulation. Thus, SGLT-2i improves cardiac output and attenuates maladaptive cardiac remodeling, chronic inflammation, oxidative stress, and endothelial dysfunction (ED) by restoring the activity of nitric oxide (NO) within the vascular endothelium. The ability of SGLT-2i to attenuate inflammation and ED strongly suggests that SGLT-2i may help prevent fluid leakage from the vascular compartment to the interstitial space and prevent edema development. Despite the fact that the direct action of SGLT-2i on cardiac tissue is unlikely since SGLT-2 protein is not expressed in the heart, the combined impact of anti-inflammatory, anti-oxidative stress, and anti-ED effects of SGLT-2i may improve ventricular and global cardiac output, reduce fibrosis, suppress edema, and overall, attenuate HF progression.

3.1. *Impact of SGLT-2i on Neurohumoral Activation toward Edema Restraining in HFrEF*

SGLT-2i might retard HF-associated chronic activation of the sympathetic nervous system and RAAS, which stimulates salt and water retention by the kidneys and crucially contributes to pulmonary and systemic edema formation [26,58,59].

Experimental data related to the association between SGLT-2i treatment and RAAS modulation are limited to translational models of type 2 diabetes mellitus [60]. Overstimulation of the sympathetic nervous system in HFrEF is associated with elevated levels of norepinephrine in circulation [59]. Neurohormonal activation was attenuated in a group of non-diabetic pigs with post-MI HFrEF treated with empagliflozin, as demonstrated by reduced plasma levels of norepinephrine catabolites in comparison with the control experimental group [61]. Persistent sympathetic activation causes tachycardia [62]. In contrast to diuretics that elevate resting heart rate, SGLT-2i treatment is not associated with heart rate elevation [49] and may even reduce it [63], therefore suppressing pathologically elevated sympathetic activity. A trial in patients with HFrEF is warranted to align this mechanism with SGLT-2i outcomes in the clinic.

3.2. *SGLT-2i Positively Affect Natriuretic Peptide System, Diuresis/Natriuresis, HF Signs, and Edema-Associated HF Plasma Biomarkers*

The natriuretic peptide system plays a central role in natriuresis, diuresis, and vasodilatation, and balances the outcomes of the sympathetic nervous system and RAAS; however, in symptomatic HFrEF, its physiologic activity is impaired [52,54,64,65]. The associations between plasma BNP/NT-pro-BNP levels and edema were incorporated into the Universal Definition of Heart Failure [5]. In pre-clinical models of HFrEF, elevated plasma and cardiac pro-BNP/BNP (and pro-ANP/NT-pro-ANP) levels were strongly associated with extracellular fluid accumulation and clinically relevant edema manifestation as pleural effusion and pulmonary edema [53,56,66–69].

In a spontaneous hypertensive rat (SHR) HF model, treatment with empagliflozin reduced expression levels of ANP/BNP and tumor necrosis factor alpha (TNFα) in the ventricular tissue that were upregulated by HF [70]. In a zebrafish model of HF (induced by aristolochic acid), treatment with empagliflozin (0.1%, 10 μg) dampened the expression of ANP/BNP and downregulated related signaling pathways [66]. In an obese rat model of spontaneous hypertensive HF, empagliflozin treatment (925 mg/kg body weight by oral gavage for 6 months) decreased hepatic congestion [71]. In a post-myocardial infarction (MI) rat model of HFrEF, empagliflozin treatment increased urinary sodium excretion associated with a substantial reduction in body weight [72]. Utilizing a TAC-induced mouse HFrEF model, the beneficial effects of empagliflozin treatment on cardiac function were not dependent on natriuresis since the diuretic effect of SGLT-2i was not associated with any significant changes in electrolyte balance (blood or urine Na^+ and K^+ concentrations). ANP/BNP levels were not provided in this study [73].

3.3. SGLT-2i May Depress Edema by Improving Cardiac Function

Left ventricular function is an important indicator of HFrEF progression [1]. Reduced cardiac output stimulates chronic activation of the sympathetic nervous system and RAAS and promotes edema. The impact of SGLT-2i on systolic and diastolic function has been evaluated in rat, mouse and pig translational models of HFrEF. Several studies have investigated cardiac function outcomes in post-MI HFrEF models of left anterior descending (LAD) coronary artery ligation. In an ischemic reperfusion (IR) model of pre-diabetic obese insulin-resistant Wistar rats on a high-fat diet, dapagliflozin improved LVEF [74]. In a post-MI HFrEF model: (1) empagliflozin treatment increased LVEF [72], improved contractility, stroke volume, and end-systolic blood pressure despite diuresis, and improved diastolic function (reduction in LV end-diastolic pressure) [75]; canagliflozin IR model treatment alleviated left ventricular (LV) systolic and diastolic dysfunction, which may be explained by the increased phosphorylation of adenosine monophosphate-activated protein kinase, eNOS, and subsequent vasodilation [76]; (2) in non-diabetic mice, dapagliflozin treatment improved LV systolic function and LV mass [77]. In an MI porcine model, empagliflozin treatment improved LV systolic function [61] and ameliorated diastolic dysfunction [78]. In a model of cardiomyopathy induced by Ang II infusion in diabetic mice, dapagliflozin treatment increased LV fractional shortening [79]. In a genetic rat model of HFrEF (inducible diabetes and hypertensive HF), treatment with empagliflozin increased EF [80]. In a mouse model of doxorubicin (DOX)-induced cardiomyopathy, treatment with empagliflozin ameliorated LV dysfunction [81]. In a mouse model of cardiac pressure overload (by transverse aortic constriction (TAC)), the treatment of empagliflozin attenuated LV systolic and diastolic dysfunction, perhaps by increasing glucose and fatty acid oxidation [82]. Treatment with empagliflozin blunted the decline in cardiac function in a mouse model of TAC-induced HFrEF [73]. In a hypertensive HF model (spontaneous hypertensive rats fed a high fat diet for 32 weeks), treatment with empagliflozin normalized end-systolic and end-diastolic volume, but LV ejection fraction was not significantly improved [70]. Through the various species and models presented, SGLT-2i—at a minimum—maintains, but in most cases shows an overwhelming objective improvement in, cardiac function. Overall outcomes of SGLT-2i in translational models of HFrEF are summarized in Table 1.

Table 1. Cardiac outcomes of treatments with SGLT-2i in pre-clinical models of HFrEF.

Preclinical Models (species)	Drug Dosage Duration	Cardiac Function	Cardiac Remodeling	Inflammation, ROS, ED
Hypertensive HF model (rat)	Empagliflozin 20 mg/kg/day 12 weeks [70]	• Normalized end diastolic, end systolic volume. • LVEF not significantly improved [70]	• Reduced cardiac fibrosis [70]	
Genetic HFrEF model (rat)	Empagliflozin 10 mg/kg/day 4 weeks	• Increased cardiac function and LVEF [80]		• Decreased infiltration by macrophages [80]
Post-MI HFrEF model (rat; mouse, pig)	Empagliflozin 10mg orally 2 months [61]	• Increased LV systolic volume [61]	• Attenuated remodeling post-MI (lower LV, dilation, sphericity) [61]	
	20 mg/kg/day 6 weeks [75]	• Improved contractility, stroke work, end-systolic blood pressure diastolic function [75]	• No improvement in interstitial fibrosis or cardiomyocyte hypertrophy [75]	
	30 mg/kg/day 2 weeks [72]	• Improved LVEF [72]	• Attenuated cardiomyocyte hypertrophy, fibrosis [72]	• Decreased inflammation alleviated oxidative stress [72,83–85];
	10 mg/day 2 months [78]	• Improved diastolic function [78]	• Ameliorated diastolic dysfunction [78]	• Reduced extracellular volume [78] • Increased eNOS activity and NO production and bioavailability associated with cGMP-PKG axis [78]
	10 mg/kg/day 2 weeks [81]		• Improved cardiac remodeling [81]	
	Dapagliflozin 1 mg/kg/day 28 days [72,74]	• Improved LVEF [72,74]		
	1.5 mg/kg/day 4 weeks [77]	• Systolic function [77]	• Inhibited cardiac apoptosis and reduced LV mass, cardiac collagen 1/3, ANP/BNP, TGF-β1 transcripts, cardiac fibrosis [77]	• Lowered levels of inflammatory cytokines [77]
	Canagliflozin 3 μg/kg 5 mins [76,82]	• Alleviated LV systolic and diastolic dysfunction [76,82]		
Ang II-induced cardiomyopathy (mouse)	Dapagliflozin 1.5 mg/kg/day 30 days [79]	• Increased LV fractional shortening [79]		• Decreased inflammation and ROS production [79]
DOX-induced cardiomyopathy (mouse)	Empagliflozin (not provided) [81]	• Ameliorated LV function [81]	• Lowered myocardial fibrosis [81]	
LPS-induced cardiomyopathy (mouse)	Empagliflozin 5 mg/kg [86]	• Preserved cardiac function [86]		• Reduced cardiac iNOS, plasma TNFα and creatine kinase MB [86]

Table 1. *Cont.*

Preclinical Models (species)	Drug Dosage Duration	Cardiac Function	Cardiac Remodeling	Inflammation, ROS, ED
TAC-induced HFrEF (mouse)	Empagliflozin 10 mg/kg/day 2 weeks post-surgery [73,82]	• Attenuated the decline in cardiac function [73,82]	• Attenuated LV remodeling [82]	• Decreased expression of markers of cardiac inflammation [73]

Heart failure (HF); Left ventricular ejection fraction (LVEF); Heart failure with reduced ejection fraction (HFrEF); Left ventricular (LV); Angiotensin II (Ang II); Reactive oxygen species (ROS); Endothelial dysfunction (ED); Doxorubicin (DOX); lipopolysaccharides (LPS); Inducible nitric oxide synthase (iNOS); Tumor necrosis factor alpha (TNFα); Transverse aortic constriction (TAC).

3.4. SGLT-2i May Block Edema by Reducing Cardiac Remodeling

Pathological cardiac remodeling contributes to the development of edema and the progression of HF [5,51]. Beneficial effects of SGLT-2i treatment were demonstrated on inducible rat and mouse models of HFrEF (Table 1). In a hypertensive HF model (spontaneous hypertensive rats on a high-fat diet for 32 weeks), the treatment of empagliflozin significantly attenuated cardiac fibrosis in atrial and ventricular tissues [70]. Empagliflozin treatment attenuated adverse LV remodeling in post-MI HFrEF in pigs [61] and reduced interstitial cardiac fibrosis in pigs HFrEF induced by 2 h balloon occlusion of the proximal left anterior descending artery [78], which was associated with reduced extracellular volume [78]. In a rat post-MI model of HFrEF, empagliflozin treatment attenuated cardiomyocyte hypertrophy and fibrosis [72], but did not show any improvement in interstitial fibrosis or cardiomyocyte hypertrophy in another study [75]. In a mouse post-MI HFrEF model: empagliflozin treatment improved cardiac remodeling by the inhibition of apoptosis, alleviated oxidative stress, restored mitochondrial membrane potential, and activated AMPK signaling [83]; dapagliflozin treatment inhibited cardiac apoptosis and reduced LV mass, the expression of cardiac collagen 1/3, atrial natriuretic peptide (ANP), B-type natriuretic peptide (BNP), and transforming growth factor-β1 (TGF-β1) transcripts of cardiac fibrosis histological staining [77]. Treatment with empagliflozin (10 mg/kg/day) administered 3 weeks before MI improved cardiac remodeling and ameliorated fibrosis and hypertrophy post-MI in both diabetic and non-diabetic rats. This is possibly due to the increase in the myocardial expression of cardiac guanosine-5′-triphosphate enzyme cyclohydrase 1 (cGCH1), which activates neuronal nitric oxide synthase (nNOS) and endothelial nitric oxide synthase (eNOS) and inhibits inducible nitric oxide synthase (iNOS) [84]. In a mouse model of DOX-induced cardiomyopathy, treatment with empagliflozin (dosage not provided) lowered myocardial fibrosis [81]. In a rat model of cardiomyopathy (salt-sensitive hypertensive rats fed a high-salt/high-fat diet) treatment with tofogliflozin (0.005% for 9 weeks) reduced cardiomyocyte hypertrophy, perivascular fibrosis and associated fibrosis genes (ANP, BNP and interleukin 6) [87]. In an Ang II-induced model of cardiomyopathy in diabetic mice, dapagliflozin treatment attenuated fibrosis [70]. Overall, SGLT-2i reduce cardiac remodeling pathology, help to maintain a healthy cardiac tissue architecture, and prevent diastolic dysfunction, thus preventing HF progression and edema development.

3.5. Impact of SGLT-2i on Cardiorenal Function Leading to Edema Suppression

Cardiac and renal function are highly interdependent and modulate sodium and water retention. Increased sodium retention by the kidneys leads to intravascular blood volume expansion, as well as interstitial fluid accumulation (extracellular water). SGLT-2i may restrict excessive sodium and water retention in the interstitial space of the kidney parenchyma and reduce edema formation in HFrEF [88]. In a rat chronic kidney disease model caused by 5/6 subtotal nephrectomy and DOX-induced dilated cardiomyopathy, empagliflozin treatment (20 mg/kg/day for 60 days) was associated with a lower kidney injury score, decreased myocardial fibrosis, inhibited LV remodeling, and decreased BNP protein level in LV (an indicator of HF/pressure overload). This may be explained by empagliflozin treatment causing downregulated autophagy, apoptosis, reduced markers of oxidative stress (NADPH oxidase, NOX-1, NOX-2) in renal tubular cells, decreased markers of DNA damage (phosphorylated histone H2AX) and mitochondrial damage (cytosolic cytochrome C), and increasing indicators of mitochondrial integrity (mitochondrial cytochrome C) [89].

3.6. SGLT-2i Block Activation of Sodium–Hydrogen Exchangers in the Heart and Kidneys That Contribute to the Clinical Progression of HFrEF Associated with Edema

SGLT-2i may slow edema development and HF progression by blocking the activity of sodium–hydrogen exchangers (NHE) expressed in the myocardium (NHE-1 isoform) and in the proximal convoluted tubule of kidneys (NHE-3 isoform) [90], as well as the activity of the late component of the cardiac sodium channel current in cardiomyocytes [91]. NHE-1 regulates cardiomyocyte pH and volume. NHE-3 is responsible for the reabsorption of approximately 70% of filtered sodium [90]. In HFrEF, chronic neurohumoral activation stimulates the activation of cardiac NHE-1 and renal NHE-3, leading to enhanced sodium retention that contributes to the physiological and clinical progression of HFrEF associated with fluid retention (edema) and increased sodium influx and intracellular calcium linked to cardiac hypertrophy, cell injury and fibrosis [90,92,93].

Excess intracellular calcium within cardiac myocytes also increases arrhythmogenicity due to increased cytosolic calcium during the relaxation phase of the cardiac cycle. The inhibition of NHE-1 activity with cariporide in animal models of HF adequately restored sodium and calcium handling, caused the regression of ventricular hypertrophy, and improved several markers of electrophysiological remodeling such as reduced QT and QRS intervals [94]. SGLT-2i antagonizes the effects of NHE in the heart and kidneys. An in silico analysis of the mechanism of action of empagliflozin showed that it binds to and inhibits the downstream signaling effects of NHE activation. The NHE blockade prevents cardiomyocyte death by increasing the expression of apoptotic inhibitors in cardiomyocytes. These findings were validated by an in vivo HF rat model, which showed that treatment with empagliflozin led to an increased expression of anti-apoptotic proteins and slowed HF progression [95]. However, several studies do not support empagliflozin as a potent inhibitor of NHE-1 in the healthy heart [96,97].

3.7. SGLT-2i May Restrain Edema by Suppressing Chronic Inflammation and ROS

Chronic inflammation and oxidative stress promote maladaptive systolic dysfunction, cardiac remodeling, and pulmonary/systemic edema, and therefore are hallmarks of HFrEF pathophysiology [36,98–100]. Treatment with SGLT-2i decreased inflammation and/or the production of reactive oxygen species (ROS) in inducible and genetic rat/mouse models of HFrEF. Empagliflozin treatment attenuated oxidative stress [72], and dapagliflozin treatment lowered cardiac transcripts of inflammatory cytokines (TNFα, TGF-β1, Vcam-1, MCP-1, Icam-1, IL-6) [77] in mouse/rat MI models. Incardiac dysfunction mouse models induced by administering lipopolysaccharide (LPS, 5 mg/kg) co-administration of empagliflozin preserved cardiac function possibly by improved AMPK phosphorylation and ATP/ADP, as well as reduced cardiac iNOS, plasma TNFα, and creatine kinase MB (isoenzyme, found mostly in cardiac and some skeletal muscles) levels [86]. In a model of cardiomyopathy by Ang II infusion in diabetic mice, dapagliflozin decreased inflammation and ROS production [70]. In a genetic rat model of HFrEF (inducible diabetes and hypertensive HF), treatment with empagliflozin decreased the infiltration of macrophages [80]. In a mouse TAC-induced model of HFrEF, empagliflozin decreased the expression of markers of sterile cardiac inflammation, possibly by attenuating the activity of the neutrophil-to-lymphocyte ratio (NLR) family pyrin domain containing 3 (NLRP3) inflammasome; this occurred in the absence of changes to ketone bodies or cardiac ATP use [73]. SGLT-2i therapy clearly attenuates cardiac inflammation and ROS associated with HFrEF.

3.8. SGLT-2i May Prevent Vascular Leakage and Edema by Improvement of Endothelial Dysfunction

The endothelium is a semi-permeable barrier that plays a crucial role in tissue fluid hemostasis by the tight control of fluid exchange from the vascular compartment to the interstitial space [101]. The proper function of the vascular endothelium is vital for preventing the inadvertent extravasation of fluid into surrounding tissues or edema. Clinical and translational HF is associated with a significant impairment of endothelium-dependent vasodilation (endothelial dysfunction (ED)), which is mediated almost entirely by the excess formation of superoxide radicals and other oxidant species that interfere with the activation of nitric oxide (NO) and the bioavailability of cyclic guanosine monophosphate (cGMP) [35,69,102]. Thus, in chronic HF, increased oxidative stress and the dysregulation of NO pathways lead to coronary/peripheral arteries and/or lung ED, which contribute to HF decompensation (NYHA class IV, pulmonary edema) and are associated with hospitalization and heart transplantation [103–105].

In HFrEF, treatment with SGLT-2i may prevent vascular leakage leading to edema by the prevention or improvement of ED. The impact of SGLT-2i on attenuating ED by restoring the activity of NO within the vascular endothelium in pre-clinical and clinical studies has been comprehensively overviewed [106–109]. Several pre-clinical studies have demonstrated that one of the off-target effects of SGLT-2i may involve a reduction in ED [106,107]. These pre-clinical studies were predominantly conducted on models of diabetes, since ED is known to be a major mediator of diabetic vascular disease. In a porcine model of HFrEF induced by 2 h balloon occlusion of proximal LAD artery, treatment with empagliflozin increased the activity of eNOS and NO production and the bioavailability associated with the activation of the cGMP–PKG axis [78]. Ex vivo studies have demonstrated that SGTL-2i may directly induce vasodilation by several mechanisms, including the modulation of cell adhesion molecules, the attenuation of inflammation, and reduced oxidative stress [106]. Treatment with empagliflozin was associated with a significant improvement in endothelial-dependent vasodilation in streptozotocin (STZ)-induced type 1 diabetes mellitus (T1DM) rat models [107]. Empagliflozin has been found to improve the enzymatic activity of eNOS, a key enzyme in the production of the one of the most important mediators of vasodilation, NO, in STZ-induced diabetic rats [107]. Thus, SGLT-2i may exert its vasodilatory effects by restoring the activity of NO within the vascular endothelium. Another possible mechanism by which SGLT-2i carriy out their vasorelaxant effect is by inhibiting glucose-mediated membrane depolarization [110]. One study found that canagliflozin and phlorizin induced membrane hyperpolarization in pulmonary artery smooth muscle cells. This finding was attributed to the activation of potassium channels in the plasma membrane of these cells by NO [110]. Empagliflozin restored the beneficial effect of cardiac microvascular endothelial cells on cardiomyocyte function in a co-culture system of human cardiac microvascular endothelial cells with adult rat ventricular cardiomyocytes by reducing mitochondrial oxidative damage and, ROS accumulation, and increasing the bioavailability of endothelial NO [109]. Still, an investigation of SGLT-2i treatment outcomes on pre-clinical model(s) of DCM-HFrEF characterized by impaired NO-cGMP bioavailability and cardiac eNOS production [69] is warranted.

3.9. Alteration of Cardiac Metabolism and Energy Utilization by SGLT-2i Improves Cardiac Structure and Function, Which May Contribute to Edema Reduction

As noted in earlier sections, the effects of SGLT-2i on cardiac tissue are largely indirect. However, cardiac metabolism is directly altered with the inclusion of SGTL-2i therapy. The failing heart is characterized by altered cardiac muscle contraction, increased oxygen demand, and high cellular turnover, resulting in an increased metabolic demand to support the pathologic condition which can manifest symptomatically as cardiac cachexia and sarcopenia [56,111]. By lowering circulating glucose levels, SGLT-2i shifts the cardiac metabolism to function under a fasting-like state (catabolic), with energy provided by gluconeogenesis and ketogenesis [112]. The energy-deficient state increases cardiac autophagy (cellular cleanup) by activating the SIRT1/PGC-1α/FGF21 pathway, thus reducing local inflammation, lowering oxidative stress, and removing dysfunctional cells that may otherwise contribute to myocardial remodeling as a consequence of cellular necrosis [112]. Ketone utilization produces more ATP than glucose or other metabolites [113], thus improving the energy supply for HFrEF. The increased availability of cardiac energetics with SGLT-2i therapy has been shown to improve cardiac structure and function in mice with and without T2DM [114,115]. These improvements should reduce cardiogenic-associated edema in HFrEF models treated with SGLT-2i therapy, though additional studies are needed to specifically investigate this correlation.

4. Limitations

Pathological RAAS overactivation and the impairment of the NP system significantly contributes to sodium retention and fluid accumulation in the interstitial space leading to edema. However, pre-clinical studies targeting the impact of SGLT-2i on these systems and overall neurohumoral activation are currently lacking. In HFrEF clinical and pre-clinical studies, treatment with SGLT-2i leads to an overall improvement of left ventricular function and the attenuation of cardiac remodeling, which are essential promoters and indicators of edema development and HF progression. Still, the direct mechanisms responsible for such beneficial action of SGLT-2i remain unclear, since cardiac tissue lacks the SGLT-2i receptor SGLT-2.

One major issue surrounding clinical and pre-clinical cardiogenic edema is the ability to detect early changes in excess fluid during the transition from the pre-symptomatic to the symptomatic phase of HF. Current diagnostic imaging modalities include echocardiography, MRI, CT, and thoracic radiography, though they often require specialized training for image collection and analysis. Quantitative magnetic resonance (QMR) has been introduced as an objective and longitudinal method for diagnosing and monitoring systemic cardiogenic edema in animal models [56,67,69,116,117], which may offer a refined method for monitoring edema therapeutic response throughout all phases of HF.

The locomotive and anatomical differences between human and animal models (mice, rats, pigs) used should also be considered. Upright (bipedal) versus horizontal (quadrupedal) orientation may have translational differences for the effects of edema development, manifestation, and clearance. Known differences include: pedal edema (humans), which is less commonly observed in the forelimbs of animals exhibiting HF due to changes in subcutaneous space [118]; the alteration of fluid lines and pulmonary patterns in thorax imaging due to gravity-dependence [119]; and evolutionary changes in baroreceptor sites and function throughout mammalian species [120]. Additional considerations for managing hospitalized HF patient positioning (horizontal recumbent, supine) when undergoing treatment and care for cardiogenic edema should be examined, as unanesthetized HF animal models rarely adopt this orientation. Newer studies suggest proning edema patients, and thus placing them in a more animal-like posture [119,121] to improve outcomes.

SGLT-2i is recommended in the outpatient setting for all patients with stable HFrEF conditions [122]. As with most pharmacological treatments, side effects and limitations for use have been reported from clinical trials and medical practice with SGLT-2i [123–125]. The most common clinical side effects might include fatigue, nausea, increased drinking, sudden urge to urinate, dry mouth, female urinary tract infections, hypotension, changes in blood pH, and, more rarely, acute renal pathology, bone fractures, and lower limb amputations. As a result of study design and inability to communicate directly with the patient, unfortunately, most pre-clinical studies are not able to account for all potential side effects prior to advancing to human administration.

5. Conclusions

Effective edema prevention and treatment are the primary goals of HF management and an unmet need to improve quality of life, reduce HF-related hospitalization rate, and prolong life. HFrEF pre-clinical studies have demonstrated that SGLT-2i treatment may attenuate edema formation through the stimulation of natriuretic and osmotic/diuretic effects, improvement in overall cardiac function, and the suppression of maladaptive cardiac remodeling, chronic inflammation, oxidative stress, and endothelial dysfunction (Figure 1). By repressing inflammation and endothelial dysfunction, SGLT-2i may prevent vascular leakage and edema development associated with extensive fluid accumulation in the interstitial space. In addition to its anti-inflammatory, anti-oxidative stress, and anti-fibrotic effects, SGLT-2i improves ventricular and global cardiac output, suppresses edema, and slows the rate of HF progression, locally in terms of cardiac function, and systemically at the kidneys, which appear to be the primary site of action.

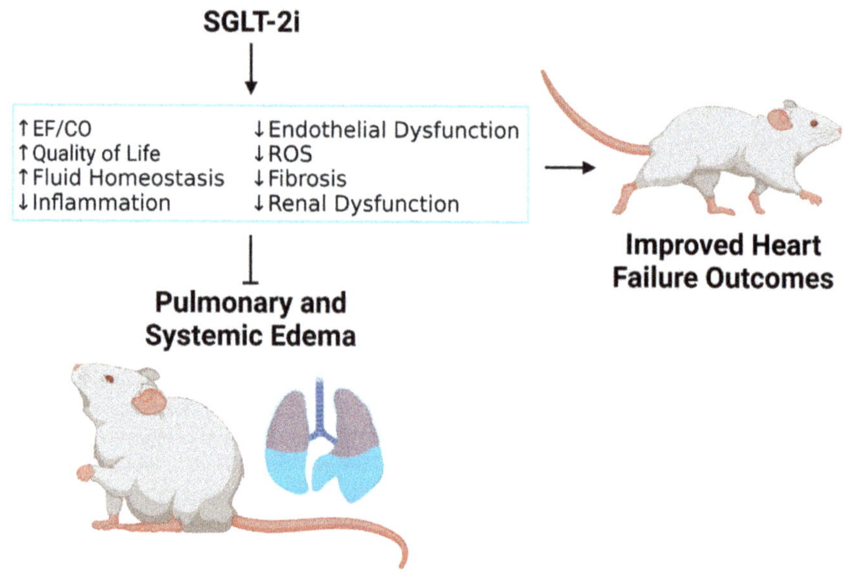

Figure 1. Mechanism by which sodium–glucose cotransporter 2 inhibitors (SGLT-2i) reduce pulmonary and systemic edema in HFrEF. Ejection fraction (EF); cardiac output (CO); reactive oxygen species (ROS). Created with BioRender.com on 18 July 2022.

Author Contributions: Conceptualization and methodology, I.P.G.; validation: R.D.S. and G.L.R.; investigation: R.D.S., M.E.M., M.H. and I.P.G.; writing—original draft preparation, R.D.S., M.E.M. and I.P.G.; writing—review and editing, R.D.S., G.L.R. and I.P.G.; formatting for publication, R.D.S.; supervision, I.P.G. All authors have read and agreed to the published version of the manuscript.

Funding: Supported by research funds from the University of Arizona, College of Medicine—Phoenix.

Institutional Review Board Statement: Not applicable.

Informed Consent Statement: Not applicable.

Data Availability Statement: Not applicable.

Conflicts of Interest: The authors declare no conflict of interest.

References

1. Groenewegen, A.; Rutten, F.H.; Mosterd, A.; Hoes, A.W. Epidemiology of Heart Failure. *Eur. J. Heart Fail.* **2020**, *22*, 1342–1356. [CrossRef]
2. Virani, S.S.; Alonso, A.; Aparicio, H.J.; Benjamin, E.J.; Bittencourt, M.S.; Callaway, C.W.; Carson, A.P.; Chamberlain, A.M.; Cheng, S.; Delling, F.N.; et al. Heart Disease and Stroke Statistics-2021 Update: A Report from the American Heart Association. *Circulation* **2021**, *143*, e254–e743. [CrossRef] [PubMed]

3. Benjamin, E.J.; Muntner, P.; Alonso, A.; Bittencourt, M.S.; Callaway, C.W.; Carson, A.P.; Chamberlain, A.M.; Chang, A.R.; Cheng, S.; Das, S.R.; et al. Heart Disease and Stroke Statistics-2019 Update: A Report from the American Heart Association. *Circulation* **2019**, *139*, e56–e528. [CrossRef] [PubMed]
4. Rosenbaum, A.N.; Agre, K.E.; Pereira, N.L. Genetics of Dilated Cardiomyopathy: Practical Implications for Heart Failure Management. *Nat. Rev. Cardiol.* **2020**, *17*, 286–297. [CrossRef] [PubMed]
5. Bozkurt, B.; Coats, A.J.S.; Tsutsui, H.; Abdelhamid, C.M.; Adamopoulos, S.; Albert, N.; Anker, S.D.; Atherton, J.; Bohm, M.; Butler, J.; et al. Universal Definition and Classification of Heart Failure: A Report of the Heart Failure Society of America, Heart Failure Association of the European Society of Cardiology, Japanese Heart Failure Society and Writing Committee of the Universal Definition of Heart Failure: Endorsed by the Canadian Heart Failure Society, Heart Failure Association of India, Cardiac Society of Australia and New Zealand, and Chinese Heart Failure Association. *Eur. J. Heart Fail.* **2021**, *23*, 352–380. [CrossRef]
6. Dzau, V.J.; Colucci, W.S.; Hollenberg, N.K.; Williams, G.H. Relation of the Renin-Angiotensin-Aldosterone System to Clinical State in Congestive Heart Failure. *Circulation* **1981**, *63*, 645–651. [CrossRef]
7. Clark, A.L.; Cleland, J.G. Causes and Treatment of Oedema in Patients with Heart Failure. *Nat. Rev. Cardiol.* **2013**, *10*, 156–170. [CrossRef]
8. Yancy, C.W.; Jessup, M.; Bozkurt, B.; Butler, J.; Casey, D.E., Jr.; Drazner, M.H.; Fonarow, G.C.; Geraci, S.A.; Horwich, T.; Januzzi, J.L.; et al. 2013 Accf/Aha Guideline for the Management of Heart Failure: A Report of the American College of Cardiology Foundation/American Heart Association Task Force on Practice Guidelines. *Circulation* **2013**, *128*, e240–e327. [CrossRef]
9. Parrinello, G.; Greene, S.J.; Torres, D.; Alderman, M.; Bonventre, J.V.; Di Pasquale, P.; Gargani, L.; Nohria, A.; Fonarow, G.C.; Vaduganathan, M.; et al. Water and Sodium in Heart Failure: A Spotlight on Congestion. *Heart Fail. Rev.* **2015**, *20*, 13–24. [CrossRef]
10. Pellicori, P.; Kaur, K.; Clark, A.L. Fluid Management in Patients with Chronic Heart Failure. *Card. Fail. Rev.* **2015**, *1*, 90–95. [CrossRef]
11. Melenovsky, V.; Andersen, M.J.; Andress, K.; Reddy, Y.N.; Borlaug, B.A. Lung Congestion in Chronic Heart Failure: Haemodynamic, Clinical, and Prognostic Implications. *Eur. J. Heart Fail.* **2015**, *17*, 1161–1171. [CrossRef] [PubMed]
12. Mentz, R.J.; Stevens, S.R.; DeVore, A.D.; Lala, A.; Vader, J.M.; AbouEzzeddine, O.F.; Khazanie, P.; Redfield, M.M.; Stevenson, L.W.; O'Connor, C.M.; et al. Decongestion Strategies and Renin-Angiotensin-Aldosterone System Activation in Acute Heart Failure. *JACC Heart Fail.* **2015**, *3*, 97–107. [CrossRef] [PubMed]
13. Metra, M.; O'Connor, C.M.; Davison, B.A.; Cleland, J.G.; Ponikowski, P.; Teerlink, J.R.; Voors, A.A.; Givertz, M.M.; Mansoor, G.A.; Bloomfield, D.M.; et al. Early Dyspnoea Relief in Acute Heart Failure: Prevalence, Association with Mortality, and Effect of Rolofylline in the Protect Study. *Eur. Heart J.* **2011**, *32*, 1519–1534. [CrossRef] [PubMed]
14. Aimo, A.; Vergaro, G.; Giannoni, A.; Emdin, M. Wet Is Bad: Residual Congestion Predicts Worse Prognosis in Acute Heart Failure. *Int. J. Cardiol.* **2018**, *258*, 201–202. [CrossRef]
15. Rubio-Gracia, J.; Demissei, B.G.; Ter Maaten, J.M.; Cleland, J.G.; O'Connor, C.M.; Metra, M.; Ponikowski, P.; Teerlink, J.R.; Cotter, G.; Davison, B.A.; et al. Prevalence, Predictors and Clinical Outcome of Residual Congestion in Acute Decompensated Heart Failure. *Int. J. Cardiol.* **2018**, *258*, 185–191. [CrossRef]
16. Palazzuoli, A.; Evangelista, I.; Nuti, R. Congestion Occurrence and Evaluation in Acute Heart Failure Scenario: Time to Reconsider Different Pathways of Volume Overload. *Heart Fail. Rev.* **2020**, *25*, 119–131. [CrossRef]
17. Lombardi, C.M.; Cimino, G.; Pellicori, P.; Bonelli, A.; Inciardi, R.M.; Pagnesi, M.; Tomasoni, D.; Ravera, A.; Adamo, M.; Carubelli, V.; et al. Congestion in Patients with Advanced Heart Failure: Assessment and Treatment. *Heart Fail. Clin.* **2021**, *17*, 575–586. [CrossRef]
18. Greene, S.J.; Fonarow, G.C.; Butler, J. Risk Profiles in Heart Failure: Baseline, Residual, Worsening, and Advanced Heart Failure Risk. *Circ. Heart Fail.* **2020**, *13*, e007132. [CrossRef]
19. Ambrosy, A.P.; Vaduganathan, M.; Huffman, M.D.; Khan, S.; Kwasny, M.J.; Fought, A.J.; Maggioni, A.P.; Swedberg, K.; Konstam, M.A.; Zannad, F.; et al. Clinical Course and Predictive Value of Liver Function Tests in Patients Hospitalized for Worsening Heart Failure with Reduced Ejection Fraction: An Analysis of the Everest Trial. *Eur. J. Heart Fail.* **2012**, *14*, 302–311. [CrossRef]
20. Boorsma, E.M.; Ter Maaten, J.M.; Damman, K.; Dinh, W.; Gustafsson, F.; Goldsmith, S.; Burkhoff, D.; Zannad, F.; Udelson, J.E.; Voors, A.A. Congestion in Heart Failure: A Contemporary Look at Physiology, Diagnosis and Treatment. *Nat. Rev. Cardiol.* **2020**, *17*, 641–655. [CrossRef]
21. Gheorghiade, M.; Gattis, W.A.; O'Connor, C.M.; Adams, K.F., Jr.; Elkayam, U.; Barbagelata, A.; Ghali, J.K.; Benza, R.L.; McGrew, F.A.; Klapholz, M.; et al. Acute Chronic Therapeutic Impact of a Vasopressin Antagonist in Congestive Heart Failure, I. Effects of Tolvaptan, a Vasopressin Antagonist, in Patients Hospitalized with Worsening Heart Failure: A Randomized Controlled Trial. *JAMA* **2004**, *291*, 1963–1971. [CrossRef] [PubMed]
22. Lala, A.; McNulty, S.E.; Mentz, R.J.; Dunlay, S.M.; Vader, J.M.; AbouEzzeddine, O.F.; DeVore, A.D.; Khazanie, P.; Redfield, M.M.; Goldsmith, S.R.; et al. Relief and Recurrence of Congestion During and after Hospitalization for Acute Heart Failure: Insights from Diuretic Optimization Strategy Evaluation in Acute Decompensated Heart Failure (Dose-Ahf) and Cardiorenal Rescue Study in Acute Decompensated Heart Failure (Caress-Hf). *Circ. Heart Fail.* **2015**, *8*, 741–748. [CrossRef] [PubMed]

23. Selvaraj, S.; Claggett, B.; Pozzi, A.; McMurray, J.J.V.; Jhund, P.S.; Packer, M.; Desai, A.S.; Lewis, E.F.; Vaduganathan, M.; Lefkowitz, M.P.; et al. Prognostic Implications of Congestion on Physical Examination among Contemporary Patients with Heart Failure and Reduced Ejection Fraction: Paradigm-Hf. *Circulation* **2019**, *140*, 1369–1379. [CrossRef] [PubMed]
24. Miller, W.L. Fluid Volume Overload and Congestion in Heart Failure: Time to Reconsider Pathophysiology and How Volume Is Assessed. *Circ. Heart Fail.* **2016**, *9*, e002922. [CrossRef]
25. Rossignol, P.; Hernandez, A.F.; Solomon, S.D.; Zannad, F. Heart Failure Drug Treatment. *Lancet* **2019**, *393*, 1034–1044. [CrossRef]
26. Sullivan, R.D.; Mehta, R.M.; Tripathi, R.; Reed, G.L.; Gladysheva, I.P. Renin Activity in Heart Failure with Reduced Systolic Function-New Insights. *Int. J. Mol. Sci.* **2019**, *20*, 3182. [CrossRef]
27. Murphy, S.P.; Ibrahim, N.E.; Januzzi, J.L., Jr. Heart Failure with Reduced Ejection Fraction: A Review. *JAMA* **2020**, *324*, 488–504. [CrossRef]
28. Pellicori, P.; Khan, M.J.I.; Graham, F.J.; Cleland, J.G.F. New Perspectives and Future Directions in the Treatment of Heart Failure. *Heart Fail. Rev.* **2020**, *25*, 147–159. [CrossRef]
29. McDonagh, T.A.; Metra, M.; Adamo, M.; Gardner, R.S.; Baumbach, A.; Bohm, M.; Burri, H.; Butler, J.; Celutkiene, J.; Chioncel, O.; et al. 2021 Esc Guidelines for the Diagnosis and Treatment of Acute and Chronic Heart Failure. *Eur. Heart J.* **2021**, *42*, 3599–3726. [CrossRef]
30. Espinoza, C.; Alkhateeb, H.; Siddiqui, T. Updates in Pharmacotherapy of Heart Failure with Reduced Ejection Fraction. *Ann. Transl. Med.* **2021**, *9*, 516. [CrossRef]
31. Patel, N.H.; Bindra, A.S. Transitioning Hfref Patients on Ace Inhibitors to Modern Medical Therapy: What Is the Next Step? *JACC Heart Fail.* **2021**, *9*, 319. [CrossRef] [PubMed]
32. Gupta, M.; Rao, S.; Manek, G.; Fonarow, G.C.; Ghosh, R.K. The Role of Dapagliflozin in the Management of Heart Failure: An Update on the Emerging Evidence. *Ther. Clin. Risk Manag.* **2021**, *17*, 823–830. [CrossRef] [PubMed]
33. Spertus, J.A.; Birmingham, M.C.; Nassif, M.; Damaraju, C.V.; Abbate, A.; Butler, J.; Lanfear, D.E.; Lingvay, I.; Kosiborod, M.N.; Januzzi, J.L. The Sglt2 Inhibitor Canagliflozin in Heart Failure: The Chief-Hf Remote, Patient-Centered Randomized Trial. *Nat. Med.* **2022**, *28*, 809–813. [CrossRef]
34. Hernandez, M.; Sullivan, R.D.; McCune, M.E.; Reed, G.L.; Gladysheva, I.P. Sodium-Glucose Cotransporter-2 Inhibitors Improve Heart Failure with Reduced Ejection Fraction Outcomes by Reducing Edema and Congestion. *Diagnostics* **2022**, *12*, 989. [CrossRef]
35. Zuchi, C.; Tritto, I.; Carluccio, E.; Mattei, C.; Cattadori, G.; Ambrosio, G. Role of Endothelial Dysfunction in Heart Failure. *Heart Fail. Rev.* **2020**, *25*, 21–30. [CrossRef]
36. Murphy, S.P.; Kakkar, R.; McCarthy, C.P.; Januzzi, J.L., Jr. Inflammation in Heart Failure: Jacc State-of-the-Art Review. *J. Am. Coll. Cardiol.* **2020**, *75*, 1324–1340. [CrossRef] [PubMed]
37. Guyton, A.C.; Granger, H.J.; Taylor, A.E. Interstitial Fluid Pressure. *Physiol. Rev.* **1971**, *51*, 527–563. [CrossRef]
38. Itkin, M.; Rockson, S.G.; Burkhoff, D. Pathophysiology of the Lymphatic System in Patients with Heart Failure: Jacc State-of-the-Art Review. *J. Am. Coll. Cardiol.* **2021**, *78*, 278–290. [CrossRef]
39. Aronson, D. The Interstitial Comartment as a Therapeutic Target in Heart Failure. *Front. Cardiovasc. Med.* **2022**, *9*, 933384. [CrossRef]
40. Hallow, K.M.; Helmlinger, G.; Greasley, P.J.; McMurray, J.J.V.; Boulton, D.W. Why Do Sglt2 Inhibitors Reduce Heart Failure Hospitalization? A Differential Volume Regulation Hypothesis. *Diabetes Obes. Metab.* **2018**, *20*, 479–487. [CrossRef]
41. Yu, H.; Basu, S.; Hallow, K.M. Cardiac and Renal Function Interactions in Heart Failure with Reduced Ejection Fraction: A Mathematical Modeling Analysis. *PLoS Comput. Biol.* **2020**, *16*, e1008074. [CrossRef] [PubMed]
42. Fujiki, S.; Tanaka, A.; Imai, T.; Shimabukuro, M.; Uehara, H.; Nakamura, I.; Matsunaga, K.; Suzuki, M.; Kashimura, T.; Minamino, T.; et al. Body Fluid Regulation Via Chronic Inhibition of Sodium-Glucose Cotransporter-2 in Patients with Heart Failure: A Post Hoc Analysis of the Candle Trial. *Clin. Res. Cardiol.* **2022**; *online ahead of print*. [CrossRef]
43. Tanaka, A.; Shimabukuro, M.; Teragawa, H.; Okada, Y.; Takamura, T.; Taguchi, I.; Toyoda, S.; Tomiyama, H.; Ueda, S.; Higashi, Y.; et al. Reduction of Estimated Fluid Volumes Following Initiation of Empagliflozin in Patients with Type 2 Diabetes and Cardiovascular Disease: A Secondary Analysis of the Placebo-Controlled, Randomized Emblem Trial. *Cardiovasc. Diabetol.* **2021**, *20*, 105. [CrossRef] [PubMed]
44. Rasalam, R.; Atherton, J.J.; Deed, G.; Molloy-Bland, M.; Cohen, N.; Sindone, A. Sodium-Glucose Cotransporter 2 Inhibitor Effects on Heart Failure Hospitalization and Cardiac Function: Systematic Review. *ESC Heart Fail.* **2021**, *8*, 4093–4118. [CrossRef]
45. Ghezzi, C.; Loo, D.D.F.; Wright, E.M. Physiology of Renal Glucose Handling Via Sglt1, Sglt2 and Glut2. *Diabetologia* **2018**, *61*, 2087–2097. [CrossRef] [PubMed]
46. Bell, R.M.; Yellon, D.M. Sglt2 Inhibitors: Hypotheses on the Mechanism of Cardiovascular Protection. *Lancet Diabetes Endocrinol.* **2018**, *6*, 435–437. [CrossRef]
47. Wojcik, C.; Warden, B.A. Mechanisms and Evidence for Heart Failure Benefits from Sglt2 Inhibitors. *Curr. Cardiol. Rep.* **2019**, *21*, 130. [CrossRef] [PubMed]
48. Nightingale, B. A Review of the Proposed Mechanistic Actions of Sodium Glucose Cotransporter-2 Inhibitors in the Treatment of Heart Failure. *Cardiol. Res.* **2021**, *12*, 60–66. [CrossRef]
49. Fathi, A.; Vickneson, K.; Singh, J.S. Sglt2-Inhibitors. More Than Just Glycosuria and Diuresis. *Heart Fail. Rev.* **2021**, *26*, 623–642. [CrossRef]
50. Sayer, G.; Bhat, G. The Renin-Angiotensin-Aldosterone System and Heart Failure. *Cardiol. Clin.* **2014**, *32*, 21–32. [CrossRef]

51. Konstam, M.A.; Kramer, D.G.; Patel, A.R.; Maron, M.S.; Udelson, J.E. Left Ventricular Remodeling in Heart Failure: Current Concepts in Clinical Significance and Assessment. *JACC Cardiovasc. Imaging* **2011**, *4*, 98–108. [CrossRef]
52. Ibebuogu, U.N.; Gladysheva, I.P.; Houng, A.K.; Reed, G.L. Decompensated Heart Failure Is Associated with Reduced Corin Levels and Decreased Cleavage of Pro-Atrial Natriuretic Peptide. *Circ. Heart Fail.* **2011**, *4*, 114–120. [CrossRef]
53. Tripathi, R.; Wang, D.; Sullivan, R.; Fan, T.H.; Gladysheva, I.P.; Reed, G.L. Depressed Corin Levels Indicate Early Systolic Dysfunction before Increases of Atrial Natriuretic Peptide/B-Type Natriuretic Peptide and Heart Failure Development. *Hypertension* **2016**, *67*, 362–367. [CrossRef]
54. Zaidi, S.S.; Ward, R.D.; Ramanathan, K.; Yu, X.; Gladysheva, I.P.; Reed, G.L. Possible Enzymatic Downregulation of the Natriuretic Peptide System in Patients with Reduced Systolic Function and Heart Failure: A Pilot Study. *Biomed. Res. Int.* **2018**, *2018*, 7279036. [CrossRef]
55. Schrier, R.W.; Abraham, W.T. Hormones and Hemodynamics in Heart Failure. *N. Engl. J. Med.* **1999**, *341*, 577–585. [CrossRef]
56. Sullivan, R.D.; Mehta, R.M.; Tripathi, R.; Gladysheva, I.P.; Reed, G.L. Normalizing Plasma Renin Activity in Experimental Dilated Cardiomyopathy: Effects on Edema, Cachexia, and Survival. *Int. J. Mol. Sci.* **2019**, *20*, 3886. [CrossRef]
57. Chiorescu, R.M.; Lazar, R.-D.; Buksa, S.-B.; Mocan, M.; Blendea, D. Biomarkers of Volume Overload and Edema in Heart Failure with Reduced Ejection Fraction. *Front. Cardiovasc. Med.* **2022**, *9*, 910100. [CrossRef]
58. Adams, K.F., Jr. Pathophysiologic Role of the Renin-Angiotensin-Aldosterone and Sympathetic Nervous Systems in Heart Failure. *Am. J. Health Syst. Pharm.* **2004**, *61*, S4–S13. [CrossRef]
59. Hartupee, J.; Mann, D.L. Neurohormonal Activation in Heart Failure with Reduced Ejection Fraction. *Nat. Rev. Cardiol.* **2017**, *14*, 30–38. [CrossRef]
60. Ansary, T.M.; Nakano, D.; Nishiyama, A. Diuretic Effects of Sodium Glucose Cotransporter 2 Inhibitors and Their Influence on the Renin-Angiotensin System. *Int. J. Mol. Sci.* **2019**, *20*, 629. [CrossRef]
61. Santos-Gallego, C.G.; Requena-Ibanez, J.A.; San Antonio, R.; Ishikawa, K.; Watanabe, S.; Picatoste, B.; Flores, E.; Garcia-Ropero, A.; Sanz, J.; Hajjar, R.J.; et al. Empagliflozin Ameliorates Adverse Left Ventricular Remodeling in Nondiabetic Heart Failure by Enhancing Myocardial Energetics. *J. Am. Coll. Cardiol.* **2019**, *73*, 1931–1944. [CrossRef]
62. Borovac, J.A.; D'Amario, D.; Bozic, J.; Glavas, D. Sympathetic Nervous System Activation and Heart Failure: Current State of Evidence and the Pathophysiology in the Light of Novel Biomarkers. *World J. Cardiol.* **2020**, *12*, 373–408. [CrossRef]
63. Shimizu, W.; Kubota, Y.; Hoshika, Y.; Mozawa, K.; Tara, S.; Tokita, Y.; Yodogawa, K.; Iwasaki, Y.K.; Yamamoto, T.; Takano, H.; et al. Effects of Empagliflozin Versus Placebo on Cardiac Sympathetic Activity in Acute Myocardial Infarction Patients with Type 2 Diabetes Mellitus: The Embody Trial. *Cardiovasc. Diabetol.* **2020**, *19*, 148. [CrossRef]
64. Wei, C.M.; Heublein, D.M.; Perrella, M.A.; Lerman, A.; Rodeheffer, R.J.; McGregor, C.G.; Edwards, W.D.; Schaff, H.V.; Burnett, J.C., Jr. Natriuretic Peptide System in Human Heart Failure. *Circulation* **1993**, *88*, 1004–1009. [CrossRef]
65. Gidlof, O. Toward a New Paradigm for Targeted Natriuretic Peptide Enhancement in Heart Failure. *Front. Physiol.* **2021**, *12*, 650124. [CrossRef]
66. Shi, X.; Verma, S.; Yun, J.; Brand-Arzamendi, K.; Singh, K.K.; Liu, X.; Garg, A.; Quan, A.; Wen, X.Y. Effect of Empagliflozin on Cardiac Biomarkers in a Zebrafish Model of Heart Failure: Clues to the Empa-Reg Outcome Trial? *Mol. Cell. Biochem.* **2017**, *433*, 97–102. [CrossRef]
67. Tripathi, R.; Sullivan, R.D.; Fan, T.M.; Houng, A.K.; Mehta, R.M.; Reed, G.L.; Gladysheva, I.P. Cardiac-Specific Overexpression of Catalytically Inactive Corin Reduces Edema, Contractile Dysfunction, and Death in Mice with Dilated Cardiomyopathy. *Int. J. Mol. Sci.* **2020**, *21*, 203. [CrossRef]
68. Ma, X.; Tannu, S.; Allocco, J.; Pan, J.; Dipiero, J.; Wong, P. A Mouse Model of Heart Failure Exhibiting Pulmonary Edema and Pleural Effusion: Useful for Testing New Drugs. *J. Pharmacol. Toxicol. Methods* **2019**, *96*, 78–86. [CrossRef]
69. Tripathi, R.; Sullivan, R.D.; Fan, T.M.; Mehta, R.M.; Gladysheva, I.P.; Reed, G.L. A Low-Sodium Diet Boosts Ang (1-7) Production and No-Cgmp Bioavailability to Reduce Edema and Enhance Survival in Experimental Heart Failure. *Int. J. Mol. Sci.* **2021**, *22*, 4035. [CrossRef]
70. Lee, H.C.; Shiou, Y.L.; Jhuo, S.J.; Chang, C.Y.; Liu, P.L.; Jhuang, W.J.; Dai, Z.K.; Chen, W.Y.; Chen, Y.F.; Lee, A.S. The Sodium-Glucose Co-Transporter 2 Inhibitor Empagliflozin Attenuates Cardiac Fibrosis and Improves Ventricular Hemodynamics in Hypertensive Heart Failure Rats. *Cardiovasc. Diabetol.* **2019**, *18*, 45. [CrossRef]
71. Abdurrachim, D.; Teo, X.Q.; Woo, C.C.; Chan, W.X.; Lalic, J.; Lam, C.S.P.; Lee, P.T.H. Empagliflozin Reduces Myocardial Ketone Utilization While Preserving Glucose Utilization in Diabetic Hypertensive Heart Disease: A Hyperpolarized (13) C Magnetic Resonance Spectroscopy Study. *Diabetes Obes. Metab.* **2019**, *21*, 357–365. [CrossRef]
72. Yurista, S.R.; Sillje, H.H.W.; Oberdorf-Maass, S.U.; Schouten, E.M.; Pavez Giani, M.G.; Hillebrands, J.L.; van Goor, H.; van Veldhuisen, D.J.; de Boer, R.A.; Westenbrink, B.D. Sodium-Glucose Co-Transporter 2 Inhibition with Empagliflozin Improves Cardiac Function in Non-Diabetic Rats with Left Ventricular Dysfunction after Myocardial Infarction. *Eur. J. Heart Fail.* **2019**, *21*, 862–873. [CrossRef] [PubMed]
73. Byrne, N.J.; Matsumura, N.; Maayah, Z.H.; Ferdaoussi, M.; Takahara, S.; Darwesh, A.M.; Levasseur, J.L.; Jahng, J.W.S.; Vos, D.; Parajuli, N.; et al. Empagliflozin Blunts Worsening Cardiac Dysfunction Associated with Reduced Nlrp3 (Nucleotide-Binding Domain-Like Receptor Protein 3) Inflammasome Activation in Heart Failure. *Circ. Heart Fail.* **2020**, *13*, e006277. [CrossRef] [PubMed]

74. Tanajak, P.; Sa-Nguanmoo, P.; Sivasinprasasn, S.; Thummasorn, S.; Siri-Angkul, N.; Chattipakorn, S.C.; Chattipakorn, N. Cardioprotection of Dapagliflozin and Vildagliptin in Rats with Cardiac Ischemia-Reperfusion Injury. *J Endocrinol* **2018**, *236*, 69–84. [CrossRef] [PubMed]
75. Connelly, K.A.; Zhang, Y.; Desjardins, J.F.; Nghiem, L.; Visram, A.; Batchu, S.N.; Yerra, V.G.; Kabir, G.; Thai, K.; Advani, A.; et al. Load-Independent Effects of Empagliflozin Contribute to Improved Cardiac Function in Experimental Heart Failure with Reduced Ejection Fraction. *Cardiovasc. Diabetol.* **2020**, *19*, 13. [CrossRef]
76. Sayour, A.A.; Korkmaz-Icoz, S.; Loganathan, S.; Ruppert, M.; Sayour, V.N.; Olah, A.; Benke, K.; Brune, M.; Benko, R.; Horvath, E.M.; et al. Acute Canagliflozin Treatment Protects against in Vivo Myocardial Ischemia-Reperfusion Injury in Non-Diabetic Male Rats and Enhances Endothelium-Dependent Vasorelaxation. *J. Transl. Med.* **2019**, *17*, 127. [CrossRef]
77. Wang, K.; Li, Z.; Sun, Y.; Liu, X.; Ma, W.; Ding, Y.; Hong, J.; Qian, L.; Xu, D. Dapagliflozin Improves Cardiac Function, Remodeling, Myocardial Apoptosis, and Inflammatory Cytokines in Mice with Myocardial Infarction. *J. Cardiovasc. Transl. Res.* **2020**; *(online ahead of print)*. [CrossRef]
78. Santos-Gallego, C.G.; Requena-Ibanez, J.A.; San Antonio, R.; Garcia-Ropero, A.; Ishikawa, K.; Watanabe, S.; Picatoste, B.; Vargas-Delgado, A.P.; Flores-Umanzor, E.J.; Sanz, J.; et al. Empagliflozin Ameliorates Diastolic Dysfunction and Left Ventricular Fibrosis/Stiffness in Nondiabetic Heart Failure: A Multimodality Study. *JACC Cardiovasc. Imaging* **2021**, *14*, 393–407. [CrossRef]
79. Arow, M.; Waldman, M.; Yadin, D.; Nudelman, V.; Shainberg, A.; Abraham, N.G.; Freimark, D.; Kornowski, R.; Aravot, D.; Hochhauser, E.; et al. Sodium-Glucose Cotransporter 2 Inhibitor Dapagliflozin Attenuates Diabetic Cardiomyopathy. *Cardiovasc. Diabetol.* **2020**, *19*, 7. [CrossRef]
80. Kraker, K.; Herse, F.; Golic, M.; Reichhart, N.; Crespo-Garcia, S.; Strauss, O.; Grune, J.; Kintscher, U.; Ebrahim, M.; Bader, M.; et al. Effects of Empagliflozin and Target-Organ Damage in a Novel Rodent Model of Heart Failure Induced by Combined Hypertension and Diabetes. *Sci. Rep.* **2020**, *10*, 14061. [CrossRef]
81. Sabatino, J.; De Rosa, S.; Tamme, L.; Iaconetti, C.; Sorrentino, S.; Polimeni, A.; Mignogna, C.; Amorosi, A.; Spaccarotella, C.; Yasuda, M.; et al. Empagliflozin Prevents Doxorubicin-Induced Myocardial Dysfunction. *Cardiovasc. Diabetol.* **2020**, *19*, 66. [CrossRef]
82. Li, X.; Lu, Q.; Qiu, Y.; do Carmo, J.M.; Wang, Z.; da Silva, A.A.; Mouton, A.; Omoto, A.C.M.; Hall, M.E.; Li, J.; et al. Direct Cardiac Actions of the Sodium Glucose Co-Transporter 2 Inhibitor Empagliflozin Improve Myocardial Oxidative Phosphorylation and Attenuate Pressure-Overload Heart Failure. *J. Am. Heart Assoc.* **2021**, *10*, e018298. [CrossRef]
83. Liu, Y.; Wu, M.; Xu, J.; Xu, B.; Kang, L. Empagliflozin Prevents from Early Cardiac Injury Post Myocardial Infarction in Non-Diabetic Mice. *Eur. J. Pharm. Sci.* **2021**, *161*, 105788. [CrossRef]
84. Asensio Lopez, M.D.C.; Lax, A.; Hernandez Vicente, A.; Saura Guillen, E.; Hernandez-Martinez, A.; Fernandez Del Palacio, M.J.; Bayes-Genis, A.; Pascual Figal, D.A. Empagliflozin Improves Post-Infarction Cardiac Remodeling through Gtp Enzyme Cyclohydrolase 1 and Irrespective of Diabetes Status. *Sci. Rep.* **2020**, *10*, 13553. [CrossRef] [PubMed]
85. Mizuno, M.; Kuno, A.; Yano, T.; Miki, T.; Oshima, H.; Sato, T.; Nakata, K.; Kimura, Y.; Tanno, M.; Miura, T. Empagliflozin Normalizes the Size and Number of Mitochondria and Prevents Reduction in Mitochondrial Size after Myocardial Infarction in Diabetic Hearts. *Physiol. Rep.* **2018**, *6*, e13741. [CrossRef] [PubMed]
86. Koyani, C.N.; Plastira, I.; Sourij, H.; Hallstrom, S.; Schmidt, A.; Rainer, P.P.; Bugger, H.; Frank, S.; Malle, E.; von Lewinski, D. Empagliflozin Protects Heart from Inflammation and Energy Depletion Via Ampk Activation. *Pharmacol. Res.* **2020**, *158*, 104870. [CrossRef] [PubMed]
87. Kimura, T.; Nakamura, K.; Miyoshi, T.; Yoshida, M.; Akazawa, K.; Saito, Y.; Akagi, S.; Ohno, Y.; Kondo, M.; Miura, D.; et al. Inhibitory Effects of Tofogliflozin on Cardiac Hypertrophy in Dahl Salt-Sensitive and Salt-Resistant Rats Fed a High-Fat Diet. *Int. Heart J.* **2019**, *60*, 728–735. [CrossRef]
88. Kuriyama, S. A Potential Mechanism of Cardio-Renal Protection with Sodium-Glucose Cotransporter 2 Inhibitors: Amelioration of Renal Congestion. *Kidney Blood Press. Res.* **2019**, *44*, 449–456. [CrossRef]
89. Yang, C.C.; Chen, Y.T.; Wallace, C.G.; Chen, K.H.; Cheng, B.C.; Sung, P.H.; Li, Y.C.; Ko, S.F.; Chang, H.W.; Yip, H.K. Early Administration of Empagliflozin Preserved Heart Function in Cardiorenal Syndrome in Rat. *Biomed. Pharmacother.* **2019**, *109*, 658–670. [CrossRef] [PubMed]
90. Packer, M. Activation and Inhibition of Sodium-Hydrogen Exchanger Is a Mechanism That Links the Pathophysiology and Treatment of Diabetes Mellitus with That of Heart Failure. *Circulation* **2017**, *136*, 1548–1559. [CrossRef]
91. Philippaert, K.; Kalyaanamoorthy, S.; Fatehi, M.; Long, W.; Soni, S.; Byrne, N.J.; Barr, A.; Singh, J.; Wong, J.; Palechuk, T.; et al. Cardiac Late Sodium Channel Current Is a Molecular Target for the Sodium/Glucose Cotransporter 2 Inhibitor Empagliflozin. *Circulation* **2021**, *143*, 2188–2204. [CrossRef]
92. Morris, K. Targeting the Myocardial Sodium-Hydrogen Exchange for Treatment of Heart Failure. *Expert. Opin. Ther. Targets* **2002**, *6*, 291–298. [CrossRef]
93. Zuurbier, C.J.; Baartscheer, A.; Schumacher, C.A.; Fiolet, J.W.T.; Coronel, R. Sglt2 Inhibitor Empagliflozin Inhibits the Cardiac Na^+/H^+ Exchanger 1: Persistent Inhibition under Various Experimental Conditions. *Cardiovasc. Res.* **2021**, *117*, 2699–2701. [CrossRef] [PubMed]
94. Baartscheer, A.; Hardziyenka, M.; Schumacher, C.A.; Belterman, C.N.; van Borren, M.M.; Verkerk, A.O.; Coronel, R.; Fiolet, J.W. Chronic Inhibition of the Na^+/H^+—Exchanger Causes Regression of Hypertrophy, Heart Failure, and Ionic and Electrophysiological Remodelling. *Br. J. Pharmacol.* **2008**, *154*, 1266–1275. [CrossRef] [PubMed]

95. Iborra-Egea, O.; Santiago-Vacas, E.; Yurista, S.R.; Lupon, J.; Packer, M.; Heymans, S.; Zannad, F.; Butler, J.; Pascual-Figal, D.; Lax, A.; et al. Unraveling the Molecular Mechanism of Action of Empagliflozin in Heart Failure with Reduced Ejection Fraction with or without Diabetes. *JACC Basic Transl. Sci.* **2019**, *4*, 831–840. [CrossRef]
96. Chung, Y.J.; Park, K.C.; Tokar, S.; Eykyn, T.R.; Fuller, W.; Pavlovic, D.; Swietach, P.; Shattock, M.J. Off-Target Effects of Sglt2 Blockers: Empagliflozin Does Not Inhibit Na^+/H^+ Exchanger-1 or Lower $[Na^+]_i$ in the Heart. *Cardiovasc. Res.* **2020**, *117*, 2794–2806. [CrossRef]
97. Chung, Y.J.; Park, K.C.; Tokar, S.; Eykyn, T.R.; Fuller, W.; Pavlovic, D.; Swietach, P.; Shattock, M.J. Sglt2 Inhibitors and the Cardiac Na^+/H^+ Exchanger-1: The Plot Thickens. *Cardiovasc. Res.* **2021**, *117*, 2702–2704. [CrossRef] [PubMed]
98. Wiig, H. Pathophysiology of Tissue Fluid Accumulation in Inflammation. *J Physiol* **2011**, *589*, 2945–2953. [CrossRef]
99. Van Linthout, S.; Tschope, C. Inflammation—Cause or Consequence of Heart Failure or Both? *Curr. Heart Fail. Rep.* **2017**, *14*, 251–265. [CrossRef]
100. Aimo, A.; Castiglione, V.; Borrelli, C.; Saccaro, L.F.; Franzini, M.; Masi, S.; Emdin, M.; Giannoni, A. Oxidative Stress and Inflammation in the Evolution of Heart Failure: From Pathophysiology to Therapeutic Strategies. *Eur. J. Prev. Cardiol.* **2020**, *27*, 494–510. [CrossRef]
101. Mehta, D.; Malik, A.B. Signaling Mechanisms Regulating Endothelial Permeability. *Physiol. Rev.* **2006**, *86*, 279–367. [CrossRef]
102. Emdin, M.; Aimo, A.; Castiglione, V.; Vergaro, G.; Georgiopoulos, G.; Saccaro, L.F.; Lombardi, C.M.; Passino, C.; Cerbai, E.; Metra, M.; et al. Targeting Cyclic Guanosine Monophosphate to Treat Heart Failure: Jacc Review Topic of the Week. *J. Am. Coll. Cardiol.* **2020**, *76*, 1795–1807. [CrossRef]
103. Fischer, D.; Rossa, S.; Landmesser, U.; Spiekermann, S.; Engberding, N.; Hornig, B.; Drexler, H. Endothelial Dysfunction in Patients with Chronic Heart Failure Is Independently Associated with Increased Incidence of Hospitalization, Cardiac Transplantation, or Death. *Eur. Heart J.* **2005**, *26*, 65–69. [CrossRef] [PubMed]
104. Kerem, A.; Yin, J.; Kaestle, S.M.; Hoffmann, J.; Schoene, A.M.; Singh, B.; Kuppe, H.; Borst, M.M.; Kuebler, W.M. Lung Endothelial Dysfunction in Congestive Heart Failure: Role of Impaired Ca^{2+} Signaling and Cytoskeletal Reorganization. *Circ. Res.* **2010**, *106*, 1103–1116. [CrossRef] [PubMed]
105. Giannitsi, S.; Bougiakli, M.; Bechlioulis, A.; Naka, K. Endothelial Dysfunction and Heart Failure: A Review of the Existing Bibliography with Emphasis on Flow Mediated Dilation. *JRSM Cardiovasc. Dis.* **2019**, *8*, 2048004019843047. [CrossRef] [PubMed]
106. Alshnbari, A.S.; Millar, S.A.; O'Sullivan, S.E.; Idris, I. Effect of Sodium-Glucose Cotransporter-2 Inhibitors on Endothelial Function: A Systematic Review of Preclinical Studies. *Diabetes Ther.* **2020**, *11*, 1947–1963. [CrossRef]
107. Ugusman, A.; Kumar, J.; Aminuddin, A. Endothelial Function and Dysfunction: Impact of Sodium-Glucose Cotransporter 2 Inhibitors. *Pharmacol. Ther.* **2021**, *224*, 107832. [CrossRef]
108. Dou, L.; Burtey, S. Reversing Endothelial Dysfunction with Empagliflozin to Improve Cardiomyocyte Function in Cardiorenal Syndrome. *Kidney Int.* **2021**, *99*, 1062–1064. [CrossRef]
109. Juni, R.P.; Al-Shama, R.; Kuster, D.W.D.; van der Velden, J.; Hamer, H.M.; Vervloet, M.G.; Eringa, E.C.; Koolwijk, P.; van Hinsbergh, V.W.M. Empagliflozin Restores Chronic Kidney Disease-Induced Impairment of Endothelial Regulation of Cardiomyocyte Relaxation and Contraction. *Kidney Int.* **2021**, *99*, 1088–1101. [CrossRef]
110. Han, Y.; Cho, Y.E.; Ayon, R.; Guo, R.; Youssef, K.D.; Pan, M.; Dai, A.; Yuan, J.X.; Makino, A. Sglt Inhibitors Attenuate No-Dependent Vascular Relaxation in the Pulmonary Artery but Not in the Coronary Artery. *Am. J. Physiol. Lung Cell. Mol. Physiol.* **2015**, *309*, L1027–L1036. [CrossRef]
111. Sala, V.; Gatti, S.; Gallo, S.; Medico, E.; Cantarella, D.; Cimino, J.; Ponzetto, A.; Crepaldi, T. A New Transgenic Mouse Model of Heart Failure and Cardiac Cachexia Raised by Sustained Activation of Met Tyrosine Kinase in the Heart. *Biomed. Res. Int.* **2016**, *2016*, 9549036. [CrossRef]
112. Packer, M. Cardioprotective Effects of Sirtuin-1 and Its Downstream Effectors: Potential Role in Mediating the Heart Failure Benefits of Sglt2 (Sodium-Glucose Cotransporter 2) Inhibitors. *Circ. Heart Fail.* **2020**, *13*, e007197. [CrossRef]
113. Mudaliar, S.; Alloju, S.; Henry, R.R. Can a Shift in Fuel Energetics Explain the Beneficial Cardiorenal Outcomes in the Empa-Reg Outcome Study? A Unifying Hypothesis. *Diabetes Care* **2016**, *39*, 1115–1122. [CrossRef] [PubMed]
114. Croteau, D.; Luptak, I.; Chambers, J.M.; Hobai, I.; Panagia, M.; Pimentel, D.R.; Siwik, D.A.; Qin, F.; Colucci, W.S. Effects of Sodium-Glucose Linked Transporter 2 Inhibition with Ertugliflozin on Mitochondrial Function, Energetics, and Metabolic Gene Expression in the Presence and Absence of Diabetes Mellitus in Mice. *J. Am. Heart Assoc.* **2021**, *10*, e019995. [CrossRef] [PubMed]
115. Xie, B.; Ramirez, W.; Mills, A.M.; Huckestein, B.R.; Anderson, M.; Pangburn, M.M.; Lang, E.Y.; Mullet, S.J.; Chuan, B.W.; Guo, L.; et al. Empagliflozin Restores Cardiac Metabolic Flexibility in Diet-Induced Obese C57bl6/J Mice. *Curr. Res. Physiol.* **2022**, *5*, 232–239. [CrossRef] [PubMed]
116. Sullivan, R.D.; Houng, A.K.; Gladysheva, I.P.; Fan, T.M.; Tripathi, R.; Reed, G.L.; Wang, D. Corin Overexpression Reduces Myocardial Infarct Size and Modulates Cardiomyocyte Apoptotic Cell Death. *Int. J. Mol. Sci.* **2020**, *21*, 3456. [CrossRef] [PubMed]
117. Reed, G.L.; Gladysheva, I.P.; Sullivan, R.D.; Mehta, R.M. Method of Personalized Treatment for Cardiomyopathy and Heart Failure and Associated Diseases by Measuring Edema and Cachexia/Sarcopenia. United States Patent Application Serial Number 17/313,904, 20210263120, 26 August 2021.
118. Weech, A.A.; Goettsch, E.; Reeves, E.B. Nutritional Edema in the Dog: I. Development of Hypoproteinemia on a Diet Deficient in Protein. *J. Exp. Med.* **1935**, *61*, 299–317. [CrossRef]

119. Kizhakke Puliyakote, A.S.; Holverda, S.; Sa, R.C.; Arai, T.J.; Theilmann, R.J.; Botros, L.; Bogaard, H.J.; Prisk, G.K.; Hopkins, S.R. Prone Positioning Redistributes Gravitational Stress in the Lung in Normal Conditions and in Simulations of Oedema. *Exp. Physiol.* **2022**, *107*, 771–782. [CrossRef] [PubMed]
120. Bagshaw, R.J. Evolution of Cardiovascular Baroreceptor Control. *Biol. Rev. Camb. Philos. Soc.* **1985**, *60*, 121–162. [CrossRef]
121. Nakos, G.; Batistatou, A.; Galiatsou, E.; Konstanti, E.; Koulouras, V.; Kanavaros, P.; Doulis, A.; Kitsakos, A.; Karachaliou, A.; Lekka, M.E.; et al. Lung and 'End Organ' Injury Due to Mechanical Ventilation in Animals: Comparison between the Prone and Supine Positions. *Crit. Care* **2006**, *10*, R38. [CrossRef]
122. Di Fusco, S.A.; Gronda, E.; Mocini, E.; Luca, F.; Bisceglia, I.; De Luca, L.; Caldarola, P.; Cipriani, M.; Corda, M.; De Nardo, A.; et al. Anmco Statement on the Use of Sodium-Glucose Cotransporter 2 Inhibitors in Patients with Heart Failure: A Practical Guide for a Streamlined Implementation. *Eur. Heart J. Suppl.* **2022**, *24*, C272–C277. [CrossRef]
123. Pittampalli, S.; Upadyayula, S.; Mekala, H.M.; Lippmann, S. Risks Vs Benefits for Sglt2 Inhibitor Medications. *Fed. Pract.* **2018**, *35*, 45–48.
124. Butler, J.; Usman, M.S.; Khan, M.S.; Greene, S.J.; Friede, T.; Vaduganathan, M.; Filippatos, G.; Coats, A.J.S.; Anker, S.D. Efficacy and Safety of Sglt2 Inhibitors in Heart Failure: Systematic Review and Meta-Analysis. *ESC Heart Fail.* **2020**, *7*, 3298–3309. [CrossRef] [PubMed]
125. Pelletier, R.; Ng, K.; Alkabbani, W.; Labib, Y.; Mourad, N.; Gamble, J.M. Adverse Events Associated with Sodium Glucose Co-Transporter 2 Inhibitors: An Overview of Quantitative Systematic Reviews. *Ther. Adv. Drug Saf.* **2021**, *12*, 2042098621989134. [CrossRef] [PubMed]

biomedicines MDPI

Article

Association of Plasma Methylglyoxal Increase after Myocardial Infarction and the Left Ventricular Ejection Fraction

Stefan Heber [1,*,†], Paul M. Haller [2,3,†], Attila Kiss [4], Bernhard Jäger [5], Kurt Huber [5,6] and Michael J. M. Fischer [1]

1 Institute of Physiology, Center for Physiology and Pharmacology, Medical University of Vienna, 1090 Vienna, Austria; michael.jm.fischer@meduniwien.ac.at
2 Department of Cardiology, University Heart & Vascular Center Hamburg, University Medical Center Hamburg-Eppendorf, 20246 Hamburg, Germany; p.haller@uke.de
3 German Center for Cardiovascular Research (DZHK), Partner Site Hamburg/Kiel/Lübeck, 20151 Hamburg, Germany
4 Center for Biomedical Research and Translational Surgery, Ludwig Boltzmann Institute for Cardiovascular Research, Medical University of Vienna, 1090 Vienna, Austria; attila.kiss@meduniwien.ac.at
5 3rd Department of Medicine, Cardiology and Intensive Care Medicine, Klinik Ottakring, 1016 Vienna, Austria; bernhard.jaeger@meduniwien.ac.at (B.J.); kurt.huber@meduniwien.ac.at (K.H.)
6 Faculty of Medicine, Sigmund Freud University, 1020 Vienna, Austria
* Correspondence: stefan.heber@meduniwien.ac.at; Tel.: +43-1-40160-31425
† These authors contributed equally to this work.

Abstract: Background: Preclinical studies suggest that methylglyoxal (MG) increases within the myocardium upon acute myocardial infarction (AMI) and thereafter contributes to adverse postinfarct remodeling. The aims of this study were to test whether MG increases in plasma of humans after AMI and whether this increase is related to the left ventricular ejection fraction (LVEF). Methods: The plasma samples of 37 patients with ST elevation AMI undergoing primary percutaneous coronary intervention (pPCI) acquired in a previously conducted randomized controlled trial testing remote ischemic conditioning (RIC) were analyzed by means of high-performance liquid chromatography. Time courses of the variables were analyzed by means of mixed linear models. Multiple regression analyses served to explore the relationship between MG levels and the LVEF. Results: Compared to the MG levels upon admission due to AMI, the levels were increased 2.4-fold (95% CI, 1.6–3.6) 0.5 h after reperfusion facilitated by pPCI, 2.6-fold (1.7–4.0) after 24 h and largely returned to the baseline after 30 d (1.1-fold, 0.8–1.5). The magnitude of the MG increase was largely independent of that of cardiac necrosis markers. Overall, the highest MG values within 24 h after AMI were associated with the lowest LVEF after 4 d. While markers of myocardial necrosis and stretch quantified within the first 24 h explained 52% of the variance of the LVEF, MG explained additional 23% of the variance ($p < 0.001$). Conclusions: Considering these observational data, it is plausible that the preclinical finding of MG generation after AMI negatively affecting the LVEF also applies to humans. Inhibition of MG generation or MG scavenging might provide a novel therapeutic strategy to target post-AMI myocardial remodeling and dysfunction.

Keywords: acute myocardial infarction; methylglyoxal; cardiac function; remodeling

Citation: Heber, S.; Haller, P.M.; Kiss, A.; Jäger, B.; Huber, K.; Fischer, M.J.M. Association of Plasma Methylglyoxal Increase after Myocardial Infarction and the Left Ventricular Ejection Fraction. *Biomedicines* **2022**, *10*, 605. https://doi.org/10.3390/biomedicines10030605

Academic Editor: Anand Prakash Singh

Received: 25 January 2022
Accepted: 1 March 2022
Published: 4 March 2022

Publisher's Note: MDPI stays neutral with regard to jurisdictional claims in published maps and institutional affiliations.

Copyright: © 2022 by the authors. Licensee MDPI, Basel, Switzerland. This article is an open access article distributed under the terms and conditions of the Creative Commons Attribution (CC BY) license (https://creativecommons.org/licenses/by/4.0/).

1. Introduction

1.1. Background

Despite significant improvements in the care of patients with acute myocardial infarction (AMI) in the recent decades, AMI remains associated with high morbidity and mortality [1]. Rapid restoration of blood flow has been proven beneficial; however, the resulting reperfusion of the ischemic myocardium acutely exacerbates tissue damage, a phenomenon known as ischemia-reperfusion injury [2,3]. Different interventions targeting ischemia-reperfusion injury have been tested in preclinical and clinical studies so far, although no intervention has yielded a convincing effect that would result in its implementation in the clinical routine, as reviewed previously [2]. One of these interventions is remote ischemic conditioning (RIC), whereby brief and repetitive ischemia and reperfusion cycles in a remote tissue or organ lead to tissue protection. In this regard, one can differentiate between pre-, per- and postconditioning, i.e., conditioning that takes place before, during or after the ischemia causing infarction [4]. Preclinical results have been promising [5,6], and preconditioning is impossible for the treatment of AMI in humans, but large clinical trials applying per- or postconditioning strategies have ultimately failed to show the cardioprotective effect of RIC in patients with ST elevation myocardial infarction [7,8].

After myocardial infarction, the main driver of morbidity and mortality is the resulting myocardial dysfunction, ultimately resulting in subsequent heart failure. An important determinant of the latter is the final infarct size [9]. However, myocardial dysfunction remains only partly explained by the loss of cardiomyocytes. As such, the concept of myocardial dysfunction following AMI involves several other processes. These include inflammatory ones that can be detected in the border zone of the infarcted myocardium, which largely contribute to adverse left ventricular remodeling resulting in cardiac dysfunction [10,11].

In this regard, advanced glycation end-products (AGEs) have been linked to worse outcomes after AMI. Specifically, an important AGE precursor methylglyoxal (MG) was shown in a preclinical study to causally affect myocardial dysfunction. Blackburn et al. [12] demonstrated that myocardial ischemia stimulates the production of MG within the myocardium. Overexpression of glyoxalase-1 (GLO1)—which constitutes the rate-limiting step in the detoxification of MG [13]—not only reduced the intramyocardial methylglyoxal levels, but also partially reversed the damage inflicted by infarction as measured with the left ventricular ejection fraction (LVEF). Notably, this functional difference occurred despite similar infarct sizes in both groups. Additionally, MG was shown as an independent predictor of the prognosis of patients with congestive heart failure [13]. This motivated us to investigate whether a negative association of MG levels after AMI and the subsequent LVEF exists in humans. In this case, translatability of the findings of Blackburn et al. [12] would seem plausible.

1.2. Objectives

Currently, it is unclear whether AMI in humans also leads to increased MG levels. Therefore, the primary objective of this study was to test whether MG levels increase over time in patients admitted to hospital due to AMI with ST elevation undergoing primary percutaneous coronary intervention (pPCI). To this end, the plasma samples previously acquired in the course of a randomized controlled trial investigating the effects of RIC on myocardial damage in patients with acute ST elevation AMI were analyzed. Thus, another exploratory objective was to test whether RIC affects MG levels, markers of myocardial damage and the myocardial function. The third objective was to investigate whether the negative impact of MG levels on the LVEF observed in mice [12] translated to a negative correlation of MG and the LVEF in patients.

2. Materials and Methods

2.1. Trial Design

The original study was a single-center, open-label, parallel-group randomized controlled trial. Patients with ST elevation AMI were randomized to RIC or sham in a 1:1 ratio. The associative research question addressed in the present report refers to observations of patients participating in the abovementioned trial. In this regard, the present analysis corresponds to an observational cohort study.

2.2. Participants, Setting and Recruitment

All the patients who presented with ST elevation AMI to the Third Medical Department (Cardiology) at Klinik Ottakring, Vienna, Austria, with intended interventional treatment by pPCI that gave written informed consent were included. We excluded patients if the symptom onset exceeded 8 h at the time of presentation to the emergency department, those with a regular intake of drugs affecting the K_{ATP} channel, patients with neurological disorders (i.e., diabetic neuropathy) and those in cardiogenic shock or other circumstances making informed consent impossible. In addition, four apparently healthy control subjects were included for assay development.

2.3. Interventions

The intervention was "remote ischemic conditioning" (RIC). It was carried out as previously described [14] using a blood pressure cuff applied to the left arm filled with a pressure of 200 mm Hg over 5 min. In case the patient's systolic blood pressure exceeded 200 mm Hg, the cuff was inflated to exceed that pressure by 15 mm Hg. The RIC protocol consisted of four cycles of 5 min inflation each. In between, the pressure was released for 5 min to allow reperfusion. RIC was performed as soon as possible after admission and following informed consent but prior to and without delaying reperfusion of the culprit lesion. The patients in the sham control group were fitted with a blood pressure cuff for 40 min without inflating it.

2.4. Outcomes

The primary outcomes for the present analysis are the levels of MG in plasma, their association with the measured myocardial necrosis markers, RIC and the LVEF. Details can be found in the Appendix B.5.

2.5. Safety Considerations

A major concern could be that RIC delays the start of myocardial reperfusion. Therefore, RIC was started as soon as possible after the arrival of each patient to the hospital. In all those cases where the RIC protocol could not be fully performed before the start of reperfusion, the missing cycles were continued throughout the routine procedures and in case of very early reperfusion in the sense of "remote ischemic postconditioning".

2.6. Sample Size

The present analysis includes all consecutive patients included from April 2016 to March 2018. No power calculation for the final sample considering the presented analyses was performed as this analysis is of exploratory character. The primary aim of the original randomized controlled trial was to confirm the previously reported benefit of RIC with respect to infarct size. White et al. [15] reported a reduction of infarct size from 24.5% in the control subjects to 18% in the RIC-treated ones. Based on their reported group difference and an assumed common standard deviation of 12%, it was calculated that 54 patients per group would be necessary to detect the relevant difference with a power of 80% accepting a type I error rate of 5%. As a dropout rate of 10% was expected, it was aimed for a total of 120 patients. The study was terminated after 37 of the intended 120 patients for the reasons provided below.

2.7. Randomization and Blinding

The allocation sequence was generated using a computer-based randomization tool generating 1:1 randomization in blocks of 8. The allocation sequence was available using serially numbered envelopes. The patients were informed that an additional intervention was offered as part of a clinical trial with a 50% chance of receiving this intervention, which might reduce damage to the heart. Due to the nature of the intervention, the patients and the physicians were not blinded regarding the group allocation. All the persons performing laboratory assessments were blinded to group allocation.

2.8. Methylglyoxal Measurements

Details on the methods, including Figures A1–A3, can be found in Appendix A.

2.9. Statistical Methods

Mixed linear models, partial correlation coefficients and multiple linear regression analyses are described in detail in Appendix B. The analyses were performed with IBM SPSS Statistics 27. They were not predefined in the study protocol, and no adjustment for multiplicity was performed; therefore, the results need to be interpreted accordingly. Only two-sided p-values are reported; p-values ≤ 0.05 were considered statistically significant.

3. Results

3.1. Participants

Between April 2016 and March 2018, 37 patients were randomized. In this period, several studies were published, which provide evidence that the beneficial effect of RIC is, at best, minimal [8]. Additionally, the recruitment of patients was far slower than expected, wherefore a decision was made to stop the study prematurely. The baseline patient characteristics were already published previously [14]. Plasma samples of 33 patients were available for MG measurements; the data of these patients are presented in this paper (Figure 1).

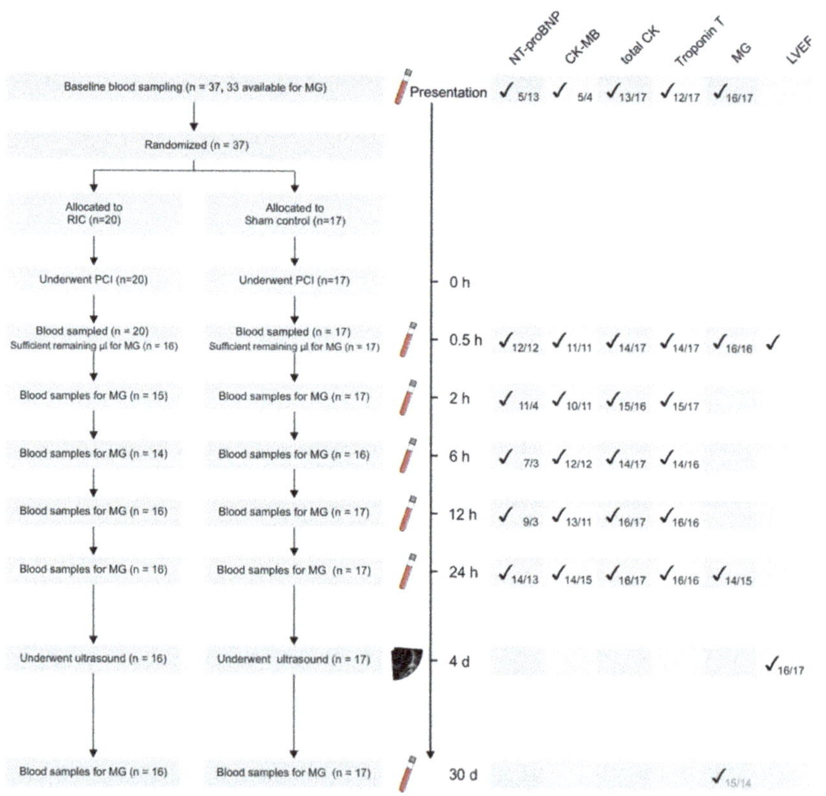

Figure 1. CONSORT flow diagram, analyzed parameters and sample size. The sample size given in the lower right corners next to the check marks indicate the sample size of the respective parameter at the timepoint. The number on the left of the slash represents the RIC patients, on the right—the sham patients.

3.2. Myocardial Infarction Is Associated with a Transient Increase in Plasma Methylglyoxal Levels Independent of RIC

Compared to the MG values at 0 h, the levels were increased 2.4-fold (95% CI, 1.6–3.6, $p < 0.001$) after 0.5 h and 2.6-fold (1.7–4.0, $p < 0.001$) after 24 h. Thirty days after admission,

MG levels did not significantly differ from the ones upon admission (1.1-fold, 0.8–1.5, $p = 0.42$). There was no evidence of an effect of RIC (interaction time * group $p = 0.15$, mixed linear model with time as a discrete variable, Figure 2A).

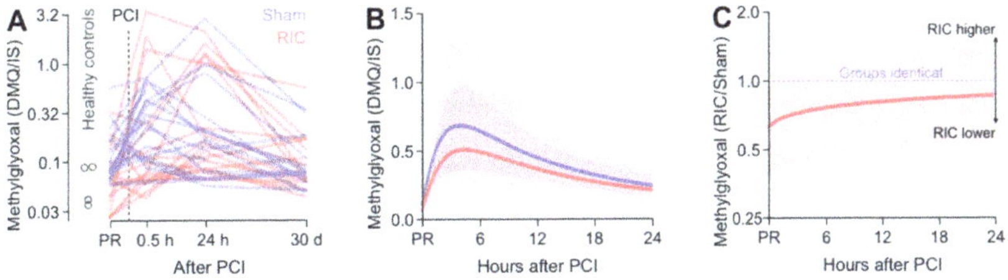

Figure 2. Time course of plasma MG levels after hospital admission due to myocardial infarction. (**A**) Individual time courses of plasma MG levels of the RIC and sham patients after presentation in the context of the plasma MG levels of four healthy control subjects as reference. Remote ischemic conditioning (RIC, red) or sham intervention (blue) were performed between 0 and 0.5 h. Logarithmic axis scaling allows visual discrimination between individual datapoints. (**B**) Estimated geometric mean time courses of methylglyoxal within the period until 24 h after admission. Error bands are 95% confidence intervals. (**C**) Estimated fold difference of MG levels between the RIC patients (red) and the sham patients (blue). Abbreviations: MG—methylglyoxal, RIC—remote ischemic conditioning, PR—presentation.

Exploration of the data using a mixed linear model with time as a continuous variable showed that they could be well-described by means of a cubic function (linear, quadratic and cubic terms $p < 0.001$ each, interactions with group $p > 0.09$ each), suggesting a RIC-independent nonlinear time course (Figure 2B,C, predicted vs. observed, Appendix B Figure A4A). The peaks of the estimated curves occur at 4 h and would correspond to a 6.7-fold (RIC: 6.1, sham: 7.2) increase compared to 0 h.

3.3. *Release of Enzymes Indicative of Myocardial Damage or Dysfunction Is Not Reduced by RIC after Myocardial Infarction*

As markers of myocardial damage, CK-MB, total CK and TnI were repeatedly determined in plasma. To repeatedly assess myocardial dysfunction, NT-proBNP was quantified. Taken together, the following analyses are hardly consistent with a beneficial effect of RIC and rather point in the opposite direction, i.e., a detrimental effect.

CK-MB data (Figure 3A) were modelled using quartic polynomials (linear, quadratic, cubic, quartic terms $p < 0.001$ each; further increasing the order of the polynomial did not improve the fit; quintic term $p = 0.067$). Interactions of these terms with the factor group suggest different time courses for each group (Figure 3B, linear term * group $p = 0.002$, quadratic term * group $p = 0.004$, cubic term * group $p = 0.009$, quartic term * group $p = 0.054$). The estimated differences on a multiplicative scale (Figure 3C) suggest higher CK-MB levels in the RIC group compared to the sham group, with the difference most pronounced at around 2 h.

The time course of the total plasma CK levels were also well-described by a quartic polynomial (linear, quadratic, cubic, quartic terms $p < 0.001$ each); however, there was no evidence for group-specific parameters (all interaction terms $p > 0.22$, Figure 3D–F). Nevertheless, point estimates of the fold differences due to RIC are constantly well above one.

Concerning TnI levels (Figure 3G), the applied statistical model suggested a group-specific time course (Figure 3H, linear, quadratic terms $p < 0.001$ each, cubic $p = 0.007$, quartic $p = 0.022$, linear term * group $p = 0.004$, quadratic term * group $p = 0.017$, cubic term * group $p = 0.038$, quartic term * group $p = 0.34$). Similar to CK-MB, the estimated differences on a multiplicative scale (Figure 3I) show nearly threefold higher TnI levels compared to the sham treatment at around 2 h. Notably, the uncertainty of this estimate, quantified using the 95% confidence error band, needs to be considered. A sensitivity analysis showed no relevant impact on the estimates by the two highest outlying datapoints (Figure A5 in Appendix B).

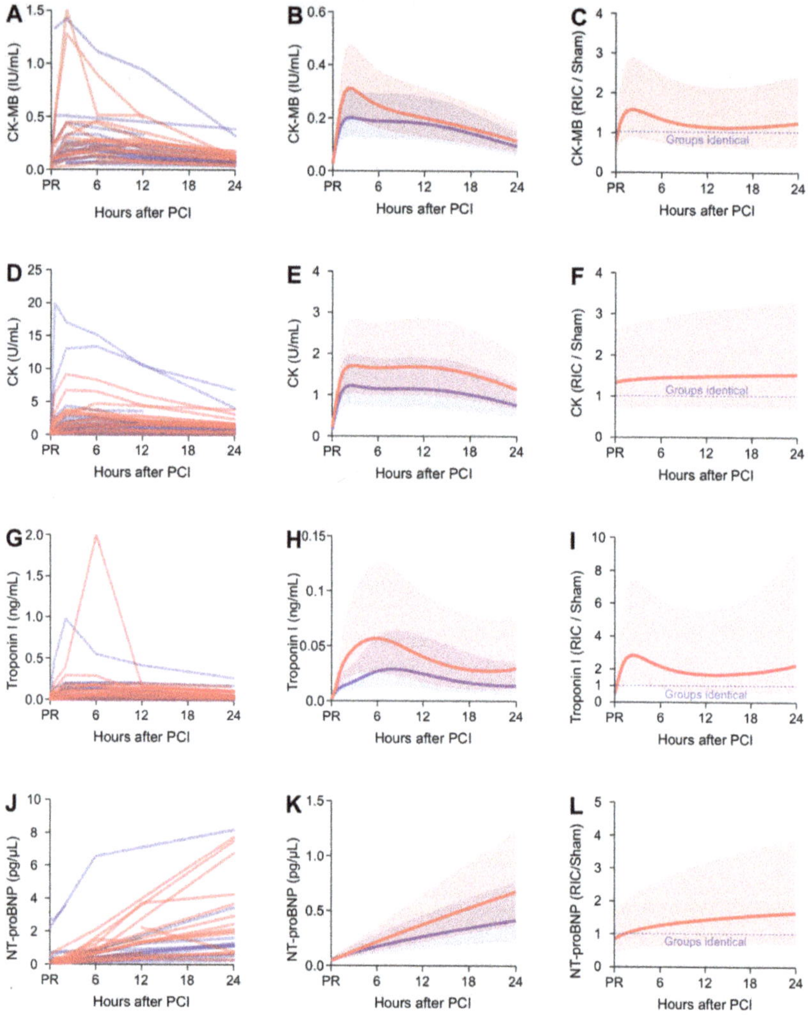

Figure 3. Biomarker levels in plasma after admission to the hospital due to myocardial infarction. The left panels show the raw values, with red lines indicating RIC and blue lines indicating sham control. The middle panels show the geometric means with 95% confidence bands estimated using mixed linear models. The right panels show the estimated fold difference of the RIC group compared to the control group with 95% confidence bands. Panels: (**A–C**) CK-MB, (**D–F**) CK, (**G–I**) TnI and (**J–L**) NT-proBNP.

Regarding NT-proBNP, the polynomial model suggested a group-specific time course as well (Figure 3J). In contrast to the abovementioned markers of myocardial damage, there was no evidence that a higher-order polynomial would fit the data better than a linear fit (linear and quadratic terms $p = 0.005$ and 0.24, Figure 3K). However, the slope of the lines differed between the groups (linear term * group $p = 0.016$), resulting in approximately 1.5-fold NT-proBNP levels increase towards the end of the day after MCI (Figure 3L).

3.4. Changes in MG Levels after AMI Are Only Weakly Dependent on Changes in CK, CK-MB, TnI or NT-proBNP, but Might Be Affected by Diabetes

As one would expect for different indices of the same underlying cause, i.e., CK-MB and TnI as markers of myocardial damage, they were strongly correlated within the patients (Figure 4A, partial r = 0.86, $p < 0.001$). TnI and NT-proBNP showed weak correlations with changes in methylglyoxal levels (NT-proBNP partial r = 0.22, $p = 0.19$, TnI partial r = 0.22, $p = 0.14$, Figure 4B,C). Myocardial damage markers CK-MB and CK were only moderately correlated with changes in methylglyoxal levels (CK-MB partial r = 0.41 $p = 0.024$, CK partial r = 0.45, $p < 0.001$, Figure 4D,E). When entered into a single multiple regression model, NT-proBNP, TnI, CK-MB and CK explained only 2.5% of the variance of methylglyoxal ($p = 0.8$). This suggests that the extent of myocardial damage cannot be the main determinant of the methylglyoxal level increase following AMI. As diabetes is associated with increased methylglyoxal levels, we explored whether the presence of diabetes type II might be an additional determinant. Although only two diabetic patients were among the included ones, these two patients indeed had comparably high methylglyoxal levels (Figure 4F), a difference that reached formal significance in the applied polynomial mixed linear model (interactions linear term * group $p = 0.008$ and quadratic term * group $p = 0.008$).

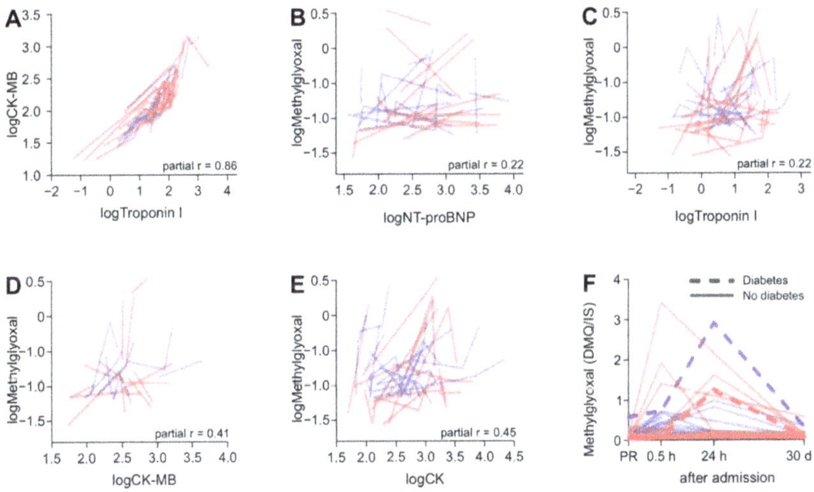

Figure 4. Correlations between myocardial damage and methylglyoxal. (**A**) Relationship between changes in TnI levels and changes in CK-MB. Each line represents measurements taken at different timepoints from a patient. Red indicates RIC, blue—sham. Correlation of changes can be appreciated by approximately parallel lines. (**B**) Correlations between intraindividual changes in methylglyoxal levels with changes in NT-proBNP, (**C**), TnI, (**D**) CK-MB and (**E**) CK. (**F**) Methylglyoxal levels of two patients with diagnosed type 2 diabetes (dotted lines) in the context of respective levels of all the nondiabetic patients.

3.5. Association of Methylglyoxal Levels within 24 h after AMI with the Myocardial Function after 4 Days

To assess the relationships between myocardial damage, methylglyoxal levels and myocardial dysfunction, the area under the curve (AUC) of CK-MB, TnI, CK and NT-proBNP was calculated for 24 h after admission for each patient. The AUC values of CK-MB, TnI and CK were highly correlated with each other (Figure 5A, $p < 0.001$ each). Higher levels of the latter two markers for myocardial damage were weakly associated with higher NT-proBNP levels (Spearman $\rho = 0.43$, $p = 0.01$ and Spearman $\rho = 0.41$, $p = 0.02$), whereas no relevant correlation between CK-MB and NT-pro-BNP was observed ($p = 0.16$). Methylglyoxal showed no relevant association with any of the abovementioned variables. There were, however, moderate negative pairwise correlations between the LVEF and each of the biomarkers CK-MB, TnI, CK and NT-proBNP ($p < 0.007$ each).

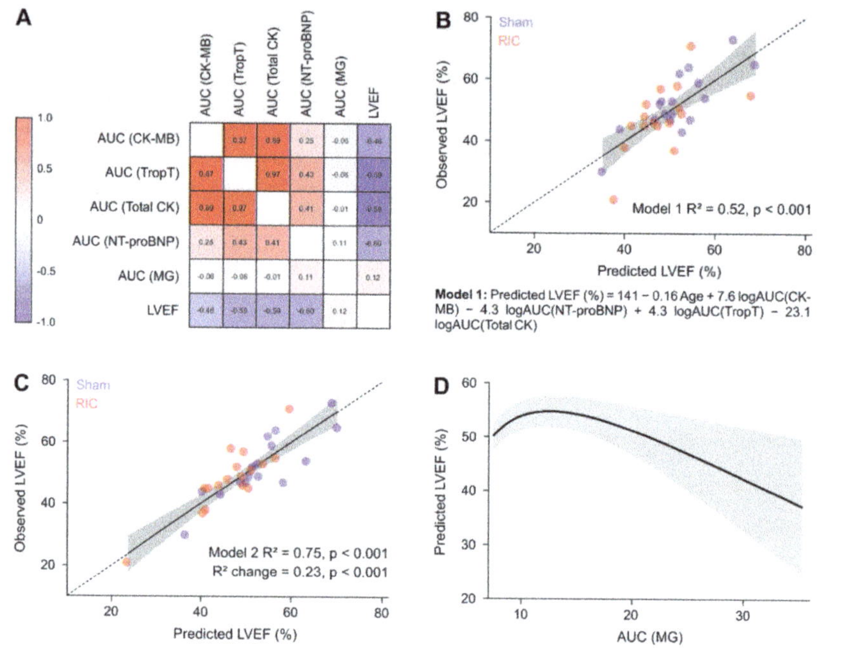

Figure 5. Methylglyoxal levels partially explain the myocardial function after myocardial infarction. (**A**) Changes in the parameters except for the LVEF were assessed by their area under the curve within the first 24 h. Heatmap of the Spearman correlation matrix including AUCs of enzymes in plasma indicating myocardial damage, AUCs of NT-proBNP, AUCs of methylglyoxal and the left ventricular ejection fraction. (**B**) LVEF values predicted without methylglyoxal. Entering each patient's values into the formula results in their predicted LVEF. The dotted line represents the perfect prediction. (**C**) Plot analogous to B, with inclusion of MG as a linear and quadratic predictor. (**D**) Relationship between AUC(MG) and the LVEF. The covariates other than MG were kept constant at their means. The AUC(MG) values in the observed range were entered into Model 2.

Next, a multivariable regression analysis was carried out to investigate whether methylglyoxal levels are associated with the LVEF independently of markers of myocardial damage and NT-proBNP. First, a model was built that predicts the LVEF (Figure 5B, Model 1) using CK, CK-MB, age, TnI and NT-proBNP. This model explained 52% of the variance of the LVEF ($p < 0.001$). Next, MG was added as a linear and quadratic term, the latter of which allows a nonlinear relationship between methylglyoxal and the LVEF. This Model 2 (Figure 5C) explained 23% more variance (R^2 change $p = 0.001$) than Model 1, namely 75%. This means that methylglyoxal levels within the 24 h after AMI contained information regarding the LVEF measured approximately three days later in addition to the information contained in NT-proBNP, CK, CK-MB, TnI and age. The relationship between methylglyoxal and the LVEF was significantly curvilinear (logAUC(MG) linear and quadratic terms $p < 0.001$ each, Figure 5D). By plotting the predicted LVEF values against the range of observed AUC(MG) values, a curvilinear relationship can be appreciated, with the highest MG levels associated with the lowest LVEF. Additionally, there was no evidence that RIC affected the LVEF (Figure A6, Appendix B).

3.6. Harms

No serious adverse events related to placing the blood pressure cuff (and inflation in case of RIC) were observed. Overall, there was one death due severely impaired cardiac function following ST-elevation AMI in the RIC group.

4. Discussion

The main finding of this exploratory study is the observation that in patients admitted to a hospital due to ST elevation AMI who have undergone reperfusion by pPCI, MG levels in plasma rise substantially within a few hours and return to levels similar to the ones upon admission within 30 days. We speculate that AMI with subsequent reperfusion causes this transient elevation. It needs to be emphasized that based on our study design, it is not possible to deduce if AMI causes higher MG levels or if reperfusion by pPCI causes them. However, given the results in mice, which clearly show that AMI alone causes increased levels of MG-derived advanced glycation end-products within the myocardium [12], it is highly plausible that in humans AMI itself causes an increase in MG. However, a clear differentiation from a potential additional effect on MG levels due to reperfusion is not possible [16].

The second central finding is an association of higher MG levels within the first day of hospitalization with lower LVEF values after 4 days. In particular, an exploratory statistical analysis showed a nonlinear association between the methylglyoxal burden within the first 24 h after admission and the LVEF four days after AMI. Thereby, the highest MG levels seemed to be associated with the lowest LVEF values. This is in line with the previous preclinical data, showing that reducing the MG burden after AMI improves the LVEF [12]. Based on this preclinical observation, our study for the first time demonstrates a similar effect in humans with AMI undergoing reperfusion, suggesting a potential new target for treatment strategies. For instance, as soon as the exact mechanism of MG generation within the myocardium during AMI and following reperfusion is elucidated, one could try to interfere with it to improve the LVEF, e.g., by increasing the glyoxylase-1 activity. Alternatively, in case generation cannot be addressed, one could aim to scavenge MG, a concept that has been implemented experimentally [17]. It might also be possible that MG exerts its negative effects on the myocardial function by the formation of advanced glycation end-products. In this case, it might constitute an additional option to interfere with the interaction of MG with proteins or lipids.

Under the assumption that MG indeed reduces the LVEF after AMI, it seems relevant which factors determine the extent of MG increase thereafter. Moreover, the extent of MG increase seems to be independent of the measured surrogate markers for myocardial damage. Thus, other factors not involved in our models seem to alter MG. Among those, inflammation seems to be an important driver as previously shown in preclinical models [12]. Hence, although AMI might trigger the MG increase, its extent is determined by an unknown factor. A possible candidate factor could be diabetes, as this metabolic state is generally associated with higher MG levels [16]. Our observation that the two diabetic patients exhibited comparatively high MG increases after AMI compared to nondiabetic patients would fit this hypothesis; however, this needs to be investigated with an adequate sample size.

As the samples from an RCT investigating RIC in the context of AMI were analyzed, a RIC effect in MG or cardiac enzymes indicative of myocardial damage could also be assessed. In this regard, we found no evidence that RIC affects MG levels or markers of myocardial necrosis. Furthermore, our study did not reveal any significant benefit of RIC on these markers, nor did RIC have any significant influence on cardiac function thereafter. On the contrary, our analyses suggest a potential higher enzyme release in the RIC group that, based on the regression model, is most pronounced at 2 h after reperfusion. However, in sight of the smaller than expected sample size and the significant width of the confidence intervals, these results should be interpreted with caution and do not necessarily imply causality. Overall, previous studies and their pooled analysis showed a very small effect in favor of RIC on infarct size and cardiac function [8]. In line, the largest clinical trial conducted demonstrated a neutral effect with respect to occurrence of death or hospitalization for heart failure following AMI [7].

Limitations and Generalizability

The results of this study seem plausible considering the published data, e.g., regarding the non-superiority of RIC compared to the control regarding myocardial damage or the MG increase after AMI and reperfusion that was already observed in preclinical studies. Nevertheless, several limitations need to be taken into consideration. Besides the effects of RIC, which were investigated in an experimental design, all other findings are of observational nature, i.e., no causal interpretation is possible. Consequently, it cannot be excluded that the MG increase is related not to AMI or reperfusion, but to an unknown confounder. Additionally, the low sample size adds some uncertainty, and validation of our results by independent researchers would be beneficial. Future studies should also include LVEF measurements at later timepoints after AMI. Another limitation concerns the multiplicity of statistical analyses. The polynomial modelling included several sequential decisions and thus might have resulted in false positive conclusions. Additionally, AUC calculations are based on the estimates derived from mixed linear models and thus are subject to some degree of uncertainty. Another limitation relates to the quantification of MG. In our view, the absolute concentration in plasma cannot be given due to matrix effects. For this purpose, a standard curve based on MG spiked, but otherwise an MG-free solution would be required. There is no method to remove MG from plasma without altering the matrix. However, the AUC of a given amount of MG obtained by HPLC heavily depends on the matrix, e.g., water, saline or plasma. The standard curve generated in H_2O clearly did not reflect the DMQ/IS relationship to MG in plasma. Of note, this does not affect the conclusions of this study, which are based on relative MG concentration changes.

5. Conclusions

In conclusion, we found a temporal relationship between the occurrence of AMI treated with reperfusion and a subsequent increase in plasma MG. We also observed an association between the magnitude of the MG increase within the first day after the infarction with the left ventricular function four days later despite statistical adjustment for the known positive association of the established cardiac necrosis markers. Based on this observation, future studies investigating the causal relationship between MG and cardiac function in humans are warranted. Overall, MG might serve as a new target for the treatment of myocardial dysfunction, reperfusion injury and associated remodeling following AMI if confirmed in future randomized controlled trials.

Author Contributions: Conceptualization, S.H., P.M.H., A.K., B.J., K.H. and M.J.M.F.; methodology, S.H., P.M.H., B.J., K.H. and M.J.M.F.; formal analysis, S.H.; investigation, P.M.H., A.K., B.J., K.H. and M.J.M.F.; data curation, S.H., P.M.H., B.J., K.H. and M.J.M.F.; writing—original draft preparation, S.H. and P.M.H.; writing—review and editing, S.H., P.M.H., A.K., B.J., K.H. and M.J.M.F.; visualization, S.H.; supervision, A.K., B.J., K.H. and M.J.M.F.; project administration, P.M.H., A.K., B.J., K.H. and M.J.M.F.; funding acquisition, P.M.H. and A.K. All authors have read and agreed to the published version of the manuscript.

Funding: This research was funded by the Ludwig Boltzmann Institute for Cardiovascular Research, the Association for Research on Arteriosclerosis, Thrombosis and Vascular Biology (ATVB) and by a grant of the Medical Scientific Fund of the Mayor of the City of Vienna (grant number 17091).

Institutional Review Board Statement: The study was performed according to the Declaration of Helsinki and Good Clinical Practice and approved by the competent ethics committee of the City of Vienna (EK 16-009-0216 on 6 April 2016).

Informed Consent Statement: All the patients gave written informed consent prior to enrolment.

Data Availability Statement: Data are available upon reasonable request after contacting the corresponding author.

Acknowledgments: The authors acknowledge Thomas Fleming for advice and providing pure methylglyoxal as a reference.

Conflicts of Interest: P.M.H. reports honoraria from Beckman Coulter and is supported by a grant of the Faculty of Medicine, University of Hamburg, and the German Centre for Cardiovascular Research (DZHK), all outside of the present study. The funders of the study had no role in the design of the study; in the collection, analyses, or interpretation of data; in the writing of the manuscript, or in the decision to publish the results.

Appendix A. Quantification of Methylglyoxal in Plasma Samples

The assay follows the principles described earlier [18,19]. Methylglyoxal can be quantified by detection of its derivative 6,7-dimethoxy-2-methylquinoxaline (DMQ), which is formed upon addition of 1,2-diamino-4,5-dimethoxybenzene (DDB). DMQ is well-detectable due to its chromophoric and fluorophoric properties. Based on availability, we relied on measuring absorption at 352 nm. Like methylglyoxal, butanedione can be derivatized by addition of DDB, which produces a suitable internal standard of the same substance class, 6,7-dimethoxy-2,3-dimethylquinoxaline (DDMQ), detectable at the same wavelength (Figure A1).

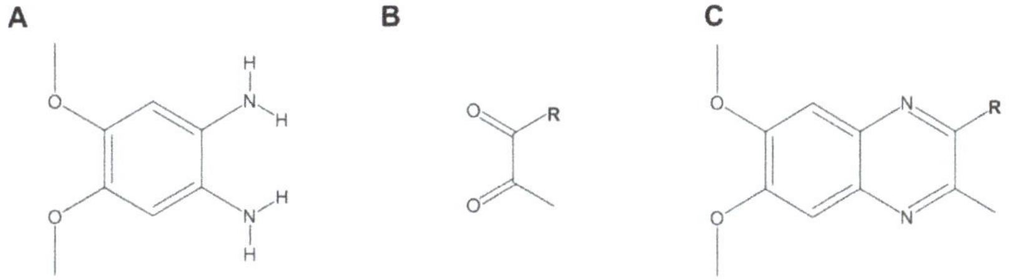

4,5-Dimethoxybenzene-1,2-diamine

R = H...Methylglyoxal
R = CH_3...Butandione

R = H...6,7-Dimethoxy-2-methylquinoxaline (DMQ, analyte)
R = CH_3...6,7-Dimethoxy-2,3-dimethylquinoxaline (DDMQ, internal standard IS)

Figure A1. Compounds relevant for methylglyoxal quantification by HPLC. 4,5-Dimethoxybenzene-1,2-diamine (**A**) is added to methylglyoxal (**B**) upon which 6,7-dimethoxy-2-methylquinoxaline (DMQ, (**C**)) is formed. DMQ can be detected by HPLC. Similarly, 4,5-dimethoxybenzene-1,2-diamine (**A**) can react with butanedione (**B**) and forms 6,7-dimethoxy-2,3-dimethylquinoxaline (DDMQ, (**C**)), which serves as the internal standard.

For analytical procedures 2,3-Butanedione and 4,5-dimethoxybenzene-1,2-diamine hydrochloride (Sigma-Aldrich, St. Louis, MO, USA), trifluoroacetic acid (Thermo Scientific, Waltham, MA, USA), and acetonitrile (Merck, Darmstadt, Germany) were purchased. Methylglyoxal was a gift from Thomas Fleming (Heidelberg, Germany).

The internal standard 6,7-dimethoxy-2,3-dimethylquinoxaline (DDMQ) was synthesized as previously described [18]. Plasma (300 μL) was spiked with 6 μL of the internal standard DDMQ (250 μM), vortexed, and then mixed with 1.2 mL acetonitrile to achieve precipitation of the proteins. The mixture was cooled on ice for 15 min and then centrifuged (20 min, 20,000× *g*). The supernatant (1300 μL) was acidified with 13 μL of trifluoroacetic acid (99%). Freshly dissolved DDB (30 μL; 12 mM) was added under nitrogen and the mixture was agitated at room temperature for 2 h, protected from light. The mixture was freeze-dried, and the residue was dissolved in 40 μL of the mobile phase (2% acetonitrile in 0.1% TFA/water), of which 35 μL were analyzed by means of HPLC. Measurements obtained from plasma samples with less than 300 μL (35 of 123 samples) were filled up to a total volume of 300 μL with saline (150 mM NaCl).

Reverse-phase high-performance liquid chromatography (RP-HPLC) was performed on a Dionex UHPLC focused UltiMate 3000 system (Sunyvale, CA, USA). The samples were run on a Phenomenex Kinetex C18 column (150 × 3.0 mm, 2.6 μm) (Torrance, CA, USA) in a mobile phase consisting of acetonitrile, water and 0.1% trifluoroacetic acid at a rate of 0.3 mL/min. An initial period of five minutes with 1.8% acetonitrile was followed by a gradient for acetonitrile from 1.8% to 66.6% within 24 min, DMQ had a peak elution after approximately 17.5 min, DDMQ—after 17.95 min (full protocol in Figure A2). Detection of the analyte was achieved at 352 nm at room temperature. For recovery of the column, acetonitrile was reduced back to 1.8% within 3 min, maintained for 10 min.

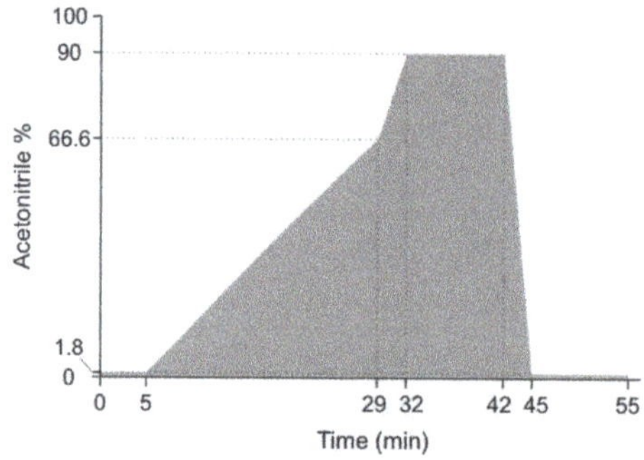

Figure A2. Acetonitrile gradient used for quantification of the analyte and the internal standard (IS).

Plasma samples of the four young and healthy control subjects with unknown methylglyoxal concentrations were spiked with methylglyoxal before adding acetonitrile. The added concentrations of methylglyoxal were 1000, 500, 350, 250, 176, 125, 87, 62, 29 and 0 nM. The assay was carried out as described above. Calibration curves were constructed by plotting the concentrations of methylglyoxal against the peak area of DMQ normalized to the peak area of the internal standard.

Spiking of plasma of the four healthy control subjects (not included in the CONSORT flow diagram above) with different amounts of MG increased the measured quotient of DMQ and IS (internal standard, Figure A3A). The linear regression lines fitted to each healthy volunteer's data resulted in DMQ/IS values of 0.090, 0.089, 0.042 and 0.037 for unspiked plasma (0 μM, intercepts). Notably, spiking of H_2O resulted in a substantially different slope. Under the assumption that the reason for the intercepts to be >0 is endogenous methylglyoxal, one can deduce a linear relation of plasma methylglyoxal and DMQ/IS, whereby DMQ/IS value of 0 represents 0 μM methylglyoxal and 1 μM increase in plasma would be reflected by an increase in the DMQ/IS value by 0.15 (Figure A3B). As it is unknown whether this assumption holds, this relationship should be interpreted with caution, but this has no impact on intraindividual time series. Subsequent statistical analyses were performed with DMQ/IS values.

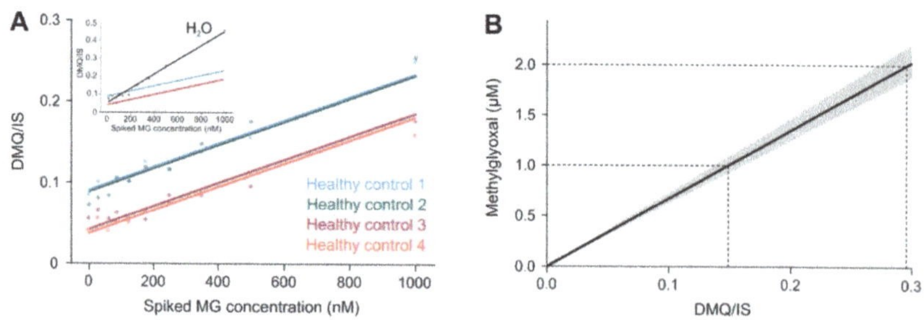

Figure A3. Relationship between DMQ/IS and methylglyoxal concentration. (**A**) DMQ/IS values resulting from spiking plasma of the four apparently healthy control subjects. (**B**) Estimated relationship with 95% CI between DMQ/IS and plasma methylglyoxal concentration.

Appendix B. Supplemental Statistical Methods

Appendix B.1. Assessment of Treatment-Specific Time Courses

To test the hypothesis that MG levels increase after AMI and reperfusion, a mixed linear model with the fixed factors time (levels: presentation, 0.5 h, 24 h and 30 d after PCI) and group (sham or RIC) was used. The patients were included as levels of a random factor, resulting in a random intercept for the patients. The duration of ischemia before reperfusion was included as continuous covariate. A diagonal covariance structure was selected based on the smallest Akaike information criterion. DMQ/IS as a dependent variable was log-transformed prior to analysis.

Time courses of MG, CK, CK-MB, TnI and NT-proBNP were modelled using the following mixed linear model approach. The time from admission with the unit hours was transformed (transformed hours = log10(hours+1). Additionally, the dependent variables were log10-transformed. In the first step, a basic model of the time course of the dependent variables was built by allowing each patient to have a random intercept and a random slope. The log10(hours+1) time variable was entered as a fixed covariate and as its quadratic, cubic, quartic and quintic form. Beginning with the highest-order polynomial, each term was removed separately until the highest-order remaining term was significant. Based on this basic model, the binary fixed factor group was introduced together with its interactions with all the polynomial terms remaining in the previous step. Following this, the group * polynomial term interactions were removed, beginning with the interaction with the highest-order polynomial until only significant interactions were left or no interaction with the factor group remained. Approximate normal distribution of residuals was checked visually, scatterplots showing predicted vs. observed values are presented in Figure A4. Least-squares means with confidence intervals were computed for timepoints closely enough together to appear as a continuous function in the graphs and back-transformed and thus represent the estimated geometric means. Group comparisons were also back-transformed and thus represent differences on a multiplicative scale (i.e., fold group differences).

Appendix B.2. Correlation between Intraindividual Changes in Methylglyoxal with Intraindividual Changes in Cardiac Enzymes and NT-proBNP

For this purpose, partial correlation coefficients were calculated. Adjustment for dependence of data originating from within patients was performed by including n–1 binary dummy variables. To quantify how much of the variance of methylglyoxal can be explained by CK, CK-MB, TnI and NT-proBNP, all the patient dummy variables were entered in the first block, the abovementioned enzymes and NT-proBNP—in the second block. The change in R^2 between the two blocks was taken as a measure of variance explained by enzymes and NT-proBNP.

Appendix B.3. Association of Methylglyoxal Levels with the Myocardial Function

To assess whether methylglyoxal might be related to the myocardial function, i.e., the LVEF, the area under the curve of the predictors methylglyoxal, CK, CK-MB, TnI and NT-proBNP was calculated. For this purpose, the predicted values of the polynomial mixed linear models were used by applying the trapezoidal rule. The predicted values were chosen instead of the raw values because there were several raw values missing. Of note, the missingness is based on organizational and technical reasons and not related to any predictors or the LVEF. Thus, bias seems unlikely. Nevertheless, this approach represents a potential limitation of the statistical approach, and the results need to be confirmed independently.

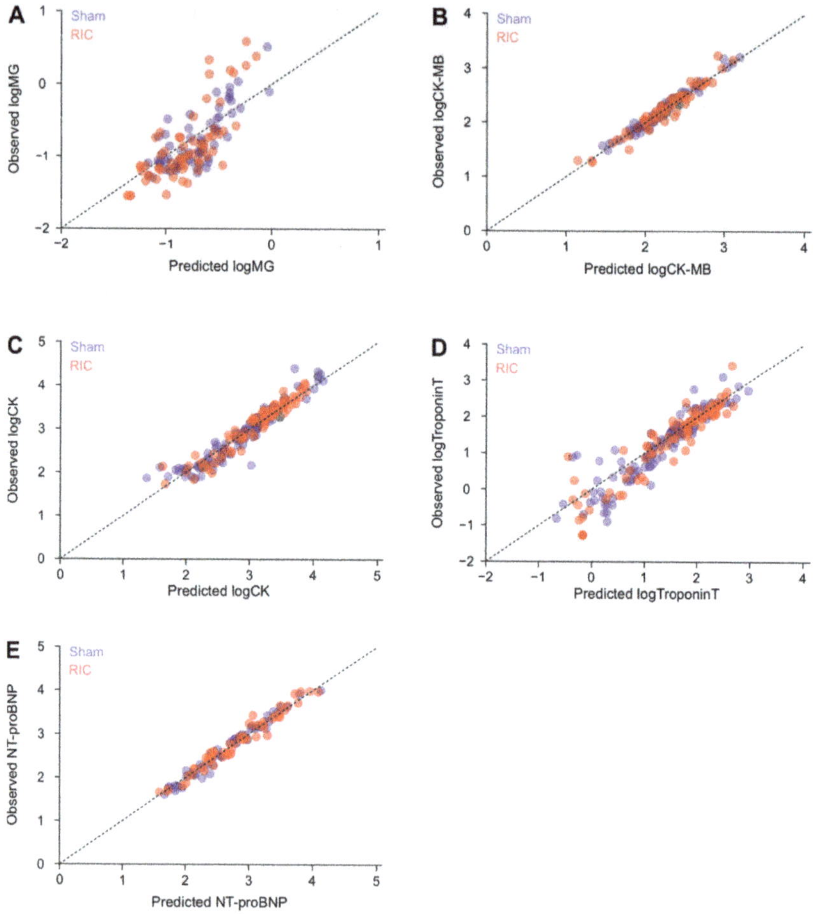

Figure A4. Predicted vs. observed values. (**A**) Methylglyoxal, (**B**) CK-MB, (**C**) CK, (**D**) TnI, (**E**) NT-proBNP.

Appendix B.4. Estimation of Methylglyoxal Levels in Healthy Subjects

A general linear model with the predictor "spiked MG concentration" as a continuous covariate and the four volunteers as levels of a factor was applied. This resulted in a parameter estimate for the covariate, which is the change in DMQ/IS for each unit in the spiked MG concentration increase. Further, this resulted in n–1 parameters indicating by how much the intercept for three volunteers is shifted in the y-direction compared to the reference subject. For the latter, the intercept of the model applies. By adding the parameters of the three subjects to the model intercept (which equals the intercept of one reference individual), each volunteer's intercept was calculated.

Appendix B.5. Additional Outcomes

The primary outcome of the initially conducted randomized controlled trial was infarct size quantified by single photon emission computer tomography one day after admission and one month, which will be reported elsewhere. Additionally, the predefined secondary outcomes included the area under the curve of the biomarkers associated with myocardial necrosis measured within 24 h after admission and the left ventricular ejection fraction (LVEF) measured four days after AMI by echocardiography. The biomarkers included creatine kinase (CK), creatine kinase myocardial band (CK-MB) and troponin I (TnI).

Appendix B.6. TnI Sensitivity Analysis

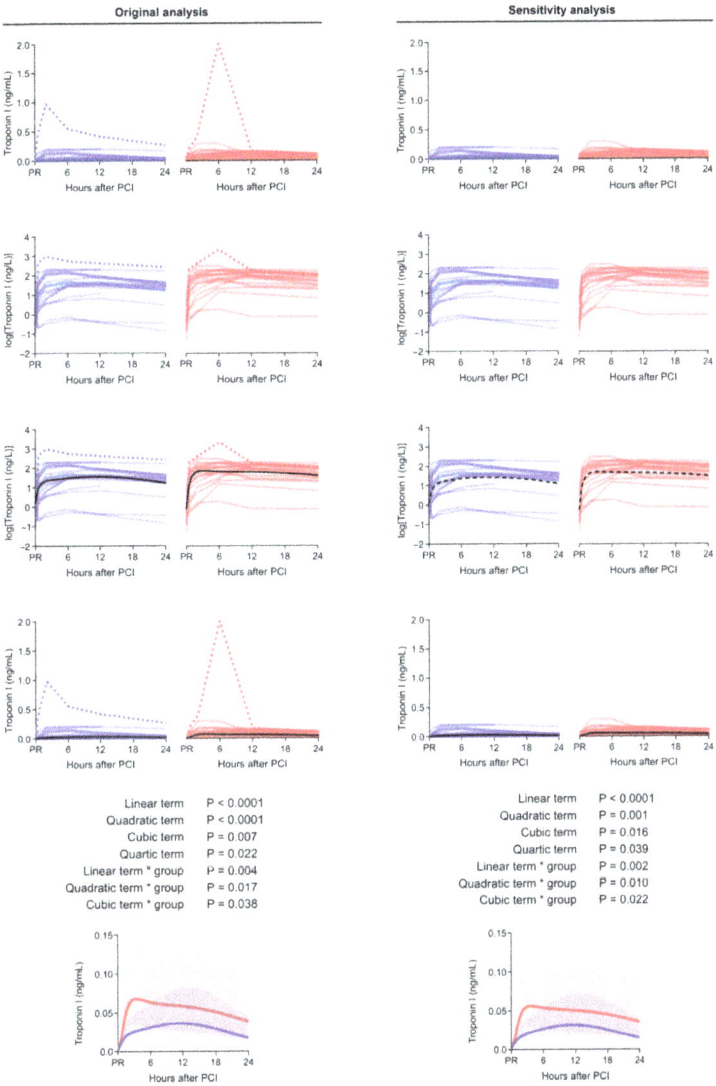

Figure A5. Sensitivity analysis of the TnI time course. The left column shows details of the analyses depicted in Figure 3G. In the first row, the raw data are depicted (sham control in blue and RIC in red). The two values visually identified as outliers are plotted as the dotted lines. As data are heavily right-skewed, they were log10-transformed prior to analysis; log10-transformed data are depicted in the second row. Notably, log transformation resulted in a distribution much closer to a normal distribution. The resulting estimated mean curve is plotted as a black line in the third row on top of the raw data. For the final plots, the data with their polynomial curve describing the central tendency was back-transformed. Thus, the black line in row four represents the estimated geometric mean on the original scale. Below, the *p*-values of the original model are listed. At the bottom, Figure 3G is reproduced to allow direct comparison with the analysis after exclusion of the outliers. The right column represents the corresponding plots of the sensitivity analysis without the dotted outliers.

Appendix B.7. No Relevant Effect of RIC on the Ejection Fraction

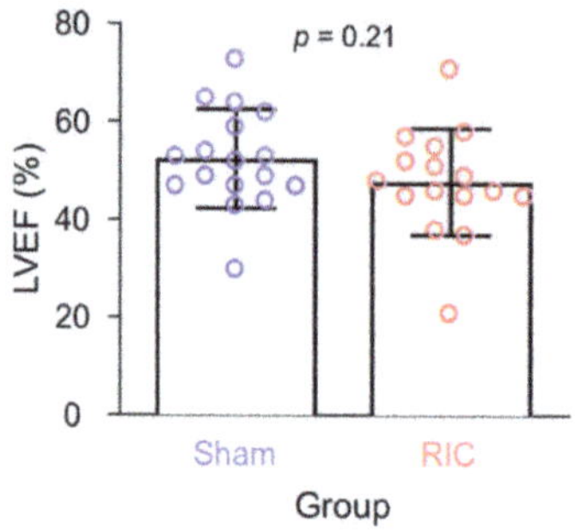

Figure A6. Effects of remote ischemic conditioning on the left ventricular ejection fraction 4 days after acute myocardial infarction.

References

1. Neumann, J.T.; Gossling, A.; Sorensen, N.A.; Blankenberg, S.; Magnussen, C.; Westermann, D. Temporal trends in incidence and outcome of acute coronary syndrome. *Clin. Res. Cardiol.* **2020**, *109*, 1186–1192. [CrossRef] [PubMed]
2. Heusch, G.; Gersh, B.J. The pathophysiology of acute myocardial infarction and strategies of protection beyond reperfusion: A continual challenge. *Eur. Heart J.* **2017**, *38*, 774–784. [CrossRef] [PubMed]
3. Ibanez, B.; James, S.; Agewall, S.; Antunes, M.J.; Bucciarelli-Ducci, C.; Bueno, H.; Caforio, A.L.P.; Crea, F.; Goudevenos, J.A.; Halvorsen, S.; et al. 2017 ESC Guidelines for the management of acute myocardial infarction in patients presenting with ST-segment elevation: The Task Force for the management of acute myocardial infarction in patients presenting with ST-segment elevation of the European Society of Cardiology (ESC). *Eur. Heart J.* **2018**, *39*, 119–177. [CrossRef] [PubMed]
4. Heusch, G.; Botker, H.E.; Przyklenk, K.; Redington, A.; Yellon, D. Remote ischemic conditioning. *J. Am. Coll. Cardiol.* **2015**, *65*, 177–195. [CrossRef] [PubMed]
5. Hausenloy, D.J.; Yellon, D.M. Ischaemic conditioning and reperfusion injury. *Nat. Rev. Cardiol.* **2016**, *13*, 193–209. [CrossRef] [PubMed]
6. Bromage, D.I.; Pickard, J.M.; Rossello, X.; Ziff, O.J.; Burke, N.; Yellon, D.M.; Davidson, S.M. Remote ischaemic conditioning reduces infarct size in animal in vivo models of ischaemia-reperfusion injury: A systematic review and meta-analysis. *Cardiovasc. Res.* **2017**, *113*, 288–297. [CrossRef] [PubMed]
7. Hausenloy, D.J.; Kharbanda, R.K.; Møller, U.K.; Ramlall, M.; Aarøe, J.; Butler, R.; Bulluck, H.; Clayton, T.; Dana, A.; Dodd, M.; et al. Effect of remote ischaemic conditioning on clinical outcomes in patients with acute myocardial infarction (CONDI-2/ERIC-PPCI): A single-blind randomised controlled trial. *Lancet* **2019**, *394*, 1415–1424. [CrossRef]
8. Haller, P.M.; Vargas, K.G.; Haller, M.C.; Piackova, E.; Wojta, J.; Gyöngyösi, M.; Gersh, B.J.; Kiss, A.; Podesser, B.K.; Huber, K. Remote ischaemic conditioning for myocardial infarction or elective PCI: Systematic review and meta-analyses of randomised trials. *Eur. Heart J. Acute Cardiovasc. Care* **2020**, *9*, 82–92. [CrossRef] [PubMed]
9. Stone, G.W.; Selker, H.P.; Thiele, H.; Patel, M.R.; Udelson, J.E.; Ohman, E.M.; Maehara, A.; Eitel, I.; Granger, C.B.; Jenkins, P.L.; et al. Relationship Between Infarct Size and Outcomes Following Primary PCI: Patient-Level Analysis From 10 Randomized Trials. *J. Am. Coll. Cardiol.* **2016**, *67*, 1674–1683. [CrossRef] [PubMed]
10. Westman, P.C.; Lipinski, M.J.; Luger, D.; Waksman, R.; Bonow, R.O.; Wu, E.; Epstein, S.E. Inflammation as a Driver of Adverse Left Ventricular Remodeling After Acute Myocardial Infarction. *J. Am. Coll. Cardiol.* **2016**, *67*, 2050–2060. [CrossRef] [PubMed]
11. Heusch, G.; Libby, P.; Gersh, B.; Yellon, D.; Böhm, M.; Lopaschuk, G.; Opie, L. Cardiovascular remodelling in coronary artery disease and heart failure. *Lancet* **2014**, *383*, 1933–1943. [CrossRef]
12. Blackburn, N.J.R.; Vulesevic, B.; McNeill, B.; Cimenci, C.E.; Ahmadi, A.; Gonzalez-Gomez, M.; Ostojic, A.; Zhong, Z.; Brownlee, M.; Beisswenger, P.J.; et al. Methylglyoxal-derived advanced glycation end products contribute to negative cardiac remodeling and dysfunction post-myocardial infarction. *Basic Res. Cardiol.* **2017**, *112*, 57. [CrossRef] [PubMed]
13. Hartog, J.W.; Voors, A.A.; Schalkwijk, C.G.; Scheijen, J.; Smilde, T.D.; Damman, K.; Bakker, S.J.; Smit, A.J.; van Veldhuisen, D.J. Clinical and prognostic value of advanced glycation end-products in chronic heart failure. *Eur. Heart J.* **2007**, *28*, 2879–2885. [CrossRef] [PubMed]
14. Haller, P.M.; Jager, B.; Piackova, E.; Sztulman, L.; Wegberger, C.; Wojta, J.; Gyongyosi, M.; Kiss, A.; Podesser, B.K.; Spittler, A.; et al. Changes in Circulating Extracellular Vesicles in Patients with ST-Elevation Myocardial Infarction and Potential Effects of Remote Ischemic Conditioning-A Randomized Controlled Trial. *Biomedicines* **2020**, *8*, 218. [CrossRef] [PubMed]
15. White, S.K.; Frohlich, G.M.; Sado, D.M.; Maestrini, V.; Fontana, M.; Treibel, T.A.; Tehrani, S.; Flett, A.S.; Meier, P.; Ariti, C.; et al. Remote ischemic conditioning reduces myocardial infarct size and edema in patients with ST-segment elevation myocardial infarction. *JACC Cardiovasc. Interv.* **2015**, *8*, 178–188. [CrossRef] [PubMed]

16. Kong, X.; Ma, M.Z.; Huang, K.; Qin, L.; Zhang, H.M.; Yang, Z.; Li, X.Y.; Su, Q. Increased plasma levels of the methylglyoxal in patients with newly diagnosed type 2 diabetes 2. *J. Diabetes* **2014**, *6*, 535–540. [CrossRef] [PubMed]
17. Brings, S.; Fleming, T.; De Buhr, S.; Beijer, B.; Lindner, T.; Wischnjow, A.; Kender, Z.; Peters, V.; Kopf, S.; Haberkorn, U.; et al. A scavenger peptide prevents methylglyoxal induced pain in mice. *Biochim. Biophys. Acta* **2017**, *1863*, 654–662. [CrossRef] [PubMed]
18. McLellan, A.C.; Thornalley, P.J. Synthesis and chromatography of 1,2-diamino-4,5-dimethoxybenzene, 6,7-dimethoxy-2-methylquinoxaline and 6,7-dimethoxy-2,3-dimethylquinoxaline for use in a liquid chromatographic fluorimetric assay of methylglyoxal. *Analytica Chimica Acta* **1992**, *263*, 137–142. [CrossRef]
19. Nemet, I.; Varga-Defterdarović, L.; Turk, Z. Preparation and quantification of methylglyoxal in human plasma using reverse-phase high-performance liquid chromatography. *Clin. Biochem.* **2004**, *37*, 875–881. [CrossRef] [PubMed]

biomedicines

MDPI

Article

Mechanisms of Regenerative Potential Activation in Cardiac Mesenchymal Cells

Pavel M. Docshin [1], Andrei A. Karpov [2,3], Malik V. Mametov [4], Dmitry Y. Ivkin [3], Anna A. Kostareva [1] and Anna B. Malashicheva [1,5,*]

1 Almazov National Medical Research Centre, Institute of Molecular Biology and Genetics, 197341 St. Petersburg, Russia; pdocshin@icloud.com (P.M.D.); kostareva_aa@almazovcentre.ru (A.A.K.)

2 Almazov National Medical Research Centre, Institute of Experimental Medicine, 194156 St. Petersburg, Russia; karpov_aa@almazovcentre.ru

3 Center of Experimental Pharmacology, Saint Petersburg State Chemical Pharmaceutical University, 197022 St. Petersburg, Russia; dmitry.ivkin@pharminnotech.com

4 Department of Pathophysiology, Pavlov First Saint Petersburg State Medical University, 197022 St. Petersburg, Russia; m_mametov@inbox.ru

5 Laboratory of Regenerative Biomedicine, Institute of Cytology, Russian Academy of Science, 194064 St. Petersburg, Russia

* Correspondence: malashicheva_ab@almazovcentre.ru

Abstract: Recovery of the contractile function of the heart and the regeneration of the myocardium after ischemic injury are contemporary issues in regenerative medicine and cell biology. This study aimed to analyze early transcriptional events in cardiac tissue after infarction and to explore the cell population that can be isolated from myocardial tissue. We induced myocardial infarction in Wistar rats by permanent ligation of the left coronary artery and showed a change in the expression pattern of Notch-associated genes and *Bmp2/Runx2* in post-MI tissues using RNA sequencing and RT-PCR. We obtained primary cardiac mesenchymal cell (CMC) cultures from postinfarction myocardium by enzymatic dissociation of tissues, which retained part of the activation stimulus and had a pronounced proliferative potential, assessed using a "xCELLigence" real-time system. Hypoxia in vitro also causes healthy CMCs to overexpress Notch-associated genes and *Bmp2/Runx2*. Exogenous activation of the Notch signaling pathway by lentiviral transduction of healthy CMCs resulted in a dose-dependent activation of the *Runx2* transcription factor but did not affect the activity of the *Bmp2* factor. Thus, the results of this study showed that acute hypoxic stress could cause short-term activation of the embryonic signaling pathways Notch and Bmp in CMCs, and this interaction is closely related to the processes of early myocardial remodeling after a heart attack. The ability to correctly modulate and control the corresponding signals in the heart can help increase the regenerative capacity of the myocardium before the formation of fibrotic conditions.

Keywords: myocardial infarction; heart regeneration; cardiac mesenchymal cells; Notch signaling pathway; early remodeling of the heart

Citation: Docshin, P.M.; Karpov, A.A.; Mametov, M.V.; Ivkin, D.Y.; Kostareva, A.A.; Malashicheva, A.B. Mechanisms of Regenerative Potential Activation in Cardiac Mesenchymal Cells. *Biomedicines* **2022**, *10*, 1283. https://doi.org/10.3390/biomedicines10061283

Academic Editors: Tânia Martins-Marques, Gonçalo F. Coutinho and Attila Kiss

Received: 30 April 2022
Accepted: 25 May 2022
Published: 31 May 2022

Publisher's Note: MDPI stays neutral with regard to jurisdictional claims in published maps and institutional affiliations.

Copyright: © 2022 by the authors. Licensee MDPI, Basel, Switzerland. This article is an open access article distributed under the terms and conditions of the Creative Commons Attribution (CC BY) license (https://creativecommons.org/licenses/by/4.0/).

1. Introduction

Myocardial infarction remains one of the major medical problems of a non-infectious character that kills and disables the world's population [1]. The search for an effective therapeutic strategy is still ongoing, and many studies are aimed at exploring the possibilities of myocardial cell therapy, but the cellular mechanisms of post-infarction cardiac recovery are still unclear. According to traditional concepts, the human heart is a definitively differentiated organ; however, foci of cell proliferative activity may occur in the heart, the intensity of which increases in the peri-infarct zone [2]. The discovery of cardiac stem cells stimulated the development of new approaches in myocardial cell therapy [3], but the ability of these cells to directly generate cardiomyocytes remains controversial.

The turnover of contractile cardiomyocytes in a healthy myocardium is approximately 0.5–2% per year [4,5], and these numbers slightly increase with heart damage [2]. The renewal of cardiomyocytes occurs mainly due to the re-entry of cells into the cell cycle, but not due to the differentiation of the cardiac stem cells themselves [5]. Mesenchymal stem cells (including cardiac stem cells) possibly participate in regenerative processes in the myocardium, releasing paracrine factors that provide myocardial protection, neovascularization, remodeling, and differentiation of the heart [6].

There is evidence that the mechanisms of myocardial functional recovery are contained in paracrine intercellular signaling, and their activation can occur precisely during the early response to damage. In this regard, the study of the mechanisms of early activation of regenerative processes in post-infarction tissue is an urgent issue. Although the existence of resident cardiac progenitor cells has been questioned [5], it is clear that myocardial regeneration exists, but the mechanisms remain undefined. This study aimed to analyze early transcriptional events in cardiac tissue after infarction and to explore the cell population that can be isolated from myocardial tissue.

Here we report that acute hypoxic stress affects the activation of the Notch signaling pathway and *Bmp2/Runx2* genes in cardiac mesenchymal cells, and the action of these pathways is associated with early myocardial remodeling processes. In this study, we induced myocardial infarction in rats and, 8/24 h after surgery, we isolated post-infarction tissues and primary CMC cultures with a pronounced proliferative potential. We showed a change in the expression pattern of Notch-associated genes and *Bmp2/Runx2* in postinfarction tissues and CMCs using RNA sequencing and RT-PCR. Hypoxia in vitro also causes healthy CMCs derived from sham-operated rats to upregulate Notch-associated genes and *Bmp2/Runx2*. Exogenous activation of the Notch signaling pathway led to dose-dependent activation of the transcription factor Runx2 but did not affect the activity of the Bmp2 factor.

2. Materials and Methods

2.1. Ethics Statements and Animals

We obtained permission from the local ethics committee of the Almazov National Medical Research Centre for conducting animal experiments. Male Wistar rats (Pushchino, Russia) of the same age and weight between 200 and 250 g were used in the experiment. Animals were kept in separate plastic cages with free access to water and standard diet food during the experiment. All experiments were performed following the Guide for the Care and Use of Laboratory Animals.

2.2. Induction of Myocardial Infarction In Vivo

Male Wistar rats (n = 16) were anesthetized with chloral hydrate (2 mg/kg intraperitoneally), intubated, and vented (SAR-830P; CWE, Inc., Ardmore, PA, USA) using room air with a tidal volume of 2 mL/100 g and a frequency of 60 breaths per minute. The core body temperature was maintained at 37.0 ± 0.5 °C using a feedback heating pad (TCAT-2LV controller; Physitemp Instruments Inc., Clifton, NJ, USA). Registration of heart rate and arrhythmias was monitored by electrocardiography. After thoracotomy, the heart was visualized through the fourth intercostal space. Further, in a blunt way, using the branches of anatomical tweezers, the pericardium was removed. At the border of the free edge of the left atrial appendage, the left coronary artery (LCA) was visualized, under which a ligature (prolene 6/0, Ethicon, Germany) was applied, directly at the edge of the left atrial appendage [7]. Myocardial ischemia was confirmed by visual examination of the anterior surface of the heart and the elevation of the ST segment on the ECG.

2.3. RNA-seq Library Preparation

We excised the post-infarction area, including the peri-infarction zone (the area was slightly visible and had a whitish hue, and was under the ligature), from the left ventricle of the ischemic heart to obtain post-infarction tissues and isolation of the primary CMC cell culture. We isolated total RNA according to the manufacturer's protocol (Eurogen,

Russia) from postinfarction tissues and CMCs ($n = 3$); RNA obtained from sham-operated rats from healthy myocardial tissues and CMCs ($n = 3$) was used as a control. The quality and quantity of the isolated RNA were checked on a NanoDrop 1000 spectrophotometer (Thermo Fisher Scientific) and an Agilent 2100 bioanalyzer (Agilent Technologies). A total of 1 μg of total RNA was used to create libraries using the TruSeq RNA sample preparation kit (Illumina) following the low sample (LS) protocol from the manufacturer's instructions.

2.4. Differential Gene Expression Analysis

Raw RNASeq reads were aligned with STAR 2.7 against Rnor_6.0 (GCA_000001895.4) and transcript annotations (ensemble 97) [8]. Differential expression analysis was performed using the DESeq2 Bioconductor package [9]. Genes with a p-value of 0.05 or less were called differentially expressed genes. Comparisons were made between ischemic tissues/cells and control tissues/cells. Data analysis and visualization (PCA Plot and Volcano Plot) were performed using Phantasus (version: 1.7.3, build: master-709) [10].

2.5. Ingenuity Pathway Analysis

Bioinformatic analysis was performed using ingenuity pathway analysis (IPA; Qiagen Silicon Valley, Redwood City, California, USA. Available online: http://www.ingenuity.com (accessed on 24 September 2020)) of differentially expressed genes to determine the interactions of genes and related networks using the default Rattus norvegicus background for settings. All the genes that passed the significance filter were identified as focus genes and uploaded to the IPA for further functional and network analysis. The specificity of connections within the network for each focus gene was calculated by the percentage of its connections with other significant genes in the database. Canonical pathway analysis identifies the paths from the IPA library that were most significant for the input data set.

2.6. Isolation of Cardiac Mesenchymal Cells

The heart was taken 8 and 24 h after surgery. Cardiac mesenchymal cells were obtained from the ischemic myocardial zone by grinding a tissue fragment followed by enzymatic treatment with type 2 collagenase (Worthington) solution (2 mg/mL) for 90 min in an incubator (37 °C, 5% CO^2, humidity 99%); a healthy myocardium obtained from rats 24 h after the sham operation was used as a negative control [11]. The cell suspension was centrifuged at 300 x g for 5 min, and cells were resuspended in growth medium (two times), seeded on the flask, and cultured in an incubator (37 °C, 5% CO^2, and 99% humidity). Medium for cardiac mesenchymal cells contains: 70% DMEM/F_{12} (Invitrogen, Waltham, MA, USA), 20% ECM (Invitrogen, USA), 10% fetal bovine serum (HyClone, Logan, UT, USA), 100 μM MEM NEAA amino acid solution (Gibco, Grand Island, New York, USA), 2 mm L-Glutamine (Gibco, USA), a mixture of penicillin (100 u/mL) and streptomycin (100 μg/mL) (Gibco, USA). In the next three days, we replaced it with a fresh culture medium once per day. On the third day, we removed large tissue debris and then continued to cultivate until the confluent state (~1 week). In the obtaining culture, live cardiomyocytes were absent, and the cell population was homogeneous. In this study, we used cells derived from rats 24 h after surgery, between 1 and 3 passages.

2.7. Assessment of Proliferative Activity of Cardiac Mesenchymal Cells

Cell proliferation was monitored in real-time using the xCELLigence RTCA DP Real-Time Cell Analyzer system. We used impedance as an indication recorded by the xCELLigence system to evaluate cell proliferative ability [12]. The system measures electrical impedance through oncoming microelectrodes embedded in the bottom of the electronic plates. The impedance measurement, which is displayed as a cell index (CI) value, provides quantitative information about the biological status of the cells, including the number of cells and their viability. Five thousand cells were sown in each well of the E-Plate (in 100 μL of cell suspension). The impedance value of each well was automatically monitored by the xCELLigence system for 72 h and expressed as the CI value. The obtained data were processed in the RTCA Software program (version number 1.0.0.1304).

2.8. In Vitro Hypoxia Induction

We used cardiac mesenchymal cells from sham-operated rats. The cells were seeded on 5 cm Petri dishes and cultured in the medium for the CMCs. The next day we transferred the cells into an incubator, where it is possible to adjust the level of oxygen in the chamber. Cardiac mesenchymal cells were hypoxic for 8 and 24 h with oxygen levels of 1% and 5% (37 °C, 5% CO^2, and 99% humidity); as a control, we used cells under normoxic conditions (37 °C, 5% CO^2, 20% O^2, and 99% humidity).

2.9. Real-Time PCR

RNA was isolated from myocardial tissues and primary cell cultures using a Trisol analog called ExtractRNA (Eurogen, Russia). Reverse transcription was performed using MMLV reverse transcriptase and the MMLV RT kit (Eurogen, Russia) according to the recommendations of the manufacturer. All samples were pretreated with DNase. We used a 5-fold reaction mixture qPCR mix-HS SYBR with intercalating dye SYBR Green I (Eurogen, Russia) for real-time PCR. cDNA (50 ng), forward and reverse primers (10 μM each) were added to the mixture, and the final volume was adjusted with sterile water to 25 μL. The sequences of the primers used: *Gapdh* (F: CCAGTATGACTCTACCCACG, R: CATTTGATGTTAGCGGGATCTC), *Notch1* (F: CAATGAGTGTGACTCACGGC, R: GCACAAGGTTCTGGCAGTTG), *Notch2* (F: CCGTGGGGCTGAAGAATCTC, R: CTTTCTTTGGCTGGGGTCCT), *Notch3* (F: GCCTAGTCCAGCAACTGCTAC, R: GGGAACAGATATGGGGTGTGG), *Dll1* (F: TAACCCCGATGGAGGCTACA, R: GCACCGTTAGAACAAGGGGA), *Dll4* (F: GCAGCTGTAAGGACCATGAGA, R: TTCACAGTGCTGGCCATAGT), *Jag1* (F: CGCCCAATGCTACAATCGTG, R: TCTTGCCCTCGTAGTCCTCA), *Hes1* (F: ACCAAAGACAGCCTCTGAGC, R: TTGGAATGCCGGGAGCTATC), *Runx2* (F: TCCCTCCGAGACCCTAAGAAA, R: GCTGCTCCCTTCTGAACCTAC), *Bmp2* (F: CTGCCATGGGGAATGTCCTT, R: TGCACTATGGCATGGTTGGT), *Hif-1a* (F: GGCGAGAACGAGAAGAAAAATAGG, R: AGATGGGAGCTCACGTTGTG), *Vegf-a* (F: GCAGCGACAAGGCAGACTAT, R: TGGCACGATTTAAGAGGGGA), *Ccnd1* (F: CTTACTTCAAGTGCGTGCAGAG, R: TTCATCTTAGAGGCCACGAACA), *Hes7* (F: CATCAACCGCAGCCTAGAAGAG, R: CACGGCGAACTCCAGTATCTCT), *Hey1* (F: CCTGGCTATGGACTATCGGAG, R: AGGCATCGAGTCCTTCAATGAT), *Myc* (F: CAGCTCGCCCAAATCCTGTA, R: TGATGGGGATGACCCTGACT). The polymerase chain reaction was carried out using a 7500 Real-Time PCR System (7500 Software v2.3, Life Technologies Ltd, Paisley, UK). Quantitative PCR was performed for 40 cycles. Data analysis was conducted using the $2^{-\Delta\Delta CT}$ method; relative gene expression was normalized on the GAPDH housekeeping gene.

2.10. Statistical Analysis

Analysis of the data was performed using Microsoft Excel and GraphPad Prism Software (version 9.3.1(350)). The significance of differences between the groups was evaluated using the unpaired nonparametric Mann–Whitney test. The differences were considered significant at $p < 0.05$. All experiments were repeated three times.

3. Results

3.1. The Gene Expression Profile Changes Significantly in Myocardial Tissues 24 h after Infarction

In order to assess transcriptional changes in the heart during acute hypoxic stress, we induced myocardial infarction in Wistar rats by permanent ligation of the left coronary artery. Twenty-four hours after the operation, we took the postinfarction area of the myocardium, including the periinfarction zone, for subsequent isolation of total RNA and preparation of libraries. As a control, we used a healthy area of the myocardium obtained from sham-operated rats. We analyzed the genetic profile of postinfarction tissues by RNA sequencing. PCA showed significant variability in data for the first and second components (Figure 1a). The transcriptional profiles of post-infarction and healthy tissues were divided and formed separate clusters. Analysis of differentially expressed genes (DEGs) showed that in postinfarction tissues, 1241 genes were upregulated and 1256 downregulated (adjusted p-value < 0.05) (Figure 1b). The top 50 DEGs include genes that determine the epithelial–mesenchymal transition, as in wound healing and fibrosis (*CD44, GPC1, SDC1, VCAN, PLAUR, SERPINE1, PMEPA1, PVR*); genes encoding proteins involved in glycolysis and gluconeogenesis (*ANGPTL4, SLC16A3*); genes upregulated by STAT5 in response to IL2 stimulation (*MYC, SLC1A5, IL1R2, SELP*), as well as genes that are regulated by NF-kB in response to TNF, in response to low oxygen levels (hypoxia), and determine the inflammatory response (Table S1).

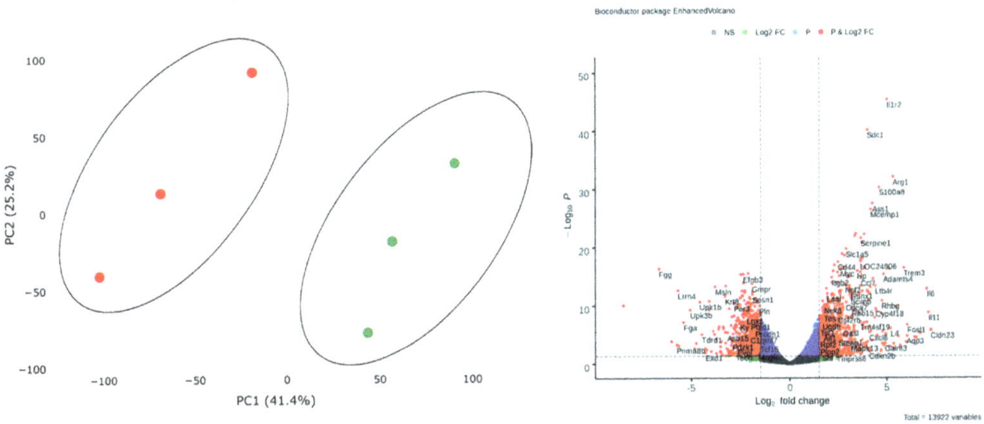

(**a**) (**b**)

Figure 1. (**a**) Principal component analysis (PCA) displays variability in gene expression in healthy (green dots, $n = 3$) and postinfarction (orange dots, $n = 3$) myocardial tissues using the Phantasus Web Tool. Major component 1 (PC1) on the x-axis and PC2 on the y-axis accounted for 41.4% and 25.2% of the total variability in gene expression, respectively. PC1 covers a significant part (41.4%) of the experimental variability and largely reflects the difference between the two states of the heart muscle. PC2 represents experimental variability (25.2%) associated with the difference in gene profile between various biological repeats (laboratory animals). The samples are visually divided into two main groups. (**b**) Volcano plots showing differentially expressed genes (adjusted p-value < 0.05) in post-infarction samples compared to controls, performed using Bioconductor software in R. The y-axis corresponds to the mean expression value of $-\log10$ (p-value), and the x-axis displays the log2 (fold-change) value. The red dots represent significantly differentially expressed genes.

3.2. Early Remodeling Processes are Enhanced in the Postinfarction Myocardium, which is Involved Components of the BMP and NOTCH Signaling Pathways

To assess the effect of acute hypoxic injury on dysregulation of canonical signaling pathways and biological functions in cardiac tissue 24 h after the onset of infarction, we performed GSEA and canonical pathway analysis of DEGs using Qiagen software. DEGs were uploaded into Qiagen Ingenuity Pathway Analysis (IPA) to identify enriched pathways in the dataset, after filtering for significance (adj. $p < 0.05$). In total, about 152 canonical signaling and metabolic pathways were identified, the significance of which was higher than $-\log$ (P)> 1.3, and only 82 signaling pathways had absolute z-scores more than 1.0 (Figure S1A).

In the heart after myocardial infarction, massive cell death occurs in the affected area and a sustained formation of an inflammatory response, in particular, aimed at providing reparative processes associated with early myocardial remodeling [13]. We observed a shift in the pattern of gene expression towards early myocardial remodeling 24 h after the induction of acute infarction. According to this, an affective *'Apelin Cardiac Fibroblast Signaling Pathway'* (ratio—0.455; z-score—−2.333; *p*-value—5.67×10^{-4}), which means the activation of cardiac fibroblasts and their differentiation into myofibroblasts and causes the formation of cardiac fibrosis, leading to heart failure, and *'Remodeling of Epithelial Adherents Junctions'* (ratio—0.382; z-score—2.449; *p*-value—1.71×10^{-5}), *'Inhibition of Matrix Metalloproteases'* (ratio—0.344; z-score—−1.265; *p*-value—4.57×10^{-3}), or *'TGF-β Signaling'* (ratio—0.218; z-score—1.5; *p*-value—4.86×10^{-2}) were found in dataset (Figure S1A). Additionally, signaling pathways involved in the regulation of the cell cycle and proliferative activity ('Cell Cycle Control of Chromosomal Replication', 'PI3K/AKT Signaling', 'Aryl Hydrocarbon Receptor Signaling', 'Cell Cycle Regulation by BTG Family Proteins', 'STAT3 Pathway', 'HIF1α Signaling', 'Cyclins and Cell Cycle Regulation') and migration activity ('Regulation of Actin-based Motility by Rho', 'Actin Cytoskeleton Signaling') were activated in postinfarction tissues.

Gene networks generated using the IPA program, which reflect forms of non-canonical signaling, were noted for various upregulated and downregulated genes involved in the growth and development of tissues and cells (Figure S1B). More than 100 molecules have been identified by gene-set enrichment analysis that are involved in the formation of the immune response, the growth and development of the cardiovascular system, the processes of cell division, and changes in cell morphology are affected. The main bio-functions and diseases are listed (Table S2). It should be noted that proliferative processes, including smooth muscle cells, differentiation of connective tissue cells, and organization of sarcomere are activated in the ischemic heart (Figure S2).

In these processes, the activity of *Bmp2* from the TGF-b subfamily and several components of the Notch signaling pathway was noted, such as the *Notch2* receptor, the target genes *Myc, Ccnd1*, and the *Runx2* transcription factor [14,15], which were upregulated, and the target gene *Hey1*, which was downregulated. In particular, *Bmp2* from the TGF-b subfamily was identified among the key regulators. A complete list of overexpressed components of the Notch signaling pathway is presented (Table S3).

3.3. Activation of Notch Signaling Pathway Components and Bmp2/Runx2 in Post-Infarction Myocardial Tissues

In order to confirm the RNA sequencing data that *Bmp2* and components of the Notch signaling pathway are activated in post-infarction tissues, we analyzed them using qPCR. Additionally, we took into account an even earlier time point after the induction of myocardial infarction in rats, 8 h, to assess the difference in gene expression between the two time intervals.

We showed (Figure 2) that acute hypoxia in vivo activates the expression of Notch signaling pathway components and Bmp2 in the ischemic myocardium compared to a healthy heart. The obtained qPCR data were in accordance with the RNA sequencing results for the 24 h point.

3.4. Postinfarction Cardiac Mesenchymal Cells had a Pronounced Ability to Proliferate

We induced myocardial infarction in Wistar rats to obtain primary cultures of cardiac mesenchymal cells from the postinfarction area (including the periinfarction zone) of the myocardium 8 and 24 h after surgery; cardiac mesenchymal cells of the ventricular myocardium of the healthy heart of sham-operated rats were used as a negative control (Figure 3a). The phenotypic characterization of this cell type is complex, and there is no specific marker or combination of markers for identifying mesenchymal stem cells (MSCs) [16,17]. The International Society for Cellular Therapy has established the following criteria for the identification of MSCs: adhesion to plastic, expression of markers, the adipogenic and osteogenic differentiation ability, and the ability to form fibroblast colony-forming units (CFU-F) [17,18]. Here, we use the term cardiac mesenchymal cells (CMC) because of some terminological differences associated with the correct classification of these cells [19,20]. The obtained CMC primary cultures clearly expressed CD90 and were positive for CD166, which was previously described as a marker of cardiac mesenchymal cells and one of the populations of cardiac stem cells obtained from cardiospheres [21]. They were also negative for endothelial markers and did not belong to hematopoietic markers, namely negative for CD45, CD31, and CD34 [17] (Figure 3b).

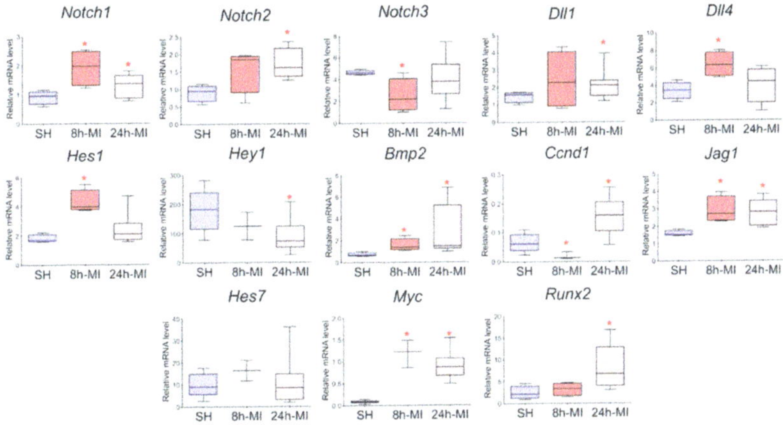

Figure 2. The dynamics of the expression of the Notch signaling pathway and *Bmp2/Runx2* in tissues by quantitative PCR analysis. SH—healthy myocardium from sham-operated rats (n = 3); 8 h-MI—tissue from the post-infarction zone of the myocardium from rats 8 h after induction (n = 4); 24 h-MI—tissue from the post-infarction zone of the myocardium from rats 24 h after induction (n = 9). Vertical—the relative amount of mRNA in each group, measured by the $2^{-\Delta\Delta CT}$ method; box plots with whiskers at min to the max are presented. * The asterisk shows significant differences between SH and MI groups at $p < 0.05$ (unpaired nonparametric Mann–Whitney test).

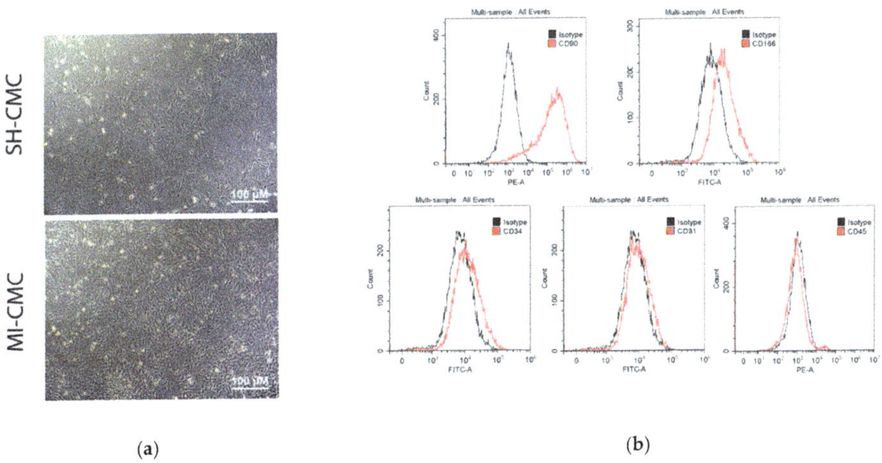

Figure 3. (**a**) Cardiac mesenchymal cells (CMCs) derived from healthy (SH-CMC) and postinfarct (MI-CMC) myocardium of rat during 5-day cultivation. CMCs were isolated 24 h after myocardial infarction. SH-CMC were obtained from rat hearts after sham operation. The primary cell culture of CMCs demonstrated the ability to form fibroblast colony-forming units (CFU-F) and had adhesion to a culture plastic. (**b**) Immunophenotyping of primary cultures of CMC using flow cytometry. The histograms show the expression levels of surface markers CD31, CD34, C45, CD90, and CD166 (red lines) concerning isotype (gray lines).

To evaluate and compare the proliferative activity of postinfarction and healthy cardiac mesenchymal cells obtained one day after surgical procedures, we used the xCELLigence system to monitor cell proliferation in real-time. We found (Figure 4) that postinfarction CMCs (orange curve) have a more pronounced potential for proliferation and, accordingly, have an activation stimulus in response to acute hypoxic stress than healthy myocardial cells (blue curve). The experiment lasted for 72 h with a frequency of measurements every 15 min.

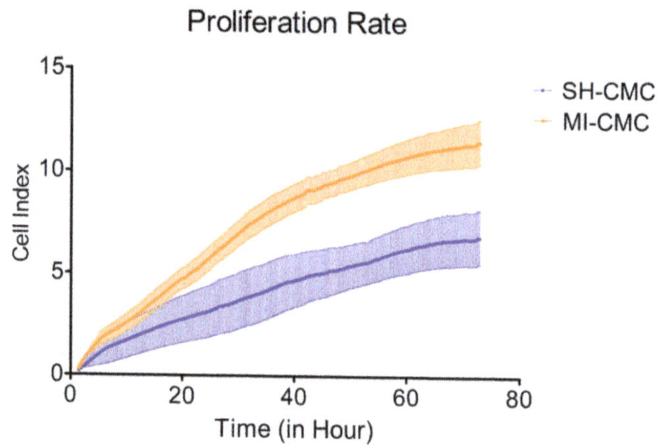

Figure 4. The proliferation rate of postinfarction cardiac mesenchymal cells (MI-CMC, orange curve, $n = 9$) and healthy cardiac mesenchymal cells (SH-CMC, blue curve, $n = 3$). Each curve represents the average values between the samples and the standard error of the mean (vertical lines). Horizontal—the time during which the measurements of proliferative activity were taken every 15 min. The experiment lasted for 72 h. Vertical—the value of the cell index, which reflects the quantitative information about the biological status of cells, including their number and viability. The seeding density was 5000 cells per well. The significance of differences between the SH-CMC and the MI-CMC is $p < 0.05$ with D'Agostino and Pearson omnibus normality test.

3.5. *The Gene Expression Profile of Postinfarction CMC is Altered*

Next, we analyzed transcriptomic profile of cardiac mesenchymal cells from the postinfarction area and from healthy hearts of sham-operated rats by RNA sequencing. Principal component analysis showed the main patterns in the resulting dataset. PC4 covers an insignificant part (14.5%, $p < 0.05$) of the experimental variability and reflects a relatively small difference between the two CMC states, which may indicate a gradual change in the expression profile during cell culture under normal conditions (Figure 5a).

Analysis of RNA sequencing data revealed 13 differentially expressed genes (adjusted p-value <0.05) in the obtained post-infarction cell cultures. Among the activated genes, the expression of *Spp1, RGD1565131, Tagap, Myh1* can be noted, while the expression of *Bmp3, Fgl2, Sfrp4,* on the contrary, was reduced in postinfarction cardiac mesenchymal cells (log2 (fold-change) $\geq$ 1.9) (Figure 5b). A complete list of differentially expressed genes (DEGs), sorted by adjusted p-value, is presented (Table S4). We collected the 50 top highly expressed genes, sorted by statistical criteria (Table S5).

We found that the expression of *Twist1* and the target genes Notch *Hey2* and *Ccnd1* are reduced in postinfarction cardiac mesenchymal cells. In addition, the activity of the *Thbs2* gene, which may be responsible for the inhibition of angiogenesis, and *Dnm1*, which is involved in vesicular transport, decreased (p-value> 0.05) (Table S6).

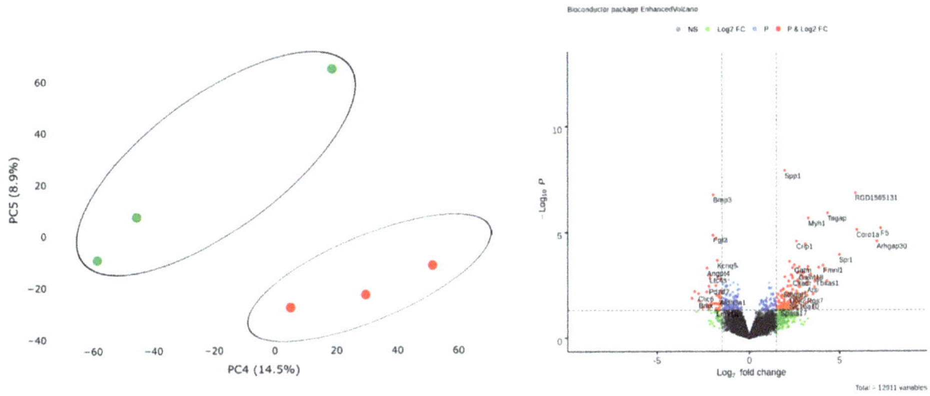

(**a**) (**b**)

Figure 5. (**a**) Principal component analysis (PCA) displays variability in gene expression in healthy (green dots, $n = 3$) and postinfarction (orange dots, $n = 3$) CMCs using the Phantasus Web Tool. Major component 1 (PC4) on the x-axis and PC5 on the y-axis accounted for 14.5% and 8.9% of the total variability in gene expression, respectively. The samples are visually divided into two main groups. (**b**) Volcano plots showing differentially expressed genes (p-value < 0.05) in post-infarction samples compared to controls, performed using Bioconductor software in R. The y-axis corresponds to the mean expression value of -log10 (p-value), and the x-axis displays the log2 (fold-change) value. The red dots represent significantly differentially expressed genes.

3.6. GSE Analysis and Canonical Pathway Analysis of DEGs Showed that the Pattern of Gene Expression Characteristic of the Ischemic Heart is Partially Retained in MI-CMC

To identify enriched signaling pathways in the dataset, genes were loaded into Qiagen Ingenuity Pathway Analysis (IPA) after significance filtering (p-value < 0.05). In total, about 14 canonical signaling pathways were identified, including Metabolic Pathways and Signaling Pathways, the significance of which was greater than −log (P)> 1.3, and only 1 STAT3 signaling pathway had absolute z-scores more than 2.0 (Figure 6). A complete list of this data is provided in Table S7.

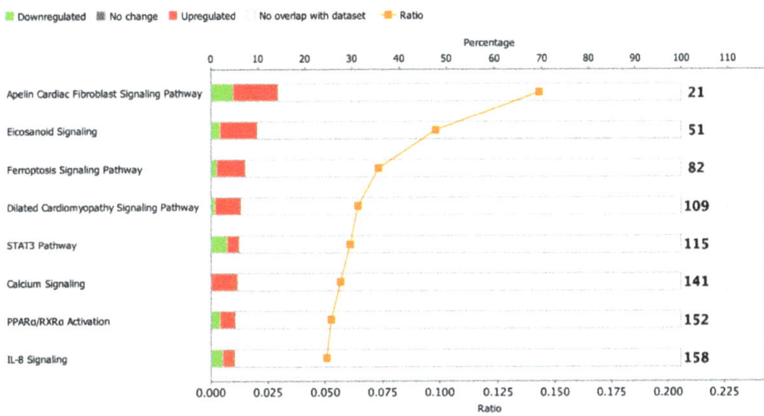

Figure 6. Dysregulated signaling pathways in postinfarction CMC samples identified by IPA analysis. The histogram represents the dysregulated canonical signaling pathways in a stacked bar chart by Fisher's Exact Test p-value (p-value > 0.05, z-score > 1). The ratio plot shows the number of significant genes expressed in the data versus the total number of genes in that particular signaling pathway.

We conducted a core analysis in the IPA program to find key diseases and bio-functions in the CMCs, and we identified about 12 categories in which processes related to diseases or functions were affected (*p*-value—0.05), and only in 3 of them, the z-score was more than 2. It was interesting to note that a partial pattern of genes associated with heart failure persisted in post-infarction cells (*'Failure of heart'*, *p*-value = 8.25×10^{-3}, z-score = 2.177), and along with this, the modulation of processes associated with neovascularization and cell migration was reflected. At the same time, the *'Expansion of cells'* parameter was suppressed (*p*-value = 6.09×10^{-3}, z-score = −2.219), which may be the result of a delay in the cell cycle, since DNA synthesis was also slightly increased (p-value = 1.31×10^{-2}, z-score = 1.077). A complete list of this data is provided (Table S8).

We constructed probabilistic networks of dysregulated genes using the IPA program based on the detected activated/repressed genes and those that were statistically filtered but present in the data set (Table S9). We combined probabilistic networks based on related processes and key regulators and found that one of the Notch *Hey2* targets is reduced in MI-CMCs.

3.7. The Expression of the Jag1 and Hes1 Genes of the Notch Signaling Pathway and Bmp2/Runx2 Factors is Preserved in Cell Culture

In order to approve the results of RNA sequencing and evaluate the level of expression of Notch signaling components using real-time PCR with reverse transcription, passages 1–3 cells were obtained from rat tissues 8 and 24 h after induction of myocardial infarction. In particular, we assessed the level of expression of *Bmp2* and *Runx2*, the dysregulation of which was noted in the transcriptome. Since the BMP signaling pathway plays an essential role in cardiogenesis, as does the Notch signaling pathway, and *Runx2* can be a target gene for both types of signaling and has also been identified in cardiac development, according to our hypothesis, *Bmp2* and *Runx2* can be early remodeling genes. We found (Figure 7) that in postinfarction cardiac mesenchymal cells, the activation potential is partially preserved, which is expressed in increased expression of the *Bmp2* and *Runx2* genes, and Jag1 and *Hes1* genes of the Notch signaling pathway compared to cells obtained from a healthy heart.

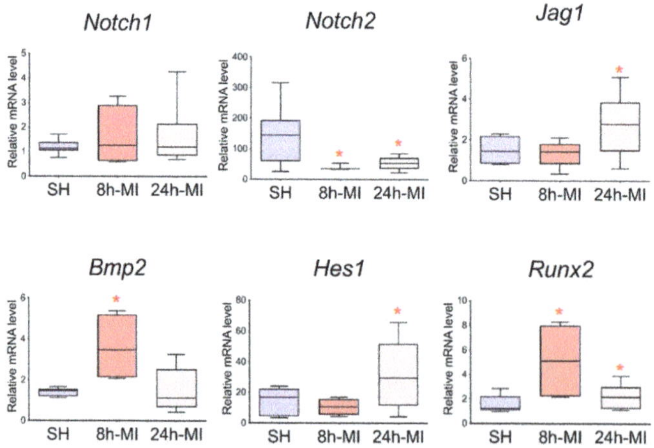

Figure 7. Dynamics of expression of Notch signaling pathway components and *Bmp2/Runx2* genes in cardiac mesenchymal cells using quantitative PCR analysis. SH—cells from a healthy myocardium of sham-operated rats (n = 3); 8 h-MI—post-infarction cells from rats 8 h after induction (n = 4); 24 h-MI—post-infarction cells from rats 24 h after induction (n = 9). Vertical—the relative amount of the mRNA in each group, measured by the $2^{-\Delta\Delta CT}$ method; box plots with whiskers at min to the max are presented. The asterisk shows significant differences between the SH and MI groups at $p < 0.05$ (unpaired nonparametric Mann–Whitney test).

3.8. Activation of Notch Signaling Pathway Components and Bmp2/Runx2 Factors in Cardiac Mesenchymal Cells during In Vitro Hypoxia Induction

To determine whether hypoxia is a sufficient factor to activate the Notch signaling pathway and *Bmp2/Runx2* factors, we took CMC from the healthy myocardium of sham-operated rats and induced hypoxia in vitro. Two different oxygen concentrations of 1% and 5% in the incubator and two time points of 8 and 24 h, during which the cells were under conditions of hypoxic stress, were chosen. We showed (Figure 8) that under conditions of hypoxia, the Notch signaling pathway components and *Bmp2/Runx2* are activated in cells compared to CMCs under normoxia conditions. *Hif-1α* and *Vegfa* were used as a control of hypoxia.

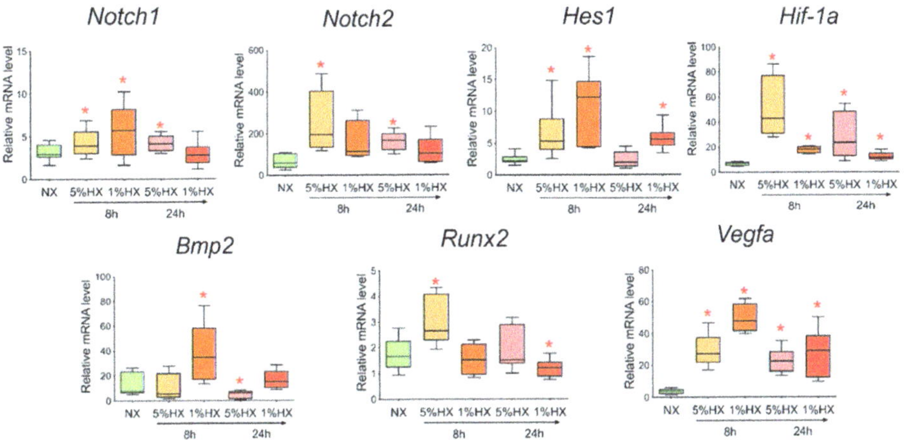

Figure 8. Dynamics of expression of Notch signaling pathway components, *Bmp2/Runx2* factors, and hypoxic stress markers in cardiac mesenchymal cells obtained from healthy myocardium of sham-operated rats by quantitative PCR analysis. NX—healthy cells in a state of normoxia ($n = 3$); 5% HX—healthy cells in a state of hypoxia with an oxygen level of 5% ($n = 3$); 1% HX—healthy cells in a state of hypoxia with an oxygen level of 1% ($n = 3$). Horizontal—the time at which the cells were in hypoxic conditions. Vertical—the relative amount of mRNA, measured by the $2^{-\Delta\Delta CT}$ method; box plots with whiskers at min to the max are presented. * The asterisk shows significant differences between the control and the 5% HX and 1% HX groups at $p < 0.05$ (unpaired nonparametric Mann–Whitney test).

3.9. Exogenous Activation of the Notch Signaling Pathway in Cardiac Mesenchymal Cells Dose-Dependently Activates Runx2

To reveal the relationship between the Notch signaling pathway and *Bmp2* and *Runx2* factors, we took CMCs from the healthy myocardium of sham-operated rats and transduced them with a lentiviral vector carrying an NICD insertion to activate the Notch signaling pathway. Viral particles were added to the culture at two different concentrations of 3 and 15 μL. Cells were cultured with the virus for 24 h. We demonstrated (Figure 9) that activation of the Notch signaling pathway led to dose-dependent activation of *Runx2*. On the contrary, the *Bmp2* did not respond to the activation of Notch.

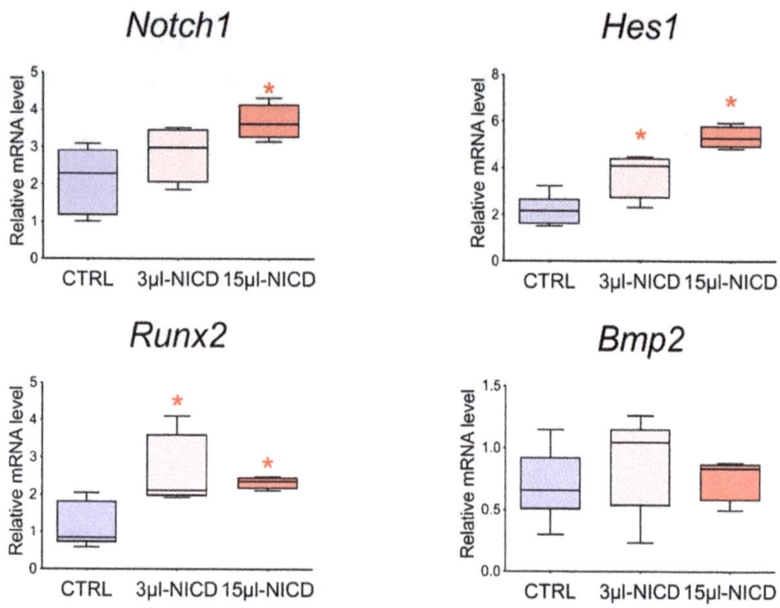

Figure 9. The dynamics of the expression of *Bmp2/Runx2* factors and Notch signaling pathways in cardiac mesenchymal cells obtained from healthy myocardium of sham-operated rats by quantitative PCR analysis. CTRL—healthy cells as a negative control ($n = 3$); 3 µL-NICD—healthy cells that had 3 µL of the virus with NICD added ($n = 3$); 15 µL-NICD—healthy cells that had 15 µL of the virus with NICD added ($n = 3$). Vertical—the relative amount of mRNA, measured by the $2^{-\Delta\Delta CT}$ method; box plots with whiskers at min to the max are presented. * The asterisk shows significant differences between CTRL and the 3 µL-NICD and 15 µL-NICD groups at $p < 0.05$ (unpaired nonparametric Mann–Whitney test).

4. Discussion

Myocardial infarction is a common acute disease that impairs heart functionality. Molecular and cellular mechanisms of cardiac early remodeling and recovery of postinfarction myocardium remain not fully understood. In this study, we induced myocardial infarction in rats to study early transcriptomic events occurring 8 and 24 h after surgical procedures. We showed a change in the expression pattern of Notch-associated genes and *Bmp2/Runx2* in postinfarction tissues using RNA sequencing and RT-PCR. We obtained primary CMC cultures from the postinfarction myocardium, which retained part of the activation stimulus and had a pronounced proliferative potential. Hypoxia in vitro also led healthy CMCs to upregulate the expression of Notch-associated genes and *Bmp2/Runx2*. Exogenous activation of the Notch signaling pathway resulted in a dose-dependent activation of the *Runx2* transcription factor but did not affect the activity of the *Bmp2* factor.

Some studies have shown that foci of proliferative activity are formed in the peri-infarction area in response to injury [2,22]. The effect of hypoxemia on myocardial recovery after a lesion has also been noted in in vivo studies in mice [23]. This study aimed to study early transcriptional events in cardiac tissue after myocardial infarction and to explore the cell population of cardiac mesenchymal cells, which can be isolated from myocardial tissue in order to analyze the molecular mechanisms of activation of the regenerative potential on these cells in vitro.

The results of RNA sequencing during induction of myocardial infarction showed significant changes in the gene expression pattern in postinfarction tissue. Myriad signaling pathways and processes associated with early cardiac remodeling, cell proliferation and migration, and immune response have been affected.

Gude et al. showed that expression of the *Notch1* receptor, the *Jagged1* ligand, and the *Hes1* target gene upregulated in interstitial cells and cardiomyocytes in the peri-infarct region [24]. Notch is a highly conserved signaling pathway involved in the embryonic development of most multicellular organisms, as well as in the regulation of tissue homeostasis, cell differentiation, and maintenance of the stem cell population in the postnatal period [25]. The role of Notch in myocardial recovery is still not fully understood [26].

We observed that components of the Notch signaling pathway are activated in postinfarction tissue. Overexpressed genes included targets from non-canonical signaling, such as *Myc*, *Ccnd1*, and *Runx2* [14,15]. The *Runx2* transcription factor has been noted in the development of the cardiovascular system [27]. BMP signaling is also involved in development [28,29], and according to our data, it is activated during myocardial infarction, in particular, *Bmp2*. It is known to play a critical role in the development of the heart, induction of differentiation of cardiac progenitor cells into cardiomyocytes, and to stimulate their contraction [30–33]. It can act with a key upregulation role and modulate the Notch signaling pathway [34], and be involved in early remodeling processes [35]. Expression of *Bmp2* has been observed in both cardiomyocytes and interstitial fibroblasts in myocardial infarction [36,37]; however, the activation mechanism also remains unknown [38]. We hypothesize that the Notch signaling pathway, together with the key factors *Bmp2* and *Runx2*, may play an important role in early myocardial events in response to injury.

In order to confirm that, we obtained a fraction of activated cardiac mesenchymal cells from the post-infarction area of the myocardium with promoted properties. The discovery and experimental use of cardiac mesenchymal cells have become a new focus in cardiovascular regenerative medicine [39]. Recently, clinical trials have been conducted using one type of these cells and their products in therapy, and mainly to improve the function of the heart with a single ventricle in patients with hypoplastic left heart syndrome [40–42]. We obtained cells 24 h after the induction of myocardial infarction, and during cultivation, they were passaged only three times to preserve their properties and perform functional tests. Proliferative activity was assessed in real-time using the xCELLigence system. We found that post-infarction cardiac mesenchymal cells have a more pronounced proliferative potential than CMCs obtained from a healthy heart. This is consistent with the data which we obtained on tissues with the analysis of the affected processes. In turn, we performed RNA sequencing of primary cultures of CMCs and found that postinfarction CMCs partially retain their transcriptional profile and reflect early events in the affected myocardium.

For comparison, we obtained an activated CMC fraction 8 h after induction of myocardial infarction. The subsequent analysis of the evaluation of the expression of the Notch signaling pathway components and putative early remodeling genes using real-time PCR showed that the Notch target gene *Hes1* and *Bmp2/Runx2* factors are activated. Moreover, the activation of the latter was expressed precisely at an earlier time point.

Hypoxia has been described as a factor modulating Notch [43], and whether in vitro hypoxic stress on healthy CMCs can lead to the same activation of the Notch signaling pathway and early remodeling genes remains a question. We induced hypoxia in vitro by placing cardiac mesenchymal cells obtained from sham-operated rats in an oxygen-controlled incubator. CMCs experienced acute hypoxic stress at 1% or 5% oxygen for 24 h. We isolated the RNA and assessed gene expression. We found that in vitro hypoxia increased the expression of Notch signaling pathway components and *Bmp2/Runx2* genes.

In embryonic development, there is crosstalk between the Notch signaling pathway and *Bmp2*, as with *Runx2*, but whether Notch can act as an activator of putative early remodeling factors in healthy CMCs remains unclear. We transduced healthy CMCs obtained from sham-operated rats, a lentiviral vector in a low and high dose containing the insertion of the Notch intracellular domain 1 (NICD1). We showed that *Runx2* is dose-dependently activated in response to increased Notch expression. *Bmp2* did not react in any way to the introduction of the vector, which, in turn, may indicate its superior position in the regulation of the gene network, but we plan to investigate this interaction in the future.

Despite these results, our work may have the following limitations. Several studies have shown [44–46] the probable effect of epicardium-derived cells on neovascularization and cardiomyogenesis by reactivating a fetal gene program in response to myocardial infarction. In our methodology, we removed the epicardium prior to isolation of RNA and primary cell cultures, but there is a possibility that the influence of the activated epicardium may have affected the myocardium, and therefore further studies are required to identify possible effects. In addition, we would like to note that under in vitro cultivation conditions, we inevitably encounter a change in the properties of cultured cells, which in turn can affect the ability of mesenchymal cells to proliferate and change their proangiogenic properties [47]. In this work, we showed that under the same conditions, postinfarction CMCs had more pronounced proliferative abilities than CMCs that were obtained from healthy animals, but whether this will affect myocardial recovery in vivo remains a question for further studies.

5. Conclusions

Recovery of the contractile function of the heart and the regeneration of the myocardium after ischemic injury are contemporary issues in regenerative medicine and cell biology. Traditional treatments such as drug therapy and revascularization only cropped symptoms but do not contribute to full recovery from developing heart failure.

Thus, the results of this study showed that acute hypoxic stress could cause short-term activation of the embryonic signaling pathways Notch and Bmp in cardiac mesenchymal cells, and this interaction is closely related to the processes of early myocardial remodeling after a heart attack. The ability to correctly modulate and control the corresponding signals in the heart can help increase the regenerative capacity of the myocardium before the formation of fibrotic conditions.

Supplementary Materials: The following supporting information can be downloaded at: https://www.mdpi.com/article/10.3390/biomedicines10061283/s1. Figure S1: A. Dysregulation of biological pathways (excluding metabolic pathways) in post-infarct tissue samples identified by IPA analysis. The histogram represents the dysregulated canonical signaling pathways in a stacked bar chart by Fisher's Exact Test p-value (adj. p-value > 0.05, z-score > 1). B. Dysregulated gene networks involved in the growth and development of tissues and cells, which were generated using the IPA software. Red means increased measurements, and green means decreased amounts. The brightness of the color means the more extreme of the transcripts in this dataset; Figure S2: Upregulated bio functions involved in the proliferation and differentiation of the heart cells, which were identified by GSEA using the IPA software. Red means increased measurements, and green means decreased amounts. The brightness of the color means the more extreme of the transcripts in this dataset; Table S1:Top 50 DEGs in postinfarction rat myocardial tissues sorted by stat using DeSeq analysis; Table S2:Top Diseases and Bio Functions in postinfarction tissues by IPA Software; Table S3: Components of the Notch signaling pathway that are differentially expressed in postinfarction myocardial tissues 24 hours after surgery; Table S4: Full list of differentially expressed genes in post-infarction cardiac mesenchymal cells; Table S5: A complete list of expression genes in post-infarction cardiac mesenchymal cells, assorted by a statistical criterion (P-Value <0.05); Table S6: Components of the Notch signaling pathway that are differentially expressed in postinfarction CMC; Table S7: A complete list of signaling pathways with involved genes is shown in the table, extracted using the IPA program; Table S8: A complete list of diseases and biofunctions with involved genes is presented in the table obtained using the IPA program; Table S9: Probabilistic networks of dysregulated genes created with the IPA software.

Author Contributions: Conceptualization, A.B.M.; methodology, P.M.D., A.A.K. (Andrey A. Karpov) and M.V.M.; software, P.M.D.; formal analysis, P.M.D.; investigation, P.M.D.; resources, A.A.K. (Anna A. Kostareva) and D.Y.I.; writing—original draft preparation, P.M.D.; writing—review and editing, A.B.M.; supervision, A.B.M. All authors have read and agreed to the published version of the manuscript.

Funding: This research was funded by Russian Foundation for Basic Research, grant number 20-015-00574.

Institutional Review Board Statement: The animal study protocol was approved by the Bioethics Committee of Saint Petersburg State Chemical Pharmaceutical University (protocol code Rats-MI-SC-2018 and date of approval—10 January 2018).

Data Availability Statement: Data is contained within the article and supplementary files. Raw and processed RNA sequencing data were uploaded to the NCBI website as specified in the rules for authors. They are available at the following link: https://www.ncbi.nlm.nih.gov/geo/query/acc.cgi?acc=GSE201888 (accessed on 2 May 2022).

Acknowledgments: We declare our deep gratitude to Artem Kiselev for preparing and conducting the RNA sequencing run and processing the raw data.

Conflicts of Interest: The authors declare no conflict of interest.

References

1. Roth, G.A.; Johnson, C.; Abajobir, A.; Abd-Allah, F.; Abera, S.F.; Abyu, G.; Ahmed, M.; Aksut, B.; Alam, T.; Alam, K.; et al. Global, Regional, and National Burden of Cardiovascular Diseases for 10 Causes, 1990 to 2015. *J. Am. Coll. Cardiol.* **2017**, *70*, 1–25. [CrossRef] [PubMed]
2. Hsieh, P.C.H.; Segers, V.F.M.; Davis, M.E.; MacGillivray, C.; Gannon, J.; Molkentin, J.D.; Robbins, J.; Lee, R.T. Evidence from a Genetic Fate-Mapping Study That Stem Cells Refresh Adult Mammalian Cardiomyocytes after Injury. *Nat. Med.* **2007**, *13*, 970–974. [CrossRef] [PubMed]
3. Mehanna, R.A.; Essawy, M.M.; Barkat, M.A.; Awaad, A.K.; Thabet, E.H.; Hamed, H.A.; Elkafrawy, H.; Khalil, N.A.; Sallam, A.; Kholief, M.A.; et al. Cardiac Stem Cells: Current Knowledge and Future Prospects. *World J. Stem Cells* **2022**, *14*, 1–40. [CrossRef] [PubMed]
4. Bergmann, O.; Bhardwaj, R.D.; Bernard, S.; Zdunek, S.; Barnabé-Heide, F.; Walsh, S.; Zupicich, J.; Alkass, K.; Buchholz, B.A.; Druid, H.; et al. Evidence for Cardiomyocyte Renewal in Humans. *Science* **2009**, *324*, 98–102. [CrossRef] [PubMed]
5. Eschenhagen, T.; Bolli, R.; Braun, T.; Field, L.J.; Fleischmann, B.K.; Frisén, J.; Giacca, M.; Hare, J.M.; Houser, S.; Lee, R.T.; et al. Cardiomyocyte Regeneration: A Consensus Statement. *Circulation* **2017**, *136*, 680–686. [CrossRef]
6. Mirotsou, M.; Jayawardena, T.M.; Schmeckpeper, J.; Gnecchi, M.; Dzau, V.J. Paracrine Mechanisms of Stem Cell Reparative and Regenerative Actions in the Heart. *J. Mol. Cell. Cardiol.* **2011**, *50*, 280–289. [CrossRef]
7. Karpov, A.A.; Ivkin, D.Y.; Dracheva, A.V.; Pitukhina, N.N.; Uspenskaya, Y.K.; Vaulina, D.D.; Uskov, I.S.; Eyvazova, S.D.; Minasyan, S.M.; Vlasov, T.D. Rat Model of Post-Infarct Heart Failure by Left Coronary Artery Occlusion: Technical Aspects, Functional and Morphological Assessment. *Biomedicine* **2014**, *1*, 32–48.
8. Dobin, A.; Davis, C.A.; Schlesinger, F.; Drenkow, J.; Zaleski, C.; Jha, S.; Batut, P.; Chaisson, M.; Gingeras, T.R. STAR: Ultrafast Universal RNA-Seq Aligner. *Bioinformatics* **2013**, *29*, 15–21. [CrossRef]
9. Love, M.I.; Huber, W.; Anders, S. Moderated Estimation of Fold Change and Dispersion for RNA-Seq Data with DESeq2. *Genome Biol.* **2014**, *15*, 550. [CrossRef]
10. Zenkova, D.; Kamenev, V.; Sablina, R.; Artyomov, M.; Sergushichev, A. Phantasus: Visual and Interactive Gene Expression Analysis 2018. Available online: https://bioconductor.riken.jp/packages/3.12/bioc/html/phantasus.html (accessed on 2 May 2022).
11. Smits, A.M.; van Vliet, P.; Metz, C.H.; Korfage, T.; Sluijter, J.P.G.; Doevendans, P.A.; Goumans, M.J. Human Cardiomyocyte Progenitor Cells Differentiate into Functional Mature Cardiomyocytes: An in Vitro Model for Studying Human Cardiac Physiology and Pathophysiology. *Nat. Protoc.* **2009**, *4*, 232–243. [CrossRef]
12. Dowling, C.M.; Herranz Ors, C.; Kiely, P.A. Using Real-Time Impedance-Based Assays to Monitor the Effects of Fibroblast-Derived Media on the Adhesion, Proliferation, Migration and Invasion of Colon Cancer Cells. *Biosci. Rep.* **2014**, *34*, 415–427. [CrossRef] [PubMed]
13. Silvestre, J.S.; Mallat, Z. Editorial: Inflammation and Reparative Process after Cardiac Injury. *Front. Cardiovasc. Med.* **2019**, *6*, 162. [CrossRef] [PubMed]
14. Garg, V.; Muth, A.N.; Ransom, J.F.; Schluterman, M.K.; Barnes, R.; King, I.N.; Grossfeld, P.D.; Srivastava, D. Mutations in NOTCH1 Cause Aortic Valve Disease. *Nature* **2005**, *437*, 270–274. [CrossRef] [PubMed]
15. Ivan, C.; Hu, W.; Bottsford-Miller, J.; Zand, B.; Dalton, H.J.; Liu, T.; Huang, J.; Nick, A.M.; Lopez-Berestein, G.; Coleman, R.L.; et al. Epigenetic Analysis of the Notch Superfamily in High-Grade Serous Ovarian Cancer. *Gynecol. Oncol.* **2013**, *128*, 506–511. [CrossRef]
16. Javazon, E.H.; Beggs, K.J.; Flake, A.W. Mesenchymal Stem Cells: Paradoxes of Passaging. *Exp. Hematol.* **2004**, *32*, 414–425. [CrossRef]
17. Dominici, M.; le Blanc, K.; Mueller, I.; Slaper-Cortenbach, I.; Marini, F.C.; Krause, D.S.; Deans, R.J.; Keating, A.; Prockop, D.J.; Horwitz, E.M. Minimal Criteria for Defining Multipotent Mesenchymal Stromal Cells. The International Society for Cellular Therapy Position Statement. *Cytotherapy* **2006**, *8*, 315–317. [CrossRef]
18. Gambini, E.; Pompilio, G.; Biondi, A.; Alamanni, F.; Capogrossi, M.C.; Agrifoglio, M.; Pesce, M. C-Kit+ Cardiac Progenitors Exhibit Mesenchymal Markers and Preferential Cardiovascular Commitment. *Cardiovasc. Res.* **2011**, *89*, 362–373. [CrossRef]

19. Sultana, N.; Zhang, L.; Yan, J.; Chen, J.; Cai, W.; Razzaque, S.; Jeong, D.; Sheng, W.; Bu, L.; Xu, M.; et al. Resident C-Kit + Cells in the Heart Are Not Cardiac Stem Cells. *Nat. Commun.* **2015**, *6*, 8701. [CrossRef]
20. Epstein, J.A. A Time to Press Reset and Regenerate Cardiac Stem Cell Biology. *JAMA Cardiol.* **2019**, *4*, 95–96. [CrossRef]
21. Oldershaw, R.; Owens, W.A.; Sutherland, R.; Linney, M.; Liddle, R.; Magana, L.; Lash, G.E.; Gill, J.H.; Richardson, G.; Meeson, A. Human Cardiac-Mesenchymal Stem Cell-like Cells, a Novel Cell Population with Therapeutic Potential. *Stem Cells Dev.* **2019**, *28*, 593–607. [CrossRef]
22. van den Borne, S.W.M.; Isobe, S.; Verjans, J.W.; Petrov, A.; Lovhaug, D.; Li, P.; Zandbergen, H.R.; Ni, Y.; Frederik, P.; Zhou, J.; et al. Molecular Imaging of Interstitial Alterations in Remodeling Myocardium After Myocardial Infarction. *J. Am. Coll. Cardiol.* **2008**, *52*, 2017–2028. [CrossRef] [PubMed]
23. Nakada, Y.; Canseco, D.C.; Thet, S.; Abdisalaam, S.; Asaithamby, A.; Santos, C.X.; Shah, A.M.; Zhang, H.; Faber, J.E.; Kinter, M.T.; et al. Hypoxia Induces Heart Regeneration in Adult Mice. *Nature* **2017**, *541*, 222–227. [CrossRef] [PubMed]
24. Gude, N.A.; Emmanuel, G.; Wu, W.; Cottage, C.T.; Fischer, K.; Quijada, P.; Muraski, J.A.; Alvarez, R.; Rubio, M.; Schaefer, E.; et al. Activation of Notch-Mediated Protective Signaling in the Myocardium. *Circ. Res.* **2008**, *102*, 1025–1035. [CrossRef] [PubMed]
25. Artavanis-Tsakonas, S.; Rand, M.D.; Lake, R.J. Notch Signaling: Cell Fate Control and Signal Integration in Development. *Science* **1999**, *284*, 770–776. [CrossRef]
26. Ferrari, R.; Rizzo, P. The Notch Pathway: A Novel Target Formyocardial Remodelling Therapy? *Eur. Heart J.* **2014**, *35*, 2140–2145. [CrossRef] [PubMed]
27. Tavares, A.L.P.; Brown, J.A.; Ulrich, E.C.; Dvorak, K.; Runyan, R.B. Runx2-I Is an Early Regulator of Epithelial-Mesenchymal Cell Transition in the Chick Embryo. *Dev. Dyn.* **2018**, *274*, 542–554. [CrossRef]
28. Niessen, K.; Karsan, A. Notch Signaling in Cardiac Development. *Circ. Res.* **2008**, *102*, 1169–1181. [CrossRef]
29. Garside, V.C.; Chang, A.C.; Karsan, A.; Hoodless, P.A. Co-Ordinating Notch, BMP, and TGF-β Signaling during Heart Valve Development. *Cell. Mol. Life Sci.* **2013**, *70*, 2899–2917. [CrossRef]
30. Zhang, H.; Bradley, A. Mice Deficient for BMP2 Are Nonviable and Have Defects in Amnion/Chorion and Cardiac Development. *Development* **1996**, *122*, 2977–2986. [CrossRef]
31. Monzen, K.; Shiojima, I.; Hiroi, Y.; Kudoh, S.; Oka, T.; Takimoto, E.; Hayashi, D.; Hosoda, T.; Habara-Ohkubo, A.; Nakaoka, T.; et al. Bone Morphogenetic Proteins Induce Cardiomyocyte Differentiation through the Mitogen-Activated Protein Kinase Kinase Kinase TAK1 and Cardiac Transcription Factors Csx/Nkx-2.5 and GATA-4. *Mol. Cell. Biol.* **1999**, *19*, 7096–7105. [CrossRef]
32. Ghosh-Choudhury, N.; Abboud, S.L.; Chandrasekar, B.; Ghosh Choudhury, G. BMP-2 Regulates Cardiomyocyte Contractility in a Phosphatidylinositol 3 Kinase-Dependent Manner. *FEBS Lett.* **2003**, *544*, 181–184. [CrossRef]
33. Ma, L.; Lu, M.F.; Schwartz, R.J.; Martin, J.F. Bmp2 Is Essential for Cardiac Cushion Epithelial-Mesenchymal Transition and Myocardial Patterning. *Development* **2005**, *132*, 5601–5611. [CrossRef] [PubMed]
34. Prados, B.; Gómez-Apiñániz, P.; Papoutsi, T.; Luxán, G.; Zaffran, S.; Pérez-Pomares, J.M.; de La Pompa, J.L. Myocardial Bmp2 Gain Causes Ectopic EMT and Promotes Cardiomyocyte Proliferation and Immaturity Article. *Cell Death Dis.* **2018**, *9*, 399. [CrossRef] [PubMed]
35. Morrell, N.W.; Bloch, D.B.; ten Dijke, P.; Goumans, M.J.T.H.; Hata, A.; Smith, J.; Yu, P.B.; Bloch, K.D. Targeting BMP Signalling in Cardiovascular Disease and Anaemia. *Nat. Rev. Cardiol.* **2016**, *13*, 106–120. [CrossRef]
36. Chang, S.; Lee, E.J.; Kang, H.; Zhang, S.; Kim, J.; Li, L.; Youn, S.; Lee, C.; Kim, K.; Won, J.; et al. Impact of Myocardial Infarct Proteins and Oscillating Pressure on the Differentiation of Mesenchymal Stem Cells: Effect of Acute Myocardial Infarction on Stem Cell Differentiation. *Stem Cells* **2008**, *26*, 1901–1912. [CrossRef]
37. Rutkovskiy, A.; Sagave, J.; Czibik, G.; Baysa, A.; Zihlavnikova Enayati, K.; Hillestad, V.; Dahl, C.P.; Fiane, A.; Gullestad, L.; Gravning, J.; et al. Connective Tissue Growth Factor and Bone Morphogenetic Protein 2 Are Induced Following Myocardial Ischemia in Mice and Humans. *Scand. J. Clin. Lab. Investig.* **2017**, *77*, 321–331. [CrossRef]
38. Hanna, A.; Frangogiannis, N.G. The Role of the TGF-β Superfamily in Myocardial Infarction. *Front. Cardiovasc. Med.* **2019**, *0*, 140. [CrossRef]
39. García, A.N. "Second-Generation" Stem Cells for Cardiac Repair. *World J. Stem Cells* **2015**, *7*, 352. [CrossRef]
40. Ishigami, S.; Ohtsuki, S.; Tarui, S.; Ousaka, D.; Eitoku, T.; Kondo, M.; Okuyama, M.; Kobayashi, J.; Baba, K.; Arai, S.; et al. Intracoronary Autologous Cardiac Progenitor Cell Transfer in Patients with Hypoplastic Left Heart Syndrome: The TICAP Prospective Phase 1 Controlled Trial. *Circ. Res.* **2015**, *116*, 653–664. [CrossRef]
41. Ishigami, S.; Ohtsuki, S.; Eitoku, T.; Ousaka, D.; Kondo, M.; Kurita, Y.; Hirai, K.; Fukushima, Y.; Baba, K.; Goto, T.; et al. Intracoronary Cardiac Progenitor Cells in Single Ventricle Physiology: The Perseus (Cardiac Progenitor Cell Infusion to Treat Univentricular Heart Disease) Randomized Phase 2 Trial. *Circ. Res.* **2017**, *120*, 1162–1173. [CrossRef]
42. Sano, T.; Ousaka, D.; Goto, T.; Ishigami, S.; Hirai, K.; Kasahara, S.; Ohtsuki, S.; Sano, S.; Oh, H. Impact of Cardiac Progenitor Cells on Heart Failure and Survival in Single Ventricle Congenital Heart Disease. *Circ. Res.* **2018**, *122*, 994–1005. [CrossRef] [PubMed]
43. Borggrefe, T.; Lauth, M.; Zwijsen, A.; Huylebroeck, D.; Oswald, F.; Giaimo, B.D. The Notch Intracellular Domain Integrates Signals from Wnt, Hedgehog, TGFβ/BMP and Hypoxia Pathways. *Biochim. Et Biophys. Acta (BBA) Mol. Cell Res.* **2016**, *1863*, 303–313. [CrossRef] [PubMed]
44. Bollini, S.; Vieira, J.M.N.; Howard, S.; Dubè, K.N.; Balmer, G.M.; Smart, N.; Riley, P.R. Re-Activated Adult Epicardial Progenitor Cells Are a Heterogeneous Population Molecularly Distinct from Their Embryonic Counterparts. *Stem Cells Dev.* **2014**, *23*, 1719–1730. [CrossRef]

45. Vieira, J.M.; Howard, S.; Villa Del Campo, C.; Bollini, S.; Dubé, K.N.; Masters, M.; Barnette, D.N.; Rohling, M.; Sun, X.; Hankins, L.E.; et al. BRG1-SWI/SNF-Dependent Regulation of the Wt1 Transcriptional Landscape Mediates Epicardial Activity during Heart Development and Disease. *Nat. Commun.* **2017**, *8*, 16034. [CrossRef] [PubMed]
46. Wagner, K.D.; Wagner, N.; Bondke, A.; Nafz, B.; Flemming, B.; Theres, H.; Scholz, H. The Wilms' Tumor Suppressor Wt1 Is Expressed in the Coronary Vasculature after Myocardial Infarction. *FASEB J. Off. Publ. Fed. Am. Soc. Exp. Biol.* **2002**, *16*, 1117–1119. [CrossRef]
47. Wagner, K.D.; Vukolic, A.; Baudouy, D.; Michiels, J.F.; Wagner, N. Inducible conditional vascular-specific overexpression of peroxisome proliferator-activated receptor beta/delta leads to rapid cardiac hypertrophy. *PPAR Res.* **2016**, *2016*, 12. [CrossRef]

biomedicines

MDPI

Article

DNMT3B System Dysregulation Contributes to the Hypomethylated State in Ischaemic Human Hearts

Estefanía Tarazón [1,2,*,†], Lorena Pérez-Carrillo [1,†], Isaac Giménez-Escamilla [1], María García-Manzanares [1,3], Luis Martínez-Dolz [1,2,4], Manuel Portolés [1,2] and Esther Roselló-Lletí [1,2,*]

1 Clinical and Translational Research in Cardiology Unit, Health Research Institute, Hospital La Fe, 46026 Valencia, Spain; lorena_perezc@iislafe.es (L.P.-C.); igies@alumni.uv.es (I.G.-E.); maria.garcia8@uchceu.es (M.G.-M.); martinez_luidol@gva.es (L.M.-D.); portoles_man@gva.es (M.P.)
2 Biomedical Research Networking Center on Cardiovascular Diseases (CIBERCV), Institute of Health Carlos III, 28029 Madrid, Spain
3 Department of Animal Medicine and Surgery, Veterinary Faculty, Cardenal Herrera-CEU University, 46115 Valencia, Spain
4 Failure and Transplantation Unit, Cardiology Department, University and Polytechnic La Fe Hospital, 46026 Valencia, Spain
* Correspondence: tarazon_est@gva.es (E.T.); esther_rosello@iislafe.es (E.R.-L.); Tel.: +34-96-1246-644 (E.T. & E.R.-L.)
† These authors contributed equally to this work.

Citation: Tarazón, E.; Pérez-Carrillo, L.; Giménez-Escamilla, I.; García-Manzanares, M.; Martínez-Dolz, L.; Portolés, M.; Roselló-Lletí, E. DNMT3B System Dysregulation Contributes to the Hypomethylated State in Ischaemic Human Hearts. *Biomedicines* **2022**, *10*, 866. https://doi.org/10.3390/biomedicines10040866

Academic Editor: Tânia Martins-Marques

Received: 11 March 2022
Accepted: 5 April 2022
Published: 7 April 2022

Publisher's Note: MDPI stays neutral with regard to jurisdictional claims in published maps and institutional affiliations.

Copyright: © 2022 by the authors. Licensee MDPI, Basel, Switzerland. This article is an open access article distributed under the terms and conditions of the Creative Commons Attribution (CC BY) license (https://creativecommons.org/licenses/by/4.0/).

Abstract: A controversial understanding of the state of the DNA methylation machinery exists in ischaemic cardiomyopathy (ICM). Moreover, its relationship to other epigenetic alterations is incomplete. Therefore, we carried out an in-depth study of the DNA methylation process in human cardiac tissue. We showed a dysregulation of the DNA methylation machinery accordingly with the genome-wide hypomethylation that we observed: specifically, an overexpression of main genes involved in the elimination of methyl groups (*TET1, SMUG1),* and underexpression of molecules implicated in the maintenance of methylation (*MBD2, UHRF1).* By contrast, we found *DNMT3B* upregulation, a key molecule in the addition of methyl residues in DNA, and an underexpression of miR-133a-3p, an inhibitor of *DNMT3B* transcription. However, we found many relevant alterations that would counteract the upregulation observed, such as the overexpression of *TRAF6*, responsible for Dnmt3b degradation. Furthermore, we showed that molecules regulating Dnmts activity were altered; specifically, SAM/SAH ratio reduction. All these results are in concordance with the Dnmts normal function that we show. Our analysis revealed genome-wide hypomethylation along with dysregulation in the mechanisms of addition, elimination and maintenance of methyl groups in the DNA of ICM. We describe relevant alterations in the DNMT3B system, which promote a normal Dnmt3b function despite its upregulation.

Keywords: ischaemic cardiomyopathy; DNMT3B; DNA methylation; non-coding RNA

1. Introduction

Ischaemic cardiomyopathy (ICM) is an important cause of mortality worldwide, associated with the development of heart failure (HF) [1]. Therefore, there is a great interest in the study of ICM to obtain a better understanding of the pathophysiology underlying the disease. One of the main research foci is epigenomics, which has allowed the identification of epigenetic alterations in ICM patients [2–4]. These epigenetic changes are associated with differential expression at the mRNA level of genes described in the development of ICM [5–7]. In addition, it has been observed that the epigenetic alterations produced during disease development could be valuable diagnostic and prognostic markers, and serve as therapeutic targets [8–11].

The most studied epigenetic modification is DNA methylation. It is a complex process, characterised by two main mechanisms: the addition and the elimination of methyl residues in DNA. The addition of methyl groups to the cytosine residues of DNA, which is associated with gene repression, is directly catalysed by three members of the DNA methyltransferase (DNMT) family. These molecules require S-adenosylmethionine (SAM), a methyl group donor, to carry out the reaction, which is then transformed into S-adenosyl-L-homocysteine (SAH). On the other hand, the removal of methyl groups from DNA is a complex process since several enzymatic reactions are required [12]. Moreover, in recent years, a large number of molecules implicated in DNA methylation have been described [13]. These key molecules can be classified based on their main function, as molecules involved in the addition, elimination and maintenance of the methyl residues on DNA or regulators of the methylation process. Other epigenetic mechanisms, such as post-translational modifications of histones and RNA-based mechanisms, mainly miRNAs, can also interact with the DNA methylation machinery. This allows complex gene regulation through the different components of the epigenome [14,15].

In the context of HF, an increase in the expression of the de novo *DNMTs*, *DNMT3A* and *DNMT3B*, has been observed [16], *DNMT3B* being the main DNA methyltransferase expressed in human and mouse hearts [17]. Controversially, greater hypomethylation has been observed in regions associated with genes that encode proteins in these patients [18]. Moreover, blood genome-wide DNA hypomethylation has been described to be a risk factor for ischaemic heart disease [19].

Therefore, due to the lack of knowledge and the controversies described related to the DNA methylation process in ICM, this study aims to elucidate the state of the DNA methylation machinery. To carry it out, we analyzed the expression of the main genes involved in DNA methylation in ischaemic hearts. Furthermore, to clarify the role of *DNMT3B* upregulation, we studied the state of the DNMT3B system and the genome-wide methylation state.

2. Materials and Methods

2.1. Cardiac Tissue Samples

Left ventricular (LV) tissue samples from 40 explanted human hearts were used in our experiments, some of them matching between different analyses. The specific sample size of each individual study is reported in Table 1. Patients were diagnosed with ICM based on the following inclusion criteria: prior documented episodes of acute myocardial infarction, an echocardiography showing normal contractility segments coexisting with other dyskinetic or akinetic segments and an electrocardiography showing signs of ischemia or myocardial necrosis. There were no signs of existence of a primary valvular disease. The ICM characteristics for inclusion in the study were obtained from clinical history, hemodynamic study, electrocardiograms and Doppler echocardiography data. All the data was collected by physicians who were blind to the subsequent analysis of the LV function. Patients were functionally classified following the New York Heart Association criteria and were receiving medical treatment in accordance with the guidelines of the European Society of Cardiology [20].

All controls (CNTs) had a normal LV function (LV ejection fraction >50%) and no history of any cardiac disease. Samples were obtained from non-diseased donor hearts that had been rejected for cardiac transplantation due to size or blood type incompatibility. All heart donors died either from a cerebrovascular accident or in a motor vehicle accident.

Tissue samples were collected from near the apex of the left ventricle, maintained in 0.9% NaCl, and preserved at 4 °C for a maximum of 4.4 $\pm$ 3 h after coronary circulation loss. The samples were stored at -80 °C until further use. Appropriate handling and rapid sample collection and storage led to high-quality samples (DNA ratios 260/280 ~1.8 and 260/230 ~2.0, RNA ratio 260/280 ~2.0 and RNA integrity number $\geq$9).

This study was approved by the Ethics Committee (Biomedical Investigation Ethics Committee of La Fe University Hospital of Valencia, Spain) and was conducted in accordance with the guidelines of the Declaration of Helsinki [21]. Signed informed consent was obtained from each patient or, in the case of CNT subjects, from their relatives.

Table 1. Clinical characteristics of ischaemic cardiomyopathy patients.

	Epigenomic Study	mRNA Sequencing	ncRNA Sequencing	*DNMT3B* Validation	Dnmt3b Nuclear Protein *	SAM/SAH Ratio
	$n = 8$	$n = 13$	$n = 22$	$n = 14$	$n = 8$	$n = 30$
Age (years)	53 ± 5	54 ± 8	55 ± 8	55 ± 8	53 ± 6	55 ± 8
Gender male (%)	100	100	100	93	100	100
NYHA class	III–IV	III–IV	III–IV	III–IV	III–IV	III–IV
BMI (kg/m^2)	28 ± 3	27 ± 4	26 ± 3	27 ± 4	28 ± 4	27 ± 4
Haemoglobin (mg/dL)	14 ± 2	14 ± 3	14 ± 2	13 ± 3	15 ± 2	14 ± 2
Haematocrit (%)	44 ± 4	41 ± 6	41 ± 6	40 ± 8	43 ± 4	41 ± 5
Total cholesterol (mg/dL)	152 ± 43	162 ± 41	174 ± 45	160 ± 40	162 ± 46	187 ± 44
Prior hypertension (%)	25	33	40	31	38	56
Prior smoking (%)	88	92	81	85	88	77
Diabetes mellitus (%)	63	42	45	38	50	55
LVEF (%)	24 ± 6	25 ± 5	24 ± 7	24 ± 6	23 ± 5	24 ± 7
LVESD (mm)	57 ± 8	57 ± 8	54 ± 8	57 ± 8	57 ± 8	54 ± 8
LVEDD (mm)	65 ± 7	65 ± 8	63 ± 9	65 ± 8	65 ± 8	63 ± 8

BMI: body mass index; ICM: ischaemic cardiomyopathy; LVEF: left ventricular ejection fraction; LVEDD: left ventricular end-diastolic diameter; LVESD: left ventricular end-systolic diameter; NYHA: New York Heart Association. Data are presented as the mean ± SD.* The same patients were used to measure Dnmts nuclear activity.

2.2. *mRNA Extraction and Sequencing*

For this analysis, 23 samples were used (ICM, $n = 13$; and CNT, $n = 10$). RNA isolation and RNA-sequencing (RNA-seq) procedures and analyses have been extensively described previously by Roselló-Lletí et al. [22]. Briefly, RNA extractions were performed using a PureLink™ Kit (Ambion Life Technologies; Waltham, MA, USA) and cDNA libraries were obtained following Illumina's recommendations. Transcriptome libraries were sequenced on the SOLiD 5500 XL (Applied Biosystems; Waltham, MA, USA) platform. All data presented in this manuscript has been deposited in the NCBI's Gene Expression Omnibus (GEO) database and are accessible through the GEO series accession number GSE55296.

2.3. *ncRNA Extraction and Sequencing*

For this analysis, 30 samples were used (ICM, $n = 22$; and CNT, $n = 8$). RNA extraction was carried out using the Quik-RNATM miniprep plus kit (Zymo Research; Irvine, CA, USA) and following the manufacturer's instructions. RNA quantification was performed using NanoDrop 1000 spectrophotometer and Qubit 3.0 fluorometer (Thermo Fisher Scientific; Horsham, UK). The purity and the integrity of the RNA samples were determined using a 0.8% agarose gel and the Agilent 2100 Bioanalyzer with RNA 6000 nano assay and small RNA assay kits (Agilent Technologies; Spain).

The cDNA libraries have been obtained following Illumina´s recommendations. Briefly, 3′ and 5′ adaptors were sequentially ligated to the RNA prior to reverse transcription and cDNA generation. The cDNA was enriched using PCR to create an indexed double-stranded cDNA library, and size selection was performed using a 6% polyacrylamide gel. The quality and quantity of the libraries were analyzed using a 4200 TapeStation (Agilent Technologies; Madrid, Spain) D1000 High-Sensitivity assay. The cDNA libraries were pooled and then sequenced using paired-end sequencing (100 × 2) in the Illumina HiSeq 2500 (Illumina; San Diego, CA, USA) sequencer.

2.4. *Validation for RT-qPCR*

Real-time quantitative polymerase chain reaction (RT-qPCR) was performed to validate the expression of *DNMT3B* in the LV samples (ICM, $n = 14$; and CNT, $n = 8$). One microgram of total RNA was reverse transcribed into cDNA using the M-MLV enzyme, as per the manufacturer's instructions (Invitrogen; Waltham, MA, USA). The resulting

56 ± 8 years, and 99% were male. They belonged to classes III–IV of the New York Heart Association functional classification and presented altered values in echocardiographic parameters. Furthermore, they were previously diagnosed with significant comorbidities, including hypertension and diabetes mellitus. CNT individuals had a mean age of 55 ± 17 years, and 65% were male. Comorbidities and other echocardiographic data were not available for the CNT group, in accordance with the Spanish Organic Law on Data Protection 15/1999.

3.2. mRNA Expressio£n of Genes Involved in DNA Methylation

Transcriptome-level differences between ICM and CNT samples were investigated using the mRNA-seq technology. We focused on the study of the genes involved in the DNA methylation machinery; these genes were classified based on their main function, relating to the addition, elimination, maintenance of methyl residues in DNA, or such as regulators of the DNA methylation process (Supplementary Table S1). We found alterations in several molecules of the methylation machinery (Figure 1a). Specifically, we observed the overexpression in *DNMT3B* gene (1.90 fold and $p < 0.001$), a molecule involved in the addition of de novo methyl groups in the DNA. Amongst the major genes analyzed associated with the elimination of methyl groups that were expressed in the heart, we observed that *TET1* (1.94 fold and $p < 0.05$) and *SMUG1* (1.36 fold and $p < 0.05$) were overexpressed in ICM patients. In addition, *MBD2* (−1.27 fold and $p < 0.01$) and *UHRF1* (−1.52 fold and $p < 0.01$) genes were underexpressed, both implicated in the maintenance of methyl groups. Finally, genes involved in the regulation of the methylation process were analyzed (Figure 1b). Specifically, we observed alterations in genes implicated in the regulation of the addition of methyl group, *AHCY* (−1.46 fold and $p < 0.05$) and *TRAF6* (1.31 fold and $p < 0.05$), and overexpression of genes related to the regulation of the elimination of methyl groups, *GADD45G* (2.33 fold and $p < 0.001$) and *OGT* (1.33 fold and $p < 0.05$).

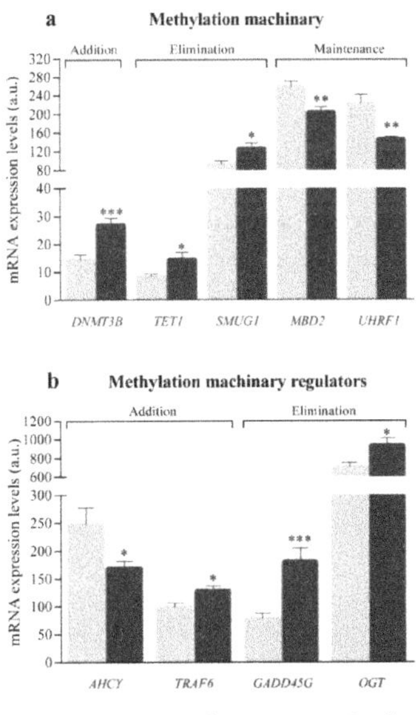

Figure 1. mRNA relative expression levels of altered genes involved in the DNA methylation process in ischaemic hearts versus controls. (**a**) Methylation machinery of DNA: genes related to addition (*DNMT3B*), elimination (*TET1* and *SMUG1*) and maintenance (*MBD2* and *UHRF1*) of methyl groups to DNA. (**b**) Methylation machinery regulators (*AHCY*, *TRAF6*, *GADD45G* and *OGT*). The results were obtained by mRNA-sequencing SOLiD 5500XL platform. Data are presented as the mean ± SEM. a.u., arbitrary units. Ischaemic cardiomyopathy patients (n = 13; blue), controls subjects (n = 10; grey). * $p < 0.05$, ** $p < 0.01$, *** $p < 0.001$.

3.3. *DNMT3B* Regulation Analysis

We validated the expression of *DNMT3B*, which plays a central role in the DNA methylation process and its overexpression contrasted with the general hypomethylation observed in these ICM patients. The results obtained by RT-qPCR corroborated the *DNMT3B* overexpression (2.22 fold and $p < 0.01$; Figure 2a) observed by RNA-seq (Figure 1a). We carried out a second RNA sequencing study, which allows the detection of ncRNAs and we observed alterations in several miRNAs previously described as regulators of *DNMT3B* expression. miR-133a-3p (Figure 2b), negative regulator of *DNMT3B* transcription, was reduced (−1.38 fold and $p < 0.001$). In addition, post-transcriptional regulators miRNAs of *DNMT3B* were altered (Figure 2c). Specifically, the miR-30d-5p (1.22 fold and $p < 0.01$) and miR-379-5p (1.41 fold and $p < 0.05$) were overexpressed, while the miR-29c-3p (−1.30 fold and $p < 0.01$) and miR-221-5p (−1.33 fold and $p < 0.01$) were underexpressed in ICM patients. Next, nuclear levels of the Dnmt3b protein were analyzed and we did not observe statistically significant changes between ICM and control individuals (Figure 2d).

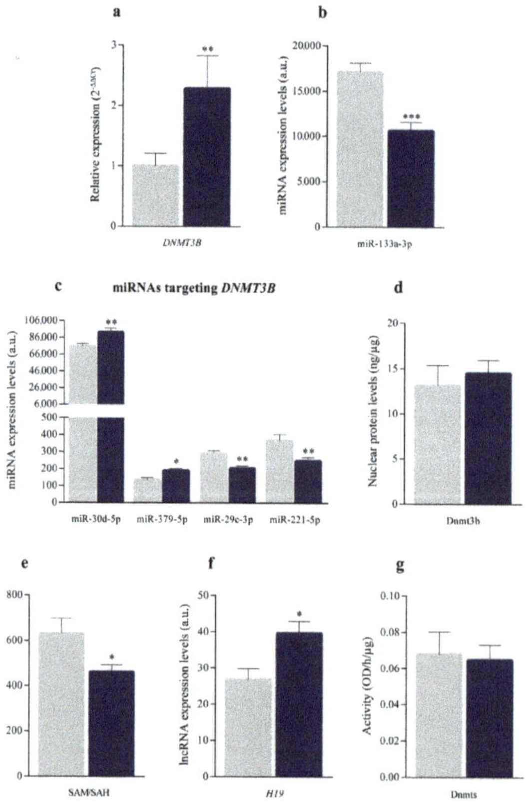

Figure 2. DNMT3B system dysregulation. (**a**) *DNMT3B* mRNA levels in ischaemic hearts ($n = 14$) and controls ($n = 8$), using RT-qPCR. (**b**) miR-133a-3p expression levels in ischaemic hearts ($n = 22$) and controls ($n = 8$), using HiSeq 2500 platform. (**c**) miRNAs expression levels in ischaemic hearts ($n = 22$) and controls (n = 8), using HiSeq 2500 platform. (**d**) Dnmt3b nuclear protein levels in ischaemic hearts ($n = 8$) and controls ($n = 8$), using ELISA. (**e**) SAM/SAH ratio in ischaemic hearts ($n = 30$) and controls ($n = 6$), using ELISA. (**f**) lncRNA *H19* expression level in ischaemic hearts ($n = 22$) and controls ($n = 8$), using the Illumina HiSeq 2500 platform. (**g**) Dnmts nuclear activity in ischaemic hearts ($n = 8$) and controls ($n = 8$), using ELISA. Data are presented as the mean ± SEM. a.u., arbitrary units. Ischaemic cardiomyopathy patients (blue), controls subjects (grey). * $p < 0.05$, ** $p < 0.01$, *** $p < 0.001$.

On the other hand, we analyzed different molecules that regulate the activity of Dnmts to determine whether their levels could favor the inhibition of Dnmts and, consequently, the global hypomethylation in ICM patients. Lower levels of SAM and higher levels of SAH are related to reduce methylation capacity, which is represented by the SAM/SAH ratio [25]. We measured SAM and SAH levels in human heart samples through ELISA analyses. The SAM/SAH ratio (Figure 2e) was reduced by 37% in ICM patients ($p < 0.05$). Specifically, we obtained a SAM/SAH ratio of 631.45 ± 163.26 in the CNT group and 461.55 ± 173.72 in ICM patients. Moreover, as previously shown, *AHCY* was underexpressed in ICM patients (Figure 1b). This gene encodes the S-adenosyl-L-homocysteine hydrolase (SAHH) protein, which is responsible for degrading SAH. Furthermore, through the sequencing of ncRNAs, we detected the levels of lncRNA *H19* (Figure 2f), which was overexpressed in ICM patients (1.68 fold and $p < 0.01$). *H19* is an SAHH inhibitor that favors the accumulation of SAH in the cells. In addition, the expression of *H19* correlated with the ejection fraction (r = 0.590 and $p < 0.01$). Next, we analysed Dnmts activity, observing absence of changes in ICM patients compared to control individuals (Figure 2g).

3.4. State of Global DNA Methylation

We analyzed global DNA methylation levels in ICM patients and CNT individuals. The β value and the *p* value were calculated for each detected CpG site, allowing the identification of 643 differentially methylated CpGs sites. We represented each CpG with a statistically significant value using a heat map (Figure 3a), which identified the ICM and CNT groups in two different methylation patterns. Specifically, the 82.9% of the CpGs sites were hypomethylated in ICM patients (Figure 3b). Furthermore, we analyzed the global methylation mean between both groups (ICM, mean β value = 0.481 ± 0.164 and CNT, mean β value = 0.569 ± 0.171; $p < 0.0001$ (Figure 3c)). These results showed genome-wide hypomethylation in ICM patients.

Next, we analyzed the position in the genome of the differentially methylated CpGs (Table 2). The 26.1% sites were distributed in the promoter proximal regions (TSS1500, TSS200 and 5´UTR), 44.3% in gene body and first exons, and 29.6% in 3´UTR and intergenic regions. In relation to CpG content and neighborhood context, the 54% of CpGs were present in open sea; the 27.2% in shores and shelves of CpG sites; and the 18.8% in CpGs sites. Although there is an evident greater number of a hypomethylated CpGs sites in ICM patients, we observed that the percentage of methylation is similar for hypermethylated and hypomethylated CpGs sites in each region, except for absence of differentially hypermethylated CpGs sites in 3´UTR regions and the marked reduction in the percentage of hypermethylation in the context of the TSS200 regions and islands.

Table 2. Comparison of statistically significant CpG site methylation ($\Delta\beta > 0.1$ and $p < 0.05$) between ischaemic cardiomyopathy patients and control individuals.

Functional Genomic Distribution						
	All CpG Sites		**Hypermethylated CpG Sites**		**Hypomethylated CpG Sites**	
	CpGs	%	CpGs	%	CpGs	%
TSS1500	67	10.4	11	10.0	56	10.5
TSS200	47	7.3	3	2.7	44	8.3
5'UTR	54	8.4	9	8.2	45	8.4
1stExon	26	4.0	5	4.5	21	3.9
Gene body	259	44.3	49	44.5	210	39.4
3'UTR	16	2.5	0	0.0	16	3.0
Intergenic	174	27.1	33	30.0	141	26.5
	643		110		533	

Table 2. *Cont.*

CpG Content and Neighbourhood Context						
	All CpG Sites		**Hypermethylated CpG Sites**		**Hypomethylated CpG Sites**	
	CpGs	%	CpGs	%	CpGs	%
North Shelf	28	4.3	8	7.3	20	3.8
Surth Shelf	19	3.0	4	3.6	15	2.8
North Shore	75	11.7	17	15.5	58	10.9
Surth Shore	53	8.2	8	7.3	45	8.4
Island	121	18.8	10	9.1	111	20.8
Open sea	347	54.0	63	57.3	284	53.3
	643		110		533	

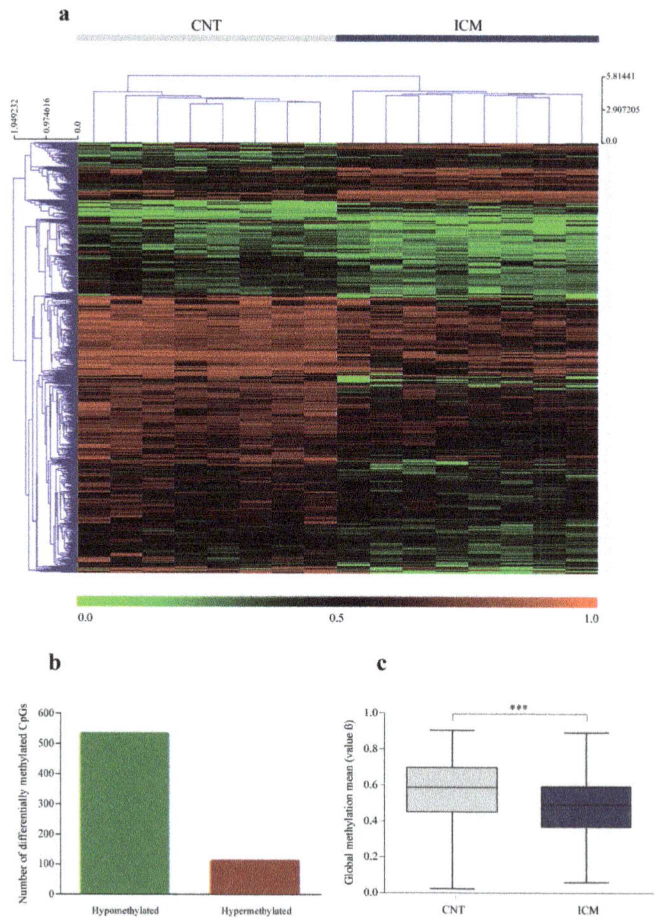

Figure 3. Global CpGs hypomethylation in ischaemic cardiomyopathy patients. (**a**) Heat map representing global methylation in ischaemic hearts ($n = 8$) and controls ($n = 8$) for each statistically significant CpGs. The results are represented as individual β values (1 = hypermethylated (red) and 0 = hypomethylated (green)). (**b**) Bar graph illustrating the number of hypermethylated and hypomethylated CpGs in ischaemic hearts. (**c**) Box plot graph with global methylation mean. Data are presented as the mean ± SEM. *** $p < 0.0001$. The results were obtained by the 850 K Infinium MethylationEPIC BeadChip platform.

4. Discussion

Gene expression is regulated by pathological mechanisms involving DNA methylation and ncRNAs function. Both processes are closely related, but incomplete knowledge of these epigenetic alterations exists in ICM [26]. Moreover, controversial results have been published regarding DNA methylation in ICM. Specifically, increased expression of methyl group addition molecules contrasts with the DNA hypomethylation observed [16,18]. Thus, this study focused on the analysis of the state of the DNA methylation machinery in these patients. Our results showed an alteration of the DNA methylation machinery in ICM patients; specifically, we found a dysregulation in addition, elimination and maintenance of methyl groups accordingly with the genome-wide hypomethylation observed.

Regarding the mechanisms of elimination of methyl groups, we observed the overexpression of *TET1* and *SMUG1* genes, both molecules favoring DNA hypomethylation [27]. *TET1* plays a major role in the initiation of the active DNA demethylation process [28,29]. Furthermore, *OGT* has been related to the post-transcriptional regulation of *TET1* in animal models [30]. Loss of Ogt in early fetal cardiomyocytes leads to multiple heart developmental defects, including hypertrabeculation, biventricular dilation, atrial septal defects, ventricular septal defects and defects in coronary vessel development [31]. In this study, we observed overexpression of the *OGT* gene. Ogt adds residues of O-GlcNAc to Tet1, thereby stimulating its activity [32]. However, the impact of Ogt on the Tet family in human cells is slightly controversial and requires additional studies [33]. In addition, we observed overexpression of the *GADD45G* gene in ICM. This gene belongs to the *GADD45* gene family, which participates in the regulation of the active demethylation process, and its expression is stimulated by cellular stress [34]. Moreover, Lucas et al. [35] demonstrated that Gadd45g is up-regulated in murine cardiomyocytes subjected to simulated ischemia, and that it promotes the development and persistence of HF by inducing cardiomyocyte apoptosis. These results indicate a dysregulation in the process of elimination of methyl groups, which could favor the demethylation of DNA.

Furthermore, the *MBD2* and *UHRF1* genes were underexpressed in this study. These genes are responsible for reading the DNA methylation pattern and maintaining the methylated state associated with transcriptional repression [12]. Genetic alterations in *MBD2* and *UHRF1* have been linked to various cardiovascular pathologies such as atherosclerosis or arterial aneurism, which can trigger HF [36,37].

On the other hand, we observed overexpression of the *DNMT3B* gene in ICM patients. This gene encodes a methyltransferase responsible for adding methyl groups to DNA, thus promoting the methylated state. Previously, *DNMT3B* overexpression has been reported in the context of HF [16], which could be favored by hypoxic conditions [38]. *DNMT3B* is a molecule subject to significant regulation. At the transcriptional level, different miRNAs regulators of its expression have been described in human and mouse cells [39–42]. Specifically, the miR-133a-3p binds to the *DNMT3B* promoter, reducing its transcription in cardiac cells [43], finding in our results the underexpression of miR-133a-3p in ICM patients. Furthermore, we observed alterations in several *DNMT3B* target miRNAs, which can induce dysregulation in the levels of mRNA, as well as at the protein level. Next, we analysed Dnmt3b nuclear protein levels. Our results did not show differences at the protein level between patients with ICM and control individuals. Yu et al. [44] have described in murine models of breast cancer that the ubiquitin E3 ligase Traf6 interacts with Dnmts, catalyzing their ubiquitination and inducing protein degradation. *TRAF6* mRNA levels were increased in ICM patients, being able to favor lysosomal degradation of Dnmt3b.

Moreover, the enzymatic activity of Dnmt3b is regulated by different molecules. Specifically, the SAH metabolite acts as an inhibitor of various methyltransferases, including Dnmt3b, thereby preventing the addition of methyl groups to DNA [45,46]. The accumulation of SAH in cells is caused by the SAM-dependent methylation, whose degradation product is SAH, and by the inactivation of SAHH, a molecule involved in the breakdown of SAH into adenosine and homocysteine. The results obtained in this study showed that the SAM/SAH ratio was reduced in ICM patients. SAH is a competitive inhibitor of Dnmts;

therefore, the increase in SAH, together with the decrease in SAM, favors the binding of this molecule to one of its main targets. In addition, the reduction in SAM/SAH ratio has been linked to the development of vascular diseases and to DNA hypomethylation of the genome [47]. These results reinforce the existence of a generalized DNA hypomethylation in ICM patients. Furthermore, we observed the underexpression of *AHCY*, gene encoding SAHH and the overexpression of lncRNA *H19* in ICM patients. *H19* acts as an SAHH inhibitor, thus preventing SAH degradation [48]. However, the Dnmts activity was similar in ICM patients and control individuals. In recent years, the role of *H19* in the development and prognosis of HF has been extensively studied [49,50]. Several authors described altered *H19* expression in cardiovascular diseases and related it with hypertrophic growth inhibition in cardiomyocytes and alleviation of myocardial ischemia reperfusion injury [51–53]. We further observed that its expression related with ejection fraction, the central measure of left ventricular function, as previously has been described [52]. These results represent a promising possibility to identify new therapeutic strategies, such as *H19*-targeted delivery for clinical ICM therapy.

To date, alterations in the methylation pattern have been identified in ICM patients in specific regions of the genome [16]. Moreover, the analysis of genome-wide DNA methylation in human ICM was carried out, observing differential methylation in the CpG sites associated with gene promoters [54], as well as a greater hypomethylation in regions associated with genes that encode proteins in these patients [18]. We also analyzed the status of the CpG sites as a whole in ICM patients compared to control individuals, and our results showed a higher degree of hypomethylated CpGs in each of the genomic regions. In addition, the results obtained in the present study indicate an increase in the regulatory mechanisms that promote DNA demethylation. Consistently, our assessment of the global DNA methylation pattern revealed that explanted hearts from individuals with ICM presented genome-wide hypomethylation. A limitation of this study includes the use of cardiac tissue samples from patients with end-stage HF, who exhibit intrinsic variability that exists between individuals. The patients who participated in this study were also treated pharmacologically, and some therapies may have influenced the results. However, our study population was etiologically homogeneous and all individuals had been receiving medical treatment according to the guidelines of the European Society of Cardiology [20].

5. Conclusions

Our analysis revealed genome-wide hypomethylation concordant with dysregulation in the mechanisms of addition, elimination and maintenance of methyl groups in DNA observed in ischaemic hearts. We describe relevant alterations in the DNMT3B system, which promote a normal Dnmt3b function despite its upregulation.

Supplementary Materials: The following supporting information can be downloaded at: https://www.mdpi.com/article/10.3390/biomedicines10040866/s1, Table S1: Genes involved in DNA methylation process in human heart.

Author Contributions: Conceptualization, M.P. and E.R.-L.; methodology, E.T. and L.P.-C.; validation, I.G.-E. and M.G.-M.; formal analysis, M.P.; investigation, E.T.; resources, L.M.-D.; writing—original draft preparation, E.T., L.P.-C., M.P. and E.R.-L.; writing—review and editing, I.G.-E., M.G.-M. and L.M.-D.; supervision, E.R.-L.; funding acquisition, E.T., L.M.-D. and E.R.-L. All authors have read and agreed to the published version of the manuscript.

Funding: This research was funded by the National Institute of Health "Fondo de Investigaciones Sanitarias del Instituto de Salud Carlos III" [PI20/01469, PI20/00071, CP18/00145, CP21/00041, FI21/00034, FI21/00186], "Consorcio Centro de Investigación Biomédica en Red, M.P." [CIBERCV, under Grant CB16/11/00261], and co-funded by European Union (ERDF, "A way to make Europe") and European Social Fund (ESF, "The ESF invests in your future").

Institutional Review Board Statement: The study was conducted according to the guidelines of the Declaration of Helsinki, and approved by the Ethics Committee (Biomedical Investigation Ethics Committee of La Fe University Hospital of Valencia, Spain; protocol code 2016/0320, 15 November 2016).

Informed Consent Statement: All patients submitted a signed informed consent prior to their inclusion in the study, this consent included permission to extract and store their samples, as well as the publication of the results obtained from the research work.

Data Availability Statement: The mRNA-seq data discussed in this publication have been deposited in NCBI's Gene Expression Omnibus and are accessible through GEO Series accession number GSE55296 (http://www.ncbi.nlm.nih.gov/geo/query/acc.cgi?acc=GSE55296, accessed on 11 March 2022).

Acknowledgments: The authors are grateful to Javier Moreno and Rosana Llobell for their technical support.

Conflicts of Interest: The authors declare no conflict of interest.

References

1. Briceno, N.; Schuster, A.; Lumley, M.; Perera, D. Ischaemic cardiomyopathy: Pathophysiology, assessment and the role of revascularisation. *Heart* **2016**, *102*, 397–406. [CrossRef]
2. Padula, S.L.; Yutzey, K.E. Epigenetic Regulation of Heart Failure: Cell Type Matters. *Circ. Res.* **2021**, *129*, 414–416. [CrossRef]
3. Di Salvo, T.G.; Haldar, S.M. Epigenetic mechanisms in heart failure pathogenesis. *Circ. Heart Fail.* **2014**, *7*, 850–863. [CrossRef]
4. Papait, R.; Greco, C.; Kunderfranco, P.; Latronico, M.V.; Condorelli, G. Epigenetics: A new mechanism of regulation of heart failure? *Basic Res. Cardiol.* **2013**, *108*, 361. [CrossRef]
5. Kimball, T.H.; Vondriska, T.M. Metabolism, Epigenetics, and Causal Inference in Heart Failure. *Trends Endocrinol. Metab.* **2020**, *31*, 181–191. [CrossRef]
6. Papait, R.; Serio, S.; Condorelli, G. Role of the Epigenome in Heart Failure. *Physiol. Rev.* **2020**, *100*, 1753–1777. [CrossRef]
7. Ortega, A.; Tarazon, E.; Gil-Cayuela, C.; Martinez-Dolz, L.; Lago, F.; Gonzalez-Juanatey, J.R.; Sandoval, J.; Portoles, M.; Rosello-Lleti, E.; Rivera, M. ASB1 differential methylation in ischaemic cardiomyopathy: Relationship with left ventricular performance in end-stage heart failure patients. *ESC Heart Fail.* **2018**, *5*, 732–737. [CrossRef]
8. Napoli, C.; Bontempo, P.; Palmieri, V.; Coscioni, E.; Maiello, C.; Donatelli, F.; Benincasa, G. Epigenetic Therapies for Heart Failure: Current Insights and Future Potential. *Vasc. Health Risk Manag.* **2021**, *17*, 247–254. [CrossRef]
9. Napoli, C.; Benincasa, G.; Donatelli, F.; Ambrosio, G. Precision medicine in distinct heart failure phenotypes: Focus on clinical epigenetics. *Am. Heart J.* **2020**, *224*, 113–128. [CrossRef]
10. Soler-Botija, C.; Galvez-Monton, C.; Bayes-Genis, A. Epigenetic Biomarkers in Cardiovascular Diseases. *Front. Genet.* **2019**, *10*, 950. [CrossRef]
11. Bain, C.R.; Ziemann, M.; Kaspi, A.; Khan, A.W.; Taylor, R.; Trahair, H.; Khurana, I.; Kaipananickal, H.; Wallace, S.; El-Osta, A.; et al. DNA methylation patterns from peripheral blood separate coronary artery disease patients with and without heart failure. *ESC Heart Fail.* **2020**, *7*, 2468–2478. [CrossRef] [PubMed]
12. Moore, L.D.; Le, T.; Fan, G. DNA methylation and its basic function. *Neuropsychopharmacology* **2013**, *38*, 23–38. [CrossRef] [PubMed]
13. Wu, H.; Zhang, Y. Reversing DNA Methylation: Mechanisms, Genomics, and Biological Functions. *Cell* **2014**, *156*, 45–68. [CrossRef]
14. Li, B.Z.; Huang, Z.; Cui, Q.Y.; Song, X.H.; Du, L.; Jeltsch, A.; Chen, P.; Li, G.; Li, E.; Xu, G.L. Histone tails regulate DNA methylation by allosterically activating de novo methyltransferase. *Cell Res.* **2011**, *21*, 1172–1181. [CrossRef] [PubMed]
15. Sinkkonen, L.; Hugenschmidt, T.; Berninger, P.; Gaidatzis, D.; Mohn, F.; Artus-Revel, C.G.; Zavolan, M.; Svoboda, P.; Filipowicz, W. MicroRNAs control de novo DNA methylation through regulation of transcriptional repressors in mouse embryonic stem cells. *Nat. Struct. Mol. Biol.* **2008**, *15*, 259–267. [CrossRef]
16. Pepin, M.E.; Drakos, S.; Ha, C.M.; Tristani-Firouzi, M.; Selzman, C.H.; Fang, J.C.; Wende, A.R.; Wever-Pinzon, O. DNA methylation reprograms cardiac metabolic gene expression in end-stage human heart failure. *Am. J. Physiol. Heart Circ. Physiol.* **2019**, *317*, H674–H684. [CrossRef]
17. Vujic, A.; Robinson, E.L.; Ito, M.; Haider, S.; Ackers-Johnson, M.; See, K.; Methner, C.; Figg, N.; Brien, P.; Roderick, H.L.; et al. Experimental heart failure modelled by the cardiomyocyte-specific loss of an epigenome modifier, DNMT3B. *J. Mol. Cell. Cardiol.* **2015**, *82*, 174–183. [CrossRef]
18. Glezeva, N.; Moran, B.; Collier, P.; Moravec, C.S.; Phelan, D.; Donnellan, E.; Russell-Hallinan, A.; O'Connor, D.P.; Gallagher, W.M.; Gallagher, J.; et al. Targeted DNA Methylation Profiling of Human Cardiac Tissue Reveals Novel Epigenetic Traits and Gene Deregulation across Different Heart Failure Patient Subtypes. *Circ. Heart Fail.* **2019**, *12*, e005765. [CrossRef]
19. Baccarelli, A.; Wright, R.; Bollati, V.; Litonjua, A.; Zanobetti, A.; Tarantini, L.; Sparrow, D.; Vokonas, P.; Schwartz, J. Ischemic heart disease and stroke in relation to blood DNA methylation. *Epidemiology* **2010**, *21*, 819–828. [CrossRef]

20. Ponikowski, P.; Voors, A.A.; Anker, S.D.; Bueno, H.; Cleland, J.G.; Coats, A.J.; Falk, V.; Gonzalez-Juanatey, J.R.; Harjola, V.P.; Jankowska, E.A.; et al. 2016 ESC Guidelines for the diagnosis and treatment of acute and chronic heart failure: The Task Force for the diagnosis and treatment of acute and chronic heart failure of the European Society of Cardiology (ESC), Developed with the special contribution of the Heart Failure Association (HFA) of the ESC. *Eur. J. Heart Fail.* **2016**, *18*, 891–975.
21. Macrae, D.J. The Council for International Organizations and Medical Sciences (CIOMS) guidelines on ethics of clinical trials. *Proc. Am. Thorac. Soc.* **2007**, *4*, 176–179. [CrossRef]
22. Rosello-Lleti, E.; Carnicer, R.; Tarazon, E.; Ortega, A.; Gil-Cayuela, C.; Lago, F.; Gonzalez-Juanatey, J.R.; Portoles, M.; Rivera, M. Human Ischemic Cardiomyopathy Shows Cardiac Nos1 Translocation and its Increased Levels are Related to Left Ventricular Performance. *Sci. Rep.* **2016**, *6*, 24060. [CrossRef] [PubMed]
23. Livak, K.J.; Schmittgen, T.D. Analysis of relative gene expression data using real-time quantitative PCR and the $2^{-\Delta\Delta CT}$ Method. *Methods* **2001**, *25*, 402–408. [CrossRef] [PubMed]
24. Tarazon, E.; Portoles, M.; Rosello-Lleti, E. Protocol for Isolation of Golgi Vesicles from Human and Animal Hearts by Flotation through a Discontinuous Sucrose Gradient. *STAR Protoc.* **2020**, *1*, 100100. [CrossRef]
25. Gao, J.; Cahill, C.M.; Huang, X.; Roffman, J.L.; Lamon-Fava, S.; Fava, M.; Mischoulon, D.; Rogers, J.T. S-Adenosyl Methionine and Transmethylation Pathways in Neuropsychiatric Diseases Throughout Life. *Neurotherapeutics* **2018**, *15*, 156–175. [CrossRef] [PubMed]
26. Handy, D.E.; Castro, R.; Loscalzo, J. Epigenetic modifications: Basic mechanisms and role in cardiovascular disease. *Circulation* **2011**, *123*, 2145–2156. [CrossRef] [PubMed]
27. Bhutani, N.; Burns, D.M.; Blau, H.M. DNA demethylation dynamics. *Cell* **2011**, *146*, 866–872. [CrossRef] [PubMed]
28. Weber, A.R.; Krawczyk, C.; Robertson, A.B.; Kusnierczyk, A.; Vagbo, C.B.; Schuermann, D.; Klungland, A.; Schar, P. Biochemical reconstitution of TET1-TDG-BER-dependent active DNA demethylation reveals a highly coordinated mechanism. *Nat. Commun.* **2016**, *7*, 10806. [CrossRef]
29. Bachman, M.; Uribe-Lewis, S.; Yang, X.; Williams, M.; Murrell, A.; Balasubramanian, S. 5-Hydroxymethylcytosine is a predominantly stable DNA modification. *Nat. Chem.* **2014**, *6*, 1049–1055. [CrossRef]
30. Hrit, J.; Goodrich, L.; Li, C.; Wang, B.A.; Nie, J.; Cui, X.; Martin, E.A.; Simental, E.; Fernandez, J.; Liu, M.Y.; et al. OGT binds a conserved C-terminal domain of TET1 to regulate TET1 activity and function in development. *Elife* **2018**, *7*, e34870. [CrossRef]
31. Mu, Y.; Yu, H.; Wu, T.; Zhang, J.; Evans, S.M.; Chen, J. O-linked beta-N-acetylglucosamine transferase plays an essential role in heart development through regulating angiopoietin-1. *PLoS Genet.* **2020**, *16*, e1008730. [CrossRef] [PubMed]
32. Shi, F.T.; Kim, H.; Lu, W.; He, Q.; Liu, D.; Goodell, M.A.; Wan, M.; Songyang, Z. Ten-eleven translocation 1 (Tet1) is regulated by O-linked N-acetylglucosamine transferase (Ogt) for target gene repression in mouse embryonic stem cells. *J. Biol. Chem.* **2013**, *288*, 20776–20784. [CrossRef] [PubMed]
33. Zhang, Q.; Liu, X.; Gao, W.; Li, P.; Hou, J.; Li, J.; Wong, J. Differential regulation of the ten-eleven translocation (TET) family of dioxygenases by O-linked beta-N-acetylglucosamine transferase (OGT). *J. Biol. Chem.* **2014**, *289*, 5986–5996. [CrossRef] [PubMed]
34. Schafer, A. Gadd45 proteins: Key players of repair-mediated DNA demethylation. *Adv. Exp. Med. Biol.* **2013**, *793*, 35–50.
35. Lucas, A.; Mialet-Perez, J.; Daviaud, D.; Parini, A.; Marber, M.S.; Sicard, P. Gadd45gamma regulates cardiomyocyte death and post-myocardial infarction left ventricular remodelling. *Cardiovasc. Res.* **2015**, *108*, 254–267. [CrossRef]
36. Rao, X.; Zhong, J.; Zhang, S.; Zhang, Y.; Yu, Q.; Yang, P.; Wang, M.H.; Fulton, D.J.; Shi, H.; Dong, Z.; et al. Loss of methyl-CpG-binding domain protein 2 enhances endothelial angiogenesis and protects mice against hind-limb ischemic injury. *Circulation* **2011**, *123*, 2964–2974. [CrossRef]
37. Elia, L.; Kunderfranco, P.; Carullo, P.; Vacchiano, M.; Farina, F.M.; Hall, I.F.; Mantero, S.; Panico, C.; Papait, R.; Condorelli, G.; et al. UHRF1 epigenetically orchestrates smooth muscle cell plasticity in arterial disease. *J. Clin. Investig.* **2018**, *128*, 2473–2486. [CrossRef]
38. Watson, C.J.; Collier, P.; Tea, I.; Neary, R.; Watson, J.A.; Robinson, C.; Phelan, D.; Ledwidge, M.T.; McDonald, K.M.; McCann, A.; et al. Hypoxia-induced epigenetic modifications are associated with cardiac tissue fibrosis and the development of a myofibroblast-like phenotype. *Hum. Mol. Genet.* **2014**, *23*, 2176–2188. [CrossRef]
39. Shiah, S.G.; Hsiao, J.R.; Chang, H.J.; Hsu, Y.M.; Wu, G.H.; Peng, H.Y.; Chou, S.T.; Kuo, C.C.; Chang, J.Y. MiR-30a and miR-379 modulate retinoic acid pathway by targeting DNA methyltransferase 3B in oral cancer. *J. Biomed. Sci.* **2020**, *27*, 46. [CrossRef]
40. Wang, J.Y.; Cheng, H.; Zhang, H.Y.; Ye, Y.Q.; Feng, Q.; Chen, Z.M.; Zheng, Y.L.; Wu, Z.G.; Wang, B.; Yao, J. Suppressing microRNA-29c promotes biliary atresia-related fibrosis by targeting DNMT3A and DNMT3B. *Cell. Mol. Biol. Lett.* **2019**, *24*, 10. [CrossRef]
41. Roscigno, G.; Quintavalle, C.; Donnarumma, E.; Puoti, I.; Diaz-Lagares, A.; Iaboni, M.; Fiore, D.; Russo, V.; Todaro, M.; Romano, G.; et al. MiR-221 promotes stemness of breast cancer cells by targeting DNMT3b. *Oncotarget* **2016**, *7*, 580–592. [CrossRef] [PubMed]
42. Gao, L.; He, R.Q.; Wu, H.Y.; Zhang, T.T.; Liang, H.W.; Ye, Z.H.; Li, Z.Y.; Xie, T.T.; Shi, Q.; Ma, J.; et al. Expression Signature and Role of miR-30d-5p in Non-Small Cell Lung Cancer: A Comprehensive Study Based on in Silico Analysis of Public Databases and in Vitro Experiments. *Cell. Physiol. Biochem.* **2018**, *50*, 1964–1987. [CrossRef] [PubMed]
43. Di Mauro, V.; Crasto, S.; Colombo, F.S.; Di Pasquale, E.; Catalucci, D. Wnt signalling mediates miR-133a nuclear re-localization for the transcriptional control of Dnmt3b in cardiac cells. *Sci. Rep.* **2019**, *9*, 9320. [CrossRef] [PubMed]

44. Yu, J.; Qin, B.; Moyer, A.M.; Nowsheen, S.; Liu, T.; Qin, S.; Zhuang, Y.; Liu, D.; Lu, S.W.; Kalari, K.R.; et al. DNA methyltransferase expression in triple-negative breast cancer predicts sensitivity to decitabine. *J. Clin. Investig.* **2018**, *128*, 2376–2388. [CrossRef] [PubMed]
45. James, S.J.; Melnyk, S.; Pogribna, M.; Pogribny, I.P.; Caudill, M.A. Elevation in S-adenosylhomocysteine and DNA hypomethylation: Potential epigenetic mechanism for homocysteine-related pathology. *J. Nutr.* **2002**, *132*, 2361S–2366S. [CrossRef] [PubMed]
46. Tehlivets, O.; Malanovic, N.; Visram, M.; Pavkov-Keller, T.; Keller, W. S-adenosyl-L-homocysteine hydrolase and methylation disorders: Yeast as a model system. *Biochim. Biophys. Acta* **2013**, *1832*, 204–215. [CrossRef]
47. Castro, R.; Rivera, I.; Struys, E.A.; Jansen, E.E.; Ravasco, P.; Camilo, M.E.; Blom, H.J.; Jakobs, C.; Tavares de Almeida, I. Increased homocysteine and S-adenosylhomocysteine concentrations and DNA hypomethylation in vascular disease. *Clin. Chem.* **2003**, *49*, 1292–1296. [CrossRef]
48. Zhou, J.; Yang, L.; Zhong, T.; Mueller, M.; Men, Y.; Zhang, N.; Xie, J.; Giang, K.; Chung, H.; Sun, X.; et al. H19 lncRNA alters DNA methylation genome wide by regulating S-adenosylhomocysteine hydrolase. *Nat. Commun.* **2015**, *6*, 10221. [CrossRef]
49. Hobuss, L.; Foinquinos, A.; Jung, M.; Kenneweg, F.; Xiao, K.; Wang, Y.; Zimmer, K.; Remke, J.; Just, A.; Nowak, J.; et al. Pleiotropic cardiac functions controlled by ischemia-induced lncRNA H19. *J. Mol. Cell. Cardiol.* **2020**, *146*, 43–59. [CrossRef]
50. Omura, J.; Habbout, K.; Shimauchi, T.; Wu, W.H.; Breuils-Bonnet, S.; Tremblay, E.; Martineau, S.; Nadeau, V.; Gagnon, K.; Mazoyer, F.; et al. Identification of Long Noncoding RNA H19 as a New Biomarker and Therapeutic Target in Right Ventricular Failure in Pulmonary Arterial Hypertension. *Circulation* **2020**, *142*, 1464–1484. [CrossRef]
51. Liu, L.; An, X.; Li, Z.; Song, Y.; Li, L.; Zuo, S.; Liu, N.; Yang, G.; Wang, H.; Cheng, X.; et al. The H19 long noncoding RNA is a novel negative regulator of cardiomyocyte hypertrophy. *Cardiovasc. Res.* **2016**, *111*, 56–65. [CrossRef] [PubMed]
52. Wang, H.; Lian, X.; Gao, W.; Gu, J.; Shi, H.; Ma, Y.; Li, Y.; Fan, Y.; Wang, Q.; Wang, L. Long noncoding RNA H19 suppresses cardiac hypertrophy through the MicroRNA-145-3p/SMAD4 axis. *Bioengineered* **2022**, *13*, 3826–3839. [CrossRef] [PubMed]
53. Li, X.; Luo, S.; Zhang, J.; Yuan, Y.; Jiang, W.; Zhu, H.; Ding, X.; Zhan, L.; Wu, H.; Xie, Y.; et al. lncRNA H19 Alleviated Myocardial I/RI via Suppressing miR-877-3p/Bcl-2-Mediated Mitochondrial Apoptosis. *Mol. Ther. Nucleic Acids* **2019**, *17*, 297–309. [CrossRef] [PubMed]
54. Li, B.; Feng, Z.H.; Sun, H.; Zhao, Z.H.; Yang, S.B.; Yang, P. The blood genome-wide DNA methylation analysis reveals novel epigenetic changes in human heart failure. *Eur. Rev. Med. Pharmacol. Sci.* **2017**, *21*, 1828–1836. [CrossRef] [PubMed]

biomedicines

MDPI

Review

Cardiac Acetylation in Metabolic Diseases

Emilie Dubois-Deruy *, Yara El Masri, Annie Turkieh, Philippe Amouyel, Florence Pinet *,† and Jean-Sébastien Annicotte *,†

Univ. Lille, Inserm, CHU Lille, Institut Pasteur de Lille, U1167-RID-AGE-Facteurs de Risque et Déterminants Moléculaires des Maladies Liées au Vieillissement, F-59000 Lille, France; yara.elmasri@pasteur-lille.fr (Y.E.M.); ani.turkieh@pasteur-lille.fr (A.T.); philippe.amouyel@pasteur-lille.fr (P.A.)

* Correspondence: emilie.deruy@pasteur-lille.fr (E.D.-D.); florence.pinet@pasteur-lille.fr (F.P.); jean-sebastien.annicotte@inserm.fr (J.-S.A.)

† These authors contributed equally to this work.

Abstract: Lysine acetylation is a highly conserved mechanism that affects several biological processes such as cell growth, metabolism, enzymatic activity, subcellular localization of proteins, gene transcription or chromatin structure. This post-translational modification, mainly regulated by lysine acetyltransferase (KAT) and lysine deacetylase (KDAC) enzymes, can occur on histone or non-histone proteins. Several studies have demonstrated that dysregulated acetylation is involved in cardiac dysfunction, associated with metabolic disorder or heart failure. Since the prevalence of obesity, type 2 diabetes or heart failure rises and represents a major cause of cardiovascular morbidity and mortality worldwide, cardiac acetylation may constitute a crucial pathway that could contribute to disease development. In this review, we summarize the mechanisms involved in the regulation of cardiac acetylation and its roles in physiological conditions. In addition, we highlight the effects of cardiac acetylation in physiopathology, with a focus on obesity, type 2 diabetes and heart failure. This review sheds light on the major role of acetylation in cardiovascular diseases and emphasizes KATs and KDACs as potential therapeutic targets for heart failure.

Keywords: acetylation; heart; obesity; diabetes; heart failure; enzymes

Citation: Dubois-Deruy, E.; El Masri, Y.; Turkieh, A.; Amouyel, P.; Pinet, F.; Annicotte, J.-S. Cardiac Acetylation in Metabolic Diseases. *Biomedicines* **2022**, *10*, 1834. https://doi.org/10.3390/biomedicines10081834

Academic Editors: Tânia Martins-Marques, Gonçalo F. Coutinho and Attila Kiss

Received: 11 July 2022
Accepted: 28 July 2022
Published: 29 July 2022

Publisher's Note: MDPI stays neutral with regard to jurisdictional claims in published maps and institutional affiliations.

Copyright: © 2022 by the authors. Licensee MDPI, Basel, Switzerland. This article is an open access article distributed under the terms and conditions of the Creative Commons Attribution (CC BY) license (https://creativecommons.org/licenses/by/4.0/).

1. Introduction

Obesity is a worldwide epidemic associated with several public health challenges including type 2 diabetes, hypertension, obstructive sleep apnea, dyslipidemia and cardiovascular disease, and notably heart failure. The prevalence of these disorders (e.g., obesity, type 2 diabetes and heart failure) is still increasing and heart failure remains the most common cause of cardiovascular morbidity and mortality in the world [1]. Interestingly, although heart failure could provoke systolic or diastolic cardiac dysfunction, obesity and type 2 diabetes are more often characterized by cardiac hypertrophy and diastolic dysfunction [2]. Moreover, diastolic heart failure remains one of the more challenging diseases to treat and, thus, a better understanding of the physio(patho)logical mechanisms involved in metabolic heart disease is necessary to propose new pharmacological therapies. In this context, epigenetic or protein post-translational modifications, including phosphorylation, methylation, O-GlcNAcylation, ubiquitylation, and acetylation, could represent interesting targets since they modulate gene or protein expression during metabolic diseases [3].

Protein acetylation is a highly conserved mechanism that was first described 60 years ago, in which an acetyl group is covalently attached to the ε-amino group of lysine residues [4]. This modification induces important changes to the protein structure at its lysine residue, by altering its charge status and adding an extra structural moiety [5]. These post-translational modifications can regulate protein–protein interactions, stability and function. Acetylation was first described to regulate chromatin structure and histone activity, associated to epigenetic-mediated gene regulation [4]. Since then, several proteomic analyses enabled to identify more than a thousand acetylation sites in both histone and non-histone proteins [6–8], and, interestingly, revealed that the subcellular distribution of acetylation is tissue dependent [9]. Focusing on cardiovascular and metabolic diseases, Table 1 indicates some of the most common proteins modified by cardiac acetylation.

Table 1. Most common proteins modified by cardiac acetylation in cardiovascular and metabolic diseases.

Class	Name	KAT	KDAC	Function of Acetylation in the Heart	References
Mitochondrial proteins	LCAD/SCAD	GCN5L1	SIRT3	Increased activity and modulated fatty acid oxidation	[5,10–13]
	β-HAD	GCN5L1	SIRT3	Increased activity and modulated fatty acid oxidation	[5,12,14]
	OPA1	Unknown	SIRT3	Decreases its activity	[15]
	PGC1α	Unknown	SIRT1	Increases its expression	[16]
	Cyclophilin D	GCN5L1	SIRT3	Induces mPTP opening	[17]
Transcription factors	TBX5	KAT2A, KAT2B	HDAC4 HDAC5	Increases transcriptional activity	[18,19]
	VGLL4	P300	Unknown	Negatively regulates its binding to TEAD1	[20]
	GATA4	P300	SIRT7	Activates its DNA binding activity	[21,22]
	MEF2A	P300 KAT2B	HDAC5	Increased hypertrophy	[23,24]
	MEF2C	KAT2B	HDAC5	Increased hypertrophy	[24,25]
Anti-oxidant proteins	SOD2	Unknown	SIRT3	Decreases SOD2 activity	[26]
				Increased mitochondrial oxidative stress and hypertrophy	[27]
	Prx1	Unknown	HDAC6	Increased peroxide-reduction activity	[28]
	Nrf2	Unknown	SIRT1	Decreases its activity	[29]
	eNOS	Unknown	SIRT1	Inactive form	[30]
Contractile proteins	β-MHC	Unknown	HDAC6	Impact myosin head positioning	[31,32]
	Titin	Unknown	HDAC6	Cardiac contraction	[33]
	CapZβ1	Unknown	HDAC3/6	Cardiac contraction	[34]
	TnI	Unknown	HDAC6	Cardiac contraction	[35]
Signaling pathway	LKB1	Unknown	SIRT2 SIRT3	Induces its activation by phosphorylation	[36,37]
	Akt	Unknown	SIRT1	Inhibition of Akt phosphorylation and activation	[38]
	SMAD2	KAT2B	SIRT1	Induced fibrosis	[39,40]
	SMAD3	Unknown	SIRT1	Induced fibrosis	[40–42]

KAT: lysine acetyltransferase; KDAC: lysine deacetylase; LCAD: long chain acyl CoA dehydrogenase; SCAD: short chain acyl CoA dehydrogenase; β-HAD: L-3-hydroxy acyl-CoA dehydrogenase; OPA1: optic atrophic 1; PGC-1α: peroxisome proliferator-activated receptor-gamma coactivator; TBX5: T-Box transcription factor 5; VGLL4: vestigial-like 4; TEAD1: TEA Domain Transcription Factor 1; GATA4: GATA-binding factor 4; MEF: myocyte enhancer factor; SOD2: superoxide dismutase 2; Prx1: peroxiredoxin 1; Nrf2: nuclear factor erythroid-2-related factor 2; eNOS: endothelial nitric oxide synthase; β-MHC: beta-myosin heavy chain; TnI: troponin I; LKB1: liver kinase B1; GCN5L1: general control of amino acid synthesis 5 like-1; SIRT: sirtuins; HDAC: histone deacetylase; mPTP: mitochondrial permeability transition pore.

2. Cardiac Acetylation

2.1. Regulation of Lysine Acetylation

Lysine acetylation is a reversible mechanism that is regulated by the dynamic actions of lysine acetyltransferase (KATs) and lysine deacetylase (KDACs) enzymes [43]. There are three categories of KATs: the MYST family, the Gcn-5-related N-acetyltransferases (GNATs) and the E1A-associated protein of 300 kDa/CREB binding protein (p300/CBP) family [44]. KDACs are broadly categorized into four classes based on function and DNA sequence similarity. Class I (reduced potassium dependency 3 family), II (histone deacetylase 1 family) and IV (HDAC11) are considered as classical KDACs with zinc-dependent active sites, whereas class III enzymes are a family of silent information regulator 2-like nicotinamide adenine dinucleotide (NAD^+)-dependent deacetylases/mono-ADP-ribosyl transferases known as sirtuins 1 to 7 [44]. Interestingly, the level of expression of KATs and KDACs follows a specific tissue distribution, suggesting specific functions in these different organs [45]. For example, in physiological conditions, KATs are weakly expressed in vascular tissue, except for the lysine acetyltransferase KAT2B, which is highly expressed [45]. Conversely, a database mining approach analyzing the expression profile of 164 enzymes in human and murine tissues highlighted heart as tissue in which KATs and KDACs are in the highest varieties [45]. In addition to tissue-specific expression, both KATs and KDACs have precise subcellular localizations related to their specific targets (Figure 1) and can regulate each other. For example, cardiomyocyte overexpression of SIRT6 decreases P300 levels, where SIRT6 promotes ubiquitination and degradation of P300 [46]. Moreover, cardiomyocyte overexpression of SIRT3 increases SIRT1 expression [27].

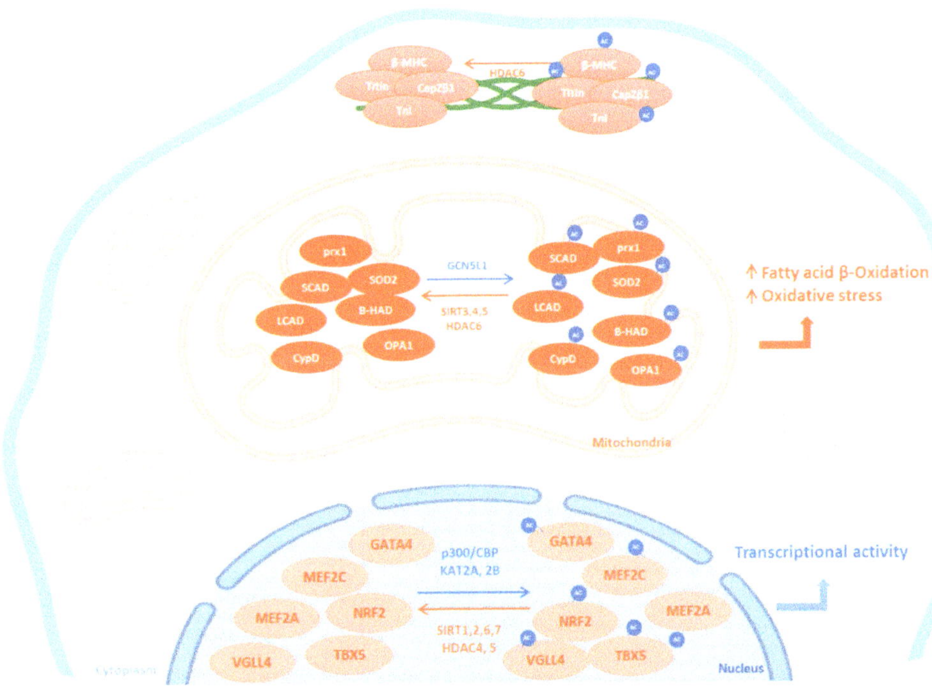

Figure 1. Subcellular localization of cardiac KATs and KDACs. This figure summarizes the subcellular localization of the most common lysine acetyltransferases (KATs, blue) and lysine deacetylases (KDAC, orange) and their cardiac targets. Ac: acetylated form; HDAC: histone deacetylase; SIRT: sirtuins; β-MHC: beta-myosin heavy chain; TnI: Troponin I; prx1: peroxiredoxin 1; LCAD: long chain acyl CoA dehydrogenase; SCAD: short chain acyl CoA dehydrogenase; β-HAD: L-3-hydroxy acyl-CoA dehydrogenase; SOD2: superoxide dismutase 2; CypD: Cyclophilin D; OPA1: optic atrophic 1; TBX5: T-Box transcription factor 5; VGLL4: vestigial-like 4; GATA4: GATA-binding factor 4; MEF: myocyte enhancer factor; Nrf2: nuclear factor erythroid-2-related factor 2; GCN5L1: general control of amino acid synthesis 5 like-1; CBP: CREB binding protein.

In addition, cofactors play an essential role to modulate acetylation. Several short-chain acyl-CoA products, such as acetyl-coenzyme A (acetyl-CoA), are important metabolite intermediates generated during catabolism of various energy fuels that modulate cardiac acetylation. Indeed, acetyl-CoA acts as a substrate for KATs. Moreover, auto-acetylation of mitochondrial proteins can occur efficiently at the pH of the matrix (pH $\geq$ 7.5) in actively respiring mitochondria [9]. On the other hand, levels of NAD^+, an important cofactor for sirtuins, may impact protein acetylation [5]. Indeed, NAD^+ levels, synthesized from metabolic pathways, connect their enzymatic activity to metabolism and provide a metabolic link between the sirtuins activity, energy homeostasis and stress responses.

2.2. *Physiological Roles of Lysine Acetylation*

Lysine acetylation can modulate cell growth, metabolism, protein–protein interactions, protein stability, subcellular localization, gene transcription, chromatin structure or enzymatic activity [6,7]. This post-translational modification plays an essential role in cardiac physiology (Figure 2).

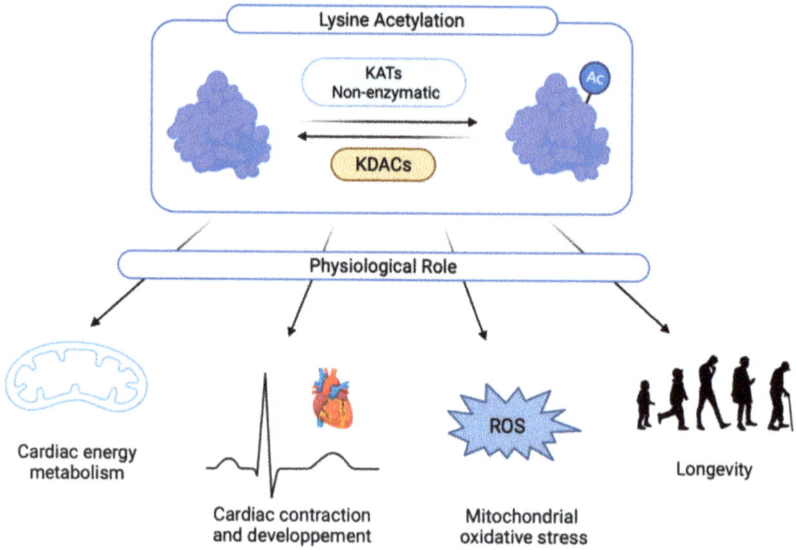

Figure 2. Physiological roles of cardiac acetylation. Ac: acetylated form; KATs: lysine acetyltransferases; KDACs: lysine deacetylase; ROS: reactive oxygen species.

2.2.1. Heart Development and Cardiac Ageing

Acetylation can regulate embryonic heart development and cardiac progenitor fate [47,48]. TBX5, a T-box family transcription factor, is involved in heart and forelimb development and is acetylated by KAT2A and KAT2B at lysine 339, leading to stimulate its transcriptional activity [18]. Knockdown of *KAT2A* and *KAT2B* in zebrafish severely impairs heart development and induces pericardial oedema by inhibiting TBX5 [18]. Besides KATs, the lysine deacetylase enzymes HDAC4 and HDAC5 are also able to interact with TBX5 to exert a repressive role on cardiac genes transcription through deacetylation [19]. Furthermore, TBX1, another T-box family transcription factor expressed in cardiac progenitors, represses myocyte enhancer factor 2c *(Mef2c)* gene expression by reducing histone 3 acetylation on lysine 27 [48]. Moreover, acetylation of vestigial-like 4 (VGLL4) at K225 by P300 negatively regulates its binding to the transcription factor TEA Domain Transcription Factor 1 (TEAD1), leading to decreased neonatal cardiomyocytes proliferation and cardiomyocytes necrosis [20]. Inhibition of HDAC1 was also described to decrease the proliferation cardiomyocyte in zebrafish [49]. Indeed, mutation in *HDAC1* gene that induces protein instability is associated with decreased cardiomyocyte proliferation, suggesting an important role of HDAC1 during heart growth [49].

In addition, sirtuins are well-described regulators of aging and have been associated with longevity [50–52]. Indeed, mice deficient for *SIRT1* [53,54], *SIRT3* [55] or *SIRT6* [56] display a shorter life span with severe cardiac damage, such as hypertrophy and fibrosis.

SIRT1 is well known to be cardioprotective by decreasing oxidative stress and inflammation, which promotes cardiomyocyte survival [54]. Moreover, SIRT3 is required for cardiomyocyte survival under several stress conditions such as serum starvation, genotoxic and oxidative stress [57]. Altogether, these studies demonstrate that KDACs and KATs exert crucial functions during heart development and cardiac ageing.

2.2.2. Cardiac Contraction

Recently, reversible acetylation of sarcomeric proteins has been described as a mechanism regulating cardiac function. The comparison of lysine acetylation patterns from rats as well as from human skeletal muscle biopsies revealed that 80% of the proteins involved in muscle contraction were acetylated [9]. Moreover, proteomic analyses suggested that HDAC6 is localized in Z-disks and acts as a sarcomeric protein desacetylase [58]. Among them, acetylation can impact the β-myosin heavy chain (lysine 34, lysine 58, lysine 213, lysine 429, lysine 951 and lysine 1195) [31], titin [33], CapZβ1 (lysine 199) [34] or cardiac troponin I [35], directly affecting cardiac contraction.

2.2.3. Role of Lysine Acetylation in Cardiac Energy Metabolism and Mitochondrial Activity

Proteomic analyses suggested that the subcellular distribution of lysine-acetylated proteins is tissue dependent [9]. For example, the heart and muscles, both high energy-demanding organs, are tissues with the largest fraction of mitochondrial protein acetylation [9]. Calorie restriction or changes in nutrition specifically affect the mitochondrial acetylome, but not the cytosol or nucleus [59], suggesting different roles for lysine acetylation in the mitochondria, nucleus and cytoplasm. Indeed, mitochondrial lysine acetylation was described to modulate cell metabolism by regulating fatty acid β-oxidation, the tricarboxylic acid cycle, the urea cycle, and oxidative phosphorylation [60].

In the early newborn period, an important change in myocardial energy substrate metabolism occurs with an increase in fatty acid β-oxidation [47], associated with an increase in long chain acyl CoA dehydrogenase (LCAD) and L-3-hydroxy acyl-CoA dehydrogenase (β-HAD) acetylation and activation [12]. At the cardiac level, mitochondrial general control of amino acid synthesis 5 like-1 (GCN5L1) triggers LCAD and β-HAD acetylation [10,11] whereas SIRT3 deacetylates both LCAD and β-HAD [13,14]. The implication of GCN5L1 in cardiac metabolism was confirmed by genetic deletion approaches. Indeed, knockdown of *GCN5L1* in H9c2 cardiomyoblasts decreases maximal mitochondrial respiration and activity of proteins involved in fatty acid oxidation [12].

However, there is no consensus about the effects of lysine acetylation on fatty acid β-oxidation. Indeed, some studies suggested an inhibitory effect [43] whereas others suggested a stimulatory effect [13,14]. As an example, it was reported that an impaired fatty acid β-oxidation was linked to a reduced acetylation and activity of LCAD in the liver [13] or heart [14] of SIRT3 knockout mice. On the other hand, cardiac mitochondrial protein hyperacetylation induced by *SIRT3* deletion is associated with an increased fatty acid β-oxidation rates [43]. This discrepancy could be due to organ, physio(patho)logical status (e.g., diabetes, obesity or newborn period) or acetylated proteins themselves, and are well described in the following review [43].

The implication of sirtuins in metabolic regulation is well detailed in a recent review [61]. Indeed, SIRT6 was described as a gatekeeper of glucose metabolism in cardiomyocytes. SIRT6 partial depletion decreased mitochondrial respiration whereas SIRT6 overexpression enhanced basal oxygen consumption [62]. Conversely, fatty acid uptake increases in *SIRT6*-deficient cardiomyocytes and is decreased in SIRT6-overexpressed cardiomyocytes [63]. Of note, proteins involved in mitochondrial biogenesis can also be acetylated, such as optic atrophy 1 (OPA1) through SIRT3 which is able to activate OPA1 by deacetylation of the residues lysine 926 and lysine 931 [15]. Finally, a cross-talk between SIRT1, peroxisome proliferator-activated receptor gamma coactivator 1 alpha (PGC-1α) and AMP-activated protein kinase (AMPK) has been described to regulate cardiac metabolism [54]. Indeed,

PGC-1α is a transcription factor playing a central role in the regulation of cellular energy metabolism, mitochondrial biogenesis and oxidative phosphorylation. SIRT1 could interact with PGC-1α and increase its expression levels [16]. Moreover, AMPK increases cellular NAD^+ levels and subsequently increases SIRT1 activity, resulting in PGC-1α deacetylation and activation [64].

2.2.4. Mitochondrial Oxidative Stress

It is well described that several mitochondrial proteins involved in oxidative stress are acetylated in the heart [60]. Oxidative stress is defined by a production of reactive oxygen species (ROS) higher than anti-oxidant capacities. Indeed, although different molecular processes may contribute to global oxidative stress, the majority of ROS originates from the mitochondrial compartment in the heart. Excessive ROS production occurs during mitochondrial dysfunction and induces irreversible damage to mitochondria, defining them as significant contributors to the development of cardiovascular disease [65,66]. In this context, we recently described that SIRT3 deacetylates and activates the mitochondrial superoxide dismutase 2 (SOD2), but there is a decreased interaction between SIRT3 and SOD2 induced SOD2 acetylation on lysine 68 and its subsequent inactivation, leading to mitochondrial oxidative stress and dysfunction in hypertrophied cardiomyocytes [27]. Moreover, SIRT3 inhibition increased oxidative stress in neonatal rat cardiomyocytes [27] or human aortic endothelial cells [26].

Other anti-oxidant enzymes were described to be acetylated in the heart, such as peroxiredoxin 1 [28] in the mitochondria or nuclear factor erythroid 2-related factor 2 [29] in the nucleus.

3. Implication of Cardiac Acetylation in Metabolic Heart Disease

Post-translational acetylation triggers modification of the activity of several proteins that occurs in obesity, diabetes and early stage of heart failure as detailed below [67,68]. Table 2 summarizes the cardiac modulation of KATs and KDACs expression levels reported in metabolic heart disease.

Table 2. Modulation of KATs and KDACs expression in heart failure and metabolic diseases.

KATs			KDACs		
Name	**Heart Failure**	**Metabolic Diseases**	**Name**	**Heart Failure**	**Metabolic Diseases**
P300	Increase [46,69]	Unknown	SIRT1	Decrease [70,71]	Decrease [30,38,72]
KAT2B	Increase [23,25,73]	Unknown	SIRT2	Decrease [37]	Decrease [74,75]
GCN5L1	Unknown	Increase in pre-diabetes [11,76]	SIRT3	Decrease [77]	Decrease [17,78,79]
		Decrease in diabetes [10]	SIRT6	Decrease [46,63]	Decrease [63]
			SIRT7	Increase [21]	Unknown
			HDAC3	Increase [80]	Increase [81]
			HDAC6	Unknown	Increase [28]

KATs: lysine acetyltransferases; KDACs: lysine deacetylases; SIRT: sirtuins; HDAC: histone deacetylase; GCN5L1: general control of amino acid synthesis 5 like-1.

3.1. *Cardiac Hypertrophy*

Cardiac hypertrophy is a consequence of genetic, mechanic or neurohormonal changes contributing to heart failure progression. Interestingly, total KAT activity was found to be increased in the hearts of mice subjected to phenylephrine (PE), an hypertrophic stimulus [23]. In this context, the increased expression of P300 and KAT2B acetylases may explain hypertrophied cardiomyocytes under PE [23,25,46,69]. In addition, PE-induced hypertrophy also decreases SIRT6 expression, leading to an increase in histone H3 acetylation on lysine 9 (H3K9) [25,46,69]. Interestingly, PE-induced hypertrophy is decreased by treatment with polyunsaturated fatty acid [69] or anacardic acid [23] in association with a decreased H3K9 acetylation. In parallel, overexpression of P300 induces an increased size of cardiomyocytes and a modification of their myofibrillar organization [69], whereas inhibition of P300 by siRNA attenuates hypertrophic response in neonatal rat cardiomyocytes [46]. On the other hand, overexpression of SIRT6 [46] or SIRT3 [27] decreased cardiomyocytes hypertrophy. As an example, NAD inhibits oxidative stress and hypertrophy induced by PE in cardiomyocytes or by angiotensin II in mice [36]. Indeed, NAD requires SIRT3 to deacetylate and activate liver kinase B1 (LKB1) and its target AMPK [36]. Furthermore, a decreased SIRT2 protein expression was reported in hypertrophic hearts from mice whereas cardiac-specific SIRT2 overexpression protected the hearts against angiotensin II–induced hypertrophy [37]. As shown for SIRT3, SIRT2 deacetylates LKB1 at lysine 48, these promote the phosphorylation of LKB1 and the subsequent activation of AMPK signaling [37]. Moreover, SIRT7 levels increase in myocardial tissues after pressure overload induced by transverse aortic constriction in mice [21]. Cardiomyocyte-specific deletion of *SIRT7* exacerbates hypertrophy and fibrosis induced by transverse aortic constriction, suggesting an antihypertrophic role of SIRT7 [21]. Indeed, SIRT7 deacetylates the hypertrophy response transcription factors such as GATA-binding factor 4 (GATA4) in cardiomyocytes [21]. Conversely, P300 acetylates GATA4 and activates its DNA binding activity, inducing cardiac hypertrophy and heart failure [22].

Finally, HDAC3 expression and activity increase during hypertrophy [80] whereas a decreased binding of HDAC3 to myofibrils is observed in PE-induced hypertrophy in neonatal rat cardiomyocytes that increases the acetylation of CapZβ1 at lysine 199 and alters myofibril growth during cardiac hypertrophy [34].

3.2. Cardiac Fibrosis

Cardiac fibrosis, defined as an excessive deposition of extracellular matrix in the cardiac muscle, is an important contributor to several heart diseases. Interestingly, in isoproterenol-injected mice, KAT2B activity was found to be increased only in cardiac fibroblasts, whereas no effects were observed in cardiomyocytes isolated from these mice [39]. In this model, KAT2B induces acetylation of SMAD2 and, subsequently, activates SMAD2 and the transforming growth factor β signaling pathway [39]. SMAD3 could also be acetylated during cardiac fibrosis [41,42]. Activation of SIRT1 by resveratrol [41], nicotinamide mononucleotide [42] or geniposide [70] could decrease SMAD3 acetylation and fibrosis. Moreover, inhibition of P300 appears to exert anti-fibrotic functions [82].

Decreased α-tubulin acetylation and increased HDAC6 were also described in cardiac fibroblasts or the heart from isoproterenol-induced fibrosis. Interestingly, treatment of cardiac fibroblasts with tubastatin A, an HDAC6 inhibitor, restored α-tubulin acetylation and decreased fibrosis [83].

3.3. Heart Failure

Cardiac hypertrophy and fibrosis are both involved in heart failure development by activation of several intracellular signaling pathways, leading to left ventricular remodeling with systolic dysfunction. Despite the best modern therapeutic management, left ventricular remodeling remains independently associated with heart failure and cardiovascular death at long-term follow-up after myocardial infarction [84]. Moreover, alterations in cardiac energy metabolism, both in terms of changes in energy substrate preference and decreased mitochondrial oxidative metabolism and ATP production, are key contributors to heart failure development [5].

An increase in P300 expression is observed during ischemia/reperfusion injury [85] and myocardial infarction [69]. Interestingly, it was described that activation of P300 acetylase, leading to an increase in acetylated H3K9 induced by myocardial infarction, is reversed by treatment with polyunsaturated fatty acid [69]. Indeed, eicosapentaenoic acid and docosahexaenoic acid directly blocks the histone acetyltransferase activity of P300 and subsequently reduces both hypertrophy and fibrosis induced by myocardial infarction in rats [69]. Other P300 inhibitors were also described to be cardioprotective, such as *Ecklonia stolonifera* Okamura extract [86], curcumin [87] and metformin [88]. Moreover, KAT2B levels are also increased by ischemia/reperfusion injury [73].

The expression and activity of sirtuins are also highly impacted during heart failure. As an example, a decrease in SIRT3 and SIRT6 expressions were observed in human failing hearts that induces a global increase in protein acetylation [63,77]. Proteomic analysis also identified an increased acetylation of mitochondrial proteins induced by transverse aortic constriction [89].

Inhibition of SIRT3 was also described to induce mitochondrial oxidative stress and hypertrophy, notably by increasing the inactive form of SOD2 (acetylated on lysine 68) in hypertrophy [27] or following ischemia/reperfusion [90]. Moreover, deletion of SIRT2 exacerbates cardiac hypertrophy and fibrosis and decreases cardiac ejection fraction and fractional shortening in angiotensin II-infused mice by inhibition of AMPK activation, whereas cardiac-specific SIRT2 overexpression reversed this phenotype [37]. SIRT1 expression and activity are decreased in the heart following ischemia/reperfusion [30]. SIRT1 inhibition was also reported to increase endoplasmic reticulum stress-induced cardiac injury by decreasing eukaryotic initiation factor 2 alpha (eIF2α) deacetylation on lysine 143 in cardiomyocytes and in adult-inducible SIRT1 knockout mice [91].

Acetylation of contractile proteins is also modified during heart failure. For example, acetylation of β-MHC on lysine 951 was decreased in both ischemic and non-ischemic failing hearts [31].

3.4. Obesity

Obesity is a major cause of disability and is often associated with cardiac hypertrophy, fibrosis, type 2 diabetes, obstructive sleep apnea and alteration of cardiac metabolism. It has notably described an increase in genes involved in fatty acid oxidation [11]. Moreover, several studies have suggested an increase in acetylation raised from the non-enzymatic reaction of high levels of acetyl-CoA generated during a high-fat diet (HFD) and obesity [5]. Furthermore, hyperacetylation of mitochondrial proteins and metabolic inflexibility was reported in response to HFD or obesity. SIRT3 expression is decreased in left ventricles of obese patients and is associated with an increased level of brain natriuretic peptide (BNP), a marker of cardiac dysfunction and of protein acetylation [17]. It was also reported that exposure of cardiomyoblasts H9c2 to palmitate led to an increase in both SIRT3 and GCN5L1 RNA levels [11]. At the metabolic level, HFD induced an increase in SCAD and LCAD acetylation and activity [11]. Moreover, the acetylation level of α-tubulin on lysine 40 is increased in the hearts of HFD mice and a pharmacological activation of α-tubulin acetylation decreases glucose transport [92].

In the other hand, cardiac specific inhibition of SIRT6 in mice exposed to HFD-induced cardiac hypertrophy and lipid accumulation [93]. In this context, SIRT6 activated the expression of endonuclease G and SOD2, that could decrease oxidative stress and hypertrophy [93].

3.5. Type 2 Diabetes

As obesity, type 2 diabetes is also associated with several cardiac dysfunctions such as cardiac hypertrophy and fibrosis, and acetylation may be involved in these mechanisms. Indeed, SIRT1 expression and activity are decreased in the heart of diabetic rats, induced by HFD and streptozotocin injection and, conversely, an up-regulation of SIRT1 by adenovirus attenuates cardiac dysfunction and oxidative stress [30]. Moreover, in a model of sucrose-fed rats, cardiac dysfunction is associated with a decreased SIRT3 protein expression (despite an increase at RNA level) and an increased GCN5L1 (protein and RNA) and mitochondrial protein acetylation [11,17]. Decreased SIRT3 levels are also associated with an increased acetylation of SOD2, and an increased oxidative stress and apoptosis in heart of diabetic mice [79]. Interestingly, exposure of cardiomyoblasts H9c2 to high glucose concentration decreased SIRT3 [78] and increased GCN5L1, oxidative stress and autophagy mediated by cytoplasmic Forkhead box O1 (FOXO1) acetylation [11,76]. Conversely, expression of GCN5L1 is decreased in hearts of diabetic ZSF1 rats, a model characterized by obese animals with hyperglycemia, hyperinsulinemia and cardiac dysfunction [10]. This decreased GCN5L1 expression and activity, with no modulation of SIRT3, induced a decrease of short- and long-chain acyl CoA dehydrogenases and a reduced respiratory capacity [10]. One explanation could be the transition from prediabetes, in which GCN5L1 becomes elevated to promote mitochondrial fatty acids oxidation, to an overt diabetic state, in which GCN5L1 expression is downregulated leading to an overall decrease in mitochondrial fuel oxidation [10].

On the other hand, OVE26 mice, a mouse model of type 1 diabetes, develop cardiac dysfunction with increased left ventricle diameters and fibrosis and decreased ejection fraction associated with an increase in HDAC3 activity, oxidative stress and inflammation [81]. Interestingly, the selective HDAC3 inhibitor RGFP966 reversed the cardiac phenotype in these mice [81]. In another model of type 1 diabetes, injection of streptozotocin in rats induced cardiac dysfunction, cardiac hypertrophy, fibrosis and inflammation [72] as well as oxidative stress [28]. This oxidative stress is due to an increased HDAC6 activity leading to a decrease in the acetylated form of peroxiredoxin 1, suggesting that acetylation of peroxiredoxin 1 is involved in this activation [28]. In this context, tubastatin A, a highly selective inhibitor of HADC6, may represent an interesting pharmacological target since treatment of diabetic rats with tubastatin A increased acetylation of peroxiredoxin 1 and decreased oxidative stress and cardiac dysfunction [28]. Conversely, SIRT1 expression and activity appear to be decreased in the left ventricles of streptozotocin-induced diabetic rats [72] and mice [38]. In these models, reduced SIRT1 expression is associated with acetylation of endothelial nitric oxide synthase [72] or Akt [38]. In addition, SIRT6 expression is also decreased in the hearts of diabetic mice [63], suggesting that sirtuin family members may play a crucial role in type 1 diabetes-associated cardiac dysfunctions.

3.6. *Pharmacological Modulation of Cardiac Acetylation*

Due to the major implication of KATs and KDACs in cardiac dysfunction induced by obesity, type 2 diabetes or heart failure, targeting these enzymes could be beneficial for patients.

First, inhibition of P300 appears to be a promising approach to inhibit cardiac hypertrophy and fibrosis induced by several stimuli. As an example, oral administration of eicosapentaenoic acid (1 g/kg) or docosahexaenoic acid (1 g/kg) one week after myocardial infarction preserved fractional shortening and decreased cardiac hypertrophy and perivascular fibrosis [68]. *Ecklonia stolonifera* Okamura extract, an algae traditionally used in Japanese foods, inhibits PE-induced hypertrophy in neonatal rat cardiomyocytes and restores fractional shortening and reduced cardiac hypertrophy and fibrosis by oral administration one week after myocardial infarction in rats [73]. Oral administration of curcumin (50 mg/kg/day) acts also as a P300 inhibitor and decreases posterior wall thickness, cardiac hypertrophy and perivascular fibrosis induced by hypertension [74]. Finally, metformin, that directly inhibits P300-mediated acetylation of H3K9, blocks PE-induced cardiomyocytes hypertrophy [75].

Based on the cardioprotective effects of SIRT3, it seems essential to develop new therapeutic strategy to increase its expression or activity. Recent reviews have well described these molecules [94,95]. For example, exogenous NAD is able to inhibit hypertrophy induced by PE in cardiomyocytes or by angiotensin II in mice [36]. Resveratrol, an activator of both SIRT1 and SIRT3, is also described to be cardioprotective [41].

4. Conclusions

Despite available therapies, cardiovascular diseases, and particularly ischemic diseases, still remain the first cause of mortality and morbidity in the world. Obesity and type 2 diabetes are some of the major risk factors of cardiovascular diseases. Acetylation is an essential mechanism involved in several processes contributing to cardiac diseases such as cell metabolism, gene transcription or enzymatic activity. In this context, targeting enzymes responsible of acetylation/deacetylation could be beneficial and is a very promising research area, as demonstrated by the development of inhibitor targeting P300 [82], modulators of SIRT6 [96] or agonists of SIRT3 [94,95] to treat cardiac dysfunctions.

Author Contributions: Writing—original draft preparation, E.D.-D., Y.E.M., A.T., F.P. and J.-S.A., funding acquisition, P.A., F.P. and J.-S.A. All authors have read and agreed to the published version of the manuscript.

Funding: E.D.D. was supported by grants from region Hauts de France, CPER "Longévité" of Institut Pasteur de Lille and "Fondation Lefoulon Delalande". F.P. and E.D.D. received a grant by the French Government, managed by the National Research Agency (ANR) under the program "Investissements d'avenir" with the reference ANR-16-RHUS-0003. Y.E.M., E.D.D. and F.P. are members of FHU-CARNAVAL. This work was supported by Programme Hubert Curien AMADEUS (France-Autriche), ANR grant BETAPLASTICITY (ANR-17-CE14-0034 to J-S.A.), INSERM, CNRS, Institut Pasteur de Lille (CPER CTRL Melodie), Université de Lille, Fondation pour la Recherche Médicale (EQU202103012732), Conseil Régional Hauts de France and Métropole Européenne de Lille and Société Francophone du Diabète.

Institutional Review Board Statement: Not applicable.

Informed Consent Statement: Not applicable.

Data Availability Statement: Data sharing not applicable.

Conflicts of Interest: The authors declare no conflict of interest.

References

1. Virani, S.S.; Alonso, A.; Aparicio, H.J.; Benjamin, E.J.; Bittencourt, M.S.; Callaway, C.W.; Carson, A.P.; Chamberlain, A.M.; Cheng, S.; Delling, F.N.; et al. Heart Disease and Stroke Statistics—2021 Update. *Circulation* **2021**, *143*, E254–E743. [CrossRef]
2. Nakamura, M.; Sadoshima, J. Cardiomyopathy in obesity, insulin resistance and diabetes. *J. Physiol.* **2020**, *598*, 2977–2993. [CrossRef]
3. Schwenk, R.W.; Vogel, H.; Schürmann, A. Genetic and epigenetic control of metabolic health. *Mol. Metab.* **2013**, *2*, 337–347. [CrossRef] [PubMed]
4. Allfrey, V.G.; Faulkner, R.; Mirsky, A.E. Acetylation and methylation of histones and their possible role in the regulation of RNA synthesis. *Proc. Natl. Acad. Sci. USA* **1964**, *51*, 786–794. [CrossRef] [PubMed]
5. Ketema, E.B.; Lopaschuk, G.D. Post-translational Acetylation Control of Cardiac Energy Metabolism. *Front. Cardiovasc. Med.* **2021**, *8*, 723996. [CrossRef] [PubMed]
6. Lundby, A.; Lage, K.; Weinert, B.T.; Bekker-Jensen, D.B.; Secher, A.; Skovgaard, T.; Kelstrup, C.D.; Dmytriyev, A.; Choudhary, C.; Lundby, C.; et al. Proteomic Analysis of Lysine Acetylation Sites in Rat Tissues Reveals Organ Specificity and Subcellular Patterns. *Cell Rep.* **2012**, *2*, 419–431. [CrossRef]
7. Kim, S.C.; Sprung, R.; Chen, Y.; Xu, Y.; Ball, H.; Pei, J.; Cheng, T.; Kho, Y.; Xiao, H.; Xiao, L.; et al. Substrate and Functional Diversity of Lysine Acetylation Revealed by a Proteomics Survey. *Mol. Cell* **2006**, *23*, 607–618. [CrossRef] [PubMed]
8. Choudhary, C.; Kumar, C.; Gnad, F.; Nielsen, M.L.; Rehman, M.; Walther, T.C.; Olsen, J.V.; Mann, M. Lysine acetylation targets protein complexes and co-regulates major cellular functions. *Science* **2009**, *325*, 834–840. [CrossRef]
9. Hosp, F.; Lassowskat, I.; Santoro, V.; De Vleesschauwer, D.; Fliegner, D.; Redestig, H.; Mann, M.; Christian, S.; Hannah, M.A.; Finkemeier, I. Lysine acetylation in mitochondria: From inventory to function. *Mitochondrion* **2017**, *33*, 58–71. [CrossRef]
10. Thapa, D.; Zhang, M.; Manning, J.R.; Guimarães, D.A.; Stoner, M.W.; Lai, Y.; Shiva, S.; Scott, I. Loss of GCN5L1 in cardiac cells limits mitochondrial respiratory capacity under hyperglycemic conditions. *Physiol. Rep.* **2019**, *7*, e14054. [CrossRef]
11. Thapa, D.; Zhang, M.; Manning, J.R.; Guimarães, D.A.; Stoner, M.W.; O'Doherty, R.M.; Shiva, S.; Scott, I. Acetylation of mitochondrial proteins by GCN5L1 promotes enhanced fatty acid oxidation in the heart. *Am. J. Physiol. Circ. Physiol.* **2017**, *313*, H265–H274. [CrossRef]
12. Fukushima, A.; Alrob, O.A.; Zhang, L.; Wagg, C.S.; Altamimi, T.; Rawat, S.; Rebeyka, I.M.; Kantor, P.F.; Lopaschuk, G.D. Acetylation and succinylation contribute to maturational alterations in energy metabolism in the newborn heart. *Am. J. Physiol. Circ. Physiol.* **2016**, *311*, H347–H363. [CrossRef] [PubMed]
13. Hirschey, M.D.; Shimazu, T.; Goetzman, E.; Jing, E.; Schwer, B.; Lombard, D.B.; Grueter, C.A.; Harris, C.; Biddinger, S.; Ilkayeva, O.R.; et al. SIRT3 regulates mitochondrial fatty-acid oxidation by reversible enzyme deacetylation. *Nature* **2010**, *464*, 121–125. [CrossRef]
14. Alrob, O.A.; Sankaralingam, S.; Ma, C.; Wagg, C.S.; Fillmore, N.; Jaswal, J.S.; Sack, M.N.; Lehner, R.; Gupta, M.P.; Michelakis, E.D.; et al. Obesity-induced lysine acetylation increases cardiac fatty acid oxidation and impairs insulin signalling. *Cardiovasc. Res.* **2014**, *103*, 485–497. [CrossRef]
15. Samant, S.A.; Zhang, H.J.; Hong, Z.; Pillai, V.B.; Sundaresan, N.R.; Wolfgeher, D.; Archer, S.L.; Chan, D.C.; Gupta, M.P. SIRT3 Deacetylates and Activates OPA1 To Regulate Mitochondrial Dynamics during Stress. *Mol. Cell. Biol.* **2014**, *34*, 807–819. [CrossRef]
16. Aquilano, K.; Vigilanza, P.; Baldelli, S.; Pagliei, B.; Rotilio, G.; Ciriolo, M.R. Peroxisome proliferator-activated receptor γ coactivator 1 α (PGC-1α) and sirtuin 1 (SIRT1) reside in mitochondria: Possible direct function in mitochondrial biogenesis. *J. Biol. Chem.* **2010**, *285*, 21590–21599. [CrossRef] [PubMed]

17. García-Rivas, G.; Castillo, E.C.; Morales, J.A.; Chapoy-Villanueva, H.; Silva-Platas, C.; Treviño-Saldaña, N.; Guerrero-Beltrán, C.E.; Bernal-Ramírez, J.; Torres-Quintanilla, A.; García, N.; et al. Mitochondrial Hyperacetylation in the Failing Hearts of Obese Patients Mediated Partly by a Reduction in SIRT3: The Involvement of the Mitochondrial Permeability Transition Pore. *Cell. Physiol. Biochem.* **2019**, *53*, 465–479.
18. Ghosh, T.K.; Aparicio-Sánchez, J.J.; Buxton, S.; Ketley, A.; Mohamed, T.; Rutland, C.S.; Loughna, S.; Brook, J.D. Acetylation of TBX5 by KAT2B and KAT2A regulates heart and limb development. *J. Mol. Cell. Cardiol.* **2018**, *114*, 185–198. [CrossRef]
19. Ghosh, T.K.; Aparicio-Sánchez, J.J.; Buxton, S.; Brook, J.D. HDAC4 and 5 repression of TBX5 is relieved by protein kinase D1. *Sci. Rep.* **2019**, *9*, 17992. [CrossRef]
20. Lin, Z.; Guo, H.; Cao, Y.; Zohrabian, S.; Zhou, P.; Ma, Q.; VanDusen, N.; Guo, Y.; Zhang, J.; Stevens, S.M.; et al. Acetylation of VGLL4 Regulates Hippo-YAP Signaling and Postnatal Cardiac Growth. *Dev. Cell* **2016**, *39*, 466–479. [CrossRef]
21. Yamamura, S.; Izumiya, Y.; Araki, S.; Nakamura, T.; Kimura, Y.; Hanatani, S.; Yamada, T.; Ishida, T.; Yamamoto, M.; Onoue, Y.; et al. Cardiomyocyte Sirt (Sirtuin) 7 Ameliorates Stress-Induced Cardiac Hypertrophy by Interacting with and Deacetylating GATA4. *Hypertension* **2020**, *75*, 98–108. [CrossRef]
22. Shimizu, S.; Sunagawa, Y.; Hajika, N.; Yorimitsu, N.; Katanasaka, Y.; Funamoto, M.; Miyazaki, Y.; Sari, N.; Shimizu, K.; Hasegawa, K.; et al. Multimerization of the GATA4 transcription factor regulates transcriptional activity and cardiomyocyte hypertrophic response. *Int. J. Biol. Sci.* **2022**, *18*, 1079–1095. [CrossRef]
23. Peng, C.; Luo, X.; Li, S.; Sun, H. Phenylephrine-induced cardiac hypertrophy is attenuated by a histone acetylase inhibitor anacardic acid in mice. *Mol. Biosyst.* **2017**, *13*, 714–724. [CrossRef]
24. Zhang, H.; Shao, Z.; Alibin, C.P.; Acosta, C.; Anderson, H.D. Liganded Peroxisome Proliferator-Activated Receptors (PPARs) Preserve Nuclear Histone Deacetylase 5 Levels in Endothelin-Treated Sprague-Dawley Rat Cardiac Myocytes. *PLoS ONE* **2014**, *9*, e115258. [CrossRef]
25. Mao, Q.; Wu, S.; Peng, C.; Peng, B.; Luo, X.; Huang, L.; Zhang, H. Interactions between the ERK1/2 signaling pathway and PCAF play a key role in PE-induced cardiomyocyte hypertrophy. *Mol. Med. Rep.* **2021**, *24*, 636. [CrossRef]
26. Winnik, S.; Gaul, D.S.; Siciliani, G.; Lohmann, C.; Pasterk, L.; Calatayud, N.; Weber, J.; Eriksson, U.; Auwerx, J.; van Tits, L.J.; et al. Mild endothelial dysfunction in Sirt3 knockout mice fed a high-cholesterol diet: Protective role of a novel C/EBP-β-dependent feedback regulation of SOD2. *Basic Res. Cardiol.* **2016**, *111*, 33. [CrossRef]
27. Peugnet, V.; Chwastyniak, M.; Mulder, P.; Lancel, S.; Bultot, L.; Fourny, N.; Renguet, E.; Bugger, H.; Beseme, O.; Loyens, A.; et al. Mitochondrial-Targeted Therapies Require Mitophagy to Prevent Oxidative Stress Induced by SOD2 Inactivation in Hypertrophied Cardiomyocytes. *Antioxidants* **2022**, *11*, 723. [CrossRef]
28. Leng, Y.; Wu, Y.; Lei, S.; Zhou, B.; Qiu, Z.; Wang, K.; Xia, Z. Inhibition of HDAC6 Activity Alleviates Myocardial Ischemia/Reperfusion Injury in Diabetic Rats: Potential Role of Peroxiredoxin 1 Acetylation and Redox Regulation. *Oxid. Med. Cell. Longev.* **2018**, *2018*, 9494052. [CrossRef]
29. Xu, J.-J.; Cui, J.; Lin, Q.; Chen, X.-Y.; Zhang, J.; Gao, E.-H.; Wei, B.; Zhao, W. Protection of the enhanced Nrf2 deacetylation and its downstream transcriptional activity by SIRT1 in myocardial ischemia/reperfusion injury. *Int. J. Cardiol.* **2021**, *342*, 82–93. [CrossRef]
30. Ding, M.; Lei, J.; Han, H.; Li, W.; Qu, Y.; Fu, E.; Fu, F.; Wang, X. SIRT1 protects against myocardial ischemia–reperfusion injury via activating eNOS in diabetic rats. *Cardiovasc. Diabetol.* **2015**, *14*, 143. [CrossRef]
31. Landim-Vieira, M.; Childers, M.C.; Wacker, A.L.; Garcia, M.R.; He, H.; Singh, R.; Brundage, E.A.; Johnston, J.R.; Whitson, B.A.; Chase, P.B.; et al. Post-translational modification patterns on β-myosin heavy chain are altered in ischemic and nonischemic human hearts. *eLife* **2022**, *11*, e74919. [CrossRef]
32. Samant, S.A.; Pillai, V.B.; Sundaresan, N.R.; Shroff, S.G.; Gupta, M.P. Histone Deacetylase 3 (HDAC3)-dependent Reversible Lysine Acetylation of Cardiac Myosin Heavy Chain Isoforms Modulates Their Enzymatic and Motor Activity. *J. Biol. Chem.* **2015**, *290*, 15559–15569. [CrossRef]
33. Loescher, C.M.; Hobbach, A.J.; Linke, W.A. Titin (TTN): From molecule to modifications, mechanics, and medical significance. *Cardiovasc. Res.* **2021**. *Online ahead of print*. [CrossRef]
34. Lin, Y.-H.; Warren, C.M.; Li, J.; McKinsey, T.A.; Russell, B. Myofibril growth during cardiac hypertrophy is regulated through dual phosphorylation and acetylation of the actin capping protein CapZ. *Cell. Signal.* **2016**, *28*, 1015–1024. [CrossRef]
35. Lin, Y.H.; Schmidt, W.; Fritz, K.S.; Jeong, M.Y.; Cammarato, A.; Foster, D.B.; Biesiadecki, B.J.; McKinsey, T.A.; Woulfe, K.C. Site-specific acetyl-mimetic modification of cardiac troponin I modulates myofilament relaxation and calcium sensitivity. *J. Mol. Cell. Cardiol.* **2020**, *139*, 135–147. [CrossRef]
36. Pillai, V.B.; Sundaresan, N.R.; Kim, G.; Gupta, M.; Rajamohan, S.B.; Pillai, J.B.; Samant, S.; Ravindra, P.V.; Isbatan, A.; Gupta, M.P. Exogenous NAD Blocks Cardiac Hypertrophic Response via Activation of the SIRT3-LKB1-AMP-activated Kinase Pathway. *J. Biol. Chem.* **2010**, *285*, 3133–3144. [CrossRef]
37. Tang, X.; Chen, X.-F.; Wang, N.-Y.; Wang, X.-M.; Liang, S.-T.; Zheng, W.; Lu, Y.-B.; Zhao, X.; Hao, D.-L.; Zhang, Z.-Q.; et al. SIRT2 Acts as a Cardioprotective Deacetylase in Pathological Cardiac Hypertrophy. *Circulation* **2017**, *136*, 2051–2067. [CrossRef] [PubMed]
38. Ding, M.; Hu, L.; Yang, H.; Gao, C.; Zeng, K.; Yu, M.; Feng, J.; Qiu, J.; Liu, C.; Fu, F.; et al. Reduction of SIRT1 blunts the protective effects of ischemic post-conditioning in diabetic mice by impairing the Akt signaling pathway. *Biochim. Biophys. Acta Mol. Basis Dis.* **2019**, *1865*, 1677–1689. [CrossRef] [PubMed]

39. Lim, Y.; Jeong, A.; Kwon, D.-H.; Lee, Y.-U.; Kim, Y.-K.; Ahn, Y.; Kook, T.; Park, W.-J.; Kook, H. P300/CBP-Associated Factor Activates Cardiac Fibroblasts by SMAD2 Acetylation. *Int. J. Mol. Sci.* **2021**, *22*, 9944. [CrossRef] [PubMed]
40. Bugyei-Twum, A.; Ford, C.; Civitarese, R.; Seegobin, J.; Advani, S.L.; Desjardins, J.-F.; Kabir, G.; Zhang, Y.; Mitchell, M.; Switzer, J.; et al. Sirtuin 1 activation attenuates cardiac fibrosis in a rodent pressure overload model by modifying Smad2/3 transactivation. *Cardiovasc. Res.* **2018**, *114*, 1629–1641. [CrossRef] [PubMed]
41. Chen, Q.; Zeng, Y.; Yang, X.; Wu, Y.; Zhang, S.; Huang, S.; Zhong, Y.; Chen, M. Resveratrol ameliorates myocardial fibrosis by regulating Sirt1/Smad3 deacetylation pathway in rat model with dilated cardiomyopathy. *BMC Cardiovasc. Disord.* **2022**, *22*, 17. [CrossRef] [PubMed]
42. Wu, K.; Li, B.; Lin, Q.; Xu, W.; Zuo, W.; Li, J.; Liu, N.; Tu, T.; Zhang, B.; Xiao, Y.; et al. Nicotinamide mononucleotide attenuates isoproterenol-induced cardiac fibrosis by regulating oxidative stress and Smad3 acetylation. *Life Sci.* **2021**, *274*, 119299. [CrossRef]
43. Fukushima, A.; Lopaschuk, G.D. Acetylation control of cardiac fatty acid β-oxidation and energy metabolism in obesity, diabetes, and heart failure. *Biochim. Biophys. Acta Mol. Basis Dis.* **2016**, *1862*, 2211–2220. [CrossRef] [PubMed]
44. Wagner, G.R.; Payne, R.M. Mitochondrial Acetylation and Diseases of Aging. *J. Aging Res.* **2011**, *2011*, 234875. [CrossRef] [PubMed]
45. Shao, Y.; Chernaya, V.; Johnson, C.; Yang, W.Y.; Cueto, R.; Sha, X.; Zhang, Y.; Qin, X.; Sun, J.; Choi, E.T.; et al. Metabolic Diseases Downregulate the Majority of Histone Modification Enzymes, Making a Few Upregulated Enzymes Novel Therapeutic Targets—"Sand Out and Gold Stays". *J. Cardiovasc. Transl. Res.* **2016**, *9*, 49–66. [CrossRef] [PubMed]
46. Shen, P.; Feng, X.; Zhang, X.; Huang, X.; Liu, S.; Lu, X.; Li, J.; You, J.; Lu, J.; Li, Z.; et al. SIRT6 suppresses phenylephrine-induced cardiomyocyte hypertrophy though inhibiting p300. *J. Pharmacol. Sci.* **2016**, *132*, 31–40. [CrossRef]
47. Yang, M.; Zhang, Y.; Ren, J. Acetylation in cardiovascular diseases: Molecular mechanisms and clinical implications. *Biochim. Biophys. Acta Mol. Basis Dis.* **2020**, *1866*, 165836. [CrossRef]
48. Pane, L.S.; Fulcoli, F.G.; Cirino, A.; Altomonte, A.; Ferrentino, R.; Bilio, M.; Baldini, A. Tbx1 represses Mef2c gene expression and is correlated with histone 3 deacetylation of the anterior heart field enhancer. *Dis. Model. Mech.* **2018**, *11*, dmm029967. [CrossRef]
49. Bühler, A.; Gahr, B.M.; Park, D.-D.; Bertozzi, A.; Boos, A.; Dalvoy, M.; Pott, A.; Oswald, F.; Kovall, R.A.; Kühn, B.; et al. Histone deacetylase 1 controls cardiomyocyte proliferation during embryonic heart development and cardiac regeneration in zebrafish. *PLOS Genet.* **2021**, *17*, e1009890. [CrossRef]
50. Cencioni, C.; Spallotta, F.; Mai, A.; Martelli, F.; Farsetti, A.; Zeiher, A.M.; Gaetano, C. Sirtuin function in aging heart and vessels. *J. Mol. Cell. Cardiol.* **2015**, *83*, 55–61. [CrossRef]
51. Watroba, M.; Szukiewicz, D. Sirtuins at the Service of Healthy Longevity. *Front. Physiol.* **2021**, *12*, 724506. [CrossRef] [PubMed]
52. Lombard, D.B.; Schwer, B.; Alt, F.W.; Mostoslavsky, R. SIRT6 in DNA repair, metabolism and ageing. *J. Intern. Med.* **2008**, *263*, 128–141. [CrossRef] [PubMed]
53. Cheng, H.-L.; Mostoslavsky, R.; Saito, S.; Manis, J.P.; Gu, Y.; Patel, P.; Bronson, R.; Appella, E.; Alt, F.W.; Chua, K.F. Developmental defects and p53 hyperacetylation in Sir2 homolog (SIRT1)-deficient mice. *Proc. Natl. Acad. Sci. USA* **2003**, *100*, 10794–10799. [CrossRef] [PubMed]
54. Packer, M. Longevity genes, cardiac ageing, and the pathogenesis of cardiomyopathy: Implications for understanding the effects of current and future treatments for heart failure. *Eur. Heart J.* **2020**, *41*, 3856–3861. [CrossRef]
55. Benigni, A.; Cassis, P.; Conti, S.; Perico, L.; Corna, D.; Cerullo, D.; Zentilin, L.; Zoja, C.; Perna, A.; Lionetti, V.; et al. Sirt3 Deficiency Shortens Life Span and Impairs Cardiac Mitochondrial Function Rescued by Opa1 Gene Transfer. *Antioxid. Redox Signal.* **2019**, *31*, 1255–1271. [CrossRef]
56. Mostoslavsky, R.; Chua, K.F.; Lombard, D.B.; Pang, W.W.; Fischer, M.R.; Gellon, L.; Liu, P.; Mostoslavsky, G.; Franco, S.; Murphy, M.M.; et al. Genomic Instability and Aging-like Phenotype in the Absence of Mammalian SIRT6. *Cell* **2006**, *124*, 315–329. [CrossRef]
57. Sundaresan, N.R.; Samant, S.A.; Pillai, V.B.; Rajamohan, S.B.; Gupta, M.P. SIRT3 Is a Stress-Responsive Deacetylase in Cardiomyocytes That Protects Cells from Stress-Mediated Cell Death by Deacetylation of Ku70. *Mol. Cell. Biol.* **2008**, *28*, 6384–6401. [CrossRef]
58. Lin, Y.-H.; Major, J.L.; Liebner, T.; Hourani, Z.; Travers, J.G.; Wennersten, S.A.; Haefner, K.R.; Cavasin, M.A.; Wilson, C.E.; Jeong, M.Y.; et al. HDAC6 modulates myofibril stiffness and diastolic function of the heart. *J. Clin. Investig.* **2022**, *132*, e148333. [CrossRef]
59. Pougovkina, O.; te Brinke, H.; Ofman, R.; van Cruchten, A.G.; Kulik, W.; Wanders, R.J.A.; Houten, S.M.; de Boer, V.C.J. Mitochondrial protein acetylation is driven by acetyl-CoA from fatty acid oxidation. *Hum. Mol. Genet.* **2014**, *23*, 3513–3522. [CrossRef]
60. Tomczyk, M.M.; Cheung, K.G.; Xiang, B.; Tamanna, N.; Fonseca Teixeira, A.L.; Agarwal, P.; Kereliuk, S.M.; Spicer, V.; Lin, L.; Treberg, J.; et al. Mitochondrial Sirtuin-3 (SIRT3) Prevents Doxorubicin-Induced Dilated Cardiomyopathy by Modulating Protein Acetylation and Oxidative Stress. *Circ. Heart Fail.* **2022**, *15*, e008547. [CrossRef]
61. Maissan, P.; Mooij, E.J.; Barberis, M. Sirtuins-mediated system-level regulation of mammalian tissues at the interface between metabolism and cell cycle: A systematic review. *Biology* **2021**, *10*, 194. [CrossRef]
62. Khan, D.; Sarikhani, M.; Dasgupta, S.; Maniyadath, B.; Pandit, A.S.; Mishra, S.; Ahamed, F.; Dubey, A.; Fathma, N.; Atreya, H.S.; et al. SIRT6 deacetylase transcriptionally regulates glucose metabolism in heart. *J. Cell. Physiol.* **2018**, *233*, 5478–5489. [CrossRef]

63. Khan, D.; Ara, T.; Ravi, V.; Rajagopal, R.; Tandon, H.; Parvathy, J.; Gonzalez, E.A.; Asirvatham-Jeyaraj, N.; Krishna, S.; Mishra, S.; et al. SIRT6 transcriptionally regulates fatty acid transport by suppressing PPARγ. *Cell Rep.* **2021**, *35*, 109190. [CrossRef]
64. Cantó, C.; Gerhart-Hines, Z.; Feige, J.N.; Lagouge, M.; Noriega, L.; Milne, J.C.; Elliott, P.J.; Puigserver, P.; Auwerx, J. AMPK regulates energy expenditure by modulating NAD + metabolism and SIRT1 activity. *Nature* **2009**, *458*, 1056–1060. [CrossRef] [PubMed]
65. Senoner, T.; Dichtl, W. Oxidative Stress in Cardiovascular Diseases: Still a Therapeutic Target? *Nutrients* **2019**, *11*, 2090. [CrossRef]
66. Dubois-Deruy, E.; Peugnet, V.; Turkieh, A.; Pinet, F. Oxidative Stress in Cardiovascular Diseases. *Antioxidants* **2020**, *9*, 864. [CrossRef] [PubMed]
67. Fukushima, A.; Lopaschuk, G.D. Cardiac fatty acid oxidation in heart failure associated with obesity and diabetes. *Biochim. Biophys. Acta Mol. Cell Biol. Lipids* **2016**, *1861*, 1525–1534. [CrossRef]
68. Hou, Y.-S.; Wang, J.-Z.; Shi, S.; Han, Y.; Zhang, Y.; Zhi, J.-X.; Xu, C.; Li, F.-F.; Wang, G.-Y.; Liu, S.-L. Identification of epigenetic factor KAT2B gene variants for possible roles in congenital heart diseases. *Biosci. Rep.* **2020**, *40*, BSR20191779. [CrossRef]
69. Sunagawa, Y.; Katayama, A.; Funamoto, M.; Shimizu, K.; Shimizu, S.; Sari, N.; Katanasaka, Y.; Miyazaki, Y.; Hosomi, R.; Hasegawa, K.; et al. The polyunsaturated fatty acids, EPA and DHA, ameliorate myocardial infarction-induced heart failure by inhibiting p300-HAT activity in rats. *J. Nutr. Biochem.* **2022**, *106*, 109031. [CrossRef]
70. Li, N.; Zhou, H.; Ma, Z.G.; Zhu, J.X.; Liu, C.; Song, P.; Kong, C.Y.; Wu, H.M.; Deng, W.; Tang, Q.Z. Geniposide alleviates isoproterenol-induced cardiac fibrosis partially via SIRT1 activation in vivo and in vitro. *Front. Pharmacol.* **2018**, *9*, 854. [CrossRef]
71. Lu, T.M.; Tsai, J.Y.; Chen, Y.C.; Huang, C.Y.; Hsu, H.L.; Weng, C.F.; Shih, C.C.; Hsu, C.P. Downregulation of Sirt1 as aging change in advanced heart failure. *J. Biomed. Sci.* **2014**, *21*, 57. [CrossRef]
72. Alshehri, A.S.; El-Kott, A.F.; Eleawa, S.M.; El-Gerbed, M.S.A.; Khalifa, H.S.; El-Kenawy, A.E.; Albadrani, G.M.; Abdel-Daim, M.M. Kaempferol protects against streptozotocin-induced diabetic cardiomyopathy in rats by a hypoglycemic effect and upregulating sirt1. *J. Physiol. Pharmacol.* **2021**, *72*, 339–355.
73. Qiu, L.; Xu, C.; Xia, H.; Chen, J.; Liu, H.; Jiang, H. Downregulation of P300/CBP-Associated Factor Attenuates Myocardial Ischemia-Reperfusion Injury Via Inhibiting Autophagy. *Int. J. Med. Sci.* **2020**, *17*, 1196–1206. [CrossRef]
74. Yuan, Q.; Zhan, L.; Zhou, Q.Y.; Zhang, L.L.; Chen, X.M.; Hu, X.M.; Yuan, X.C. SIRT2 regulates microtubule stabilization in diabetic cardiomyopathy. *Eur. J. Pharmacol.* **2015**, *764*, 554–561. [CrossRef]
75. Turdi, S.; Li, Q.; Lopez, F.L.; Ren, J. Catalase alleviates cardiomyocyte dysfunction in diabetes: Role of Akt, Forkhead transcriptional factor and silent information regulator 2. *Life Sci.* **2007**, *81*, 895–905. [CrossRef]
76. Lai, Y.C.; Tabima, D.M.; Dube, J.J.; Hughan, K.S.; Vanderpool, R.R.; Goncharov, D.A.; St. Croix, C.M.; Garcia-Ocanã, A.; Goncharova, E.A.; Tofovic, S.P.; et al. SIRT3-AMP-Activated Protein Kinase Activation by Nitrite and Metformin Improves Hyperglycemia and Normalizes Pulmonary Hypertension Associated with Heart Failure with Preserved Ejection Fraction. *Circulation* **2016**, *133*, 717–731.
77. Zhang, X.; Ji, R.; Liao, X.; Castillero, E.; Kennel, P.J.; Brunjes, D.L.; Franz, M.; Möbius-Winkler, S.; Drosatos, K.; George, I.; et al. MicroRNA-195 Regulates Metabolism in Failing Myocardium Via Alterations in Sirtuin 3 Expression and Mitochondrial Protein Acetylation. *Circulation* **2018**, *137*, 2052–2067. [CrossRef]
78. Li, L.; Zeng, H.; He, X.; Chen, J. Sirtuin 3 Alleviates Diabetic Cardiomyopathy by Regulating TIGAR and Cardiomyocyte Metabolism. *J. Am. Heart Assoc.* **2021**, *10*, e018913. [CrossRef]
79. Dai, L.; Xie, Y.; Zhang, W.; Zhong, X.; Wang, M.; Jiang, H.; He, Z.; Liu, X.; Zeng, H.; Wang, H. Weighted Gene Co-Expression Network Analysis Identifies ANGPTL4 as a Key Regulator in Diabetic Cardiomyopathy via FAK/SIRT3/ROS Pathway in Cardiomyocyte. *Front. Endocrinol.* **2021**, *12*, 705154. [CrossRef]
80. Wang, Y.-Y.; Gao, B.; Yang, Y.; Jia, S.-B.; Ma, X.-P.; Zhang, M.-H.; Wang, L.-J.; Ma, A.-Q.; Zhang, Q.-N. Histone deacetylase 3 suppresses the expression of SHP-1 via deacetylation of DNMT1 to promote heart failure. *Life Sci.* **2022**, *292*, 119552. [CrossRef]
81. Xu, Z.; Tong, Q.; Zhang, Z.; Wang, S.; Zheng, Y.; Liu, Q.; Qian, L.; Chen, S.; Sun, J.; Cai, L. Inhibition of HDAC3 prevents diabetic cardiomyopathy in OVE26 mice via epigenetic regulation of DUSP5-ERK1/2 pathway. *Clin. Sci.* **2017**, *131*, 1841–1857. [CrossRef] [PubMed]
82. Rai, R.; Sun, T.; Ramirez, V.; Lux, E.; Eren, M.; Vaughan, D.E.; Ghosh, A.K. Acetyltransferase p300 inhibitor reverses hypertension-induced cardiac fibrosis. *J. Cell. Mol. Med.* **2019**, *23*, 3026–3031. [CrossRef] [PubMed]
83. Tao, H.; Yang, J.-J.; Shi, K.-H.; Li, J. Epigenetic factors MeCP2 and HDAC6 control α-tubulin acetylation in cardiac fibroblast proliferation and fibrosis. *Inflamm. Res.* **2016**, *65*, 415–426. [CrossRef] [PubMed]
84. Bauters, C.; Dubois, E.; Porouchani, S.; Saloux, E.; Fertin, M.; de Groote, P.; Lamblin, N.; Pinet, F. Long-term prognostic impact of left ventricular remodeling after a first myocardial infarction in modern clinical practice. *PLoS ONE* **2017**, *12*, e0188884. [CrossRef] [PubMed]
85. Yang, L.; Chen, X.; Bi, Z.; Liao, J.; Zhao, W.; Huang, W. Curcumin attenuates renal ischemia reperfusion injury via JNK pathway with the involvement of p300/CBP-mediated histone acetylation. *Korean J. Physiol. Pharmacol. Off. J. Korean Physiol. Soc. Korean Soc. Pharmacol.* **2021**, *25*, 413–423. [CrossRef] [PubMed]
86. Katagiri, T.; Sunagawa, Y.; Maekawa, T.; Funamoto, M.; Shimizu, S.; Shimizu, K.; Katanasaka, Y.; Komiyama, M.; Hawke, P.; Hara, H.; et al. Ecklonia stolonifera Okamura Extract Suppresses Myocardial Infarction-Induced Left Ventricular Systolic Dysfunction by Inhibiting p300-HAT Activity. *Nutrients* **2022**, *14*, 580. [CrossRef] [PubMed]

87. Sunagawa, Y.; Funamoto, M.; Shimizu, K.; Shimizu, S.; Sari, N.; Katanasaka, Y.; Miyazaki, Y.; Kakeya, H.; Hasegawa, K.; Morimoto, T. Curcumin, an inhibitor of p300-hat activity, suppresses the development of hypertension-induced left ventricular hypertrophy with preserved ejection fraction in dahl rats. *Nutrients* **2021**, *13*, 2608. [CrossRef] [PubMed]
88. Sunagawa, Y.; Shimizu, K.; Katayama, A.; Funamoto, M.; Shimizu, K.; Nurmila, S.; Shimizu, S.; Miyazaki, Y.; Katanasaka, Y.; Hasegawa, K.; et al. Metformin suppresses phenylephrine-induced hypertrophic responses by inhibiting p300-HAT activity in cardiomyocytes. *J. Pharmacol. Sci.* **2021**, *147*, 169–175. [CrossRef] [PubMed]
89. Horton, J.L.; Martin, O.J.; Lai, L.; Riley, N.M.; Richards, A.L.; Vega, R.B.; Leone, T.C.; Pagliarini, D.J.; Muoio, D.M.; Bedi, K.C., Jr.; et al. Mitochondrial protein hyperacetylation in the failing heart. *JCI Insight* **2016**, *1*, e84897. [CrossRef]
90. Porter, G.A.; Urciuoli, W.R.; Brookes, P.S.; Nadtochiy, S.M. SIRT3 deficiency exacerbates ischemia-reperfusion injury: Implication for aged hearts. *Am. J. Physiol. Heart Circ. Physiol.* **2014**, *306*, H1602. [CrossRef]
91. Prola, A.; Pires Da Silva, J.; Guilbert, A.; Lecru, L.; Piquereau, J.; Ribeiro, M.; Mateo, P.; Gressette, M.; Fortin, D.; Boursier, C.; et al. SIRT1 protects the heart from ER stress-induced cell death through eIF2α deacetylation. *Cell Death Differ.* **2017**, *24*, 343–356. [CrossRef]
92. Renguet, E.; De Loof, M.; Fourny, N.; Ginion, A.; Bouzin, C.; Poüs, C.; Horman, S.; Beauloye, C.; Bultot, L.; Bertrand, L. α-Tubulin acetylation on lysine 40 controls cardiac glucose uptake. *Am. J. Physiol. Heart Circ. Physiol.* **2022**, *322*, H1032–H1043. [CrossRef]
93. Gao, S.; Yang, Q.; Peng, Y.; Kong, W.; Liu, Z.; Li, Z.; Chen, J.; Bao, M.; Li, X.; Zhang, Y.; et al. SIRT6 regulates obesity-induced oxidative stress via ENDOG/SOD2 signaling in the heart. *Cell Biol. Toxicol.* **2022**. *Online ahead of print*. [CrossRef]
94. Cao, M.; Zhao, Q.; Sun, X.; Qian, H.; Lyu, S.; Chen, R.; Xia, H.; Yuan, W. Sirtuin 3: Emerging therapeutic target for cardiovascular diseases. *Free Radic. Biol. Med.* **2022**, *180*, 63–74. [CrossRef] [PubMed]
95. Chen, J.; Chen, S.; Zhang, B.; Liu, J. SIRT3 as a potential therapeutic target for heart failure. *Pharmacol. Res.* **2021**, *165*, 105432. [CrossRef] [PubMed]
96. Saiyang, X.; Deng, W.; Qizhu, T. Sirtuin 6: A potential therapeutic target for cardiovascular diseases. *Pharmacol. Res.* **2021**, *163*, 105214. [CrossRef]

biomedicines MDPI

Review

Cardiotoxicity of Selected Vascular Endothelial Growth Factor Receptor Tyrosine Kinase Inhibitors in Patients with Renal Cell Carcinoma

Beata Franczyk [1,†], Jacek Rysz [1], Janusz Ławiński [2], Aleksandra Ciałkowska-Rysz [3] and Anna Gluba-Brzózka [1,*,†]

1 Department of Nephrology, Hypertension and Family Medicine, Medical University of Lodz, 113 Żeromskiego Street, 90-549 Lodz, Poland
2 Department of Urology, Institute of Medical Sciences, Medical College of Rzeszow University, 35-055 Rzeszow, Poland
3 Palliative Medicine Unit, Oncology Department, Medical University of Lodz, 90-419 Lodz, Poland
* Correspondence: anna.gluba-brzozka@umed.lodz.pl
† These authors contributed equally to this work.

Abstract: Renal cell carcinoma (RCC) is one of the most frequent malignant neoplasms of the kidney. The therapeutic options available for the treatment of advanced or metastatic RCC include vascular endothelial growth factor receptor (VEGFR)-targeted molecules, for example, tyrosine kinase inhibitors (TKI). Various VEGFR-TKIs proved to be effective in the treatment of patients with solid tumours. The combination of two drugs may prove most beneficial in the treatment of metastatic RCC; however, it also enhances the risk of toxicity compared to monotherapy. Specific VEGFR-TKIs (e.g., sunitinib, sorafenib or pazopanib) may increase the rate of cardiotoxicity in metastatic settings. VEGF inhibitors modulate multiple signalling pathways; thus, the identification of the mechanism underlying cardiotoxicity appears challenging. VEGF signalling is vital for the maintenance of cardiomyocyte homeostasis and cardiac function; therefore, its inhibition can be responsible for the reported adverse effects. Disturbed growth factor signalling pathways may be associated with endothelial dysfunction, impaired revascularization, the development of dilated cardiomyopathy, cardiac hypertrophies and altered peripheral vascular load. Patients at high cardiovascular risk at baseline could benefit from clinical follow-up in the first 2–4 weeks after the introduction of targeted molecular therapy; however, there is no consensus concerning the surveillance strategy.

Keywords: cardiotoxicity; RCC; metastasis; VEGFR-TKIs

Citation: Franczyk, B.; Rysz, J.; Ławiński, J.; Ciałkowska-Rysz, A.; Gluba-Brzózka, A. Cardiotoxicity of Selected Vascular Endothelial Growth Factor Receptor Tyrosine Kinase Inhibitors in Patients with Renal Cell Carcinoma. *Biomedicines* **2023**, *11*, 181. https://doi.org/10.3390/biomedicines11010181

Academic Editor: Tânia Martins-Marques

Received: 2 November 2022
Revised: 28 December 2022
Accepted: 3 January 2023
Published: 11 January 2023

Copyright: © 2023 by the authors. Licensee MDPI, Basel, Switzerland. This article is an open access article distributed under the terms and conditions of the Creative Commons Attribution (CC BY) license (https://creativecommons.org/licenses/by/4.0/).

1. Introduction

Renal cell carcinoma (RCC) is among the most frequent malignant neoplasms of the kidney (present in 90% of cases) [1]. At the same time, RCC-related mortality is the highest among urological neoplasms. Since this type of tumour does not give easily recognizable alarming symptoms, in nearly one-third of patients it is diagnosed at the metastatic stage of the disease. The estimated overall 5-year survival is 76%; however, such survival is significantly reduced in patients with stage IV disease [2]. The treatment of this tumour is challenging since RCC comprises highly heterogeneous cancers with different underlying genetic and epigenetic mechanisms and molecular pathways, including clear cell (accounting for 70–75% of all cases and caused by the loss of tumour suppression gene VHL (von Hippel–Lindau) on the short arm of chromosome 3) and non-clear cell subtypes: papillary (15% of all kidney cancers), chromophobe RCC (occurring in 5–10% of all cases, typically a slowly growing type), collecting duct carcinoma and renal medullary carcinoma (aggressive types accounting for less than 5% of all RCC cases, resistant to most systemic treatment options), as well as sarcomatoid RCC (occurring in about 5% of all RCC cases, generally symptomatic and very aggressive) [3–7]. Approximately 20–50% of patients with localized disease will ultimately relapse and advance to the metastatic stage despite treatment [2,8]. Features of various subtypes of RCC are summarized in Table 1 (based on data from [9]).

Table 1. Summary of features of various subtypes of RCC.

Features	Clear Cell RCC (ccRCC)	Papillary RCC (pRCC)	Chromophobe RCC (cRCC)	Clear Cell Papillary RCC (ccpRCC)	Multilocular Cystic *
Occurrence	~70–75% of RCC	~10–20% of RCC	~5–7% of RCC	1–4% of RCC	~1–4% of RCC
Age of onset	>50 years	>50 years	>50 years		40–50 years
Morphology	Clear or pale appearance, sometimes yellow Acinar, nested, alveolar, tubular architecture with small cysts Capillary vessels intimately associated with the tumour	Cells with finger-like projections. Papillary architecture, tubulo-papillary, glomeruloid, and dense papillary simulating solid areas also occur Slender papillae lined with a single layer of cuboidal cells with scant, basophilic cytoplasm and inconspicuous nucleoli	Solid architecture with extension to adjacent renal parenchymal entrapping pre-existent tubules Large eosinophilic cells and vegetal-like cells with distinct cytoplasmic membrane. Eosinophilic or finely granular cytoplasm with perinuclear halos and nuclei showing "raisinoid" morphology	Cells with "piano-key-like" pattern Clear cells with tubular (predominant), papillary, acinar, cystic, cords and solid growth	Cysts lined with cells with clear cytoplasm and low-grade nuclei with variable sizes separated by delicate septae that may contain cords of clear cell
Characteristic traits	Originates from proximal nephron/ tubular epithelium Aggressiveness related to the stage and Furham grade	2 types Originates from distal nephron/tubular epithelium typically low grade; cases with high-grade nuclear features (prominent nucleoli, non-basophilic cytoplasm), sometimes invasive growth	Originates from distal nephron/intercalated cells of distal tubules Less aggressive compared to ccRCC and pRCC (mortality ~10% patients) Rhabdoid/sarcomatoid and tumour necrosis—more aggressive behaviour	Indolent, not metastatic	More prevalent in males Indolent Not metastatic Good prognosis
Features	**Medullary Carcinoma**	**Mucinous Tubular and Spindle Cell Carcinoma**	**Collecting Duct Carcinoma (CDC)**	**Xp11 Translocation RCC**	**Unclassified RCC**
Occurrence	~1% of RCC	Rare	~1–2% of RCC		3–6% of RCC
Age of onset	20–30 years	40–50 years	20–30 years	childhood	Variable
Morphology	Similar to CDC, tubular, papillary and infiltrative architectures Adenoid cystic, reticular and microcystic patterns	Tightly packed tubular component lined with cuboidal cells transitioning into a bland spindle cell component Variable amount of mucinous/myxoid stroma Low-grade tubules, spindle cell component and mucin	Predominant tubular morphology (tubulo-papillary and papillary patterns are also common) Desmoplastic stromal reaction High-grade nuclei Absence of other RCC subtypes	Papillae lined with epithelioid clear and eosinophilic cells with abundant psammoma bodies	More than one cell type visible under a microscope
Characteristic traits	Originates from distal nephron Extremely aggressive centred in renal medulla Associated with sickle cell trait/disease/related hemoglobinopathies	Originates from distal nephron/tubular cells Indolent, usually low-grade; rarely high-grade nuclei and sarcomatoid, rarely metastatic	Originates from distal nephron High-grade adenocarcinoma Medullary involvement High aggressiveness (2-year mortality 70%)	Overall prognosis comparable with clear cell RCC	Aggressive High mortality

* multilocular cystic renal neoplasm of low malignant potential.

The therapeutic options available for the treatment of advanced or metastatic RCC have changed over the years [3]. Initially, high-dose interleukin-2 [IL-2] and interferon-α were used to affect the immune system signalling cascade and intracellular oncogenic pathways. The introduction of molecular targeted agents and immunotherapies, including vascular endothelial growth factor receptor (VEGFR)-targeted molecules (e.g., tyrosine kinase inhibitors; TKI), immune checkpoint inhibitors (ICIs) and inhibitors of the mechanistic target of rapamycin (mTOR) has improved the prognosis and survival of patients with metastatic RCC (mRCC) [5,8,10]. Tyrosine kinases (TK) activate numerous proteins via phosphorylation, thus affecting signal transduction and regulating cellular activity [4]. In turn, TK inhibitors (TKI) attach to the ATP-binding pocket of these kinases, thus hampering their activity. Various VEGFR-TKIs proved to be effective in the treatment of patients with solid tumours [11]. Current guidelines suggest that the combination of two of the aforementioned drugs may prove most beneficial in the treatment of metastatic RCC [12–14]. However, the combination of two treatment options may increase the risk of toxicity in comparison to monotherapy. Cancer-treatment-related toxicities affect many organs; however, it appears that cardiovascular adverse effects are the most important since they worsen patients' prognosis and quality of life [15]. Many studies and clinical trials have suggested that specific anti-VEGF tyrosine kinase inhibitors (TKI) such as sunitinib, sorafenib or pazopanib may possibly be responsible for a higher rate of cardiotoxicity in metastatic settings. Cardio-oncology is a relatively new area of knowledge which focuses on the treatment of cancer taking into consideration the risk of cardiovascular adverse effects of the therapies used [16]. This review will focus on the cardiotoxicity risk associated with VEGFR-TKIs treatment in patients with RCC.

We conducted a PubMed search to identify articles that are suitable for inclusion in this narrative review. We did not perform a systematic review. Terms that were searched for included: "cardiotoxicity", "vascular endothelial growth factor receptor tyrosine kinase inhibitors", "VEGFR-TKIs", "renal cell carcinoma", "RCC", "prognosis", "metastasis".

2. Vascular Endothelial Growth Factor Receptor Tyrosine Kinase Inhibitors (VEGFR-TKIs)

Five tyrosine kinase inhibitors (TKIs) of the vascular endothelial growth factor receptor (VEGFR) pathway (VEGFR-TKI) have been approved by the Food and Drug Administration (FDA) in the treatment of metastatic RCC [17]. Pazopanib has been approved as the first-line treatment in patients with RCC3 [17]. These small-molecule tyrosine kinase inhibitors, including sorafenib, axitinib, sunitinib, cabozantinib and pazopanib, target vascular endothelial growth factor receptors (VEGFRs) [18]. VEGFR-TKIs act at the intracellular sites of the VEGF receptor, which affects the survival of microvascular endothelial cells [19]. Moreover, the less the VEGFR-TKI is specific, the more the effects are pronounced [20]. Inhibitors of VEGF signalling have been used with success in the treatment of various cancers since they prevent tumour angiogenesis, thus limiting its growth [17]. Inhibitors of VEGFR also enhance T lymphocyte trafficking into tumours, improving in consequence malignant cells' responsiveness to immunotherapy [21,22]. The results of randomized phase III trials demonstrated their clinical benefits in the treatment of metastatic renal cell carcinoma [23]. The introduction of small-molecule targeted VEGFR-TKIs has increased median progression-free survival and overall survival in advanced/mRCC compared to previous treatment modes by 6 and 14 months, respectively [24,25]. Clinical data indicate that the use of TKIs is frequently associated with systemic adverse effects in patients. Nausea, diarrhoea, fatigue, rhabdomyolysis, hypertension, neutropenia, renal failure, QT prolongation and heart failure are among the frequent adverse effects [26]. Common adverse events related to therapy with sorafenib and sunitinib include hand-foot skin reaction, reversible skin rashes, haemorrhage, diarrhoea, leukopenia, increased pancreatic enzymes levels, hypophosphatemia and proteinuria [27,28]. Growing evidence suggests that despite being beneficial in terms of cancer, such treatment exerts cardiotoxic effects of VEGFRs-TKIs, including asymptomatic left ventricular (LV) dysfunction, hypertension and even congestive heart failure (CHF) [29,30].

Sunitinib, sorafenib, pazopanib and cabozantinib are used in the treatment of RCC. Sunitinib (known also as SU11248 or Sutent), an orally active multi-targeted receptor tyrosine kinase inhibitor, was approved by the FDA in 2006 for the treatment of, i.a., advanced renal-cell carcinoma (RCC), pancreatic cancer (PC), chronic myeloid leukaemia and imatinib-resistant gastrointestinal stromal tumour (GIST) [31,32]. This first-line therapy for metastatic RCC blocks VEGFR 1,2,3, platelet-derived growth factor (PDGF), colony stimulating factor-1, Fms-related receptor tyrosine kinase 3 (FLT-3) as well as tyrosine-protein kinase KIT (c-kit) [33,34]. Sunitinib's actions involve the inhibition of angiogenesis and the limitation of blood supply to the tumour cells. The impact on blood supply, impairment of signal transduction, cellular metabolism and transcription is associated with elevated cardiovascular risk in cancer patients treated with such drugs [35]. Due to its relatively cDNA was used as a template for RT-qPCR, which was performed in duplicate using the high-performance ViiATM 7 Real-Time PCR System thermal cycler, according to the manufacturer's instructions (Applied Biosystems; Waltham, MA, USA) and using TaqMan® (Thermo Fisher Scientific; Horsham, UK) probes: *DNMT3B* (Hs00171876_m1) and *GAPDH* (Hs99999905_m1). *GAPDH* was used as the reference gene. The relative expression of the DNMT3B gene was calculated according to the Livak method of $2^{-\Delta\Delta Ct}$ [23].

2.5. Enzyme-Linked Immunosorbent Assay

Dnmt3b protein levels, Dnmts activity and the concentrations of S-adenosylmethionine (SAM) and S-adenosylhomocysteine (SAH) were determined via a specific sandwich enzyme-linked immunosorbent assay, as per the manufacturer´s instructions (DNMT3B Assay Kit ab113471 from Abcam (Cambridge, UK), DNMT Activity Quantification Kit (Colorimetric) ab113467 from Abcam, Human S-adenosylmethionine SAM ELISA kit MBS2605308 and Human S-adenosylhomocysteine SAH ELISA kit MBS2602645 from MyBioSource (San Diego, CA, USA)). The tests were quantified at 450 nm in a dual-wavelength microplate reader (Sunrise; Tecan, Tecan Ibérica Instrumentación S.L Barcelona, Spain) using Magellan version 2.5 software (Tecan).

Dnmt3b protein levels and Dnmts activity were measured in nuclear extracts from homogenate heart tissue (ICM, $n = 8$; CNT, $n = 8$). The isolation method of the nuclear fraction is detailed by Tarazón et al. [24]. Briefly, 100–150 mg of frozen left ventricle was used for the extraction, which was homogenized in an extraction buffer. The homogenates were subjected to a sucrose gradient and to successive ultracentrifugations until the isolation of the nuclear fraction. The concentrations of SAM and SAH were measured in cardiac tissue homogenates (ICM, $n = 30$; and CNT, $n = 6$). For the extraction, we used 25 mg of the frozen left ventricle. These were then homogenized with an extraction buffer in a FastPrep-24 homogenizer (MP Biomedicals; Irvine, CA, USA) with specifically designed Lysing Matrix D tubes. The homogenates were centrifuged, and the supernatant was aliquoted. The SAM and SAH tests have a limit of detection up to 5 ng/mL and 0.06 ng/mL, respectively. No significant cross-reactivity or interference between these metabolites and their analogues were observed.

2.6. DNA Extraction and Infinium MethylationEPIC BeadChip

For this analysis, 16 samples were used (ICM, $n = 8$; and CNT, $n = 8$). DNA isolation and DNA methylation analyses—using the Infinium MethylationEPIC BeadChip platform (Illumina; San Diego, CA, USA) to interrogate over 850,000 CpG sites—were performed and extensively described by Ortega et al. [7].

2.7. Statistical Methods

Data are expressed as the mean ± standard deviation (SD) of the mean for continuous variables and as percentage values for discrete variables. The Kolmogorov–Smirnov test was used to analyze the data distribution. Significant differences between the means of groups with a normal distribution were analyzed using the Student's *t*-test, while the non-parametric Mann–Whitney U test was used for data that was non-normally distributed. Finally, Pearson's correlation coefficient was calculated to determine the statistical relationship between variables. The molecules with $p < 0.05$ and fold > 1.20 were considered statistically significant. CpGs with a $\Delta\beta \geq \pm 0.1$ and $p < 0.05$ were considered as differentially methylated. All statistical analyses were performed using the SPSS software (version 20.0) for Windows (IBM SPSS Inc.; Endicott, NY, USA).

3. Results

3.1. Clinical Characteristics of Patients

This study included explanted heart tissue samples from CNT individuals and HF patients from ischaemic etiology. The six study populations with ICM were homogeneous with regard to their clinical characteristics (Table 1). ICM patients had a mean age of nonselective binding to the intracellular catalytic site of receptors, sunitinib inhibits a wide range of tyrosine kinases [36]. Its low specificity raised hope that it would inhibit angiogenesis and tumour growth but at the same time, it would be less vulnerable to drug resistance [37]. However, the inhibition of various growth factor pathways, particularly those involved in cardiac functioning, may be associated with cardiotoxicity. Sorafenib is a multikinase inhibitor affecting transmembrane VEGFR-2, VEGFR-3, FLT-3, PDGFR-B and KIT receptors, as well as intracellular serine/threonine-protein kinase B-raf (BRAF) and RAF proto-oncogene serine/threonine-protein kinase (CRAF) receptors, which is used in the treatment of patients with RCC [38]. The aforementioned kinases are involved in intracellular signalling pathways and angiogenesis; thus, their inhibition translates into hampered tumour growth [39]. A randomized, double-blind, placebo-controlled phase III trial (TARGET) enrolling over 900 patients previously resistant to therapy showed a significant benefit of sorafenib (vs. placebo) in terms of progression-free survival (PFS) and a 28% decrease in the risk of death in patients receiving sorafenib [40,41]. The National Comprehensive Cancer Network (NCCN) Task Force report [27] suggests that sorafenib may also be appropriate for certain naïve patients with clear cell mRCC. Pazopanib therapy is associated with the inhibition of VEGFR 1–3 and subsequent hampering of angiogenesis and RCC regression [42]. This drug also hinders the actions of stem cell factor KIT receptors and platelet-derived growth factor receptors. Finally, cabozantinib is an oral TKI of MET, VEGFRs and Anexelekto (AXL). This therapy was used in the phase III METEOR trial in pre-treated patients with advanced RCC. It was found to increase progression-free survival (PFS) and objective response rate improvements [43,44]. Currently, cabozantinib has received the approval of the FDA for RCC [45].

3. Cardiotoxicity and Involved Mechanisms

Therapy with VEGFR inhibitors has been demonstrated to improve overall survival and progression-free survival (PFS) in patients with metastatic renal cell cancer. However, tyrosine kinase inhibitors (VEGFR-TKIs) can induce adverse cardiovascular (CV) toxicities [25,40,46]. Many pathways inhibited by tyrosine kinase inhibitors play crucial roles in the preservation of cardiovascular development, CV function and response to CV stress [47–49]. Such therapies interfere with key cardiovascular signalling pathways; thus, they may induce considerable cardiovascular toxicities [50].

The term cardiotoxicity refers to cardiovascular complications of therapies which result in higher morbidity and mortality [51]. The occurrence of this phenomenon differs between oncological therapies, some of which are associated with early clinical manifestation of cardiotoxicity, while in the case of others the adverse effects appear years after the initiation of treatment. Since patients with cancers are administered many drugs, the prediction of cardiotoxicity seems challenging. According to estimations, the incidence of VEGFR-TKI cardiotoxicity is in the range of 3–30% and depends on the drug, study population and diagnostic criteria [48,51–53]. Cancer survivors suffer from late CV risk due to prior cardiotoxic exposure and the appearance of new cardiovascular risk factors with advancing age. Cardiovascular mortality rates in paediatric cancer survivors have been found to be up to ten times higher compared to age-matched controls [54]. Late CV risk is also considerably higher among long-term adult cancer survivors and the risk is particularly pronounced in adult-onset cancer survivors with underlying CV risk factors [55]. Therefore, it is important to recognise potential long-term CV toxicity in cancer survivors in order to implement aggressive correction of CV risk factors in this population and treatment of incident CV dysfunction [56].

3.1. Possible Mechanisms

Since VEGF inhibitors affect multiple signalling pathways, the identification of the underlying mechanism that causes cardiotoxicity can be challenging [51]. Cardiotoxicity may be ascribed to the inhibition of tyrosine kinases that are normally expressed in non-neoplastic tissues, including blood vessels and the myocardium; however, the exact mechanism remains not fully understood. VEGF signalling is of importance for the maintenance of cardiomyocyte homeostasis and cardiac function; therefore its inhibition, especially accompanied by the blockade of PDGFR, RSK and AMPK kinases participating in cardiomyocytes energy metabolism and survival, can be responsible for the observed adverse effects [46].

The results of studies have demonstrated that disturbances related to these growth factors may be associated with endothelial dysfunction, impaired revascularization, the development of dilated cardiomyopathy, cardiac hypertrophies and altered peripheral vascular load [57–61]. The family of VEGFR (VEGFR-1, -2 and -3) is involved in numerous vascular functions important for the preservation of proper functioning of the cardiovascular system [36,62,63]. Since VEGF promotes endothelial cell proliferation and survival, thus contributing to vascular integrity, the inhibition of VEGF signalling is associated with a diminished regenerative capacity of endothelial cells, enhanced proliferation of vascular smooth muscle cells, increased haematocrit and blood viscosity promoting pro-coagulant changes and favouring thrombosis [27,64]. The loss of VEGF signalling has been suggested to cause oxidative stress, vascular rarefaction, the inhibition of nitric oxide pathway and glomerular injury and result in the development of hypertension [65]. Moreover, VEGF inhibition can induce renal thrombotic microangiopathy [66]. Available data indicate that the sequestration of VEGF results in impaired adaptive cardiac hypertrophy in response to pressure overload [59].

The evidence concerning the cardiovascular toxicity of sunitinib appears to be most convincing [29,67,68]. The majority of data concerning the mechanisms of sunitinib-induced cardiotoxicity come from animal studies. Sunitinib-related cardiac side effects are associated with coronary microvascular dysfunction, the triggering of the endothelin-1 system, the inhibition of adenosine 5′-monophosphate-activated protein kinase (AMPK) resulting in the impairment of normal mitochondrial function and consequent cellular energy homeostasis compromise within the heart, as well as hindering of mast/stem cell growth factor receptor [36]. Indeed, the study of sunitinib's influence on cardiac mitochondrial function in the culture of cardiomyocytes revealed considerable abnormalities in mitochondrial structure [29]. This finding was confirmed in another study in which the incubation of rat neonatal cardiomyocytes with a high dose of sunitinib triggered the activation of a caspase-9-related mitochondrial apoptotic pathway as well as the loss of mitochondrial membrane potential and energy rundown as a result of the inhibition of AMP-activated protein kinase [69]. The impairment of mitochondria was presented also in other animal studies. The administration of 40 mg/kg per day of sunitinib for 12 days to mice was associated with the appearance of aberrantly shaped and swollen mitochondria with disrupted cristae [29]. However, cardiomyocyte apoptosis was not triggered until severe hypertension was induced with the use of phenylephrine in the studied animals. The occurrence of cardiac apoptosis was seven times higher in mice treated with sunitinib (10 mg/kg per day) + phenylephrine in comparison to animals fed only with phenylephrine. Thus, it seems that sunitinib-related mitochondrial dysfunction and apoptosis are facilitated by the presence of additional cardiac stress [36]. Since AMPK is involved in the maintenance of cardiac energy homeostasis in a state of enhanced cardiac stress, the impairment of this pathway by sunitinib may result in cardiac dysfunction. AMPK may participate in the hindering of anabolic pathways and inducing energy generation under energy stress via the regulation of acetyl-CoA-carboxylase (ACC) activity and subsequent uptake and metabolism of prime cardiomyocytes energy source-free fatty acids. The role of altered AMPK signalling in heart failure was demonstrated for the first time in patients with the familial form of hypertrophic cardiomyopathy, a rare disease associated with missense SNP within $\gamma 2$ regulatory subunit of AMPK (PRKAG2) [70]. The mechanism of AMPK-inhibition-related sunitinib-induced cardiotoxicity under stress conditions is slowly emerging, but the picture is still not complete. It appears that AICAR (5-aminoimidazole-4-carboxamide-1-β-D-ribofuranoside; acadesine) may be involved in this process. According to studies, AICAR, which activates AMPK, may diminish myocardial ischemic injury via the limitation of oxidative stress, leukocyte plugging and platelet aggregation [71]. However, the blockage of AMPK was found to reverse the beneficial effects of acadesine on the apoptosis of rat cardiomyocytes exposed to hypoxic stress [72]. Also, the role of AMPK signalling in the settings of pressure overload was assessed. A study of AMPKα2 knockout mice (lacking the catalytic unit of AMPK expressed predominantly in the heart) which underwent transverse aortic constriction (TAC) showed a more pronounced loss of LV function, accelerated LV hypertrophy and considerably increased mortality in comparison to wild-type animals after 3 weeks of TAC. In an animal model, the administration of sunitinib reduced the phosphorylation of acetyl-CoA-carboxylase (ACC) in heart tissue which translated into the loss of the activity of AMPK [69]. In turn, the delivery of constitutively active AMPK into cardiomyocytes was associated with their partial resistance to sunitinib-triggered apoptosis. These findings may suggest that the impairment of AMPK signalling may be associated with disturbed adaptation to systolic pressure overload and consequent severe cardiac dysfunction. This mechanism may be potentially responsible for cardiotoxicity observed in hypertensive patients treated with sunitinib. However, in vitro study of sunitinib's effect on isolated rat heart mitochondria and intact rat myoblast cells failed to demonstrate the direct adverse impact of treatment on mitochondrial function [73]. According to the authors, sunitinib-induced mitochondrial abnormalities are due to the inhibition of RSK, which can stimulate proapoptotic factor Bad, leading to the release of cytochrome and apoptosis. Another study confirmed that sunitinib acts as a strong inhibitor of RSK [74].

Also, the impact of sunitinib on platelet-derived growth factor receptor (PDGFR) and AMP-activated protein kinase (AMPK) may affect cardiomyocyte function and survival [46,75]. Some authors have hypothesized that apart from the greater release of endothelin-1, lower production of nitric oxide in the arteriole wall as well as microvascular rarefaction (involving the apoptosis of endothelial cells and the remodelling of capillary beds) could also be responsible for VEGF-inhibitor-related hypertension [76–78]. The observed rise in resistive load may support the role of the reduced number of microvessels in the development of sunitinib-related hypertension [50]. The results of other studies also demonstrated that TKI-induced hypertension may be associated with greater systemic afterload following VEGF inhibition as well as the destruction of endothelial cells, VEGF-receptor-inhibition-related disturbances in vasoconstrictor–vasodilator balance, lower survival of mesangial cells and impaired glomerular function and filtration [79–82]. Moreover, Catino et al. [50] suggested that the worsening of arterial stiffness may also contribute to the pathophysiology of hypertension induced by sunitinib. The authors suggested that the combination of calcium channel blockers with inorganic nitrates may prove useful in the management of hypertension in sunitinib-treated patients due to their vasodilating properties and ability to ameliorate conduit artery function, which appears to be impaired in this group of patients [50]. The aforementioned increased arterial stiffness is also an important risk factor for coronary and cerebrovascular disease [29,50].

The cardiotoxicity of sorafenib has been suggested to be related to the imbalance of pro-survival factor RAF1 (probably also involved in cardiac functioning) and pro-apoptotic factors MST2 (serine/threonine kinase 3) and ASK1 (apoptosis signal-regulating kinase 1) [46]. RAF1 was suggested to inhibit apoptosis-signal-regulating kinase 1 (ASK1) and mammalian sterile 20 kinase 2 (MST2), which exert apoptotic, ERK-independent effects and participate in oxidant-stress-induced injury [46]. Sorafenib-induced impairment of RAF1–ASK1 and/or RAF1–MST2 interactions may cause higher cardiotoxicity compared to solely ERK cascade hindering [46]. A study in an animal model demonstrated that a cardiac-muscle-specific Raf-1-knockout (Raf CKO) mirroring Raf-1 inhibition caused LV systolic dysfunction and heart dilatation induced by considerable elevation of apoptotic cardiomyocyte amounts and the promotion of fibrosis [83]. TKI drugs may also impair angiogenesis by affecting the Src family and downstream RAF1 [84]. According to studies, TKI could target the c-kit, thus negatively affecting the expression of AT2 (angiotensin II receptor type 2) involved in the repair of ischemic injury [85]. However, a vast range of cardiotoxic phenotypes and degrees of toxicity cannot be ascribed only to the abovementioned mechanisms. According to another theory, this class of drugs may exert a cardiotoxic effect via acting "off-target" [20]. It has also been suggested that TKI may trigger mitochondrial toxicity, thus damaging metabolically active cardiomyocytes of the heart [46]. Will et al. [73] provided evidence that sorafenib may act as an inhibitor of Complex V and mitochondrial uncoupler. The results of an animal (rat) study revealed that sorafenib disturbed mitochondrial cristae [86]. In turn, studies of cell cultures indicated that treatment of cancerous cell lines with sorafenib increased the levels of the autophagy markers Beclin1 and LC3 (microtubule-associated protein 1 light chain 3) [87]. In general, autophagy is considered a protective mechanism; low levels of autophagy in hearts appear to maintain the homeostasis and turnover of organelles; however, its enhanced activation may result in cellular death [88–90]. Also, apoptosis has been suggested to contribute to cardiomyocyte loss and subsequent heart failure or hypertensive cardiomyopathy [91,92]. The activation of the pro-apoptotic protein BAD is associated with the downregulation of the anti-apoptotic protein Bcl-2's expression and the subsequent triggering of the initiator and effector caspases, caspase-9 and -3, and the promotion of death of cardiomyocytes [93]. Moreover, according to some authors, the cardiotoxic effects of sorafenib are associated with an increase in metabolites such as urea and fatty acid levels in plasma [94].

Another theory concerning cardiotoxicity related to VEGF inhibitors states that perfusion–contraction match may play a role in this phenomenon [95,96]. According to this thesis, these drugs may contribute to the decrease in myocardial perfusion that is associated with the additive effects of diffuse, non-significant reductions in the coronary arteries' luminal diameter [97]. Evolving aberrations of the coronary microcirculation may result in the impairment of myocardial perfusion. Since patients with cardiovascular diseases display a lower tolerability margin in this area, cardiotoxicity in this group is more common [98]. Chintalgattu et al. [99] confirmed that sunitinib can diminish coronary flow reserve, impair the integrity of the coronary microcirculation and worsen cardiac function. They suggested that the inhibition of the PDGF signalling pathway could be responsible for these effects via a reduction in the pericyte population, resulting in the destabilization of endothelial cells, coronary microcirculation and eventually cardiac function. In the opinion of the authors, systemic hypertension combined with impaired PDGFR signalling could be responsible for the development of heart failure in patients treated with sunitinib [100]. PDGF is a vital growth factor for various cell types, such as cardiomyocytes, smooth muscle cells, endothelial cells and stromal cells, that promote angiogenesis and the maintenance of endothelial function [46,101]. It was also demonstrated to mediate the signalling between heart myocytes and adjacent endothelial cells [102]. Both receptor subtypes of PDGF are inhibited by sunitinib. The inhibition of PDGFR receptor-ß tyrosine kinase by sunitinib is associated with decreased myocardial pericytes, myocardial microvascular density and worsened cardiac function [99]. The inhibition of PDGFR signalling has been found to adversely affect cardiac function, particularly in the stressed heart [36]. The aforementioned pericyte–endothelial–myocardial coupling appears to partly explain the role of VEGF signalling inhibition in cardiotoxicity [103]. Suggested mechanisms of cardiotoxicity of VEGF signalling inhibition are presented in Figure 1.

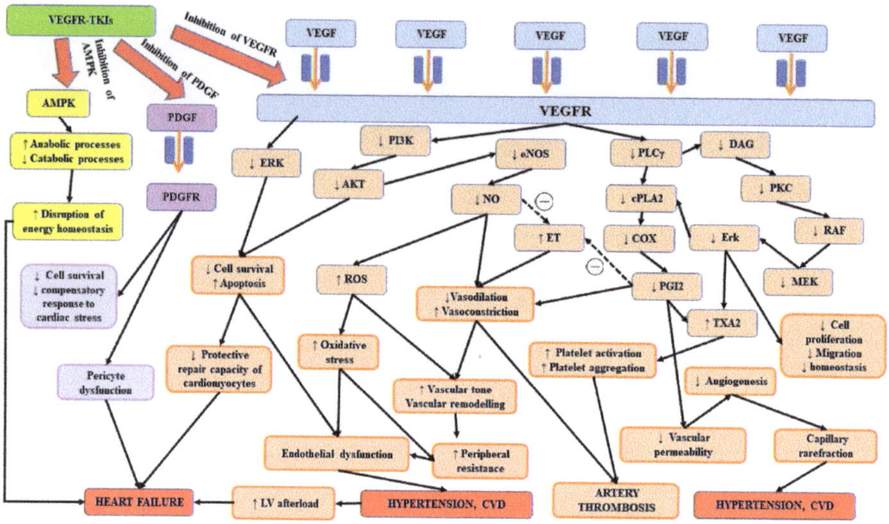

Figure 1. The results of VEGFR, AMPK and PDGFR inhibition leading to cardiotoxicity of VEGF signalling inhibition.

3.2. The Results of the Studies

Clinical data show that VEGF inhibitors, especially TKI, can trigger either reversible or irreversible cardiac side effects [51]. According to estimations, heart dysfunction is observed in 3–15% of patients treated with sunitinib, pazopanib and axitinib, while symptomatic heart failure occurs in 1–10% of these patients [104–106]. However, clinical data from large clinical trials are not available for sorafenib and vandetanib, both of which can also promote cardiac dysfunction. VEGF-inhibitor therapy-related hypertension develops in approximately 19 to 47% of patients [76]. Meta-analyses of randomized clinical trials of sorafenib, pazopanib and sunitinib demonstrated that such treatment is associated with a higher risk of decline in left ventricular ejection fraction (LVEF), hypertension and arterial thrombotic events [66,107–111]. In turn, a recent meta-analysis assessing treatment with sunitinib, axitinib, ponatinib, vandetanib, cabozantinib, sorafenib, pazopanib and regorafenib demonstrated a 2.69-fold rise in the risk of congestive HF (all grades) [52,53]. Therapy with either sorafenib or sunitinib was found to be associated with hypertension, LV systolic and diastolic dysfunction, heart failure (HF) and myocardial ischemia [33,47,48,67]. Sorafenib therapy appeared to affect resistive and pulsatile load [112]. Other clinical events associated with cardiotoxicity include congestive heart failure (CHF) and arterial thromboembolic events (ATE) [27,113,114]. Moreover, adjuvant sunitinib and sorafenib could trigger arrhythmia and cardiac ischemia [33]. Abdel-Qadir et al. [115] demonstrated a higher risk of arterial thromboembolism (odds ratio [OR] = 1.52, 95% confidence interval (CI) 1.17–1.98) in patients treated with VEGF inhibitors. The results of another meta-analysis revealed a considerably greater risk of all-grade bleeding, all-grade and high-grade hypertension as well as all-grade cardiac dysfunction in patients with tumours receiving VEGFR-TKIs [116]. Also, other studies indicated a significantly greater risk of high-grade (RR 4.60, 95% CI 3.92–5.40) and all-grade (RR 3.85, 95% CI 3.37–4.40) hypertension [108]. VEGFR-TKI therapy was also suggested to trigger QTc interval prolongation [117]. The most pronounced effect was reported in the case of sunitinib and vandetanib. Bayesian network meta-analysis of nine FDA-approved VEGFR-TKIs demonstrated their impact on the occurrence of all grades and grade 3 or higher cardiovascular events, hypertension and cardiac damage [11]. Totzeck et al. [118] found a higher relative risk of cardiac ischemia (RR 1.69, 95% CI 1.12–2.57), LV systolic dysfunction (RR 2.53, 95% CI 1.79–3.57) QT corrected interval prolongation (RR 6.25, 95% CI 3.44–11.38) and arterial hypertension (RR 3.78, 95%

CI 3.15–4.54), especially in sunitinib-treated patients. However, a re-analysis of the aforementioned meta-analysis including 71 trials and eight different VEGFR-TKIs suggested that the previously observed effect was exaggerated since, according to the authors, such treatment was associated with a slight elevation in the risk of bleeding, hypertension, thrombocytopenia and arterial thrombotic damage [119].

The results of a recent meta-analysis indicated that the incidence of sunitinib-induced HF might amount to 4.1%, while the prevalence of asymptomatic LVEF deterioration could be even higher in patients with metastatic disease [113,120]. The worsening of LV ejection fraction appears usually in the first cycle of treatment with sunitinib [50]. A multi-centre, longitudinal prospective cohort study confirmed early worsening of vascular function following sunitinib exposure [50]. Such treatment was associated with higher markers of resistive load, total peripheral resistance and arterial elastance. Catino et al. [50] reported the worsening of diastolic dysfunction (E/e') and LV filling pressures (BNP) in sunitinib-treated patients. Sunitinib's impact on cardiac function was reported in a phase III trial, in which either sunitinib or interferon alfa was used in the treatment of patients with advanced RCC [25]. In that trial, the sunitinib-related decline in left ventricular ejection fraction (LVEF) was observed in 10% of patients, including 2% of subjects who developed a grade 3 decline in LVEF. The discontinuation of treatment or dose reduction appeared to reverse this condition. A retrospective study enrolling patients with RCC or imatinib-resistant gastrointestinal stromal tumours (GISTs) demonstrated symptomatic grade 3 or 4 LV dysfunction in 15% of sunitinib-treated patients [48]. The results of other studies have suggested a higher, up to 27%, incidence of cardiac problems [121]. Based on the results of a prospective study, Narayan et al. [68] suggested that the declines in LVEF occurring in approximately 9.7% of patients, even the substantial ones observed in 1.9% of patients, returned to near baseline values despite the continuation of sunitinib therapy in a reduced dose when careful cardiovascular management was provided. Moreover, according to the authors, routine cardiac monitoring in asymptomatic individuals after the third cycle of therapy will not bring great clinical benefit since at that time cardiac dysfunction rates are low.

A meta-analysis of 16 clinical trials including 6935 patients provided evidence for a higher risk of congestive heart disease in patients treated with sunitinib (risk ratio 1.81; 95% CI 1.30–2.50; $p < 0.001$). The 1.5% incidence of high-grade CHF in this analysis translated into a three-times-increased risk of developing serious cardiovascular events (risk ratio 3.30; 95% CI 1.29–8.45; $p < 0.01$). Chu et al. [29] indicated that patients treated with sunitinib experience continuous, gradual worsening of cardiac function. This effect is also observed in many patients with stabilized hypertension dynamics (BP < 140/90 mmHg) treated with beta-blocker and angiotensin-converting enzyme inhibitors. The authors suggested that perhaps more aggressive BP control may prove beneficial. Increased resting systemic blood pressure and higher systemic and coronary vascular resistance were also reported in animals treated with sunitinib [122]. Chronically elevated afterload led to hypertrophic heart remodelling. One animal study indicated that the process of cardiomyocyte apoptosis occurred as a result of the combined effect of sunitinib treatment and the increase in blood pressure [29]. The results of retrospective studies have found that the incidence of sunitinib-induced cardiotoxicity in mRCC patients ranges from 3% to 30% [47,48,52,123]. Such discrepancies in results could stem from the fact that cardiac monitoring protocols are not standardized and different definitions of cardiotoxicity were used in studies [68].

According to one of the studies, sorafenib also markedly decreased left ventricular (LV) pressure, indexes of myocardial contractility and relaxation as well as prolonged systolic and diastolic periods [94]. Such therapy was also found to trigger vasospasm [124]. Sorafenib-related elevation in mid to late systolic load of the LV can potentially lead to myocardial hypertrophy, fibrosis and heart failure [125–128]. In turn, one of the most common side effects of cabozantinib is the development of hypertension with an incidence of 37% in the METEOR trial (15% of grade 3–4) and 81% in the CABOSUN study (28% of grade 3–4) [43,44,129,130]. Such a high frequency of hypertension related to cabozantinib treatment was confirmed in a meta-analysis. Compared to other VEGFR-TKIs, the occurrence of such adverse effects was considerably higher in the case of cabozantinib [129]. A prospective study focusing on the chances of LV systolic dysfunction in cabozantinib-treatment patients revealed an extremely modest risk of developing this disorder [18]. A decline in LVEF by more than 10% was observed in 11.1% of cases after 3 months of therapy; however, it did not translate into LV systolic dysfunction or the appearance of clinical symptoms. This mode of cancer management was not associated with an elevation in cardiac biomarkers, such as proBNP or hsTnI. Iacovelli et al. [18] reported that cabozantinib did not significantly increase the risk of cardiac dysfunction even in patients with cardiovascular comorbidities.

Also, the use of pazopanib has been reported to be associated with cardiotoxicity in the form of hypertension (HTN), thrombotic events, heart failure (HF) with a reduced left ventricular ejection fraction (LVEF) and myocardial ischemia in many patients [52,131,132]. The incidence of hypertension reaches up to 40% and an elevation of N-terminal B-type natriuretic peptide (NT-pro-BNP) levels is observed in up to 26% of patients while HF is reported in 2.4% [52,53]. Another meta-analysis revealed a slightly higher incidence of all-grade HF (3.2%) related to therapy with VEGFR-TKIs [106]. The cardiotoxicity of pazopanib was confirmed in animal studies. One of them demonstrated that the administration of pazopanib considerably increased blood pressure (BP) and diminished CO. The latter finding may suggest early cardiomyocyte stress and probable remodelling in the presence of higher mean arterial pressure [17]. However, another study demonstrated that pazopanib did not negatively affect mouse growth or survival [133].

In contrast to the aforementioned studies, the results of other research have demonstrated a very low risk of cardiac ischemia/infarction related to treatment with angiogenesis inhibitors or TKIs [38,67,114,134]. It reached 2.9% in patients on sorafenib therapy, 1.5% in those on bevacizumab therapy and <1.0% in patients treated with sunitinib. A meta-analysis of the impact of pazopanib, sunitinib, sorafenib and vandetanib on the occurrence of venous thromboembolic events failed to demonstrate a significant relationship [109]. Also, a meta-analysis concerning the risk of arterial thromboembolic events (ATE) and VTEs associated with various VEGFR-TKIs demonstrated a lack of marked increase in the risk of developing all-grade and high-grade VTEs [135]. It seems that a higher occurrence of cardiotoxicity may be associated with a relatively high presence of pre-existing CV disease and/or cardiovascular risk factors in RCC patients [136]. Numerous studies have indicated that hypertension and cardiovascular disease at baseline are crucial predictors of major adverse cardiac events (congestive heart failure, cardiovascular death, myocardial infarction) after the VEGF-TKI sunitinib therapy [29,98,123]. Also, the prevalence of sorafenib-related LVEF dysfunction and/or CHF is higher in patients with a history of hypertension or coronary artery disease [137–139]. More than 70% of hypertensive patients (grade 3 HTN) developed LV systolic dysfunction following the initiation of sunitinib treatment. Khakoo et al. [47] hypothesized that acute rises in blood pressure together with the adverse impact of sunitinib on cardiomyocytes may result in the impairment of cardiac response to BP elevation and subsequent heart failure. The randomized, double-blinded phase III ECOG 2805 trial of adjuvant sunitinib, sorafenib or placebo demonstrated a low incidence of treatment-related significant LVEF decline in formerly untreated patients with completely resected RCC at high risk for recurrence and without baseline cardiovascular comorbidities. The authors suggested that the prevalence of cardiac dysfunction could be higher in the population of symptomatic patients. Therefore, it appears there is a need for close CV monitoring combined with immediate hypertension therapy in RCC patients treated with sunitinib or sorafenib [33]. The control of hypertension can potentially decrease the risk of heart failure. In turn, appropriate HF management can ameliorate already-developed cardiac dysfunction [140].

The occurrence of cardiotoxicity could be underestimated in some studies since many clinical trials exclude patients with cardiovascular diseases [141]. A real-life setting study of major adverse cardiovascular events (MACE) incidence in patients treated with TKI revealed that arterial thrombotic events occurred in 3.99% of study participants, rhythm disorders (atrial fibrillation, atrioventricular block) in 2.66%, and pulmonary embolism and heart failure in 1.57% at 1 year of follow-up [141]. According to the authors, a high incidence of atrial fibrillation in the early period of therapy was associated with an anti-VEGFR treatment-related increase in blood pressure and diastolic dysfunction of the left ventricle. After the initial period, AF could be associated with early remodelling of the atrial ventricles.

According to the recommendations of the European Society for Medical Oncology echocardiography at baseline and every 3 months for the first 6 months and optionally the evaluation of global longitudinal strain should be performed during the first months of TKI therapy in order to carefully monitor arrhythmias, the development of heart failure and pulmonary embolism [141,142]. Considering the increasing prevalence of thrombotic events, high-risk patients should obtain adequate therapy to prevent thrombotic adverse effects [141]. Patients at high cardiovascular risk at baseline could benefit from clinical follow-up in the first 2–4 weeks after the introduction of targeted molecular therapy with, e.g., sorafenib, sunitinib or pazopanib [51]. Patients should undergo periodic reassessment of cardiac function to detect early symptoms of developing cardiac complications. There is no consensus concerning the surveillance strategy. It seems rational to perform periodic echocardiography until the stabilization of LVEF values. Moreover, the determination of values of cardiac biomarkers, such as troponin or N-terminal pro-B-type natriuretic peptide (NT-proBNP), is recommended to increase the number of diagnosed complications. The timing of the aforementioned assessments should be adjusted to the needs of a given patient, his baseline cardiovascular risk and antitumour regimen [143]. The lack of optimal surveillance negatively affects clinical outcomes. However, the optimal timing of biomarker testing has still not been established. One systematic review suggested that the use of ACE inhibitors, angiotensin II receptor blockers (ARBs) and beta-blockers appears beneficial in patients who have developed develop asymptomatic LV dysfunction or HF during the treatment of cancers [144]. The results of studies presenting benefits and cardiotoxicity risk related to VEGF-TKI are presented in Table 2.

The potential risk of cardiotoxicity of TKI has resulted in the addition of special warnings on product labelling [93]. However, sunitinib remains the gold standard in the treatment of some tumours despite the current focus on its cardiotoxicity. Many studies have demonstrated its effectiveness and safety and it seems that a better understanding of the underlying mechanisms would enable the reduction of this risk [32,151]. The inclusion of studies in this narrative review lacked a systematic approach which could affect our conclusions concerning this field.

Table 2. The results of studies presenting benefits and cardiotoxicity risk related to VEGF TKIs.

Name of Drug/Mechanism of Action	Anti-Tumour Effects	Possible Cardiotoxicity
Sorafenib Inhibitor of VEGFR-2, -3, FLT-3, PDGFR-B and KIT, BRAF and CRAF receptors	• Hampers tumour growth [38]. • 28% reduction in the risk of death among patients receiving a dose of 400 mg twice daily compared with placebo. • Median progression-free survival of 5.5 mo compared to 2.8 mo in the placebo group (HR 0.44; 95% CI 0.35 to 0.55; $p < 0.01$). • PFS shorter in second- vs. first-line treatment [145]. • In phase 1 clinical studies: sorafenib used in advanced, refractory solid tumours demonstrated disease stabilization with acceptable toxicity [146]. • A phase 2 randomized discontinuation trial of metastatic RCC: tumour shrinkage, 50% of patients were progression free vs. 18% on placebo, significantly increased PFS (24 weeks vs. 6 weeks on placebo; $p = 0.0087$) [147]. • Phase 3 TARGET trial (ccRCC): interim analysis: median PFS: 5.5 vs. 2.8 mo on placebo, $p < 0.001$) [40]. No statistically significant difference in OS between study arms. • Survival advantage over placebo when patients crossing over were censored (17.8 vs. 14.3 months, respectively; $p = 0.029$) [41].	• Hypertension occurred in 12% of patients receiving 400 mg twice daily [105]. • Significantly increased risk of all-grade hypertension with RR of 6.11 (2.44–15.32], $p < 0.001$) compared with controls [66]. • In meta-analysis, RR of ATEs associated with sorafenib and sunitinib was 3.03 (95% CI, 1.25 to 7.37; $p = 0.015$) compared with control patients [107]. • Increased risk of all-grade hypertension (RR 1.99; 95% CI 1.73–2.29) and high-grade hypertension (RR 0.98; 95% CI 0.75–1.30 [111]. • LVEF decline >15% from baseline and below the institutional lower limit of normal reported in 1.4% of patients [33]. • Increase by ≥10 mmHg in SBP in 75% of patients and by ≥20 mmHg in 60% of patients from baseline value; mean change of 20.6 mmHg ($p < 0.0001$) after 3 weeks of therapy [112]. • Cardiac ischemia or infarction occurred in 3% of patients receiving a dose of 400 mg twice daily in 6-week cycles for the first 24 weeks and in 8-week cycles thereafter [40].
Sunitinib Blocks VEGFR 1, 2, 3, PDGF, CSF-1, FLT-3, c-KIT	• Inhibition of angiogenesis [37]. • Limitation of blood supply to the tumour cells [37]. • Longer overall survival compared with IFN-α. • Increases progression-free survival in the first-line treatment of patients with metastatic RCC vs. INF-α (randomized, phase III trial) [24]. • Increases progression-free survival (11 mo vs. 5 mo INFα). • Improves objective response rate (31% vs. 6% INFα, $p < 0.001$). • Improves quality of life compared to INFα ($p < 0.001$) [25].	• Cardiac dysfunction occurred in 11% of patients receiving once-daily dose of 50 mg for 4 weeks, followed by 2 weeks without treatment [104]. • Myocardial infarction or ischemia occurred in 4% of patients [104]. • In meta-analysis, RR of ATEs associated with sorafenib and sunitinib was 3.03 (95% CI, 1.25 to 7.37; $p = 0.015$) compared with control patients [107]. • Increased risk of all-grade hypertension (RR 2.20; 95% CI 1.92–2.52) and high-grade hypertension (RR 0.81; 95% CI 0.62–1.06) [111]. • LVEF decline >15% from baseline and below the institutional lower limit of normal reported in 1.8% of patients [33]. • 2.7% of patients receiving sunitinib malate developed HF which resulted in substantial morbidity and mortality. Symptomatic HF occurred after a mean of 22 days from treatment initiation. Decline in cardiac function and elevations in BP pressure were not completely reversible [47]. • 15% of patients developed symptomatic grade 3/4 HF [48]. • Increased mean SBP by 9.5 mm Hg (95% CI 2.0–17.1; $p = 0.02$) and DBP by 7.2 mm Hg (95% CI 4.3–10.0; $p < 0.001$) in all participants. Increased large-artery stiffness and resistive load ($p < 0.05$) and changes in pulsatile load [50].
Pazopanib Targets VEGFR-1, -2, -3, PDGFR-α and -β, c-KIT	• Inhibitor of angiogenesis and RCC regression [41]. • PFS of 10.6. Median OS of 14.5 mo [148]. • PARACHUTE, phase IV trial: 39% of patients remained progression free (at 12 months); median PFS was 10 months (95% CI: 8.48–11.83) [149]. • 19% of patients were long-term responders. CR/PR in 24%, stable disease in 44% and PD in 31% patients [149]. • Phase III COMPARZ study: Median time to response—11.9 weeks, CR/PR ≥ 10 months in 14% of patients, PFS ≥ 10 months—31% [150].	*Animal study* • 30 mg/kg of pazopanib twice daily—significant elevation in blood pressure after 2 weeks which persists for the duration of dosing [17]. • Decrease in CO suggestive of early cardiomyocyte stress and possible remodelling [17]. *Human trial* • Hypertension—one of most frequent AE • Cardiac dysfunction present in 13% of patients treated with once-daily dose of 800 mg (continuous dosing) [104]. • Myocardial infarction or ischemia occurred in 2% of patients [104]. • Increased risk of developing all-grade (RR 4.97, 95% CI 3.38–7.30, $p < 0.001$) and high-grade hypertension (RR 2.87, 95% CI 1.16–7.12, $p = 0.023$) [111].

Table 2. *Cont.*

Name of Drug/Mechanism of Action	Anti-Tumour Effects	Possible Cardiotoxicity
Cabozantinib Inhibitor of c-MET, VEGFR2, Ret, KIT, FLT-1/3/4, Tie2, AXL	• Increased PFS compared with sunitinib in CABOSUN trial (median of 8.6 vs. 5.3 mo; HR 0.48, 95% CI 0.31 to 0.74; two-sided p = 0.0008) [42]. • Higher median OS of 26.6 mo for cabozantinib vs. 21.2 mo for sunitinib (HR 0.79, 95% CI 0.53 to 1.2; two-sided p = 0.27) [42]. • Overall survival of 21.4 months (95% CI 18.7–not estimable) vs. 16.5 months (14.7–18.8) with everolimus (HR 0·66 [95% CI 0.53–0.83]; p = 0.00026) [44]. • Improved PFS (HR 0.51 [95% CI 0.41–0.62]; p < 0.0001) and objective response (17% [13–22] [44].	• Hypertension occurred in 28% of patients receiving dose of 60 mg once per day [42]. • Hypertension as the most common grade 3 or 4 adverse event in 15% of patients. • Significantly increased risk of developing all-grade (RR 5.48; 95%CI, 3.76–7.99; p < 0.001) and high-grade (5.09; 95% CI: 2.71–9.54, p < 0.001) hypertension in comparison with controls [129]. • Substantially higher risk of high-grade hypertension compared with sorafenib, sunitinib, vandetanib and pazopanib [129]. • Modest risk of developing left ventricular systolic dysfunction [18].

AE, adverse event; ATE, arterial thromboembolic events; AXL, Anexelekto; BRAF, serine/threonine-protein kinase B-raf; c-KIT, tyrosine-protein kinase KIT; c-MET, tyrosine-protein kinase Met; CR, complete response; CRAF, RAF proto-oncogene serine/threonine-protein kinase; CSF-1; colony-stimulating factor 1; DBP, diastolic blood pressure; FLT-3, Fms-related receptor tyrosine kinase 3; HF, heart failure; HR, hazard ratio; INFα, interferon α; LVEF, left ventricular ejection fraction; mo, months; OS, overall survival; PD, progressive disease; PDGF, platelet-derived growth factor; PDGFR, platelet-derived growth factor receptor; PFS, progression-free survival; PR, partial response; RR, risk ratio; SBP, systolic blood pressure; Tie2, endothelial-enriched tunica interna endothelial cell kinase 2.

Author Contributions: All authors have read and agreed to the published version of the manuscript.

Funding: This study received no external funding.

Institutional Review Board Statement: Not applicable.

Informed Consent Statement: Not applicable.

Data Availability Statement: Not applicable.

Conflicts of Interest: The authors declare no conflict of interest.

References

1. American Cancer Society. What is kidney cancer? Available online: https://www.cancer.org/cancer/kidney-cancer/about/what-is-kidney-cancer.html (accessed on 22 June 2022).
2. Padala, S.A.; Barsouk, A.; Thandra, K.C.; Saginala, K.; Mohammed, A.; Vakiti, A.; Rawla, P.; Barsouk, A. Epidemiology of renal cell carcinoma. *World J. Oncol.* **2020**, *11*, 79–87. [CrossRef]
3. Fontes-Sousa, M.; Magalhães, H.; Oliveira, A.; Carneiro, F.; dos Reis, F.P.; Madeira, P.S.; Meireles, S. Reviewing Treatment Options for Advanced Renal Cell Carcinoma: Is There Still a Place for Tyrosine Kinase Inhibitor (TKI) Monotherapy? *Adv. Ther.* **2022**, *39*, 1107–1125. [CrossRef] [PubMed]
4. Choueiri, T.K.; Motzer, R.J. Systemic therapy for metastatic renal-cell carcinoma. *N. Engl. J. Med.* **2017**, *376*, 354–366. [CrossRef] [PubMed]
5. Jonasch, E.; Gao, J.; Rathmell, W.K. Renal cell carcinoma. *BMJ Br. Med. J.* **2014**, *349*, g4797. [CrossRef] [PubMed]
6. Linehan, W.M.; Ricketts, C.J. The Cancer Genome Atlas of renal cell carcinoma: Findings and clinical implications. *Nat. Rev. Urol.* **2019**, *16*, 539–552. [CrossRef]
7. Akhtar, M.; Al-Bozom, I.A.; Al Hussain, T. Papillary renal cell carcinoma (PRCC): An update. *Adv. Anat. Pathol.* **2019**, *26*, 124–132. [CrossRef]
8. Barata, P.C.; Rini, B.I. Treatment of renal cell carcinoma: Current status and future directions. *CA Cancer J. Clin.* **2017**, *67*, 507–524. [CrossRef]
9. Athanazio, D.A.; Amorim, L.S.; da Cunha, I.W.; Leite, K.R.M.; da Paz, A.R.; de Paula Xavier Gomes, R.; Tavora, F.R.F.; Faraj, S.F.; Cavalcanti, M.S.; Bezerra, S.M. Classification of renal cell tumors—Current concepts and use of ancillary tests: Recommendations of the Brazilian Society of Pathology. *Surg. Exp. Pathol.* **2021**, *4*, 4. [CrossRef]
10. Huang, J.; Shi, G.; Wang, Y.; Wang, P.; Zhang, J.; Kong, W.; Huang, Y.; Wang, S.; Xue, W. Second-line treatment with axitinib plus toripalimab in metastatic renal cell carcinoma: A retrospective multicenter study. *Future Oncol.* **2022**, *18*, 1461–1471. [CrossRef]
11. Hou, W.; Ding, M.; Li, X.; Zhou, X.; Zhu, Q.; Varela-Ramirez, A.; Yi, C. Comparative evaluation of cardiovascular risks among nine FDA-approved VEGFR-TKIs in patients with solid tumors: A Bayesian network analysis of randomized controlled trials. *J. Cancer Res. Clin. Oncol.* **2021**, *147*, 2407–2420. [CrossRef] [PubMed]
12. Ljungberg, B.; Albiges, L.; Abu-Ghanem, Y.; Bedke, J.; Capitanio, U.; Dabestani, S.; Fernández-Pello, S.; Giles, R.H.; Hofmann, F.; Hora, M.; et al. European Association of Urology Guidelines on Renal Cell Carcinoma: The 2022 Update. *Eur. Urol.* **2022**, *82*, 399–410. [CrossRef]
13. Powles, T.; Albiges, L.; Bex, A.; Grünwald, V.; Porta, C.; Procopio, G.; Schmidinger, M.; Suárez, C.; De Velasco, G. ESMO Clinical Practice Guideline update on the use of immunotherapy in early stage and advanced renal cell carcinoma. *Ann. Oncol.* **2021**, *32*, 1511–1519. [CrossRef] [PubMed]
14. Wood, D.E. National Comprehensive Cancer Network (NCCN) clinical practice guidelines for lung cancer screening. *Thorac. Surg. Clin.* **2015**, *25*, 185–197. [CrossRef] [PubMed]
15. Lobenwein, D.; Kocher, F.; Dobner, S.; Gollmann-Tepeköylü, C.; Holfeld, J. Cardiotoxic mechanisms of cancer immunotherapy—A systematic review. *Int. J. Cardiol.* **2021**, *323*, 179–187. [CrossRef]
16. Nhola, L.F.; Abdelmoneim, S.S.; Villarraga, H.R.; Kohli, M.; Grothey, A.; Bordun, K.A.; Cheung, M.; Best, R.; Cheung, D.; Huang, R.; et al. Echocardiographic Assessment for the Detection of Cardiotoxicity Due to Vascular Endothelial Growth Factor Inhibitor Therapy in Metastatic Renal Cell and Colorectal Cancers. *J. Am. Soc. Echocardiogr.* **2019**, *32*, 267–276. [CrossRef]
17. Kempton, A.; Justice, C.; Guo, A.; Cefalu, M.; Makara, M.; Janssen, P.; Ho, T.H.; Smith, S.A. Pazopanib for renal cell carcinoma leads to elevated mean arterial pressures in a murine model. *Clin. Exp. Hypertens.* **2018**, *40*, 524–533. [CrossRef] [PubMed]
18. Iacovelli, R.; Ciccarese, C.; Fornarini, G.; Massari, F.; Bimbatti, D.; Mosillo, C.; Rebuzzi, S.E.; Di Nunno, V.; Grassi, M.; Fantinel, E.; et al. Cabozantinib-related cardiotoxicity: A prospective analysis in a real-world cohort of metastatic renal cell carcinoma patients. *Br. J. Clin. Pharmacol.* **2019**, *85*, 1283–1289. [CrossRef]
19. Lee, S.; Chen, T.T.; Barber, C.L.; Jordan, M.C.; Murdock, J.; Desai, S.; Ferrara, N.; Nagy, A.; Roos, K.P.; Iruela-Arispe, M.L. Autocrine VEGF signaling is required for vascular homeostasis. *Cell* **2007**, *130*, 691–703. [CrossRef]
20. Hasinoff, B.B.; Patel, D. The lack of target specificity of small molecule anticancer kinase inhibitors is correlated with their ability to damage myocytes in vitro. *Toxicol. Appl. Pharmacol.* **2010**, *249*, 132–139. [CrossRef]
21. Vanneman, M.; Dranoff, G. Combining immunotherapy and targeted therapies in cancer treatment. *Nat. Rev. Cancer* **2012**, *12*, 237–251. [CrossRef]

22. Shrimali, R.K.; Yu, Z.; Theoret, M.R.; Chinnasamy, D.; Restifo, N.P.; Rosenberg, S.A. Antiangiogenic Agents Can Increase Lymphocyte Infiltration into Tumor and Enhance the Effectiveness of Adoptive Immunotherapy of Cancer Adoptive Cell Therapy and Antiangiogenesis. *Cancer Res.* **2010**, *70*, 6171–6180. [CrossRef] [PubMed]
23. Escudier, B.; Porta, C.; Schmidinger, M.; Algaba, F.; Patard, J.; Khoo, V.; Eisen, T.; Horwich, A. Renal cell carcinoma: ESMO Clinical Practice Guidelines for diagnosis, treatment and follow-up. *Ann. Oncol.* **2014**, *25*, iii49–iii56. [CrossRef] [PubMed]
24. Motzer, R.J.; Hutson, T.E.; Tomczak, P.; Michaelson, M.D.; Bukowski, R.M.; Oudard, S.; Negrier, S.; Szczylik, C.; Pili, R.; Bjarnason, G.A.; et al. Overall survival and updated results for sunitinib compared with interferon alfa in patients with metastatic renal cell carcinoma. *J. Clin. Oncol.* **2009**, *27*, 3584–3590. [CrossRef] [PubMed]
25. Motzer, R.J.; Hutson, T.E.; Tomczak, P.; Michaelson, M.D.; Bukowski, R.M.; Rixe, O.; Oudard, S.; Negrier, S.; Szczylik, C.; Kim, S.T.; et al. Sunitinib versus interferon alfa in metastatic renal-cell carcinoma. *N. Engl. J. Med.* **2007**, *356*, 115–124. [CrossRef] [PubMed]
26. Chrisoulidou, A.; Mandanas, S.; Margaritidou, E.; Mathiopoulou, L.; Boudina, M.; Georgopoulos, K.; Pazaitou-Panayiotou, K. Treatment compliance and severe adverse events limit the use of tyrosine kinase inhibitors in refractory thyroid cancer. *Onco Targets Ther.* **2015**, *8*, 2435–2442. [CrossRef] [PubMed]
27. Hudes, G.R.; Carducci, M.A.; Choueiri, T.K.; Esper, P.; Jonasch, E.; Kumar, R.; Margolin, K.A.; Michaelson, M.D.; Motzer, R.J.; Pili, R.; et al. NCCN Task Force report: Optimizing treatment of advanced renal cell carcinoma with molecular targeted therapy. *J. Natl. Compr. Cancer Netw.* **2011**, *9*, S-1–S-29. [CrossRef]
28. Stadler, W.M.; Figlin, R.A.; McDermott, D.F.; Dutcher, J.P.; Knox, J.J.; Miller, W.H., Jr.; Hainsworth, J.D.; Henderson, C.A.; George, J.R.; Hajdenberg, J.; et al. Safety and efficacy results of the advanced renal cell carcinoma sorafenib expanded access program in North America. *Cancer Interdiscip. Int. J. Am. Cancer Soc.* **2010**, *116*, 1272–1280. [CrossRef]
29. Chu, T.F.; Rupnick, M.A.; Kerkela, R.; Dallabrida, S.M.; Zurakowski, D.; Nguyen, L.; Woulfe, K.; Pravda, E.; Cassiola, F.; Desai, J.; et al. Cardiotoxicity associated with tyrosine kinase inhibitor sunitinib. *Lancet* **2007**, *370*, 2011–2019. [CrossRef]
30. Buti, S.; Bersanelli, M. Is cabozantinib really better than sunitinib as first-line treatment of metastatic renal cell carcinoma? *J. Clin. Oncol.* **2017**, *35*, 1858–1859. [CrossRef]
31. Blumenthal, G.M.; Cortazar, P.; Zhang, J.J.; Tang, S.; Sridhara, R.; Murgo, A.; Justice, R.; Pazdur, R. FDA approval summary: Sunitinib for the treatment of progressive well-differentiated locally advanced or metastatic pancreatic neuroendocrine tumors. *Oncologist* **2012**, *17*, 1108–1113. [CrossRef]
32. Akaza, H.; Naito, S.; Ueno, N.; Aoki, K.; Houzawa, H.; Pitman Lowenthal, S.; Lee, S.Y. Real-world use of sunitinib in Japanese patients with advanced renal cell carcinoma: Efficacy, safety and biomarker analyses in 1689 consecutive patients. *Jpn. J. Clin. Oncol.* **2015**, *45*, 576–583. [CrossRef]
33. Haas, N.B.; Manola, J.; Ky, B.; Flaherty, K.T.; Uzzo, R.G.; Kane, C.J.; Jewett, M.; Wood, L.; Wood, C.G.; Atkins, M.B.; et al. Effects of Adjuvant Sorafenib and Sunitinib on Cardiac Function in Renal Cell Carcinoma Patients without Overt Metastases: Results from ASSURE, ECOG 2805. *Clin. Cancer Res.* **2015**, *21*, 4048–4054. [CrossRef] [PubMed]
34. SUTENT (Sunitinib Malate) Capsules, Oral. 2018. Available online: https://www.accessdata.fda.gov/drugsatfda_docs/label/2018/021938s035lbl.pdf (accessed on 10 July 2022).
35. Hamo, C.E.; Bloom, M.W. Getting to the Heart of the Matter: An Overview of Cardiac Toxicity Related to Cancer Therapy. *Clin. Med. Insights Cardiol.* **2015**, *9*, 47–51. [CrossRef]
36. Greineder, C.F.; Kohnstamm, S.; Ky, B. Heart failure associated with sunitinib: Lessons learned from animal models. *Curr. Hypertens. Rep.* **2011**, *13*, 436–441. [CrossRef] [PubMed]
37. Desai, J.; Shankar, S.; Heinrich, M.C.; Fletcher, J.A.; Fletcher, C.D.; Manola, J.; Morgan, J.A.; Corless, C.L.; George, S.; Tuncali, K.; et al. Clonal evolution of resistance to imatinib in patients with metastatic gastrointestinal stromal tumors. *Clin. Cancer Res.* **2007**, *13*, 5398–5405. [CrossRef] [PubMed]
38. Pantaleo, M.A.; Mandrioli, A.; Saponara, M.; Nannini, M.; Erente, G.; Lolli, C.; Biasco, G. Development of coronary artery stenosis in a patient with metastatic renal cell carcinoma treated with sorafenib. *BMC Cancer* **2012**, *12*, 231. [CrossRef]
39. Wilhelm, S.M.; Carter, C.; Tang, L.; Wilkie, D.; McNabola, A.; Rong, H.; Chen, C.; Zhang, X.; Vincent, P.; McHugh, M.; et al. BAY 43-9006 exhibits broad spectrum oral antitumor activity and targets the RAF/MEK/ERK pathway and receptor tyrosine kinases involved in tumor progression and angiogenesis. *Cancer Res.* **2004**, *64*, 7099–7109. [CrossRef]
40. Escudier, B.; Eisen, T.; Stadler, W.M.; Szczylik, C.; Oudard, S.; Siebels, M.; Negrier, S.; Chevreau, C.; Solska, E.; Desai, A.A.; et al. Sorafenib in advanced clear-cell renal-cell carcinoma. *N. Engl. J. Med.* **2007**, *356*, 125–134. [CrossRef] [PubMed]
41. Escudier, B.; Eisen, T.; Stadler, W.M.; Szczylik, C.; Oudard, S.; Staehler, M.; Negrier, S.; Chevreau, C.; Desai, A.A.; Rolland, F.; et al. Sorafenib for treatment of renal cell carcinoma: Final efficacy and safety results of the phase III treatment approaches in renal cancer global evaluation trial. *J. Clin. Oncol.* **2009**, *27*, 3312–3318. [CrossRef]
42. Hamberg, P.; Verweij, J.; Sleijfer, S. (Pre-)clinical pharmacology and activity of pazopanib, a novel multikinase angiogenesis inhibitor. *Oncologist* **2010**, *15*, 539–547. [CrossRef] [PubMed]
43. Choueiri, T.; Hessel, C.; Halabi, S.; Sanford, B.; Hahn, O.; Michaelson, M.; Walsh, M.; Olencki, T.; Picus, J.; Small, E.; et al. Progression-free survival (PFS) by independent review and updated overall survival (OS) results from Alliance A031203 trial (CABOSUN): Cabozantinib versus sunitinib as initial targeted therapy for patients (pts) with metastatic renal cell carcinoma (mRCC). *Ann. Oncol.* **2017**, *28*, v623. [CrossRef]

44. Choueiri, T.K.; Escudier, B.; Powles, T.; Tannir, N.M.; Mainwaring, P.N.; Rini, B.I.; Hammers, H.J.; Donskov, F.; Roth, B.J.; Peltola, K.; et al. Cabozantinib versus everolimus in advanced renal cell carcinoma (METEOR): Final results from a randomised, open-label, phase 3 trial. *Lancet Oncol.* **2016**, *17*, 917–927. [CrossRef] [PubMed]
45. Atkins, M.B.; Philips, G.K. Emerging monoclonal antibodies for the treatment of renal cell carcinoma (RCC). *Expert Opin. Emerg. Drugs* **2016**, *21*, 243–254. [CrossRef] [PubMed]
46. Force, T.; Krause, D.S.; Van Etten, R.A. Molecular mechanisms of cardiotoxicity of tyrosine kinase inhibition. *Nat. Rev. Cancer* **2007**, *7*, 332–344. [CrossRef]
47. Khakoo, A.Y.; Kassiotis, C.M.; Tannir, N.; Plana, J.C.; Halushka, M.; Bickford, C.; Trent, J., 2nd; Champion, J.C.; Durand, J.B.; Lenihan, D.J. Heart failure associated with sunitinib malate: A multitargeted receptor tyrosine kinase inhibitor. *Cancer Interdiscip. Int. J. Am. Cancer Soc.* **2008**, *112*, 2500–2508. [CrossRef] [PubMed]
48. Telli, M.; Witteles, R.; Fisher, G.; Srinivas, S. Cardiotoxicity associated with the cancer therapeutic agent sunitinib malate. *Ann. Oncol.* **2008**, *19*, 1613–1618. [CrossRef]
49. Force, T.; Kolaja, K.L. Cardiotoxicity of kinase inhibitors: The prediction and translation of preclinical models to clinical outcomes. *Nat. Rev. Drug Discov.* **2011**, *10*, 111–126. [CrossRef]
50. Catino, A.B.; Hubbard, R.A.; Chirinos, J.A.; Townsend, R.; Keefe, S.; Haas, N.B.; Puzanov, I.; Fang, J.C.; Agarwal, N.; Hyman, D.; et al. Longitudinal Assessment of Vascular Function with Sunitinib in Patients With Metastatic Renal Cell Carcinoma. *Circ. Heart Fail.* **2018**, *11*, e004408. [CrossRef]
51. Zamorano, J.L.; Lancellotti, P.; Rodriguez Muñoz, D.; Aboyans, V.; Asteggiano, R.; Galderisi, M.; Habib, G.; Lenihan, D.J.; Lip, G.Y.H.; Lyon, A.R.; et al. 2016 ESC Position Paper on cancer treatments and cardiovascular toxicity developed under the auspices of the ESC Committee for Practice Guidelines: The Task Force for cancer treatments and cardiovascular toxicity of the European Society of Cardiology (ESC). *Eur. Heart J.* **2016**, *37*, 2768–2801. [CrossRef]
52. Hall, P.S.; Harshman, L.C.; Srinivas, S.; Witteles, R.M. The frequency and severity of cardiovascular toxicity from targeted therapy in advanced renal cell carcinoma patients. *JACC Heart Fail.* **2013**, *1*, 72–78. [CrossRef]
53. Ghatalia, P.; Morgan, C.J.; Je, Y.; Nguyen, P.L.; Trinh, Q.-D.; Choueiri, T.K.; Sonpavde, G. Congestive heart failure with vascular endothelial growth factor receptor tyrosine kinase inhibitors. *Crit. Rev. Oncol. Hematol.* **2015**, *94*, 228–237. [CrossRef]
54. Tukenova, M.; Guibout, C.; Oberlin, O.; Doyon, F.; Mousannif, A.; Haddy, N.; Guérin, S.; Pacquement, H.; Aouba, A.; Hawkins, M.; et al. Role of cancer treatment in long-term overall and cardiovascular mortality after childhood cancer. *J. Clin. Oncol.* **2010**, *28*, 1308–1315. [CrossRef] [PubMed]
55. Armenian, S.H.; Xu, L.; Ky, B.; Sun, C.; Farol, L.T.; Pal, S.K.; Douglas, P.S.; Bhatia, S.; Chao, C. Cardiovascular disease among survivors of adult-onset cancer: A community-based retrospective cohort study. *J. Clin. Oncol.* **2016**, *34*, 1122–1130. [CrossRef]
56. Narayan, V.; Ky, B. Common Cardiovascular Complications of Cancer Therapy: Epidemiology, Risk Prediction, and Prevention. *Annu. Rev. Med.* **2018**, *69*, 97–111. [CrossRef] [PubMed]
57. Carmeliet, P. Angiogenesis in life, disease and medicine. *Nature* **2005**, *438*, 932–936. [CrossRef]
58. Pepe, M.; Mamdani, M.; Zentilin, L.; Csiszar, A.; Qanud, K.; Zacchigna, S.; Ungvari, Z.; Puligadda, U.; Moimas, S.; Xu, X.; et al. Intramyocardial VEGF-B167 gene delivery delays the progression towards congestive failure in dogs with pacing-induced dilated cardiomyopathy. *Circ. Res.* **2010**, *106*, 1893–1903. [CrossRef]
59. Izumiya, Y.; Shiojima, I.; Sato, K.; Sawyer, D.B.; Colucci, W.S.; Walsh, K. Vascular endothelial growth factor blockade promotes the transition from compensatory cardiac hypertrophy to failure in response to pressure overload. *Hypertension* **2006**, *47*, 887–893. [CrossRef] [PubMed]
60. Zhao, Q.; Ishibashi, M.; Hiasa, K.-i.; Tan, C.; Takeshita, A.; Egashira, K. Essential role of vascular endothelial growth factor in angiotensin II–induced vascular inflammation and remodeling. *Hypertension* **2004**, *44*, 264–270. [CrossRef]
61. Abraham, D.; Hofbauer, R.; Schäfer, R.; Blumer, R.; Paulus, P.; Miksovsky, A.; Traxler, H.; Kocher, A.; Aharinejad, S. Selective downregulation of VEGF-A165, VEGF-R1, and decreased capillary density in patients with dilative but not ischemic cardiomyopathy. *Circ. Res.* **2000**, *87*, 644–647. [CrossRef]
62. Hsieh, P.C.; Davis, M.E.; Lisowski, L.K.; Lee, R.T. Endothelial-cardiomyocyte interactions in cardiac development and repair. *Annu. Rev. Physiol.* **2006**, *68*, 51–66. [CrossRef]
63. Heloterä, H.; Alitalo, K. The VEGF family, the inside story. *Cell* **2007**, *130*, 591–592. [CrossRef] [PubMed]
64. Kamba, T.; McDonald, D. Mechanisms of adverse effects of anti-VEGF therapy for cancer. *Br. J. Cancer* **2007**, *96*, 1788–1795. [CrossRef]
65. Milan, A.; Puglisi, E.; Ferrari, L.; Bruno, G.; Losano, I.; Veglio, F. Arterial hypertension and cancer. *Int. J. Cancer* **2014**, *134*, 2269–2277. [CrossRef]
66. Wu, S.; Chen, J.J.; Kudelka, A.; Lu, J.; Zhu, X. Incidence and risk of hypertension with sorafenib in patients with cancer: A systematic review and meta-analysis. *Lancet Oncol.* **2008**, *9*, 117–123. [CrossRef]
67. Schmidinger, M.; Zielinski, C.C.; Vogl, U.M.; Bojic, A.; Bojic, M.; Schukro, C.; Ruhsam, M.; Hejna, M.; Schmidinger, H. Cardiac toxicity of sunitinib and sorafenib in patients with metastatic renal cell carcinoma. *J. Clin. Oncol.* **2008**, *26*, 5204–5212. [CrossRef]
68. Narayan, V.; Keefe, S.; Haas, N.; Wang, L.; Puzanov, I.; Putt, M.; Catino, A.; Fang, J.; Agarwal, N.; Hyman, D. Prospective Evaluation of Sunitinib-Induced Cardiotoxicity in Patients with Metastatic Renal Cell CarcinomaSunitinib Cardiotoxicity. *Clin. Cancer Res.* **2017**, *23*, 3601–3609. [CrossRef] [PubMed]

69. Kerkela, R.; Woulfe, K.C.; Durand, J.B.; Vagnozzi, R.; Kramer, D.; Chu, T.F.; Beahm, C.; Chen, M.H.; Force, T. Sunitinib-induced cardiotoxicity is mediated by off-target inhibition of AMP-activated protein kinase. *Clin. Transl. Sci.* **2009**, *2*, 15–25. [CrossRef] [PubMed]
70. Arad, M.; Seidman, C.E.; Seidman, J. AMP-activated protein kinase in the heart: Role during health and disease. *Circ. Res.* **2007**, *100*, 474–488. [CrossRef]
71. Gruber, H.E.; Hoffer, M.E.; McAllister, D.R.; Laikind, P.K.; Lane, T.A.; Schmid-Schoenbein, G.W.; Engler, R.L. Increased adenosine concentration in blood from ischemic myocardium by AICA riboside. Effects on flow, granulocytes, and injury. *Circulation* **1989**, *80*, 1400–1411. [CrossRef]
72. Terai, K.; Hiramoto, Y.; Masaki, M.; Sugiyama, S.; Kuroda, T.; Hori, M.; Kawase, I.; Hirota, H. AMP-activated protein kinase protects cardiomyocytes against hypoxic injury through attenuation of endoplasmic reticulum stress. *Mol. Cell. Biol.* **2005**, *25*, 9554–9575. [CrossRef]
73. Will, Y.; Dykens, J.A.; Nadanaciva, S.; Hirakawa, B.; Jamieson, J.; Marroquin, L.D.; Hynes, J.; Patyna, S.; Jessen, B.A. Effect of the multitargeted tyrosine kinase inhibitors imatinib, dasatinib, sunitinib, and sorafenib on mitochondrial function in isolated rat heart mitochondria and H9c2 cells. *Toxicol. Sci.* **2008**, *106*, 153–161. [CrossRef]
74. Hasinoff, B.B.; Patel, D.; O'Hara, K.A. Mechanisms of myocyte cytotoxicity induced by the multiple receptor tyrosine kinase inhibitor sunitinib. *Mol. Pharmacol.* **2008**, *74*, 1722–1728. [CrossRef] [PubMed]
75. Rainer, P.P.; Doleschal, B.; Kirk, J.A.; Sivakumaran, V.; Saad, Z.; Groschner, K.; Maechler, H.; Hoefler, G.; Bauernhofer, T.; Samonigg, H.; et al. Sunitinib causes dose-dependent negative functional effects on myocardium and cardiomyocytes. *BJU Int.* **2012**, *110*, 1455–1462. [CrossRef]
76. Hahn, V.S.; Lenihan, D.J.; Ky, B. Cancer therapy–induced cardiotoxicity: Basic mechanisms and potential cardioprotective therapies. *J. Am. Heart Assoc.* **2014**, *3*, e000665. [CrossRef] [PubMed]
77. de Jesus-Gonzalez, N.; Robinson, E.; Moslehi, J.; Humphreys, B.D. Management of antiangiogenic therapy-induced hypertension. *Hypertension* **2012**, *60*, 607–615. [CrossRef]
78. van der Veldt, A.A.; de Boer, M.P.; Boven, E.; Eringa, E.C.; van den Eertwegh, A.J.; van Hinsbergh, V.W.; Smulders, Y.M.; Serné, E.H. Reduction in skin microvascular density and changes in vessel morphology in patients treated with sunitinib. *Anti-Cancer Drugs* **2010**, *21*, 439–446. [CrossRef] [PubMed]
79. Eremina, V.; Cui, S.; Gerber, H.; Ferrara, N.; Haigh, J.; Nagy, A.; Ema, M.; Rossant, J.; Jothy, S.; Miner, J.H.; et al. Vascular endothelial growth factor a signaling in the podocyte-endothelial compartment is required for mesangial cell migration and survival. *J. Am. Soc. Nephrol.* **2006**, *17*, 724–735. [CrossRef] [PubMed]
80. Inai, T.; Mancuso, M.; Hashizume, H.; Baffert, F.; Haskell, A.; Baluk, P.; Hu-Lowe, D.D.; Shalinsky, D.R.; Thurston, G.; Yancopoulos, G.D.; et al. Inhibition of vascular endothelial growth factor (VEGF) signaling in cancer causes loss of endothelial fenestrations, regression of tumor vessels, and appearance of basement membrane ghosts. *Am. J. Pathol.* **2004**, *165*, 35–52. [CrossRef] [PubMed]
81. Kappers, M.H.; van Esch, J.H.; Sluiter, W.; Sleijfer, S.; Danser, A.J.; van den Meiracker, A.H. Hypertension induced by the tyrosine kinase inhibitor sunitinib is associated with increased circulating endothelin-1 levels. *Hypertension* **2010**, *56*, 675–681. [CrossRef] [PubMed]
82. Ku, D.D.; Zaleski, J.K.; Liu, S.; Brock, T.A. Vascular endothelial growth factor induces EDRF-dependent relaxation in coronary arteries. *Am. J. Physiol.—Heart Circ. Physiol.* **1993**, *265*, H586–H592. [CrossRef]
83. Yamaguchi, O.; Watanabe, T.; Nishida, K.; Kashiwase, K.; Higuchi, Y.; Takeda, T.; Hikoso, S.; Hirotani, S.; Asahi, M.; Taniike, M.; et al. Cardiac-specific disruption of the c-raf-1 gene induces cardiac dysfunction and apoptosis. *J. Clin. Investig.* **2004**, *114*, 937–943. [CrossRef]
84. Chou, M.T.; Wang, J.; Fujita, D.J. Src kinase becomes preferentially associated with the VEGFR, KDR/Flk-1, following VEGF stimulation of vascular endothelial cells. *BMC Biochem.* **2002**, *3*, 32. [CrossRef] [PubMed]
85. Altarche-Xifró, W.; Curato, C.; Kaschina, E.; Grzesiak, A.; Slavic, S.; Dong, J.; Kappert, K.; Steckelings, M.; Imboden, H.; Unger, T.; et al. Cardiac c-kit+ AT2+ cell population is increased in response to ischemic injury and supports cardiomyocyte performance. *Stem Cells* **2009**, *27*, 2488–2497. [CrossRef] [PubMed]
86. French, K.J.; Coatney, R.W.; Renninger, J.P.; Hu, C.X.; Gales, T.L.; Zhao, S.; Storck, L.M.; Davis, C.B.; McSurdy-Freed, J.; Chen, E.; et al. Differences in effects on myocardium and mitochondria by angiogenic inhibitors suggest separate mechanisms of cardiotoxicity. *Toxicol. Pathol.* **2010**, *38*, 691–702. [CrossRef]
87. Prieto-Domínguez, N.; Ordóñez, R.; Fernández, A.; García-Palomo, A.; Muntané, J.; González-Gallego, J.; Mauriz, J.L. Modulation of autophagy by sorafenib: Effects on treatment response. *Front. Pharmacol.* **2016**, *7*, 151. [CrossRef]
88. Nakai, A.; Yamaguchi, O.; Takeda, T.; Higuchi, Y.; Hikoso, S.; Taniike, M.; Omiya, S.; Mizote, I.; Matsumura, Y.; Asahi, M.; et al. The role of autophagy in cardiomyocytes in the basal state and in response to hemodynamic stress. *Nat. Med.* **2007**, *13*, 619–624. [CrossRef]
89. Knaapen, M.W.; Davies, M.J.; De Bie, M.; Haven, A.J.; Martinet, W.; Kockx, M.M. Apoptotic versus autophagic cell death in heart failure. *Cardiovasc. Res.* **2001**, *51*, 304–312. [CrossRef]
90. Nishida, K.; Kyoi, S.; Yamaguchi, O.; Sadoshima, J.; Otsu, K. The role of autophagy in the heart. *Cell Death Differ.* **2009**, *16*, 31–38. [CrossRef] [PubMed]

91. González, A.; Fortuño, M.a.A.; Querejeta, R.; Ravassa, S.; López, B.; López, N.; Díez, J. Cardiomyocyte apoptosis in hypertensive cardiomyopathy. *Cardiovasc. Res.* **2003**, *59*, 549–562. [CrossRef]
92. Kunapuli, S.; Rosanio, S.; Schwarz, E.R. "How do cardiomyocytes die?" apoptosis and autophagic cell death in cardiac myocytes. *J. Card. Fail.* **2006**, *12*, 381–391. [CrossRef]
93. Vaidya, T.; Kamta, J.; Chaar, M.; Ande, A.; Ait-Oudhia, S. Systems pharmacological analysis of mitochondrial cardiotoxicity induced by selected tyrosine kinase inhibitors. *J. Pharmacokinet. Pharmacodyn.* **2018**, *45*, 401–418. [CrossRef]
94. Abdelgalil, A.A.; Mohamed, O.Y.; Ahamad, S.R.; Al-Jenoobi, F.I. The protective effect of losartan against sorafenib induced cardiotoxicity: Ex-vivo isolated heart and metabolites profiling studies in rat. *Eur. J. Pharmacol.* **2020**, *882*, 173229. [CrossRef]
95. Heusch, G.; Schulz, R. Perfusion-contraction match and mismatch. *Basic Res. Cardiol.* **2001**, *96*, 1–10. [CrossRef]
96. Ross, J., Jr. Myocardial perfusion-contraction matching. Implications for coronary heart disease and hibernation. *Circulation* **1991**, *83*, 1076–1083. [CrossRef]
97. Gould, K.L. Does coronary flow trump coronary anatomy? *JACC Cardiovasc. Imaging* **2009**, *2*, 1009–1023. [CrossRef]
98. Touyz, R.M.; Herrmann, J. Cardiotoxicity with vascular endothelial growth factor inhibitor therapy. *NPJ Precis. Oncol.* **2018**, *2*, 13. [CrossRef]
99. Chintalgattu, V.; Rees, M.L.; Culver, J.C.; Goel, A.; Jiffar, T.; Zhang, J.; Dunner, K., Jr.; Pati, S.; Bankson, J.A.; Pasqualini, R.; et al. Coronary microvascular pericytes are the cellular target of sunitinib malate-induced cardiotoxicity. *Sci. Transl. Med.* **2013**, *5*, 187ra69. [CrossRef] [PubMed]
100. Chintalgattu, V.; Ai, D.; Langley, R.R.; Zhang, J.; Bankson, J.A.; Shih, T.L.; Reddy, A.K.; Coombes, K.R.; Daher, I.N.; Pati, S.; et al. Cardiomyocyte PDGFR-β signaling is an essential component of the mouse cardiac response to load-induced stress. *J. Clin. Investig.* **2010**, *120*, 472–484. [CrossRef] [PubMed]
101. Force, T.; Kerkelä, R. Cardiotoxicity of the new cancer therapeutics–mechanisms of, and approaches to, the problem. *Drug Discov. Today* **2008**, *13*, 778–784. [CrossRef] [PubMed]
102. Edelberg, J.M.; Lee, S.H.; Kaur, M.; Tang, L.; Feirt, N.M.; McCabe, S.; Bramwell, O.; Wong, S.C.; Hong, M.K. Platelet-derived growth factor-AB limits the extent of myocardial infarction in a rat model: Feasibility of restoring impaired angiogenic capacity in the aging heart. *Circulation* **2002**, *105*, 608–613. [CrossRef] [PubMed]
103. Herrmann, J.; Yang, E.H.; Iliescu, C.A.; Cilingiroglu, M.; Charitakis, K.; Hakeem, A.; Toutouzas, K.; Leesar, M.A.; Grines, C.L.; Marmagkiolis, K. Vascular Toxicities of Cancer Therapies: The Old and the New—An Evolving Avenue. *Circulation* **2016**, *133*, 1272–1289. [CrossRef]
104. Motzer, R.J.; Hutson, T.E.; Cella, D.; Reeves, J.; Hawkins, R.; Guo, J.; Nathan, P.; Staehler, M.; de Souza, P.; Merchan, J.R.; et al. Pazopanib versus sunitinib in metastatic renal-cell carcinoma. *N. Engl. J. Med.* **2013**, *369*, 722–731. [CrossRef]
105. Motzer, R.J.; Escudier, B.; Tomczak, P.; Hutson, T.E.; Michaelson, M.D.; Negrier, S.; Oudard, S.; Gore, M.E.; Tarazi, J.; Hariharan, S.; et al. Axitinib versus sorafenib as second-line treatment for advanced renal cell carcinoma: Overall survival analysis and updated results from a randomised phase 3 trial. *Lancet Oncol.* **2013**, *14*, 552–562. [CrossRef] [PubMed]
106. Qi, W.X.; Shen, Z.; Tang, L.N.; Yao, Y. Congestive heart failure risk in cancer patients treated with vascular endothelial growth factor tyrosine kinase inhibitors: A systematic review and meta-analysis of 36 clinical trials. *Br. J. Clin. Pharmacol.* **2014**, *78*, 748–762. [CrossRef] [PubMed]
107. Choueiri, T.K.; Schutz, F.; Je, Y.; Rosenberg, J.E.; Bellmunt, J. Risk of arterial thromboembolic events with sunitinib and sorafenib: A systematic review and meta-analysis of clinical trials. *J. Clin. Oncol.* **2010**, *28*, 2280–2285. [CrossRef] [PubMed]
108. Liu, B.; Ding, F.; Liu, Y.; Xiong, G.; Lin, T.; He, D.; Zhang, Y.; Zhang, D.; Wei, G. Incidence and risk of hypertension associated with vascular endothelial growth factor receptor tyrosine kinase inhibitors in cancer patients: A comprehensive network meta-analysis of 72 randomized controlled trials involving 30,013 patients. *Oncotarget* **2016**, *7*, 67661. [CrossRef]
109. Qi, W.X.; Min, D.L.; Shen, Z.; Sun, Y.J.; Lin, F.; Tang, L.N.; He, A.N.; Yao, Y. Risk of venous thromboembolic events associated with VEGFR-TKIs: A systematic review and meta-analysis. *Int. J. Cancer* **2013**, *132*, 2967–2974. [CrossRef]
110. Liu, B.; Ding, F.; Zhang, D.; Wei, G.-h. Risk of venous and arterial thromboembolic events associated with VEGFR-TKIs: A meta-analysis. *Cancer Chemother. Pharmacol.* **2017**, *80*, 487–495. [CrossRef]
111. Qi, W.-X.; Lin, F.; Sun, Y.-j.; Tang, L.-N.; He, A.-N.; Yao, Y.; Shen, Z. Incidence and risk of hypertension with pazopanib in patients with cancer: A meta-analysis. *Cancer Chemother. Pharmacol.* **2013**, *71*, 431–439. [CrossRef]
112. Veronese, M.L.; Mosenkis, A.; Flaherty, K.T.; Gallagher, M.; Stevenson, J.P.; Townsend, R.R.; O'Dwyer, P.J. Mechanisms of hypertension associated with BAY 43-9006. *J. Clin. Oncol.* **2006**, *24*, 1363–1369. [CrossRef]
113. Richards, C.J.; Je, Y.; Schutz, F.; Heng, D.; Dallabrida, S.M.; Moslehi, J.J.; Choueiri, T.K. Incidence and risk of congestive heart failure in patients with renal and nonrenal cell carcinoma treated with sunitinib. *J. Clin. Oncol.* **2011**, *29*, 3450–3456. [CrossRef]
114. Ranpura, V.; Hapani, S.; Chuang, J.; Wu, S. Risk of cardiac ischemia and arterial thromboembolic events with the angiogenesis inhibitor bevacizumab in cancer patients: A meta-analysis of randomized controlled trials. *Acta Oncol.* **2010**, *49*, 287–297. [CrossRef] [PubMed]
115. Abdel-Qadir, H.; Ethier, J.-L.; Lee, D.S.; Thavendiranathan, P.; Amir, E. Cardiovascular toxicity of angiogenesis inhibitors in treatment of malignancy: A systematic review and meta-analysis. *Cancer Treat. Rev.* **2017**, *53*, 120–127. [CrossRef] [PubMed]
116. Li, J.; Gu, J. Cardiovascular toxicities with vascular endothelial growth factor receptor tyrosine kinase inhibitors in cancer patients: A meta-analysis of 77 randomized controlled trials. *Clin. Drug Investig.* **2018**, *38*, 1109–1123. [CrossRef] [PubMed]

117. Ghatalia, P.; Je, Y.; Kaymakcalan, M.; Sonpavde, G.; Choueiri, T. QTc interval prolongation with vascular endothelial growth factor receptor tyrosine kinase inhibitors. *Br. J. Cancer* **2015**, *112*, 296–305. [CrossRef]
118. Totzeck, M.; Mincu, R.-I.; Mrotzek, S.; Schadendorf, D.; Rassaf, T. Cardiovascular diseases in patients receiving small molecules with anti-vascular endothelial growth factor activity: A meta-analysis of approximately 29,000 cancer patients. *Eur. J. Prev. Cardiol.* **2018**, *25*, 482–494. [CrossRef]
119. Furuya-Kanamori, L.; Doi, S.A.; Onitilo, A.; Akhtar, S. Is there truly an increase in risk of cardiovascular and hematological adverse events with vascular endothelial growth factor receptor tyrosine kinase inhibitors? *Expert Opin. Drug Saf.* **2020**, *19*, 223–228. [CrossRef] [PubMed]
120. Yancy, C.W.; Jessup, M.; Bozkurt, B.; Butler, J.; Casey, D.E.; Drazner, M.H.; Fonarow, G.C.; Geraci, S.A.; Horwich, T.; Januzzi, J.L. 2013 ACCF/AHA guideline for the management of heart failure: A report of the American College of Cardiology Foundation/American Heart Association Task Force on Practice Guidelines. *J. Am. Coll. Cardiol.* **2013**, *62*, e147–e239. [CrossRef]
121. Highlights of Prescribing Information: SUTENT® (sunitinib malate) capsules, oral. Available online: https://hemonc.org/docs/packageinsert/sunitinib.pdf (accessed on 10 June 2022).
122. Kappers, M.H.; de Beer, V.J.; Zhou, Z.; Danser, A.J.; Sleijfer, S.; Duncker, D.J.; van den Meiracker, A.H.; Merkus, D. Sunitinib-induced systemic vasoconstriction in swine is endothelin mediated and does not involve nitric oxide or oxidative stress. *Hypertension* **2012**, *59*, 151–157. [CrossRef]
123. Di Lorenzo, G.; Autorino, R.; Bruni, G.; Carteni, G.; Ricevuto, E.; Tudini, M.; Ficorella, C.; Romano, C.; Aieta, M.; Giordano, A.; et al. Cardiovascular toxicity following sunitinib therapy in metastatic renal cell carcinoma: A multicenter analysis. *Ann. Oncol.* **2009**, *20*, 1535–1542. [CrossRef]
124. Lestuzzi, C.; Vaccher, E.; Talamini, R.; Lleshi, A.; Meneguzzo, N.; Viel, E.; Scalone, S.; Tartuferi, L.; Buonadonna, A.; Ejiofor, L.; et al. Effort myocardial ischemia during chemotherapy with 5-fluorouracil: An underestimated risk. *Ann. Oncol.* **2014**, *25*, 1059–1064. [CrossRef]
125. Townsend, R.R.; Wilkinson, I.B.; Schiffrin, E.L.; Avolio, A.P.; Chirinos, J.A.; Cockcroft, J.R.; Heffernan, K.S.; Lakatta, E.G.; McEniery, C.M.; Mitchell, G.F. Recommendations for improving and standardizing vascular research on arterial stiffness: A scientific statement from the American Heart Association. *Hypertension* **2015**, *66*, 698–722. [CrossRef]
126. Hashimoto, J.; Westerhof, B.E.; Westerhof, N.; Imai, Y.; O'Rourke, M.F. Different role of wave reflection magnitude and timing on left ventricular mass reduction during antihypertensive treatment. *J. Hypertens.* **2008**, *26*, 1017–1024. [CrossRef]
127. Kobayashi, S.; Yano, M.; Kohno, M.; Obayashi, M.; Hisamatsu, Y.; Ryoke, T.; Ohkusa, T.; Yamakawa, K.; Matsuzaki, M. Influence of aortic impedance on the development of pressure-overload left ventricular hypertrophy in rats. *Circulation* **1996**, *94*, 3362–3368. [CrossRef]
128. Chirinos, J.A.; Kips, J.G.; Jacobs, D.R.; Brumback, L.; Duprez, D.A.; Kronmal, R.; Bluemke, D.A.; Townsend, R.R.; Vermeersch, S.; Segers, P. Arterial wave reflections and incident cardiovascular events and heart failure: MESA (Multiethnic Study of Atherosclerosis). *J. Am. Coll. Cardiol.* **2012**, *60*, 2170–2177. [CrossRef] [PubMed]
129. Zhang, X.; Shao, Y.; Wang, K. Incidence and risk of hypertension associated with cabozantinib in cancer patients: A systematic review and meta-analysis. *Expert Rev. Clin. Pharmacol.* **2016**, *9*, 1109–1115. [CrossRef] [PubMed]
130. Schmidinger, M.; Danesi, R. Management of adverse events associated with cabozantinib therapy in renal cell carcinoma. *Oncol.* **2018**, *23*, 306–315. [CrossRef] [PubMed]
131. Sternberg, C.N.; Hawkins, R.E.; Wagstaff, J.; Salman, P.; Mardiak, J.; Barrios, C.H.; Zarba, J.J.; Gladkov, O.A.; Lee, E.; Szczylik, C.; et al. A randomised, double-blind phase III study of pazopanib in patients with advanced and/or metastatic renal cell carcinoma: Final overall survival results and safety update. *Eur. J. Cancer* **2013**, *49*, 1287–1296. [CrossRef]
132. Duffaud, F.; Sleijfer, S.; Litiere, S.; Ray-Coquard, I.; Le Cesne, A.; Papai, Z.; Judson, I.; Schöffski, P.; Chawla, S.; Dewji, R.; et al. Hypertension (HTN) as a potential biomarker of efficacy in pazopanib-treated patients with advanced non-adipocytic soft tissue sarcoma. A retrospective study based on European Organisation for Research and Treatment of Cancer (EORTC) 62043 and 62072 trials. *Eur. J. Cancer* **2015**, *51*, 2615–2623. [CrossRef]
133. Kumar, R.; Knick, V.B.; Rudolph, S.K.; Johnson, J.H.; Crosby, R.M.; Crouthamel, M.-C.; Hopper, T.M.; Miller, C.G.; Harrington, L.E.; Onori, J.A.; et al. Pharmacokinetic-pharmacodynamic correlation from mouse to human with pazopanib, a multikinase angiogenesis inhibitor with potent antitumor and antiangiogenic activity. *Mol. Cancer Ther.* **2007**, *6*, 2012–2021. [CrossRef]
134. Lim, T.J.; Lee, J.H.; Chang, S.-G.; Lee, C.H.; Min, G.E.; Yoo, K.H.; Jeon, S.H. Life-threatening complications associated with the tyrosine kinase inhibitor sunitinib malate. *Urol. Int.* **2010**, *85*, 475–478. [CrossRef]
135. Liu, J.; Li, L.; Deng, K.; Xu, C.; Busse, J.W.; Vandvik, P.O.; Li, S.; Guyatt, G.H.; Sun, X. Incretin based treatments and mortality in patients with type 2 diabetes: Systematic review and meta-analysis. *BMJ* **2017**, *357*, j2499. [CrossRef] [PubMed]
136. Maharsy, W.; Aries, A.; Mansour, O.; Komati, H.; Nemer, M. Ageing is a risk factor in imatinib mesylate cardiotoxicity. *Eur. J. Heart Fail.* **2014**, *16*, 367–376. [CrossRef] [PubMed]
137. Porto, I.; Leo, A.; Miele, L.; Pompili, M.; Landolfi, R.; Crea, F. A case of variant angina in a patient under chronic treatment with sorafenib. *Nat. Rev. Clin. Oncol.* **2010**, *7*, 476–480. [CrossRef] [PubMed]
138. Arima, Y.; Oshima, S.; Noda, K.; Fukushima, H.; Taniguchi, I.; Nakamura, S.; Shono, M.; Ogawa, H. Sorafenib-induced acute myocardial infarction due to coronary artery spasm. *J. Cardiol.* **2009**, *54*, 512–515. [CrossRef]
139. Naib, T.; Steingart, R.; Chen, C. Sorafenib-associated multivessel coronary artery vasospasm. *Herz* **2011**, *36*, 348–351. [CrossRef]

140. Shelburne, N.; Adhikari, B.; Brell, J.; Davis, M.; Desvigne-Nickens, P.; Freedman, A.; Minasian, L.; Force, T.; Remick, S.C. Cancer treatment-related cardiotoxicity: Current state of knowledge and future research priorities. *J. Natl. Cancer Inst.* **2014**, *106*, dju232. [CrossRef]
141. Vallerio, P.; Orenti, A.; Tosi, F.; Maistrello, M.; Palazzini, M.; Cingarlini, S.; Colombo, P.; Bertuzzi, M.; Spina, F.; Amatu, A.; et al. Major adverse cardiovascular events associated with VEGF-targeted anticancer tyrosine kinase inhibitors: A real-life study and proposed algorithm for proactive management. *ESMO Open* **2022**, *7*, 100338. [CrossRef]
142. Curigliano, G.; Lenihan, D.; Fradley, M.; Ganatra, S.; Barac, A.; Blaes, A.; Herrmann, J.; Porter, C.; Lyon, A.; Lancellotti, P.; et al. Management of cardiac disease in cancer patients throughout oncological treatment: ESMO consensus recommendations. *Ann. Oncol.* **2020**, *31*, 171–190. [CrossRef]
143. Thakur, A.; Witteles, R.M. Cancer therapy-induced left ventricular dysfunction: Interventions and prognosis. *J. Card. Fail.* **2014**, *20*, 155–158. [CrossRef]
144. Polk, A.; Vistisen, K.; Vaage-Nilsen, M.; Nielsen, D.L. A systematic review of the pathophysiology of 5-fluorouracil-induced cardiotoxicity. *BMC Pharmacol. Toxicol.* **2014**, *15*, 47. [CrossRef] [PubMed]
145. Fishman, M.N.; Tomshine, J.; Fulp, W.J.; Foreman, P.K. A systematic review of the efficacy and safety experience reported for sorafenib in advanced renal cell carcinoma (RCC) in the post-approval setting. *PLoS ONE* **2015**, *10*, e0120877. [CrossRef] [PubMed]
146. Strumberg, D.; Richly, H.; Hilger, R.A.; Schleucher, N.; Korfee, S.; Tewes, M.; Faghih, M.; Brendel, E.; Voliotis, D.; Haase, C.G.; et al. Phase I clinical and pharmacokinetic study of the Novel Raf kinase and vascular endothelial growth factor receptor inhibitor BAY 43-9006 in patients with advanced refractory solid tumors. *J. Clin. Oncol.* **2005**, *23*, 965–972. [CrossRef] [PubMed]
147. Ratain, M.J.; Eisen, T.; Stadler, W.M.; Flaherty, K.T.; Kaye, S.B.; Rosner, G.L.; Gore, M.; Desai, A.A.; Patnaik, A.; Xiong, H.Q.; et al. Phase II placebo-controlled randomized discontinuation trial of sorafenib in patients with metastatic renal cell carcinoma. *J. Clin. Oncol.* **2006**, *24*, 2505–2512. [CrossRef] [PubMed]
148. Topal, A.; Sayın, S.; Gokyer, A.; Kucukarda, A.; Kostek, O.; Bekir Hacıoglu, M.; Uzunoglu, S.; Erdogan, B.; Cicin, I. Real-life data on first-line Sunitinib and Pazopanib therapy in metastatic renal cell carcinoma patients: A single center experience. *J. BUON* **2021**, *26*, 1628–1634. [PubMed]
149. Erman, M.; Biswas, B.; Danchaivijitr, P.; Chen, L.; Wong, Y.F.; Hashem, T.; Lim, C.S.; Karabulut, B.; Chung, H.J.; Chikatapu, C.; et al. Prospective observational study on Pazopanib in patients treated for advanced or metastatic renal cell carcinoma in countries in Asia Pacific, North Africa, and Middle East regions: PARACHUTE study. *BMC Cancer* **2021**, *21*, 1021. [CrossRef]
150. Sternberg, C.N.; Motzer, R.J.; Hutson, T.E.; Choueiri, T.K.; Kollmannsberger, C.; Bjarnason, G.A.; Nathan, P.; Porta, C.; Grünwald, V.; Dezzani, L.; et al. COMPARZ Post Hoc Analysis: Characterizing Pazopanib Responders With Advanced Renal Cell Carcinoma. *Clin. Genitourin Cancer* **2019**, *17*, 425–435.e4. [CrossRef] [PubMed]
151. Yang, Y.; Bu, P. Progress on the cardiotoxicity of sunitinib: Prognostic significance, mechanism and protective therapies. *Chem. Biol. Interact.* **2016**, *257*, 125–131. [CrossRef] [PubMed]

Disclaimer/Publisher's Note: The statements, opinions and data contained in all publications are solely those of the individual author(s) and contributor(s) and not of MDPI and/or the editor(s). MDPI and/or the editor(s) disclaim responsibility for any injury to people or property resulting from any ideas, methods, instructions or products referred to in the content.

biomedicines

Review

Mitochondrial Determinants of Anti-Cancer Drug-Induced Cardiotoxicity

Carmine Rocca [1], Ernestina Marianna De Francesco [2], Teresa Pasqua [3], Maria Concetta Granieri [1], Anna De Bartolo [1], Maria Eugenia Gallo Cantafio [4], Maria Grazia Muoio [2], Massimo Gentile [5], Antonino Neri [6,7], Tommaso Angelone [1,8,*], Giuseppe Viglietto [4] and Nicola Amodio [4,*]

1 Laboratory of Cellular and Molecular Cardiovascular Pathophysiology, Department of Biology, Ecology and Earth Sciences (DiBEST), University of Calabria, Arcavacata di Rende, 87036 Cosenza, Italy; carmine.rocca@unical.it (C.R.); mariaconcetta.granieri@unical.it (M.C.G.); anna.de_bartolo@unical.it (A.D.B.)
2 Unit of Endocrinology, Department of Clinical and Experimental Medicine, University of Catania, Garibaldi-Nesima Hospital, 95122 Catania, Italy; ernestina.defrancesco@unict.it (E.M.D.F.); mariagrazia.muoio@unict.it (M.G.M.)
3 Department of Health Science, University Magna Graecia of Catanzaro, 88100 Catanzaro, Italy; teresa.pasqua@unicz.it
4 Department of Experimental and Clinical Medicine, Magna Graecia University of Catanzaro, 88100 Catanzaro, Italy; mariaeugenia.gallocantafio@unicz.it (M.E.G.C.); viglietto@unicz.it (G.V.)
5 Hematology Unit, "Annunziata" Hospital of Cosenza, 87100 Cosenza, Italy; m.gentile@aocs.it
6 Department of Oncology and Hemato-Oncology, University of Milan, 20122 Milan, Italy; antonino.neri@unimi.it
7 Hematology Fondazione Cà Granda, IRCCS Policlinico, 20122 Milan, Italy
8 National Institute of Cardiovascular Research (I.N.R.C.), 40126 Bologna, Italy
* Correspondence: tommaso.angelone@unical.it (T.A.); amodio@unicz.it (N.A.)

Citation: Rocca, C.; De Francesco, E.M.; Pasqua, T.; Granieri, M.C.; De Bartolo, A.; Gallo Cantafio, M.E.; Muoio, M.G.; Gentile, M.; Neri, A.; Angelone, T.; et al. Mitochondrial Determinants of Anti-Cancer Drug-Induced Cardiotoxicity. *Biomedicines* **2022**, *10*, 520. https://doi.org/10.3390/biomedicines10030520

Academic Editors: Tânia Martins-Marques, Gonçalo F. Coutinho and Attila Kiss

Received: 23 January 2022
Accepted: 19 February 2022
Published: 22 February 2022

Publisher's Note: MDPI stays neutral with regard to jurisdictional claims in published maps and institutional affiliations.

Copyright: © 2022 by the authors. Licensee MDPI, Basel, Switzerland. This article is an open access article distributed under the terms and conditions of the Creative Commons Attribution (CC BY) license (https://creativecommons.org/licenses/by/4.0/).

Abstract: Mitochondria are key organelles for the maintenance of myocardial tissue homeostasis, playing a pivotal role in adenosine triphosphate (ATP) production, calcium signaling, redox homeostasis, and thermogenesis, as well as in the regulation of crucial pathways involved in cell survival. On this basis, it is not surprising that structural and functional impairments of mitochondria can lead to contractile dysfunction, and have been widely implicated in the onset of diverse cardiovascular diseases, including ischemic cardiomyopathy, heart failure, and stroke. Several studies support mitochondrial targets as major determinants of the cardiotoxic effects triggered by an increasing number of chemotherapeutic agents used for both solid and hematological tumors. Mitochondrial toxicity induced by such anticancer therapeutics is due to different mechanisms, generally altering the mitochondrial respiratory chain, energy production, and mitochondrial dynamics, or inducing mitochondrial oxidative/nitrative stress, eventually culminating in cell death. The present review summarizes key mitochondrial processes mediating the cardiotoxic effects of anti-neoplastic drugs, with a specific focus on anthracyclines (ANTs), receptor tyrosine kinase inhibitors (RTKIs) and proteasome inhibitors (PIs).

Keywords: anticancer therapy; cardiotoxicity; heart failure; mitochondrial function

1. Introduction

Despite the great energy consumption needed for contraction and ion transport, the human heart is characterized by a limited content of endogenous high-energy phosphate, able to support cardiac activity only for a very short time [1]. For this reason, adenosine triphosphate (ATP) is constantly produced, especially by mitochondria which, beside representing one third of myocyte volume, account for more than 95% of the cardiac ATP [2]. Mitochondria not only produce ATP by oxidative phosphorylation (OXPHOS), but are also involved in the balance of the redox status, in Ca^{2+} homeostasis, and in the modulation of nuclear gene expression that may result in the regulation of crucial pathways involved in cell survival [3]. Hence, it is not surprising that disorders of these organelles may disrupt cardiac physiology, leading to cardiovascular diseases (CVDs), as convincingly demonstrated by different comprehensive studies [4,5]. Over the past decades, further information has described mitochondria as dynamic organelles undergoing a finely tuned process, known as mitochondrial dynamics, which contributes to cellular homeostasis, allowing the generation of an appropriate response to environmental changes [6–9]. Moreover, to accomplish their activities, mitochondria exploit a selective quality control machinery whose purpose is to target and remove misfolded proteins or aberrant organelles which could impair cardiac homeostasis [4,10].

Because of the dominant role of mitochondria in calcium signaling, redox homeostasis, and thermogenesis, as well as in dictating the fate of a cell, mitochondrial disorders represent a major challenge in medicine [11,12]. Mitochondrial impairment—in terms of defective apoptosis, cytoplasmic and mitochondrial matrix calcium regulation, reactive oxygen species (ROS) generation and detoxification, ATP generation, metabolite synthesis, and intracellular metabolite transport—has been implicated in diverse pathological conditions. Specifically, mitochondria predominantly contribute to maintaining the heart's homeostasis; thus, structural and functional alterations in this organelle lead to contractile dysfunction, and underlie the pathophysiology of several cardiovascular diseases (CVDs), including ischemic cardiomyopathy, heart failure, and stroke [4,13].

Recently, mitochondrial targets have also emerged as important determinants in the cardiotoxic effects triggered by an increasing number of chemotherapeutic agents [14,15], which clinically present as a dose-dependent cardiomyopathy leading to chronic heart failure (CHF), significantly impacting morbidity and mortality [16]. Given the increasing number of long-term cancer survivors and the clinical impact of chemotherapy-related cardiotoxicity, standardizing risk stratification, evaluating the multifactorial processes relying on the interaction between genetic and environmental factors during anticancer therapy, and improving the knowledge of the mechanisms underlying anticancer-drug cardiotoxicity and cardiovascular adverse effects (CVAEs) still represent major challenges in the field of cardio-oncology [13,16].

In this perspective, the present review aims to provide a comprehensive analysis of the key role played by mitochondria in cardiac patho-physiology, focusing on mitochondrial processes implicated in normal cardiac homeostasis, and on their perturbations upon treatment with those cardiotoxic anti-neoplastic drugs which are relevant from a cardio-oncology viewpoint, namely anthracyclines (ANTs), receptor tyrosine kinase inhibitors (RTKIs) and proteasome inhibitors (PIs).

2. Mitochondria and Heart Physio-Pathology

Energy supply in cardiac cells. To cope with the energy demands of the heart, mitochondria produce ATP from a wide range of substrates, such as carbohydrates, fatty acids, amino acids and ketone bodies; however, under basal conditions, energy is mainly drawn from fats (60–90% of cardiac energy supply) [1]. Specifically, while fatty acids (FAs) are directly subjected to β-oxidation in the mitochondria, glucose is preliminarily subjected to glycolysis in the cytosol to produce pyruvate, which in turn is transferred to the mitochondria for oxidation. Usually, glucose and FAs establish a reciprocal relationship described by the Randle cycle, i.e., a dynamic adaptation that induces cardiomyocytes to use these energetic substrates depending on their availability [17,18]. Altered mitochondria result in impaired ATP production and defective energy metabolism that may predispose a higher risk for developing cardiac diseases [10,19].

Redox homeostasis. The oxidative phosphorylation that leads to ATP synthesis is accompanied by electron shift, as visible in the electron transport chain (ETC) by the contribution of electron carriers such as $FADH_2$ and NADH. During this process, a small number of electrons (0.2–2%) slip and are transferred to O_2 to form superoxide [2]. This phenomenon helps explain why mitochondria represent the main cellular source of ROS, as byproducts of electron transfer, whose accumulation not only causes mitochondrial injury but also can lead to the development of cardiovascular diseases. To regulate oxidative stress, mitochondria employ efficient networks, able to scavenge ROS [2,20], which importantly supports the general antioxidant activity of cardiac cells, mitigating oxidative stress [21–23]. The first defense against mitochondrial ROS is represented by superoxide dismutase (SOD), which transforms the superoxide anion into hydrogen peroxide; the latter is then detoxified by catalase, glutathione peroxidase (GSH-PX), and peroxiredoxin/thioredoxin (PRX/Trx) systems. Catalase is a crucial element of the intracellular ROS detoxification process, and is localized not only in peroxisome but also in cardiac mitochondria [24], indicating a role in controlling the ROS pool of these organelles; these enzymes act on hydrogen peroxide, generating water and oxygen. GSH-PX1 and GSH-PX4 are confined in the mitochondria and, by using reduced glutathione (GSH), convert hydrogen peroxide into water and produce oxidized glutathione (GSSG), which is next reconverted into GSH by glutathione reductase with the support of NADPH [25]. In addition, GSH represents a non-enzymatic antioxidant, able to directly neutralize the hydroxyl radical [26]. In this context, it is important to underline that the GSH/GSSG ratio can be considered a useful indicator of oxidative stress [27]. Of note, even if both catalase and GSH-PX are able to reduce hydrogen peroxide, they show important catalytic differences. GPX-PX reduces hydrogen peroxide by making use of glutathione, while catalase mainly acts through the Fenton reaction [28]. Moreover, a differential role of these enzymes in their scavenging activity has been postulated, indicating catalase as a primary defense against low hydrogen peroxide concentrations and GSH-PX as a protective system under high hydrogen peroxide levels [29].

Ionic balance. A fine regulated ion balance, obtained by the presence of selective channels and appropriate exchangers, ensures the physiological potential of the mitochondrial membrane that, in turn, contributes to correct redox regulation and ATP production. In particular, the mitochondrial membrane potential ($\Delta\Psi m$) and the negative charge detectable in the matrix are generated by the flow of electrons in the respiratory chain, and act as a crucial driving force for ATP synthesis [30]. Accordingly, $\Delta\Psi m$ represents a useful indicator of cardiac cell health, and its preservation is vital for cardiomyocytes [31,32]. Calcium channels and transporters are localized on both the outer (OMM) and the inner (IMM) mitochondrial membranes [33–35] and make mitochondria able to detect calcium cytosolic signaling and eventually mediate its sequestration [36]. It is well established that the amount of intracellular calcium (100 nM) is more than 10,000-fold less than the extracellular [37], and that in the mitochondrial matrix calcium levels range from 100 to 200 nM under resting conditions [38]. When several stresses induce an increase of intracellular Ca^{2+} levels, mitochondria act as efficient Ca^{2+} buffering organelles [39]. A rise in intracellular Ca^{2+} increases mitochondrial uptake [40], causing an elevation of intra-mitochondrial Ca^{2+} and a drop in $\Delta\Psi m$ that enhances ROS production and oxidative stress.

Programmed cell death. A wide range of stimuli may activate mitochondrial-related apoptosis, as in the case of ischemia/reperfusion (I/R), loss of nutrients, oxidative stress, increased Ca^{2+} levels, chemotherapeutics, and targeted cancer therapies [41]. The main event in the mitochondria-driven apoptotic process is the permeabilization of the OMM, which allows several apoptogens to move towards the cytosol and activate procaspases. The whole mechanism is strictly regulated by the BCL-2 (B cell lymphoma-2) proteins [42], a protein family including three subfamilies, which are grouped according to their function and to the BCL-2 homology (BH) domains: (i) pro-survival proteins (containing BH1-4), such as BCLW, MCL-1, BCL-xL, and BCL-2 itself; (ii) pro-cell death proteins (containing BH1-3, or rarely BH1-4), such as BAX, BAK, and BOK; and (iii) pro-cell-death proteins (containing only BH3) such as BIM, BID, PUMA, and NOXA [41]. BH3 proteins are able to physically bind BAX and BAK, inducing their conformational activation, which results in their homo- or hetero-oligomerization within the OMM [43]. This critical step produces OMM permeabilization and the leak of apoptogens [44–46] from the mitochondria with the activation of cytosolic pro-caspases, which in turn trigger apoptosis [47]. In particular, the released cytochrome c induces the assembly of the apoptosome, a multiprotein complex that activates caspase-9 by the cleavage of pro-caspase-9, then inducing other apoptotic effectors [48–50]. Conversely, BCL-2 is able to both sequester BH-3 proteins and bind BAX/BAK, inhibiting this death process and promoting cell survival [41,51].

2.1. Mitochondrial Quality Control

Cardiac homeostasis strictly depends on healthy mitochondria, and for this reason they exploit a selective quality control machinery that, by targeting damaged mitochondria or mitochondrial proteins, drives them to degradative and/or removal processes [4]. Indeed, several cardiomyopathies are characterized by the presence of abnormal mitochondria clusters [4,10,19]. Two main pathways intervene to support the quality control of mitochondria: (i) the ubiquitin-proteasome system (UPS), which degrades damaged mitochondrial proteins; and (ii) the autophagy-lysosomal pathway (i.e., mitophagy), which degrades the whole mitochondrion [52,53]. UPS and mitophagy share a common key element, namely ubiquitin, which covalently binds the substrates which are thus targeted for degradation and removal [54].

2.1.1. Ubiquitin Proteasome System (UPS)

UPS promotes ubiquitination, a multistep and ATP-dependent mechanism, through the activity of three enzymes: E1, which activates ubiquitin; E2, which conjugates ubiquitin; and E3, ubiquitin ligases. A polyubiquitin chain, created by successive ubiquitination reactions, is then able to interact with the proteasome leading the substrate degradation [55]. Deubiquitinating enzymes ensure the reversibility of the entire process [56,57]. UPS dynamically regulates the mitochondrial proteome, which depends on both the importation of newly synthesized proteins from the cytosol and their degradation. Indeed, this quality control system extracts ubiquitinated proteins from the OMM and/or IMM, and degrades non-imported mitochondrial proteins [58]. In the specific case of cytosolic UPS, it controls the delivery of functional proteins to the mitochondria. Accordingly, cardiac diseases that involve the perturbation of protein homeostasis, i.e., proteostasis, alter mitochondrial function and activate death processes [59,60]. Moreover, data obtained from animal models and from human patients demonstrates that a proteasomal inefficiency, together with increased levels of protein ubiquitination, correlates with cardiomyopathies [61,62]. Accessible proteins of the OMM may be degraded by UPS, after ubiquitination, extraction from the OMM, and delivery to the proteasome, producing significative effects not only on mitochondrial morphology but also on apoptosis. For instance, when UPS induces the degradation of MCL-1, an anti-apoptotic molecule, the apoptotic proteins BAX/BAK are activated [63]. The turnover of mitochondrial proteins is also guaranteed by the translocase of the OMM, involved in the exportation of proteins localized in the intermembrane space [11,64]. Furthermore, UPS also controls nuclear-encoded mitochondrial proteins before their transport into the organelle by TOM/TIM complexes [64]. Since nuclear-encoded mitochondrial proteins are transported in an unfolded state, mitochondria possess an intrinsic quality control system, composed of chaperones and proteases, able to avoid the accumulation of misfolded or damaged proteins [65,66]. When these quality control systems fail to compensate for the excessive generation/accumulation of misfolded proteins, the mitochondrial unfolded protein response (URPmt) is activated. URPmt activates a nuclear transcriptional program that aims to restore mitochondrial homeostasis, inducing both proteases and chaperones [67].

2.1.2. Mitophagy

When the total protein injury overcomes the restorative ability of the URPmt and UPS quality control systems, mitochondria are driven to mitophagy. The importance of mitophagy as a crucial cardiac mitochondrial quality control mechanism has been widely reported [68]. In general, autophagy represents the main degradation mechanism in cells and uses autophagosome vesicles to deliver cytoplasmic elements to the lysosomes. In this context, mitophagy is a fine-tuned process that supports the previously mentioned mitochondrial quality control systems, selectively removing damaged mitochondria. Compared to non-selective autophagy, mitophagy shows a complex organization that relies on two main events: (i) identification and labeling of mitochondria that have to be degraded; and (ii) generation of vesicular structures that transport mitochondria to lysosomes [69]. The leading processes that drive mitophagy are the PTEN-induced putative kinase 1 (PINK1)/Parkin pathway and the OMM mitophagy receptors. Especially in the heart, where inefficient mitochondria need to be degraded in order to prevent cardiomyocyte death and cardiac diseases, the PINK1/Parkin-dependent mitophagy plays a pivotal role. For instance, in the hearts of mice that were fed a high-fat diet, mitophagy increased and Parkin deficiency worsened diabetic cardiomyopathy [70]. Additionally, the PINK1/Parkin pathway is stimulated by cardiac pressure overload [71,72], during I/R [73], and under myocardial infarction [74]. The Parkin gene encodes an E3 ubiquitin ligase that interacts with E2 ubiquitin, the enzyme promoting the ubiquitination and the final removal and degradation of targeted proteins [68,75]. Mitofusin (MFN2), which will be discussed later, seems to be necessary for this mitochondrial quality control process, and is supposed to act as a mitochondrial receptor for Parkin [76]. Gong et al. elegantly demonstrated that when PINK1, located on the mitochondria, phosphorylates MFN2, it recruits cytosolic Parkin, which, in turn, ubiquitinates outer membrane proteins which are then able to interact, via protein p62, with the autophagosomal LC-3 [77]. Notably, LC-3, i.e., the microtubule-associated protein 1 light chain, has been identified by Kabeya et al. as the first mammalian protein associated with the membranes of autophagosomes [78]. A few years later, LC-3 was characterized as a crucial protein involved in the binding of PINK1 during mitophagy [79].

PINK1 promotes Parkin translocation into the mitochondria by its phosphorylation, a fundamental step for its recruitment and for the resulting ubiquitination of additional proteins, such as mitofusin 2 (MFN2), which will be discussed later [75–77]. Ubiquitination represents the key signal for the binding of mitophagy proteins such as sequestosome 1 (p62/SQSTM1), a so-called autophagy adaptor, providing a molecular link able to concurrently bind ubiquitin and specific proteins located on the autophagosome [80]. Autophagy adaptor proteins are characterized by a ubiquitin binding domain (UBD) and by the presence of an LC-3-interacting region (LIR), both needed to address mitochondria to their autophagosome sequestration and subsequent elimination through the lysosome intervention [53,81].

2.2. Mitochondrial Dynamics

Despite the fact that mitochondria were previously considered independent, static, and isolated organelles, it is now accepted that they form a dynamic network inside the cell, maintained by "mitochondrial dynamics". Mitochondrial dynamics refers to the ability of mitochondria to undergo continuous cycles of fusion, during which segregated mitochondria join; and fission, during which the mitochondria divide [82]. Accordingly, mitochondria are highly dynamic organelles, whose function is dynamically regulated by their fusion and fission, movement along the cytoskeleton, and mitophagy. These processes are essential to maintaining normal mitochondrial morphology, distribution, and function—including mitochondrial respiration, mitochondrial metabolism, and ROS production—as well as normal cell metabolism [83].

Selective mitochondrial fusion proteins known as membrane-anchored dynamin family members, which are abundantly expressed in the adult heart, mediate the fusion of two adjacent mitochondria to form a more elongated mitochondrion; in particular, fusion is promoted by mitofusin-1 (MFN1) and MFN2 proteins, whose normal functions rely on the activity of guanosine triphosphatases (GTPases), by forming stable homo-oligomeric and hetero-oligomeric complexes through their GTPase domain at the outer mitochondrial membrane, and optic atrophy 1 (OPA1), which is located in the IMM and in the intermembrane space; OPA1 is a dynamin-like GTPase that is anchored to the IMM by an N-terminal transmembrane domain, and mediates IMM fusion, enhancing the interconnection of the mitochondrial network [84,85]. Mitochondrial fusion allows the exchange of intramitochondrial material (i.e., mitochondrial DNA (mtDNA), proteins, lipids, and metabolites), necessary for maintaining a balanced pool of mitochondrial protein, as well as a genetic and biochemical homogeneity within the mitochondrial population [83].

On the other hand, mitochondrial fission proteins participate in mitochondrial fission, a multistep and complex process that divides a single mitochondrion into two mitochondria; the key factor mediating mitochondrial fission is dynamin-related protein 1 (Drp1), a homologous protein of GTPase power protein, which is recruited from the cytosol to the OMM by various OMM-anchored adapter proteins, including fission protein 1 (Fis1) and mitochondrial fission factor (MFF), which act as Drp1 receptors [8,86]. Mitochondrial fission is necessary to replicate the mitochondria during cell division, to facilitate the transport and distribution of mitochondria, and to permit the isolation of damaged mitochondria for mitophagy. Alterations of mitochondrial dynamics lead to cardiac mitochondrial integrity and mtDNA damage, and cell death ultimately occurs [87].

In the case of prolonged exposure of the heart to stressful conditions, such as hypoxia, ischemia/reperfusion, oxidative and nitrosative stress, and hyperglycemia, the profound alterations of mitochondrial dynamics and mitophagy lead to irreversible damage of the mtDNA and excessive ROS released by damaged mitochondria, ultimately leading to cardiotoxicity [87,88].

3. Cardiac Mitochondrial Dysfunction Secondary to Anti-Cancer Drug Treatments

It is well-established that the cardiotoxic side effects of several anti-cancer therapies are frequently mediated by mitochondrial damage [89]. This evidence was first demonstrated through the detrimental effects of chemotherapy on skeletal muscle, a tissue in which the number of mitochondria is very high, although lower than cardiomyocytes [90]. Accordingly, skeletal muscle weakness, together with persistent fatigue, are common in cancer patients undergoing chemotherapy, and some of the skeletal-muscle-specific symptoms are due to mitochondrial dysfunction [12,91,92]. At the molecular level, different processes, including but not limited to oxidative stress, inflammation, immunometabolism, pyroptosis, and autophagy, act together, promoting chemotherapy-induced multifactorial cardiotoxicity [93].

In this context, growing evidence highlights the involvement of diverse mechanisms that mainly converge on mitochondrial dysfunction. There are a number of potential reasons why cardiac mitochondria represent a major target of antineoplastic drugs. Firstly, cardiomyocytes show a high susceptibility to oxidative stress because they are rich in mitochondria and possess relatively low endogenous antioxidant defense systems [94]; additionally, they use enormous amounts of ATP, whose production occurs in mitochondria and is maintained, as discussed above, by mitochondrial biogenesis, replication, and autophagy/mitophagy [95]. Overall, mechanisms that induce mitochondrial toxicity via anti-tumor agents are many, and mostly related to the alterations occurring in ROS/redox system regulation, the mitochondrial calcium homeostasis system, mitochondrial dynamics, and endoplasmic reticulum (ER) stress signaling, all processes linked by a vicious cycle that disrupts cardiac cell homeostasis and induces cell death [96,97].

In the following paragraphs, we will analyze the main mitochondrial determinants of cardiotoxicity secondary to three major classes of antineoplastic drugs widely reported as cardiotoxic, represented by ANTs, RTKIs and PIs.

3.1. *Anthracyclines (ANTs)*

ANTs, primarily doxorubicin (DOX), are antibiotics that exert their anti-tumor activity by inducing single- and double-strand breaks in DNA, preventing DNA synthesis, intercalating with DNA base pairs and stabilizing the topoisomerase (Top) 2α complex after DNA cleavage [98,99]. ANTs still represent the cornerstone of treatment in many malignancies, including lymphomas, leukemias, sarcomas, advanced and early breast cancer, and small cell lung cancer [100,101]. However, the clinical use of ANTs is seriously hampered by dose-related cardiomyocyte injury and death, leading to left ventricular dysfunction and heart failure, representing the most clinically-limiting adverse feature of ANTs [94,100–102].

The most relevant ANT-related cardiac dysfunction from a cardio-oncological point of view involves the myocardium, and is manifested by a decreased left-ventricular ejection fraction, which may progress to congestive heart failure [103]. Mechanically, cardiac dysfunction induced by ANTs relies on alteration in iron metabolism and ROS, and reactive nitrogen species (RNS) overproduction; however, intriguing evidence emerged in recent years indicating that ANTs may use alternative damaging mechanisms, such as Top 2β inhibition, inflammation, immunometabolism, pyroptosis, and autophagy, which explains, at least in part, the complexity of iatrogenic ANT-induced progressive cardiomyopathy and heart failure (Figure 1) [104]. On the other hand, ANTs typically associate with an irreversible form of cardiac dysfunction (known as type I cardiotoxicity) characterized by evident ultrastructural myocardial abnormalities, as evinced by vacuoles, myofibrillar disarray and dropout, and myocyte necrosis at higher cumulative doses [103].

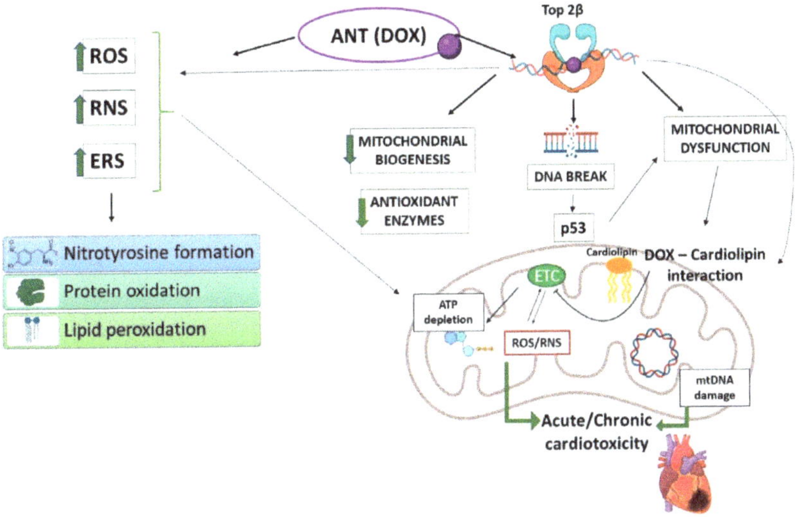

Figure 1. Schematic representation of major events leading to mitochondrial dysfunction during ANT (DOX)-induced cardiotoxicity. ANT: anthracycline; DOX: doxorubicin; ROS: reactive oxygen species; RNS: reactive nitrogen species; ERS: endoplasmic reticulum stress; Top2β: topoisomerase 2β; ETC: electron transport chain; mtDNA: mitochondrial DNA.

Although the pathogenetic mechanisms accounting for ANT-dependent cardiotoxicity remain complex and multifactorial, mitochondrial oxidative stress, in addition to the redox cycling secondary to ANT-iron complex formation, and targeting of Top 2β (one of the two types of Top2 present in quiescent non-proliferating cells, including cardiomyocytes), are the most relevant. The inhibition of Top 2β by ANTs causes double-stranded DNA breaks and the consequent activation of the tumor suppressor protein p53, strongly contributing to the development of cardiotoxicity [93,105,106]. Importantly, over the last decades, a large number of studies reported sub-chronic/chronic mitochondrial cardiac alterations, in terms of disrupted mitochondrial calcium homeostasis [107,108] and mitochondrial respiration alteration [109,110] during DOX exposure in both pre-clinical and human models. The primary effect of DOX on mitochondrial activity is related to its capacity to interfere with oxidative phosphorylation and inhibit ATP synthesis. In particular, DOX can inhibit mitochondrial Complex I by diverting electrons from NADH to molecular oxygen, leading to DOX recycling and generating a futile cycle, a major ROS production site in DOX-induced toxicity [111,112]. Other evidence subsequently demonstrated that DOX also interferes with Complexes III and IV, the phosphate carrier and the adenine nucleotide translocator [112]. Free radicals derived from DOX redox cycling are responsible for many of the secondary effects of oxidative stress induced by DOX; these include alteration of macromolecules as well as depletion of GSH and pyridine nucleotide reducing equivalents [113]. The generation of excessive ROS and RNS overcomes the endogenous capacity in producing antioxidant enzymes, including mitochondrial antioxidant systems, leading to the typical redox modifications of macromolecules, including nitrotyrosine formation, protein carbonylation, and lipid peroxidation (Figure 1) [93,102,114]. In addition to the lower antioxidant surplus in the heart respective to other tissues, the ability of DOX to accumulate primarily in mitochondria and nuclei [115] can explain the cardio-selective toxicity of the drug. In this context, it should also be noted that ANTs are able to selectively bind the phospholipid cardiolipin, localized in the IMM, in close proximity to the mitochondrial electron-transport chain, leading to mitochondrial accumulation of the drug (Figure 1) [116]. Cardiolipin is an acidic phospholipid that plays a crucial role in the regulation of mitochondrial function, structure, and dynamics, and mitochondrial dysfunction in different CVDs correlate with cardiolipin remodeling; in particular, cardiolipin peroxidation induces mitochondrial impairments and CVD progression [117]. In this regard, several studies on animal models demonstrated that the ANT-cardiolipin interaction alters cardiolipin function since, in this condition, cardiolipin is not able to anchor cytochrome c or lipid-protein interfaces for the other important mitochondrial proteins [118]; the oxidized cardiolipin can disrupt the electron transport chain, stimulating additional ROS/RNS production and inducing mitochondrial DNA damage (Figure 1) [112].

As elegantly reviewed by Wallace et al., mechanistic studies showed that the acute inhibition of mitochondrial oxidative phosphorylation induced by DOX may induce compensatory selective cardiomyocyte adaptations [119]. For instance, as indicated in an acute in vitro model (i.e., H9c2 rat cardiac myoblasts), a major cellular defense mechanism secondary to DOX exposure concerns the activation of the Keap1 (kelch-like ECH-associated protein 1)/Nrf2 (Nfe2l2, nuclear factor erythroid derived 2 like 2)-antioxidant response element (ARE) signaling pathway [120]. Other in vitro reports suggest that acute DOX exposure can induce, in cardiomyocytes, the nuclear up-regulation of p66Shc, an adaptor protein modulating cellular redox status and serving as an oxidative stress sensor, in order to modulate FoxO (Forkhead box subgroup O) nuclear transcription factors, inducing cell death in order to eliminate damaged cells [121].

Both experimental and clinical evidence supports the hypothesis that specific antioxidants may be effective in protecting the heart from ANT toxicity, in terms of HF prevention or cardiac damage mitigation. Clinical trials and meta-analytical studies have been conducted to determine the protective effect of specific antioxidants, such as carvedilol, L-carnitine, and dexrazoxane in ANT-induced cardiomyopathy [122–127]. However, it is still unclear whether these antioxidants exert cardioprotective effects in humans without impairing the anticancer activity of ANTs; moreover, most of these studies evaluated the effects of ANTs alone, not in combination with other therapies. Therefore, larger multicenter trials are required to effectively evaluate the beneficial activity of antioxidant agents in co-administration with ANTs and other anticancer drugs [128,129].

Notably, mitochondrial alteration secondary to ANTs is profoundly interconnected with Top 2β targeting and ROS/RNS generation, since indirect effects on mitochondrial function can also occur through nuclear-mediated effects related to the inhibition of Top 2β in cardiomyocytes. Accordingly, after DNA breaks secondary to DOX-Top 2β binding, p53 stimulation also induces defective mitochondria biogenesis and metabolic impairment by decreasing the transcription of crucial genes involved in mitochondrial biogenesis and function, such as peroxisome proliferator-activated receptor gamma coactivator 1-α (PGC-1α), which is also a key regulator of SOD, and peroxisome proliferator-activated receptor gamma coactivator 1-β (PGC-1β), and alteration of oxidative phosphorylation [105]. DOX is also able to downregulate uncoupling protein 2 (UCP-2) and uncoupling protein 3 (UCP-3), members of the superfamily of mitochondrial transport proteins which regulate mitochondrial ROS production, predisposing the failing heart to oxidative stress [130]. These data are of particular interest since it has been reported that polymorphisms in the human UCP genes can affect the expression/function of the protein [131]; thus, genetic variations in human UCP-2 and/or UCP-3 may affect the susceptibility of patients to DOX-related cardiotoxicity.

Human studies and pre-clinical models indicate that the redox and metabolic alterations, as well as mitochondrial impairment secondary to a DOX regimen, persist after therapy completion (one to five weeks following the last of six drug treatments) and that the toxic effects of DOX can propagate to successive generations of mitochondria, leading to cumulative dose-dependent and progressive mitochondrial dysfunction [132,133]. This can correlate with DOX cardiotoxicity memory, according to which myocardial mass reduction following DOX administration may predispose the heart to further alterations after subsequent DOX treatments [119,134] (Figure 1).

There is also growing evidence that ANTs can disrupt mitochondrial dynamics, which is increasingly recognized as a major process driving ANT-dependent heart dysfunction, so that several therapeutic interventions targeting mitochondrial dynamics have shown promising effects in attenuating DOX cardiac toxicity in both cell and animal models (Figure 2).

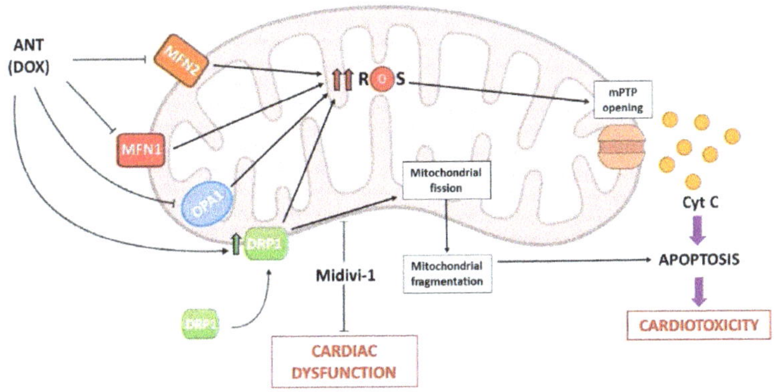

Figure 2. Schematic representation of mitochondrial dynamics alterations induced by ANT (DOX) leading to cardiotoxicity. ANT: anthracycline; DOX: doxorubicin; ROS: reactive oxygen species; MFN1: mitofusin-1; MFN2: mitofusin-2; OPA1: optic atrophy 1; DRP1: dynamin-related protein 1; Midivi-1: mitochondrial division inhibitor-1; mPTP: mitochondrial permeability transition pore; cyt c: cytochrome c.

In vitro evidence on cultured neonatal rat cardiomyocytes demonstrated that DOX negatively affects levels of MFN2, thus promoting mitochondrial fission and ROS production, while increasing MFN2 levels counteracted these processes [135]. Similarly, other studies indicate that MFN1 and OPA1 are downregulated in response to apoptotic stimulation following DOX exposure in cardiomyocytes [136]. Conversely, DOX can upregulate the expression of mitochondrial fission protein 1 in HL-1 cardiac myocytes, while its lessening reduces DOX-dependent apoptosis, preventing dynamin 1-like accumulation in mitochondria [137]. In vivo, sub-chronic DOX treatment in rats increased mitochondrial permeability transition pore (mPTP) susceptibility and induced apoptosis, decreasing the expression of MFN1, MFN2, and OPA1, and increasing Drp1, activating autophagy and mitophagy signaling [138]. Moreover, Xia et al. (2017) demonstrated in H9c2 cardiomyocytes, as well as in a mouse model of DOX-induced cardiomyopathy, that DOX exposure augmented Drp1 and its Ser 616 phosphorylation [139]. These findings were corroborated by the ability of both LCZ696, a novel angiotensin receptor-neprilysin inhibitor, and of mitochondrial division inhibitor-1 (Midivi-1), a specific inhibitor of Drp1, to mitigate the DOX-dependent mitochondrial dynamics alterations and cardiac dysfunction (Figure 2). On the other hand, the overexpression of Drp1 antagonized the beneficial effect of LCZ696 in vitro [139]. The crucial involvement of Drp1 in DOX-dependent cardiotoxicity was further demonstrated by Zhuang et al. (2021) [140], who confirmed that the expression of Drp1 increased following DOX treatment both in vitro and in vivo, leading to apoptosis of cardiomyocytes. In this study, the authors also found that an overexpression of Klotho (an anti-aging protein whose defects in its gene expression accelerated cardiac hypertrophy and remodeling in mice and human vascular calcification) [141,142] or Midivi-1 can trigger cardioprotection through inhibition of cell death and reversal of mitochondrial dynamics perturbation.

Consistently, other in vitro and in vivo reports strongly support a key role for Drp1-dependent mitochondrial fragmentation in DOX-dependent cardiomyopathy. Catanzaro et al. (2019) indicated that a short interference-RNA-mediated knockdown of Drp1 prevents DOX-induced mitochondrial fragmentation, mitophagy flux, and apoptosis in H9c2 cells, while Drp1-deficient mice were protected from DOX-induced cardiac dysfunction [143]. Various studies reported that Drp1 can be reversibly phosphorylated at its serine residues, and that this phosphorylation strongly affects both the localization and activation of cardiac Drp1 [144]. Specifically, when Drp1 is phosphorylated at Ser 637, its translocation to mitochondria is prevented and mitochondrial fission is inhibited [145]. In this regard, a very recent study identified the cardiomyocyte mitochondrial dynamic-related lncRNA 1 (CMDL-1) as the most significantly downregulated long non-coding RNA (lncRNA) in cardiomyocytes after DOX exposure, and demonstrated that CMDL-1 can inhibit Drp1 translocation to mitochondria by promoting Drp1 Ser 637 phosphorylation, thereby preventing mitochondrial fission and apoptosis [146].

Among the different OMM proteins that promote mitochondrial fission by recruiting Drp1 to the mitochondrial surface, it has also been shown that mitochondrial dynamics proteins of 49 kDa (MiD49, MIEF2) can participate in the regulation of cardiac mitochondrial dynamics during DOX treatment. Accordingly, recent studies identified MIEF2 as a transcriptional target of the transcription factor FoxO3a, and reported that FoxO3a can prevent DOX-induced mitochondrial fission, apoptosis, and cardiotoxicity by suppressing MIEF2 expression [147].

Overall, these data indicate that DOX displays inhibitory effects on mitochondrial fusion while promoting mitochondrial fission; in particular, the increased Drp1 expression, whose protein levels were previously found increased in patients with ischemic cardiomyopathy and dilated cardiomyopathy [148], represents a key factor also promoting the shift toward mitochondrial fission during DOX exposure.

Taken together, these observations suggest that preventing mitochondrial fission and targeting mitochondrial dynamics could represent a promising strategy in saving cardiomyocyte loss due to DOX-induced cardiotoxicity (Figure 2).

3.2. RTK Inhibitors (RTKIs)

Receptor tyrosine kinases (RTKs) are cell surface transmembrane proteins activated in response to ligand binding, an event conveying downstream stimulatory signals towards cell proliferation, migration, invasion, differentiation, and angiogenesis [149]. Aberrant RTK signaling, which may occur in response to genome amplification, gain of function mutations, or chromosome rearrangements, has been shown to contribute to tumor development and progression, as well as to anti-cancer treatment failure [149,150]. Most of the known human RTKs share a similar protein structure, with an extracellular ligand-binding (N)-terminal domain, a single spanning transmembrane helix, and an intracellular carboxyl(C)-terminal domain [151,152]. A number of pharmacological approaches have been proposed to block aberrant RTK signaling in cancer, including the use of monoclonal antibodies targeting either specific receptors or their ligands, as well as the use of RTKIs' small molecules.

RTKIs mainly act by preventing receptor autophosphorylation through interference with the ATP binding site within the kinase catalytic domain of the protein; nevertheless, certain RTKIs are non-ATP competitors [153]. One of the clinical advantages of targeting aberrant RTK signaling is that fewer off-target effects are to be expected when using targeted therapies compared with chemo- and radiotherapy. Despite the risk of developing cardiovascular effects appearing to be generally low, long-term use of certain RTKIs can significantly increase the risk of cardiovascular events. Such effects appear to be highly variable among the class of RTKIs, although it is generally accepted that pre-existing cardiac pathological conditions, such as hypertension, hyperlipidemia, and diabetes, as well as both the genetic background and immune status of the patient, may influence the risk and severity of RTKI-associated cardiovascular toxicity [154].

RTKI-triggered cardiovascular side effects range from asymptomatic left ventricular dysfunction to symptomatic congestive heart failure, arrhythmia/QT prolongation, hypertension, and acute coronary syndrome [155]. Despite the fact that the mechanisms are various and drug-specific side effects are observed, a general model of toxicity involves both on-target and off-target effects.

The most important pharmacological strategy aimed at blocking tumor angiogenesis is the targeting of the vascular endothelial growth factor (VEGF)/VEGFR transduction pathway. Both anti-VEGF monoclonal antibodies and VEGFR small molecule inhibitors have been shown to induce left ventricular dysfunction, ischemia, and thromboembolic events [156]. Commonly, the most strongly observed effect in response to anti-VEGF therapies is hypertension, which is due to unbalanced production in blood pressure regulators (i.e., increased endothelin-1 and decreased nitric oxide production, respectively), as well as reduced capillary density [157]. It is worth mentioning that certain detrimental cardiovascular effects induced by RTKIs are directly attributable to loss of RTK function and therefore compromised cardiomyocyte biology. This is the case for anticancer therapies that target the ERBB family of RTKs [158].

As ERBB family members play a crucial role in the maintenance of cardiomyocytes' homeostasis and cell response to stress and injury, the disruption of their transduction network results in myocyte dysfunction. For instance, interfering with ERBB-mediated signaling may promote the mitochondrial release of cytochrome c [159], together with the inhibition of antiapoptotic pathways, the induction of caspase activation, and the subsequent activating of apoptotic cell death [160]. Additional studies have shown that the monoclonal antibody trastuzumab, which targets ERBB family members, may compromise the ability of cardiomyocytes to cope with stress, including pressure overload and/or ANT injury, thus providing a rationale for the increased risk of cardiotoxicity of the drug combination (trastuzumab plus ANT) compared to single agent treatment [161].

Interestingly, cardiac toxicity has also been detected after inhibition of non-receptor TKs. For instance, imatinib mesylate, which mainly targets the fusion protein bcr-Abl and represents the drug of choice in chronic myelogenous leukemia (CML) and Philadelphia chromosome-positive B-acute lymphoblastic leukemia (Ph+ B-ALL), induces myocyte dysfunctions resulting in severe CHF [162]. The analysis of endomyocardial bioptic tissue obtained from patients who developed CHF after treatment with imatinib mesylate revealed profound ultrastructural mitochondrial changes and abnormalities, including pleomorphisms, swelling, and erosions of cristae, together with intense cytosolic signs of cell stress, like formation of vacuoles [162]. Cardiomyocytes cultured with imatinib mesylate had high ER stress, deep alterations of mitochondrial membrane potential, reduction of ATP production, release of cytochrome c into the cytosol, and activation of cell death programs (Figure 3) [157,162]. Of note, myocytes' mitochondrial damage and subsequent energy rundown may also be attributable to the impaired activity of the energy-restoring AMP-activated protein kinase (AMPK), a frequently observed off-target effect of RTKIs [163].

Further corroborating these findings, deranged mitochondrial energetics were also observed in response to clinically relevant concentrations of sorafenib, which compromised oxidative phosphorylation by inhibiting complexes V, II, and III of the electron transport chain [164,165], thereby halting ATP production necessary for myocyte contractility (Figure 3).

Of note, promising clinical effects of the multi-targeting TKI ponatinib, approved for the treatment of CML and Ph+ B-ALL [166], have been mitigated by the cardiac-specific toxicity induced by this drug, including myocardial infarction, severe congestive heart failure, and cardiac arrhythmias.

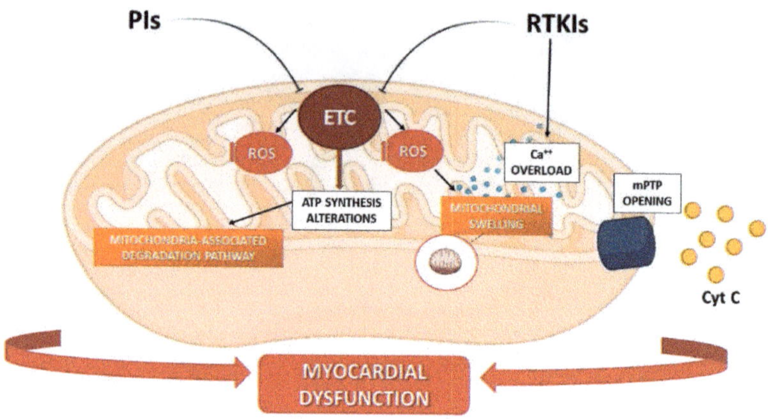

Figure 3. Proposed mechanism of cardiac mitochondrial alterations secondary to PIs and RTKIs exposure. PIs: Proteasome inhibitors; RTKIs: Receptor tyrosine kinase inhibitors; ETC: electron transport chain; ROS: reactive oxygen species; mPTP: mitochondrial permeability transition pore; cyt c: cytochrome c.

A well-designed approach by Talbert et al. demonstrated that the cardiac toxicity potential of ponatinib is reflected by dramatic changes in ROS generation and lipid formation, consistent with mitochondrial impairment and metabolic imbalances [167]. In addition, the authors developed a comprehensive in vitro screening tool based on the use of human-induced pluripotent stem cell-derived cardiomyocytes (hiPSC-CM), which was able to accurately predict human cardiac toxicity by evaluating several indices, including signs of mitochondrial stress [167].

Likewise, enhanced ROS generation and oxidative stress are largely implicated in the initiation of mitochondrial dysfunction, which triggers cell damage in a broad range of cellular components. It should be mentioned that certain RTKIs promote mitochondrial dysfunctions in an indirect fashion. This is the case for regorafenib, a drug approved for metastatic colorectal cancer and advanced gastrointestinal stromal tumors, which disrupts calcium homeostasis, thereby inducing mitochondrial swelling due to calcium overload [168].

On the other hand, abnormalities in mitochondrial structures and function may result as a consequence of RTKIs' action on several off-target kinases, including c-Jun N-terminal kinase, protein kinase A and pyruvate dehydrogenase kinase (PDK); moreover, PDK, a mitochondrial enzyme acting with pyruvate dehydrogenase phosphatase to regulate pyruvate dehydrogenase complex, has been shown a promising therapeutic target in complex diseases including diabetes, heart failure, and cancer, as well as in the mitochondrial toxicity induced by RTKIs [169,170]. Accordingly, the inhibition of these signaling pathways may disrupt oxidative phosphorylation, and facilitate the establishment of both morphological abnormalities consistent with hypertrophic responses and the shift of energetic metabolism toward anaerobic dependency [171].

Clearly, the disruption of mitochondrial structure and function represents the main trigger for cardiomyocytes' metabolic reprogramming, as nicely shown by Wang et al., who performed a systems-level analysis of human cardiomyocytes differentiated from hiPSCs and exposed to different RTKIs [172]. Results showed a parallel inhibition of mitochondrial ATP production and an increase in glycolysis after treatment with RTKIs [172]. The effect on mitochondrial functionality appeared to be reversible upon drug withdrawal, and the metabolic remodeling toward the glycolytic pathway served as an alternate route to cope with metabolic stress. Likewise, an increased tendency to rely on glycolysis is a peculiar feature of hypertrophic myocardium and myocardial ischemia, as well as heart failure [173]. Despite the fact that the mechanisms involved in RTKI cardiotoxicity are an active topic under investigation, and less-known than other anti-cancer drugs like

ANTs, a relative lack of adequate pre-clinical platforms to predict, detect and hamper drug-associated cardiovascular effects still represents a challenge to basic researchers and clinicians in this field. Therefore, additional effort has to be implemented to minimize the detrimental cardiac effects of RTK inhibition, taking into account the complexity of the RTK signaling networks. For instance, the inhibition of EGFR by gefitinib (mainly used for the treatment of non-small cell lung cancer), has been shown to induce mitochondrial membrane potential alteration, cellular plasma membrane permeabilization, and activation of apoptosis in cardiomyocytes [174]. These effects are triggered by the CYP1A1-dependent formation of toxic reactive metabolites within myocytes' microsomes. It is worth recalling that in various contexts, EGFR cooperates with other non-RTK transduction partners to promote biological responses. This is the case for the G-protein coupled receptor 30, namely GPER, which serves as an alternate receptor for estrogens [175,176]. Numerous studies have demonstrated that GPER activation elicits beneficial cardiovascular effects by regulating myocyte cell response to stressful conditions, including ischemia, inflammation, and hypertension [177,178]. Additionally, GPER activation has been shown to reduce DOX cardiotoxicity [179]. Table 1 summarizes the main RTKIs and their cardiovascular toxicity.

Table 1. List of main RTKIs and their cardiovascular toxicity.

Tyrosine Kinase Inhibitor	Molecular Target	Type of Study	Type of Cancer	Cardiotoxic Effect	Ref.
Sunitinib	Multi-tyrosine kinases (VEGFR, PDGFR, c-KIT)	Phase I/II clinical trial Multicenter prospective study	Imatinib-resistant, metastatic, gastrointestinal stromal tumors metastatic renal cell carcinoma	Left ventricular dysfunction congestive heart failure hypertension	[157,180]
Pazopanib	Multi-tyrosine kinases (VEGFR, PDGFR, c-KIT)	Randomized, double-blind, placebo-controlled study	Advanced solid tumors	Hypertension reduction in heart rate small prolongation of the QTc interval	[181]
Sorafenib	Multi-tyrosine kinases (VEGFR, PDGFR, FLT3)	Systematic review and meta-analysis	Renal cell carcinoma melanoma	Hypertension myocardial infarction ischemia acute coronary syndrome rarely heart failure	[182]
Regorafenib	Multi-tyrosine kinases (VEGFR1-3, PDGFR-β, FGFR)	Meta-analysis of 45 RTCs	Solid tumors	Hypertension generally few cardiovascular side effects	[183]
Ponatinib	Multi-tyrosine kinases FGFR, PDGFR, and VEGFR	Phase II clinical trial Review	Chronic myeloid leukemia; Philadelphia chromosome-positive leukemias	Arterial thrombotic events	[184,185]
Cabozantinib	Flt-3, RET, MET	Multicenter prospective study Review	Metastatic renal cell carcinoma medullary thyroid cancer	Modest risk of developing left ventricular systolic dysfunction hypertension	[186,187]
Nilotinib	PDGFR, CSF-1R,	Retrospective study	Chronic myeloid leukemia	Accelerated atherosclerosis peripheral arterial occlusive disease (PAOD) QTc prolongation.	[188]
Axitinib	VEGFR	Clinical trial	Metastatic renal cell carcinoma	Hypertension myocardial infarction	[189]

3.3. Proteasome Inhibitors (PIs)

As mentioned above, UPS, a crucial mechanism for protein degradation, regulates protein turnover, thus affecting various cellular functions [190]. UPS is a relevant therapeutic target in cancer, especially in hematological malignancies like multiple myeloma (MM), a cancer of terminally differentiated plasma cells accumulating in the bone marrow [191,192]. Since plasma cells produce high amounts of immunoglobulins, they are very sensitive to the deregulation of protein degradation; moreover, malignant plasma cells appear even more susceptible to proteasomal inhibition than the healthy ones, due to constitutive activation of the oncogenic NF-κB pathway [193]. In fact, PIs act by blocking IκB degradation and thus, indirectly, inhibiting NF-κB signaling, although other processes are emerging, which contribute to the antitumor effects of PIs, and include inhibition of altered cell cycle control and apoptosis, ER stress, angiogenesis, and DNA repair [194], as well as epigenetic modulating effects [195,196].

The striking sensitivity of malignant cells to PIs has led to their approval for MM treatment, with three drugs being routinely used in a clinical setting [197] in association with other anti-MM therapies such as dexamethasone, and immunomodulatory drugs (lenalidomide), chemotherapy (DOX, mephalan, or cyclophosphamide), antibodies (elotuzumab or daratumumab), or histone deacetylase inhibitors (panobinostat) [198]. The first-in-class PI was bortezomib, a boronic acid derivative acting as a slowly reversible inhibitor of the β5 catalytic proteasomal subunit. Next, the irreversible inhibitor of β5 site carfilzomib and the first oral PI, ixazomib, were approved [197].

Although the toxicity of PIs is well-controlled in a clinical setting, distinct adverse profiles (such as peripheral neuropathy and cardiotoxicity) frequently arise and can lead to early discontinuation of the therapy [199].

The cardiotoxicity of bortezomib is still under debate, and likely depends on whether the drug is administered in patients with significant cardiovascular disease risk factors or previously treated with known cardiotoxic chemotherapeutics [200].

Molecular mechanisms involved in bortezomib-induced cardiovascular toxicity remain to be fully elucidated. In rat cardiomyoblast H9c2 cells, bortezomib causes the accumulation of polyubiquitinated proteins which, in turn, leads to ER stress and compensatory autophagy [201]. MG262, another boronic acid-based PI, promotes the translocation of the nuclear factor of activated T-cells (NFAT) in neonatal rat ventricular myocytes through the activation of the calcineurin-NFAT pathway, with significant changes in the cell morphology [202]; moreover, the inhibition of the proteasome by bortezomib in primary neonatal rat ventricular myocytes activates caspase-3 and caspase-7, triggering apoptosis [203]. Notably, mitochondria have been identified as a relevant target of cardiotoxicity because bortezomib inhibits complex V of the respiratory chain, resulting in a drop in ATP synthesis in the hearts of treated rats, and in a decreased cell shortening of primary rat left ventricular myocytes [204]. Functional and reversible changes accompanied the structural alterations of the mitochondria, which become pleomorphic and enlarged with concentric cristae and electron-dense inclusions, and showing misalignment of the myofibrillar network [201]. Moreover, bortezomib-mediated mitochondrial dysfunction might also be explained by the recently described process of extraction of misfolded proteins from mitochondria, and their subsequent degradation in proteasomes, called mitochondria-associated degradation (MAD) [205]; inhibition of proteasome leads to accumulation of misfolded and damaged proteins in mitochondria, resulting in their dysfunction (Figure 3).

The cardiovascular effects of bortezomib have been also addressed in several in vivo preclinical models that led to contradictory results. In fact, left ventricular systolic and diastolic function was preserved and no morphological myocardial abnormalities were detectable in adult male rabbits upon exposure of bortezomib [206]; conversely, male Wistar rats treated with bortezomib developed a reversible cardiac dysfunction with a significant decrease in left ventricular ejection fraction [201].

In cancer patients, the cardiovascular AEs associated with bortezomib treatment so far include heart failure, conduction disorders such as complete atrioventricular block, arrhythmias including atrial fibrillation, ischemic heart disease, pericardial effusion, and orthostatic hypotension [207]. A systematic review and meta-analysis of 25 prospective phase II/III trials evaluating bortezomib in different malignancies indicated that it does not significantly increase the risk of cardiac AEs as compared to control medications [208]. The overall cardiac safety profile of bortezomib was confirmed in a later retrospective analysis of patients included in the phase II registration study for US and EU regulatory approval, and in all phase III studies that led to US and EU approval of the drug [209], reporting no significant differences in the incidence of cardiovascular toxicities between bortezomib- and non-bortezomib-based arms [207].

Carfilzomib, which binds irreversibly to β5 (chymotryptic-like activity) and β5i immunoproteasome, was found to have greater selectivity for β5 subunits, with minimal affinity to β1 and β2 subunits when compared with bortezomib [210]. Carfilzomib induced proteasome inhibition in excess of 80% of patients [211], and its efficacy in bortezomib-resistant cells was likely due to prolonged and sustained inhibition of the proteasome. Carfilzomib received FDA approval in 2012 for use in relapsed and refractory MM (RRMM) patients who had previously received at least two therapies. Overall, several studies of carfilzomib noted an increased risk of cardiovascular AEs. A pooled analysis of phase II studies with carfilzomib showed 22% of patients developing cardiac side effects, such as arrhythmias, heart failure, treatment-associated cardiomyopathy, and ischemic heart disease [212]. In 2015, a carfilzomib combination regimen with lenalidomide and dexamethasone (KRd) was approved by the FDA for RRMM with one or more prior lines of treatment, based on significantly improved PFS and improved quality of life in a phase III trial [213,214]. However, this trial (ASPIRE) reported that the combination with the immunomodulatory drug lenalidomide increased cases of CVAEs, such as hypertension rates, heart failure rate, and ischemic heart disease rates [212,215]. The higher potency and irreversible inhibition by carfilzomib, along with dose-limiting neuropathy associated with bortezomib, may be the link between carfilzomib and higher incidences of CVAEs. In a systemic review and meta-analysis of 24 prospective clinical trials that included 2594 patients, a large range of reported CVAEs, with all grades of CVAE ranging from 0 to 52% and high-grade CVAEs ranging from 0 to 45% [216] was found. In an effort to better define risk factors and outcomes in patients who receive PI therapy, a prospective, observational study (PROTECT), was conducted [217], in which patients underwent baseline assessments over 6 months of bortezomib or carfilzomib; cardiac biomarkers included troponin I or T, BNP, NT-proBNP, ECG, and echocardiography. Of the CVAEs, 51% were in patients treated with carfilzomib, and 17% of those were treated with bortezomib, confirming the superior cardiotoxicity profile of carfilzomib. The study also demonstrated an association between BNP and NT-proBNP rise and increased CVAE risk. Overall, this trial reported a much higher incidence of CVAEs than prior studies, possibly due to its prospective nature as well as to the fact that CVAEs were captured as primary endpoint, showing that cardiotoxicity mainly occurred in patients with cardiac comorbidities.

Ixazomib (MLN9708), like bortezomib, acts as a reversible inhibitor on the β5 (chymotrypsin-like) and β5i subunits of the immunoproteasome, with additional inhibition of β1 and β2 subunits at higher concentrations [218,219]; it was the first orally bioavailable drug approved by the FDA in 2015 for RRMM, used in combination with lenalidomide and dexamethasone for MM patients in which one or more prior lines of treatment failed. It showed a pattern of cardiovascular AEs similar to bortezomib, although the trial excluded patients with cardiac comorbidities [220,221].

To overcome the cardiotoxicity of PIs like carfilzomib, mitochondrial functions affected by PIs are being dissected, and novel PIs devoid of cardiotoxicity are also being developed and analyzed in preclinical studies [222]. Combination strategies reducing PI doses are currently being evaluated in clinical trials to counteract dose-dependent CVAEs [223].

Table 2 recapitulates the main PIs used in clinical settings and their relative cardiotoxic effects, as well as the potential preventive/cardioprotective strategies to reduce their CVAEs.

Table 2. Main PIs, associated CVAEs, and potential preventive/cardioprotective strategies to reduce cardiotoxicity.

<table>
<tr><th>Proteasome Inhibitors</th><th>Mechanism of Action</th><th>Type of Study</th><th>Type of Cancer</th><th>Cardiotoxic Effects</th><th>Potential Preventive/Cardioprotective Strategies</th><th>Ref.</th></tr>
<tr><td>Bortezomib</td><td>Slowly-reversible inhibitor of β5 and β5i subunits</td><td>Systematic review and meta-analysis of 25 prospective phase II/III trials</td><td>Untreated multiple myeloma</td><td>Heart failure, conduction disorders, arrhythmias, ischemic heart disease, pericardial effusion and orthostatic hypotension</td><td rowspan="3">Assessment of cardiac function, evaluation of serum biomarkers of heart failure; Evaluation of atrial fibrillation history; Identification of cardiovascular risk factors; Use of β-blockers, ACE inhibitors, angiotensin II receptor blockers, apremilast (PDE4 inhibitor), metformin, PKG activator</td><td>[201,203–205,207,221,224–228]</td></tr>
<tr><td>Carfilzomib</td><td>Irreversible inhibitor of β5 and β5i subunits</td><td>Phase III trial (ASPIRE trial) Prospective, observational study (PROTECT trial)</td><td>Relapsed and refractory multiple myeloma</td><td>Arrhythmias, heart failure, cardiomyopathy, ischemic heart disease</td><td>[212,217,221,224–228]</td></tr>
<tr><td>Ixazomib</td><td>Reversible inhibitor of β5 and β5i subunits, inhibition of β1 and β2 subunits at high concentration</td><td>Randomized phase III trial (TOURMALINE-MM1 trial)</td><td>Relapsed and refractory multiple myeloma</td><td>Heart failure</td><td>[220,221,224–229]</td></tr>
</table>

4. Conclusions

Cardiotoxicity associated with widely used anticancer drugs, such as ANTs, RTKIs, and PIs, still represents a significant clinical challenge that compromises the quality of life and overall survival of cancer patients. Although the mechanisms driving the cardiotoxicity of these anticancer drugs is multifactorial, and different pathways seem implicated, a growing line of evidence strongly suggests that the cardiac AEs from these anticancer therapeutics involve direct or indirect mitochondria-related toxicity. In addition to the ability of the anticancer drugs to affect mitochondrial bioenergetics, mitochondrial DNA replication, mitochondrial oxidative/nitrative stress, and cell death, emerging evidence also underscores dysregulated mitochondrial dynamics as determinant of anticancer-drug-dependent cardiotoxicity. A thorough understanding of the mitochondrial processes underlying cardiovascular toxicity is therefore fundamental to rationally develop effective strategies preventing cardiomyocyte dysfunction or loss elicited by several chemotherapeutic regimens.

Author Contributions: Conceptualization: C.R., T.P., T.A. and N.A.; data curation: C.R, E.M.D.F., M.C.G., A.D.B., M.E.G.C., M.G.M., M.G., A.N., G.V. and N.A.; writing—original draft preparation: C.R., E.M.D.F., T.P., T.A. and N.A.; writing—review and editing: C.R. and N.A. All authors have read and agreed to the published version of the manuscript.

Funding: This work was supported by the Associazione Italiana per la Ricerca sul Cancro (AIRC), Milan, Italy (Investigator Grant-24449 to N.A.) and by a Fondazione Associazione Italiana per la Ricerca sul Cancro (AIRC) Start Up Reintegration Grant (21651) to E.M.D.F.

Institutional Review Board Statement: Not applicable.

Informed Consent Statement: Not applicable.

Data Availability Statement: Not applicable.

Acknowledgments: C.R. acknowledges POR Calabria (Italy) FESR-FSE 2014/2020-Azione 10.5.12-Linea B (DR n. 683 del 21 May 2019) for financial support for an RTDa position.

Conflicts of Interest: The authors declare no conflict of interest.

Abbreviations

AMPK	AMP-activated protein kinase
ANTs	anthracyclines
ARE	antioxidant response element
BCL-2	B cell lymphoma-2
CHF	chronic heart failure
CMDL-1	cardiomyocyte mitochondrial dynamic-related lncRNA 1
CML	chronic myelogenous leukemia
CVAEs	cardiovascular adverse events
CVDs	cardiovascular diseases
DOX	doxorubicin
Drp1	dynamin-related protein 1
ER	endoplasmic reticulum
Fis1	fission protein 1
FoxO	Forkhead box subgroup O
GPER	G-protein coupled receptor 30
GSH	glutathione
hiPSC-CM	human-induced pluripotent stem cell-derived cardiomyocytes
I/R	ischemia/reperfusion
IMM	inner mitochondrial membrane
Keap1	kelch-like ECH-associated protein 1
LIR	LC-3-interacting region
MAD	mitochondria-associated degradation
MFF	mitochondrial fission factor
MFN1	mitofusin-1
MFN2	mitofusin 2
Midivi-1	mitochondrial division inhibitor-1
MM	multiple myeloma
mtDNA	mitochondrial DNA
NFAT	nuclear factor of activated T-cells
Nfe2l2	nuclear factor erythroid derived 2 like 2
Nrf2	nuclear factor erythroid 2–related factor 2
OMM	outer mitochondrial membrane
OPA1	optic atrophy 1
PDK	pyruvate dehydrogenase kinase
PGC-1α	peroxisome proliferator-activated receptor gamma coactivator 1-α
PGC-1β	peroxisome proliferator-activated receptor gamma coactivator 1-β
Ph+ B-ALL	Philadelphia chromosome-positive B-acute lymphoblastic leukemia
PINK1	PTEN-induced putative kinase 1
PIs	proteasome inhibitors
RNS	reactive nitrogen species
ROS	reactive oxygen species
RTKIs	receptor tyrosine kinase inhibitors
RTKs	Receptor tyrosine kinases
SOD	superoxide dismutase
Top	topoisomerase
TXNRD	thioredoxin reductase
UBD	ubiquitin binding domain
UCP-2	uncoupling protein 2
UCP-3	uncoupling protein 3
UPS	ubiquitin–proteasome system
URPmt	mitochondrial unfolded protein response
VEGF	vascular endothelial growth factor

References

1. Pasqua, T.; Rocca, C.; Giglio, A.; Angelone, T. Cardiometabolism as an Interlocking Puzzle between the Healthy and Diseased Heart: New Frontiers in Therapeutic Applications. *J. Clin. Med.* **2021**, *10*, 721. [CrossRef]
2. Nguyen, B.Y.; Ruiz-Velasco, A.; Bui, T.; Collins, L.; Wang, X.; Liu, W. Mitochondrial function in the heart: The insight into mechanisms and therapeutic potentials. *J. Cereb. Blood Flow Metab.* **2019**, *176*, 4302–4318. [CrossRef]
3. Cadete, V.J.; Vasam, G.; Menzies, K.J.; Burelle, Y. Mitochondrial quality control in the cardiac system: An integrative view. *Biochim. Biophys. Acta (BBA) Mol. Basis Dis.* **2019**, *1865*, 782–796. [CrossRef]
4. Chistiakov, D.A.; Shkurat, T.P.; Melnichenko, A.A.; Grechko, A.V.; Orekhov, A.N. The role of mitochondrial dysfunction in cardiovascular disease: A brief review. *Ann. Med.* **2018**, *50*, 121–127. [CrossRef]
5. Marín-García, J.; Akhmedov, A.T. Mitochondrial dynamics and cell death in heart failure. *Heart Fail. Rev.* **2016**, *21*, 123–136. [CrossRef]
6. Liesa, M.; Palacín, M.; Zorzano, A. Mitochondrial Dynamics in Mammalian Health and Disease. *Physiol. Rev.* **2009**, *89*, 799–845. [CrossRef]
7. Westermann, B. Mitochondrial fusion and fission in cell life and death. *Nat. Rev. Mol. Cell Biol.* **2010**, *11*, 872–884. [CrossRef]
8. Youle, R.J.; van der Bliek, A.M. Mitochondrial fission, fusion, and stress. *Science* **2012**, *337*, 1062–1065. [CrossRef]
9. Friedman, J.R.; Nunnari, J. Mitochondrial form and function. *Nature* **2014**, *505*, 335–343. [CrossRef]
10. Rosca, M.G.; Hoppel, C.L. Mitochondrial dysfunction in heart failure. *Heart Fail. Rev.* **2012**, *18*, 607–622. [CrossRef]
11. Zhou, B.; Tian, R. Mitochondrial dysfunction in pathophysiology of heart failure. *J. Clin. Investig.* **2018**, *128*, 3716–3726. [CrossRef]
12. Cardinale, D.; Colombo, A.; Lamantia, G.; Colombo, N.; Civelli, M.; De Giacomi, G.; Rubino, M.; Veglia, F.; Fiorentini, C.; Cipolla, C.M. Anthracycline-Induced Cardiomyopathy: Clinical Relevance and Response to Pharmacologic Therapy. *J. Am. Coll. Cardiol.* **2010**, *55*, 213–220. [CrossRef]
13. Randle, P.; Garland, P.; Hales, C.; Newsholme, E. The glucose fatty-acid cycle its role in insulin sensitivity and the metabolic disturbances of diabetes mellitus. *Lancet* **1963**, *281*, 785–789. [CrossRef]
14. Varga, Z.; Ferdinandy, P.; Liaudet, L.; Pacher, P. Drug-induced mitochondrial dysfunction and cardiotoxicity. *Am. J. Physiol. Circ. Physiol.* **2015**, *309*, H1453–H1467. [CrossRef]
15. Zamorano, J.L.; Lancellotti, P.; Rodriguez Munoz, D.; Aboyans, V.; Asteggiano, R.; Galderisi, M.; Habib, G.; Lenihan, D.J.; Lip, G.Y.H.; Lyon, A.R.; et al. ESC position paper on cancer treatments and cardiovascular toxicity developed under the auspices of the ESC committee for practice guidelines: The task force for cancer treatments and cardiovascular toxicity of the European Society of Cardiology (ESC). *Eur. Heart J.* **2016**, *37*, 2768–2801. [CrossRef]
16. Natarajan, V.; Chawla, R.; Mah, T.; Vivekanandan, R.; Tan, S.Y.; Sato, P.Y.; Mallilankaraman, K. Mitochondrial Dysfunction in Age-Related Metabolic Disorders. *Proteomics* **2020**, *20*, e1800404. [CrossRef]
17. Koklesova, L.; Liskova, A.; Samec, M.; Zhai, K.; Al-Ishaq, R.K.; Bugos, O.; Šudomová, M.; Biringer, K.; Pec, M.; Adamkov, M.; et al. Protective Effects of Flavonoids against Mitochondriopathies and Associated Pathologies: Focus on the Predictive Approach and Personalized Prevention. *Int. J. Mol. Sci.* **2021**, *22*, 8649. [CrossRef]
18. Randle, P.J.; Priestman, D.A.; Mistry, S.C.; Halsall, A. Glucose fatty acid interactions and the regulation of glucose disposal. *J. Cell. Biochem.* **1994**, *55*, 1–11. [CrossRef]
19. Holmgren, D.; Wahlander, H.; Eriksson, B.; Oldfors, A.; Holme, E.; Tulinius, M. Cardiomyopathy in children with mitochondrial disease Clinical course and cardiological findings. *Eur. Heart J.* **2003**, *24*, 280–288. [CrossRef]
20. Murphy, E.; Ardehali, H.; Balaban, R.S.; DiLisa, F.; Dorn, G.W., 2nd; Kitsis, R.N.; Otsu, K.; Ping, P.; Rizzuto, R.; Sack, M.N.; et al. Mitochondrial Function, Biology, and Role in Disease: A Scientific Statement from the American Heart Association. *Circ. Res.* **2016**, *118*, 1960–1991. [CrossRef]
21. Rocca, C.; Pasqua, T.; Boukhzar, L.; Anouar, Y.; Angelone, T. Progress in the emerging role of selenoproteins in cardiovascular disease: Focus on endoplasmic reticulum-resident selenoproteins. *Cell Mol. Life Sci.* **2019**, *76*, 3969–3985. [CrossRef]
22. Rocca, C.; Boukhzar, L.; Granieri, M.C.; Alsharif, I.; Mazza, R.; Lefranc, B.; Tota, B.; Leprince, J.; Cerra, M.C.; Anouar, Y.; et al. A selenoprotein T-derived peptide protects the heart against ischaemia/reperfusion injury through inhibition of apoptosis and oxidative stress. *Acta Physiol.* **2018**, *223*, e13067. [CrossRef]
23. Peoples, J.N.; Saraf, A.; Ghazal, N.; Pham, T.T.; Kwong, J.Q. Mitochondrial dysfunction and oxidative stress in heart disease. *Exp. Mol. Med.* **2019**, *51*, 1–13. [CrossRef]
24. Radi, R.; Turrens, J.; Chang, L.; Bush, K.; Crapo, J.; Freeman, B. Detection of catalase in rat heart mitochondria. *J. Biol. Chem.* **1991**, *266*, 22028–22034. [CrossRef]
25. Andreyev, A.Y.; Kushnareva, Y.E.; Murphy, A.N.; Starkov, A. Mitochondrial ROS metabolism: 10 Years later. *Biochemistry* **2015**, *80*, 517–531. [CrossRef]
26. Marí, M.; Morales, A.; Colell, A.; García-Ruiz, C.; Fernández-Checa, J.C. Mitochondrial Glutathione, a Key Survival Antioxidant. *Antioxid. Redox Signal.* **2009**, *11*, 2685–2700. [CrossRef]
27. Bhandari, S. Impact of intravenous iron on cardiac and skeletal oxidative stress and cardiac mitochondrial function in experimental uraemia chronic kidney disease. *Front. Biosci.* **2021**, *26*, 442. [CrossRef]
28. Fenton, H.J.H. Oxidation of tartaric acid in presence of iron. *J. Chem. Soc. Trans.* **1894**, *65*, 899–910. [CrossRef]
29. Masaki, H.; Okano, Y.; Sakurai, H. Differential role of catalase and glutathione peroxidase in cultured human fibroblasts under exposure of H_2O_2 or ultraviolet B light. *Arch. Dermatol. Res.* **1998**, *290*, 113–118. [CrossRef]

30. Logan, A.; Pell, V.R.; Shaffer, K.J.; Evans, C.; Stanley, N.; Robb, E.L.; Prime, T.A.; Chouchani, E.T.; Cochemé, H.M.; Fearnley, I.M.; et al. Assessing the Mitochondrial Membrane Potential in Cells and In Vivo using Targeted Click Chemistry and Mass Spectrometry. *Cell Metab.* **2016**, *23*, 379–385. [CrossRef]
31. Jonckheere, A.I.; Smeitink, J.A.M.; Rodenburg, R.J.T. Mitochondrial ATP synthase: Architecture, function and pathology. *J. Inherit. Metab. Dis.* **2011**, *35*, 211–225. [CrossRef]
32. Perry, S.W.; Norman, J.P.; Barbieri, J.; Brown, E.B.; Gelbard, H.A. Mitochondrial membrane potential probes and the proton gradient: A practical usage guide. *BioTechniques* **2011**, *50*, 98–115. [CrossRef]
33. Arruda, A.P.; Hotamisligil, G.S. Calcium homeostasis and organelle function in the pathogenesis of obesity and diabetes. *Cell Metab.* **2015**, *22*, 381–397. [CrossRef]
34. Baughman, J.M.; Perocchi, F.; Girgis, H.S.; Plovanich, M.; Belcher-Timme, C.A.; Sancak, Y.; Bao, X.R.; Strittmatter, L.; Goldberger, O.; Bogorad, R.L.; et al. Integrative genomics identifies MCU as an essential component of the mitochondrial calcium uniporter. *Nature* **2011**, *476*, 341–345. [CrossRef]
35. De Stefani, D.; Raffaello, A.; Teardo, E.; Szabò, I.; Rizzuto, R. A forty-kilodalton protein of the inner membrane is the mitochondrial calcium uniporter. *Nature* **2011**, *476*, 336–340. [CrossRef]
36. Luongo, T.S.; Lambert, J.P.; Gross, P.; Nwokedi, M.; Lombardi, A.A.; Shanmughapriya, S. The mitochondrial Na^+/Ca^{2+} exchanger is essential for Ca^{2+} homeostasis and viability. *Nature* **2017**, *545*, 93–97. [CrossRef]
37. Eisner, D.A.; Caldwell, J.L.; Kistamás, K.; Trafford, A.W. Calcium and Excitation-Contraction Coupling in the Heart. *Circ. Res.* **2017**, *121*, 181–195. [CrossRef]
38. Sheu, S.S.; Jou, M.J. Mitochondrial free Ca^{2+} concentration in living cells. *J. Bioenerg. Biomembr.* **1994**, *26*, 487–493. [CrossRef]
39. Olson, M.L.; Chalmers, S.; McCarron, J.G. Mitochondrial organization and Ca^{2+} uptake. *Biochem. Soc. Trans.* **2012**, *40*, 158–167. [CrossRef]
40. Patergnani, S.; Suski, J.M.; Agnoletto, C.; Bononi, A.; Bonora, M.; De Marchi, E.; Giorgi, C.; Marchi, S.; Missiroli, S.; Poletti, F.; et al. Calcium signaling around Mitochondria Associated Membranes (MAMs). *Cell Commun. Signal.* **2011**, *9*, 19. [CrossRef]
41. Del Re, D.P.; Amgalan, D.; Linkermann, A.; Liu, Q.; Kitsis, R.N. Fundamental Mechanisms of Regulated Cell Death and Implications for Heart Disease. *Physiol. Rev.* **2019**, *99*, 1765–1817. [CrossRef]
42. Chipuk, J.E.; Moldoveanu, T.; Llambi, F.; Parsons, M.J.; Green, D.R. The BCL-2 Family Reunion. *Mol. Cell* **2010**, *37*, 299–310. [CrossRef]
43. Gavathiotis, E.; Reyna, D.E.; Davis, M.L.; Bird, G.H.; Walensky, L.D. BH3-Triggered Structural Reorganization Drives the Activation of Proapoptotic BAX. *Mol. Cell* **2010**, *40*, 481–492. [CrossRef]
44. Du, C.; Fang, M.; Li, Y.; Li, L.; Wang, X. Smac, a Mitochondrial Protein that Promotes Cytochrome c–Dependent Caspase Activation by Eliminating IAP Inhibition. *Cell* **2000**, *102*, 33–42. [CrossRef]
45. Verhagen, A.M.; Ekert, P.G.; Pakusch, M.; Silke, J.; Connolly, L.M.; Reid, G.E.; Moritz, R.L.; Simpson, R.J.; Vaux, D.L. Identification of DIABLO, a Mammalian Protein that Promotes Apoptosis by Binding to and Antagonizing IAP Proteins. *Cell* **2000**, *102*, 43–53. [CrossRef]
46. Suzuki, Y.; Imai, Y.; Nakayama, H.; Takahashi, K.; Takio, K.; Takahashi, R. A Serine Protease, HtrA2, Is Released from the Mitochondria and Interacts with XIAP, Inducing Cell Death. *Mol. Cell* **2001**, *8*, 613–621. [CrossRef]
47. Chipuk, J.E.; McStay, G.P.; Bharti, A.; Kuwana, T.; Clarke, C.J.; Siskind, L.J.; Obeid, L.M.; Green, D.R. Sphingolipid Metabolism Cooperates with BAK and BAX to Promote the Mitochondrial Pathway of Apoptosis. *Cell* **2012**, *148*, 988–1000. [CrossRef]
48. Li, P.; Nijhawan, D.; Budihardjo, I.; Srinivasula, S.M.; Ahmad, M.; Alnemri, E.S.; Wang, X. Cytochrome c and dATP-dependent formation of Apaf-1/caspase-9 complex initiates an apoptotic protease cascade. *Cell* **1997**, *91*, 479–489. [CrossRef]
49. Zou, H.; Henzel, W.; Liu, X.; Lutschg, A.; Wang, X. Apaf-1, a Human Protein Homologous to C. elegans CED-4, Participates in Cytochrome c–Dependent Activation of Caspase-3. *Cell* **1997**, *90*, 405–413. [CrossRef]
50. Liu, X.; Kim, C.N.; Yang, J.; Jemmerson, R.; Wang, X. Induction of Apoptotic Program in Cell-Free Extracts: Requirement for dATP and Cytochrome c. *Cell* **1996**, *86*, 147–157. [CrossRef]
51. Edlich, F.; Banerjee, S.; Suzuki, M.; Cleland, M.M.; Arnoult, D.; Wang, C.; Neutzner, A.; Tjandra, N.; Youle, R.J. Bcl-xL Retrotranslocates Bax from the Mitochondria into the Cytosol. *Cell* **2011**, *145*, 104–116. [CrossRef] [PubMed]
52. Minoia, M.; Boncoraglio, A.; Vinet, J.; Morelli, F.F.; Brunsting, J.F.; Poletti, A.; Krom, S.; Reits, E.; Kampinga, H.H.; Carra, S. BAG3 induces the sequestration of proteasomal clients into cytoplasmic puncta: Implications for a proteasome-toautophagy switch. *Autophagy* **2014**, *10*, 1603–1621. [CrossRef]
53. Hammerling, B.C.; Gustafsson, Å.B. Mitochondrial quality control in the myocardium: Cooperation between protein degradation and mitophagy. *J. Mol. Cell. Cardiol.* **2014**, *75*, 122–130. [CrossRef] [PubMed]
54. Bragoszewski, P.; Turek, M.; Chacinska, A. Control of mitochondrial biogenesis and function by the ubiquitin–proteasome system. *Open Biol.* **2017**, *7*, 7. [CrossRef] [PubMed]
55. Ciechanover, A.; Stanhill, A. The complexity of recognition of ubiquitinated substrates by the 26S proteasome. *Biochim. Biophys. Acta* **2014**, *1843*, 86–96. [CrossRef]
56. Kwon, Y.T.; Ciechanover, A. The Ubiquitin Code in the Ubiquitin-Proteasome System and Autophagy. *Trends Biochem. Sci.* **2017**, *42*, 873–886. [CrossRef]
57. Kodroń, A.; Mussulini, B.H.; Pilecka, I.; Chacińska, A. The ubiquitin-proteasome system and its crosstalk with mitochondria as therapeutic targets in medicine. *Pharmacol. Res.* **2021**, *163*, 105248. [CrossRef]

58. Karbowski, M.; Youle, R.J. Regulating mitochondrial outer membrane proteins by ubiquitination and proteasomal degradation. *Curr. Opin. Cell Biol.* **2011**, *23*, 476–482. [CrossRef]
59. Quiles, J.M.; Gustafsson, Å.B. Mitochondrial Quality Control and Cellular Proteostasis: Two Sides of the Same Coin. *Front. Physiol.* **2020**, *11*, 515. [CrossRef]
60. Szczepanowska, K.; Maiti, P.; Kukat, A.; Hofsetz, E.; Nolte, H.; Senft, K.; Becker, C.; Ruzzenente, B.; Hornig-Do, H.-T.; Wibom, R.; et al. CLPP coordinates mitoribosomal assembly through the regulation of ERAL 1 levels. *EMBO J.* **2016**, *35*, 2566–2583. [CrossRef]
61. Weinhäupl, K.; Lindau, C.; Hessel, A.; Wang, Y.; Schütze, C.; Jores, T.; Melchionda, L.; Schönfisch, B.; Kalbacher, H.; Bersch, B.; et al. Structural Basis of Membrane Protein Chaperoning through the Mitochondrial Intermembrane Space. *Cell* **2018**, *175*, 1365–1379.e25. [CrossRef] [PubMed]
62. Lavie, J.; De Belvalet, H.; Sonon, S.; Ion, A.M.; Dumon, E.; Melser, S.; Lacombe, D.; Dupuy, J.-W.; Lalou, C.; Bénard, G. Ubiquitin-Dependent Degradation of Mitochondrial Proteins Regulates Energy Metabolism. *Cell Rep.* **2018**, *23*, 2852–2863. [CrossRef] [PubMed]
63. Hofmann, C.; Katus, H.A.; Doroudgar, S. Protein Misfolding in Cardiac Disease. *Circulation* **2019**, *139*, 2085–2088. [CrossRef] [PubMed]
64. Ranek, M.J.; Zheng, H.; Huang, W.; Kumarapeli, A.R.; Li, J.; Liu, J.; Wang, X. Genetically induced moderate inhibition of 20S proteasomes in cardiomyocytes facilitates heart failure in mice during systolic overload. *J. Mol. Cell. Cardiol.* **2015**, *85*, 273–281. [CrossRef] [PubMed]
65. Predmore, J.M.; Wang, P.; Davis, F.; Bartolone, S.; Westfall, M.; Dyke, D.B.; Pagani, F.; Powell, S.R.; Day, S.M. Ubiquitin Proteasome Dysfunction in Human Hypertrophic and Dilated Cardiomyopathies. *Circulation* **2010**, *121*, 997–1004. [CrossRef]
66. Zhong, Q.; Gao, W.; Du, F.; Wang, X. Mule/ARF-BP1, a BH3-Only E3 Ubiquitin Ligase, Catalyzes the Polyubiquitination of Mcl-1 and Regulates Apoptosis. *Cell* **2005**, *121*, 1085–1095. [CrossRef]
67. Münch, C.; Harper, J.W. Mitochondrial unfolded protein response controls matrix pre-RNA processing and translation. *Nature* **2016**, *534*, 710–713. [CrossRef]
68. Tahrir, F.G.; Langford, D.; Amini, S.; Mohseni Ahooyi, T.; Khalili, K. Mitochondrial quality control in cardiac cells: Mechanisms and role in cardiac cell injury and disease. *J. Cell. Physiol.* **2019**, *234*, 8122–8133. [CrossRef]
69. Gustafsson, Å.B.; Dorn, G.W., 2nd. Evolving and Expanding the Roles of Mitophagy as a Homeostatic and Pathogenic Process. *Physiol. Rev.* **2019**, *99*, 853–892. [CrossRef]
70. Tong, M.; Saito, T.; Zhai, P.; Oka, S.-I.; Mizushima, W.; Nakamura, M.; Ikeda, S.; Shirakabe, A.; Sadoshima, J. Mitophagy Is Essential for Maintaining Cardiac Function During High Fat Diet-Induced Diabetic Cardiomyopathy. *Circ. Res.* **2019**, *124*, 1360–1371. [CrossRef]
71. Billia, F.; Hauck, L.; Konecny, F.; Rao, V.; Shen, J.; Mak, T.W. PTENinducible kinase 1 (PINK1)/Park6 is indispensable for normal heart function. *Proc. Natl. Acad. Sci. USA* **2011**, *108*, 9572–9577. [CrossRef] [PubMed]
72. Shirakabe, A.; Zhai, P.; Ikeda, Y.; Saito, T.; Maejima, Y.; Hsu, C.-P.; Nomura, M.; Egashira, K.; Levine, B.; Sadoshima, J. Drp1-Dependent Mitochondrial Autophagy Plays a Protective Role Against Pressure Overload–Induced Mitochondrial Dysfunction and Heart Failure. *Circulation* **2016**, *133*, 1249–1263. [CrossRef] [PubMed]
73. Siddall, H.K.; Yellon, D.M.; Ong, S.-B.; Mukherjee, U.A.; Burke, N.; Hall, A.R.; Angelova, P.R.; Ludtmann, M.H.; Deas, E.; Davidson, S.M.; et al. Loss of PINK1 Increases the Heart's Vulnerability to Ischemia-Reperfusion Injury. *PLoS ONE* **2013**, *8*, e62400. [CrossRef]
74. Kubli, D.A.; Zhang, X.; Lee, Y.; Hanna, R.A.; Quinsay, M.N.; Nguyen, C.K.; Jimenez, R.; Petrosyan, S.; Murphy, A.N.; Gustafsson, A.B. Parkin Protein Deficiency Exacerbates Cardiac Injury and Reduces Survival following Myocardial Infarction. *J. Biol. Chem.* **2013**, *288*, 915–926. [CrossRef]
75. Truban, D.; Hou, X.; Caulfield, T.R.; Fiesel, F.C.; Springer, W. PINK1, Parkin, and Mitochondrial Quality Control: What can we Learn about Parkinson's Disease Pathobiology? *J. Park. Dis.* **2017**, *7*, 13–29. [CrossRef]
76. Chen, Y.; Dorn, G.W. PINK1-Phosphorylated Mitofusin 2 Is a Parkin Receptor for Culling Damaged Mitochondria. *Science* **2013**, *340*, 471–475. [CrossRef]
77. Gong, G.; Song, M.; Csordas, G.; Kelly, D.P.; Matkovich, S.J.; Dorn, G.W., 2nd. Parkin-mediated mitophagy directs perinatal cardiac metabolic maturation in mice. *Science* **2015**, *350*, aad2459. [CrossRef]
78. Kabeya, Y.; Mizushima, N.; Ueno, T.; Yamamoto, A.; Kirisako, T.; Noda, T.; Kominami, E.; Ohsumi, Y.; Yoshimori, T. LC3, a mammalian homologue of yeast Apg8p, is localized in autophagosome membranes after processing. *EMBO J.* **2000**, *19*, 5720–5728. [CrossRef]
79. Kawajiri, S.; Saiki, S.; Sato, S.; Sato, F.; Hatano, T.; Eguchi, H.; Hattori, N. PINK1 is recruited to mitochondria with parkin and associates with LC3 in mitophagy. *FEBS Lett.* **2010**, *584*, 1073–1079. [CrossRef]
80. Pankiv, S.; Clausen, T.H.; Lamark, T.; Brech, A.; Bruun, J.A.; Outzen, H.; Øvervatn, A.; Bjørkøy, G.; Johansen, T. p62/SQSTM1 binds directly to Atg8/LC3 to facilitate degradation of ubiquitinated protein aggregates by autophagy. *J. Biol. Chem.* **2007**, *282*, 24131–24145. [CrossRef]
81. Narendra, D.; Tanaka, A.; Suen, D.-F.; Youle, R.J. Parkin is recruited selectively to impaired mitochondria and promotes their autophagy. *J. Cell Biol.* **2008**, *183*, 795–803. [CrossRef] [PubMed]

82. Shah, M.; Chacko, L.A.; Joseph, J.P.; Ananthanarayanan, V. Mitochondrial dynamics, positioning and function mediated by cytoskeletal interactions. *Cell Mol. Life Sci.* **2021**, *78*, 3969–3986. [CrossRef] [PubMed]
83. El-Hattab, A.W.; Suleiman, J.; Almannai, M.; Scaglia, F. Mitochondrial dynamics: Biological roles, molecular machinery, and related diseases. *Mol. Genet. Metab.* **2018**, *125*, 315–321. [CrossRef] [PubMed]
84. Gerald, W.D., II. Evolving Concepts of Mitochondrial Dynamics. *Ann. Rev. Physiol.* **2019**, *81*, 1–17.
85. Cipolat, S.; De Brito, O.M.; Dal Zilio, B.; Scorrano, L. OPA1 requires mitofusin 1 to promote mitochondrial fusion. *Proc. Natl. Acad. Sci. USA* **2004**, *101*, 15927–15932. [CrossRef]
86. Smirnova, E.; Griparic, L.; Shurland, D.-L.; van der Bliek, A.M. Dynamin-related Protein Drp1 Is Required for Mitochondrial Division in Mammalian Cells. *Mol. Biol. Cell* **2001**, *12*, 2245–2256. [CrossRef]
87. Li, A.; Gao, M.; Jiang, W.; Qin, Y.; Gong, G. Mitochondrial Dynamics in Adult Cardiomyocytes and Heart Diseases. *Front. Cell Dev. Biol.* **2020**, *8*. [CrossRef]
88. Yin, Y.; Shen, H. Advances in Cardiotoxicity Induced by Altered Mitochondrial Dynamics and Mitophagy. *Front. Cardiovasc. Med.* **2021**, *8*. [CrossRef]
89. Gorini, S.; De Angelis, A.; Berrino, L.; Malara, N.; Rosano, G.; Ferraro, E. Chemotherapeutic Drugs and Mitochondrial Dysfunction: Focus on Doxorubicin, Trastuzumab, and Sunitinib. *Oxid. Med. Cell. Longev.* **2018**, *2018*, 7582730, Correction in *Oxid. Med. Cell. Longev.* **2019**, *2019*, 9601435. [CrossRef]
90. Larsen, S.; Nielsen, J.; Hansen, C.N.; Nielsen, L.B.; Wibrand, F.; Stride, N.; Schrøder, H.D.; Boushel, R.; Helge, J.W.; Dela, F.; et al. Biomarkers of mitochondrial content in skeletal muscle of healthy young human subjects. *J. Physiol.* **2012**, *590*, 3349–3360. [CrossRef]
91. Gilliam, L.A.; St Clair, D.K. Chemotherapy-Induced Weakness and Fatigue in Skeletal Muscle: The Role of Oxidative Stress. *Antioxid. Redox Signal.* **2011**, *15*, 2543–2563. [CrossRef] [PubMed]
92. Sorensen, J.C.; Cheregi, B.D.; Timpani, C.; Nurgali, K.; Hayes, A.; Rybalka, E. Mitochondria: Inadvertent targets in chemotherapy-induced skeletal muscle toxicity and wasting? *Cancer Chemother. Pharmacol.* **2016**, *78*, 673–683. [CrossRef] [PubMed]
93. Rocca, C.; Pasqua, T.; Cerra, M.C.; Angelone, T. Cardiac Damage in Anthracyclines Therapy: Focus on Oxidative Stress and Inflammation. *Antioxid. Redox Signal.* **2020**, *32*, 1081–1097. [CrossRef] [PubMed]
94. Minotti, G.; Menna, P.; Salvatorelli, E.; Cairo, G.; Gianni, L. Anthracyclines: Molecular Advances and Pharmacologic Developments in Antitumor Activity and Cardiotoxicity. *Pharmacol. Rev.* **2004**, *56*, 185–229. [CrossRef]
95. Fischer, F.; Hamann, A.; Osiewacz, H.D. Mitochondrial quality control: An integrated network of pathways. *Trends Biochem. Sci.* **2012**, *37*, 284–292. [CrossRef]
96. Kim, C.-W.; Choi, K.-C. Effects of anticancer drugs on the cardiac mitochondrial toxicity and their underlying mechanisms for novel cardiac protective strategies. *Life Sci.* **2021**, *277*, 119607. [CrossRef]
97. Egea, J.; Fabregat, I.; Frapart, Y.M.; Ghezzi, P.; Görlach, A.; Kietzmann, T.; Kubaichuk, K.; Knaus, U.G.; Lopez, M.G.; Olaso-Gonzalez, G.; et al. European contribution to the study of ROS: A summary of the findings and prospects for the future from the COST action BM1203 (EU-ROS). *Redox Biol.* **2017**, *13*, 94–162. [CrossRef]
98. Tewey, K.M.; Rowe, T.C.; Yang, L.; Halligan, B.D.; Liu, L.F. Adriamycin-Induced DNA Damage Mediated by Mammalian DNA Topoisomerase II. *Science* **1984**, *226*, 466–468. [CrossRef]
99. Henriksen, P.A. Anthracycline cardiotoxicity: An update on mechanisms, monitoring and prevention. *Heart* **2018**, *104*, 971–977. [CrossRef]
100. Gianni, L.; Herman, E.H.; Lipshultz, S.E.; Minotti, G.; Sarvazyan, N.; Sawyer, D.B. Anthracycline Cardiotoxicity: From Bench to Bedside. *J. Clin. Oncol.* **2008**, *26*, 3777–3784. [CrossRef]
101. Eschenhagen, T.; Force, T.; Ewer, M.S.; De Keulenaer, G.; Suter, T.M.; Anker, S.D.; Avkiran, M.; de Azambuja, E.; Balligand, J.-L.; Brutsaert, D.L.; et al. Cardiovascular side effects of cancer therapies: A position statement from the Heart Failure Association of the European Society of Cardiology. *Eur. J. Heart Fail.* **2011**, *13*, 1–10. [CrossRef] [PubMed]
102. Han, X.; Zhou, Y.; Liu, W. Precision cardio-oncology: Understanding the cardiotoxicity of cancer therapy. *NPJ Precis. Oncol.* **2017**, *1*, 31. [CrossRef] [PubMed]
103. Ewer, M.S.; Lippman, S.M. Type II Chemotherapy-Related Cardiac Dysfunction: Time to Recognize a New Entity. *J. Clin. Oncol.* **2005**, *23*, 2900–2902. [CrossRef] [PubMed]
104. Tocchetti, C.G.; Cadeddu, C.; Di Lisi, D.; Femminò, S.; Madonna, R.; Mele, D.; Monte, I.; Novo, G.; Penna, C.; Pepe, A.; et al. From Molecular Mechanisms to Clinical Management of Antineoplastic Drug-Induced Cardiovascular Toxicity: A Translational Overview. *Antioxid. Redox Signal.* **2019**, *30*, 2110–2153. [CrossRef] [PubMed]
105. Zhang, S.; Liu, X.; Bawa-Khalfe, T.; Lu, L.-S.; Lyu, Y.L.; Liu, L.F.; Yeh, E.T.H. Identification of the molecular basis of doxorubicin-induced cardiotoxicity. *Nat. Med.* **2012**, *18*, 1639–1642. [CrossRef]
106. Vejpongsa, P.; Yeh, E.T. Prevention of anthracycline-induced cardiotoxicity: Challenges and opportunities. *J. Am. Coll. Cardiol.* **2014**, *64*, 938–945. [CrossRef]
107. Borowitz, J.; Rathinavelu, A.; Kanthasamy, A.; Wilsbacher, J.; Isom, G. Accumulation of Labeled Cyanide in Neuronal Tissue. *Toxicol. Appl. Pharmacol.* **1994**, *129*, 80–85. [CrossRef]
108. Montaigne, D.; Marechal, X.; Preau, S.; Baccouch, R.; Modine, T.; Fayad, G.; Lancel, S.; Neviere, R. Doxorubicin induces mitochondrial permeability transition and contractile dysfunction in the human myocardium. *Mitochondrion* **2011**, *11*, 22–26. [CrossRef]

109. Ferrero, M.; Ferrero, E.; Gaja, G.; Bernelli-Zazzera, A. Adriamycin: Energy metabolism and mitochondrial oxidations in the heart of treated rabbits. *Biochem. Pharmacol.* **1976**, *25*, 125–130. [CrossRef]
110. Zhou, S.; Starkov, A.; Froberg, M.K.; Leino, R.L.; Wallace, K.B. Cumulative and irreversible cardiac mitochondrial dysfunction induced by doxorubicin. *Cancer Res.* **2001**, *61*, 771–777.
111. Davies, K.J.; Doroshow, J.H. Redox cycling of anthracyclines by cardiac mitochondria. I. Anthracycline radical formation by NADH dehydrogenase. *J. Biol. Chem.* **1986**, *261*, 3060–3067. [CrossRef]
112. Pereira, G.C.; Pereira, S.P.; Tavares, L.C.; Carvalho, F.S.; Magalhães-Novais, S.; Barbosa, I.A.; Santos, M.S.; Bjork, J.; Moreno, A.J.; Wallace, K.B.; et al. Cardiac cytochrome c and cardiolipin depletion during anthracycline-induced chronic depression of mitochondrial function. *Mitochondrion* **2016**, *30*, 95–104. [CrossRef] [PubMed]
113. Wallace, K.B. Adriamycin-induced interference with cardiac mitochondrial calcium homeostasis. *Cardiovasc. Toxicol.* **2007**, *7*, 101–107. [CrossRef] [PubMed]
114. Rocca, C.; Scavello, F.; Colombo, B.; Gasparri, A.M.; Dallatomasina, A.; Granieri, M.C.; Amelio, D.; Pasqua, T.; Cerra, M.C.; Tota, B.; et al. Physiological levels of chromogranin A prevent doxorubicin-induced cardiotoxicity without impairing its anticancer activity. *FASEB J.* **2019**, *33*, 7734–7747. [CrossRef]
115. Cova, D.; De Angelis, L.; Monti, E.; Piccinini, F. Subcellular Distribution of Two Spin Trapping Agents in Rat Heart: Possible Explanation for Their Different Protective Effects against Doxorubicin-Induced Cardiotoxicity. *Free Radic. Res. Commun.* **1992**, *15*, 353–360. [CrossRef]
116. Goormaghtigh, E.; Huart, P.; Praet, M.; Brasseur, R.; Ruysschaert, J.-M. Structure of the adriamycin-cardiolipin complex: Role in mitochondrial toxicity. *Biophys. Chem.* **1990**, *35*, 247–257. [CrossRef]
117. El-Hafidi, M.; Correa, F.; Zazueta, C. Mitochondrial dysfunction in metabolic and cardiovascular diseases associated with cardiolipin remodeling. *Biochim. Biophys. Acta (BBA) Mol. Basis Dis.* **2020**, *1866*, 165744. [CrossRef]
118. Goormaghtigh, E.; Huart, P.; Brasseur, R.; Ruysschaert, J.-M. Mechanism of inhibition of mitochondrial enzymatic complex I–III by adriamycin derivatives. *Biochim. Biophys. Acta (BBA) Biomembr.* **1986**, *861*, 83–94. [CrossRef]
119. Wallace, K.B.; Sardão, V.A.; Oliveira, P.J. Mitochondrial Determinants of Doxorubicin-Induced Cardiomyopathy. *Circ. Res.* **2020**, *126*, 926–941. [CrossRef]
120. Nordgren, K.K.; Wallace, K.B. Keap1 redox-dependent regulation of doxorubicin-induced oxidative stress response in cardiac myoblasts. *Toxicol. Appl. Pharmacol.* **2014**, *274*, 107–116. [CrossRef]
121. Sampaio, S.F.; Branco, A.F.; Wojtala, A.; Vega-Naredo, I.; Wieckowski, M.R.; Oliveira, P.J. p66Shc signaling is involved in stress responses elicited by anthracycline treatment of rat cardiomyoblasts. *Arch. Toxicol.* **2016**, *90*, 1669–1684. [CrossRef] [PubMed]
122. Nabati, M.; Janbabai, G.; Baghyari, S.; Esmaili, K.; Yazdani-Charati, J. Cardioprotective Effects of Carvedilol in Inhibiting Doxorubicin-induced Cardiotoxicity. *J. Cardiovasc. Pharmacol.* **2017**, *69*, 279–285. [CrossRef] [PubMed]
123. Waldner, R.; Laschan, C.; Lohninger, A.; Gessner, M.; Tüchler, H.; Huemer, M.; Spiegel, W.; Karlic, H. Effects of doxorubicin-containing chemotherapy and a combination with l-carnitine on oxidative metabolism in patients with non-Hodgkin lymphoma. *J. Cancer Res. Clin. Oncol.* **2005**, *132*, 121–128. [CrossRef] [PubMed]
124. Carrasco, R.; Ramirez, M.C.; Nes, K.; Schuster, A.; Aguayo, R.; Morales, M.; Ramos, C.; Hasson, D.; Sotomayor, C.G.; Henriquez, P.; et al. Prevention of doxorubicin-induced Cardiotoxicity by pharmacological non-hypoxic myocardial preconditioning based on Docosahexaenoic Acid (DHA) and carvedilol direct antioxidant effects: Study protocol for a pilot, randomized, double-blind, controlled trial (CarDHA trial). *Trials* **2020**, *21*, 137. [CrossRef]
125. Macedo, A.V.; Hajjar, L.A.; Lyon, A.R.; Nascimento, B.R.; Putzu, A.; Rossi, L.; Costa, R.B.; Landoni, G.; Nogueira-Rodrigues, A.; Ribeiro, A.L. Efficacy of Dexrazoxane in Preventing Anthracycline Cardiotoxicity in Breast Cancer. *JACC CardioOncol.* **2019**, *1*, 68–79. [CrossRef]
126. Kheiri, B.; Abdalla, A.; Osman, M.; Haykal, T.; Chahine, A.; Ahmed, S.; Osman, K.; Hassan, M.; Bachuwa, G.; Bhatt, D.L. Meta-Analysis of Carvedilol for the Prevention of Anthracycline-Induced Cardiotoxicity. *Am. J. Cardiol.* **2018**, *122*, 1959–1964. [CrossRef]
127. Bosch, X.; Rovira, M.; Sitges, M.; Domènech, A.; Ortiz-Pérez, J.T.; de Caralt, T.M.; Morales-Ruiz, M.; Perea, R.J.; Monzó, M.; Esteve, J. Enalapril and carvedilol for preventing chemotherapy-induced left ventricular systolic dysfunction in patients with malignant hemopathies: The OVERCOME trial (preventiOn of left Ventricular dysfunction with Enalapril and caRvedilol in patients submitted to intensive ChemOtherapy for the treatment of Malignant hEmopathies). *J. Am. Coll. Cardiol.* **2013**, *61*, 2355–2362.
128. Vincent, D.T.; Ibrahim, Y.F.; Espey, M.G.; Suzuki, Y.J. The role of antioxidants in the era of cardio-oncology. *Cancer Chemother. Pharmacol.* **2013**, *72*, 1157–1168. [CrossRef]
129. Fabiani, I.; Aimo, A.; Grigoratos, C.; Castiglione, V.; Gentile, F.; Saccaro, L.F.; Arzilli, C.; Cardinale, D.; Passino, C.; Emdin, M. Oxidative stress and inflammation: Determinants of anthracycline cardiotoxicity and possible therapeutic targets. *Heart Fail. Rev.* **2021**, *26*, 881–890. [CrossRef]
130. Bugger, H.; Guzman, C.; Zechner, C.; Palmeri, M.; Russell, K.S.; Russell, R.R. Uncoupling protein downregulation in doxorubicin-induced heart failure improves mitochondrial coupling but increases reactive oxygen species generation. *Cancer Chemother. Pharmacol.* **2010**, *67*, 1381–1388. [CrossRef]

131. Walder, K.; Norman, R.A.; Hanson, R.; Schrauwen, P.; Neverova, M.; Jenkinson, C.P.; Easlick, J.; Warden, C.H.; Pecqueur, C.; Raimbault, S.; et al. Association between uncoupling protein polymorphisms (UCP2-UCP3) and energy metabolism/obesity in Pima indians. *Hum. Mol. Genet.* **1998**, *7*, 1431–1435. [CrossRef] [PubMed]
132. Carvalho, R.A.; Sousa, R.P.; Cadete, V.J.; Lopaschuk, G.D.; Palmeira, C.M.; Bjork, J.A.; Wallace, K.B. Metabolic remodeling associated with subchronic doxorubicin cardiomyopathy. *Toxicology* **2010**, *270*, 92–98. [CrossRef] [PubMed]
133. Berthiaume, J.M.; Wallace, K.B. Persistent Alterations to the Gene Expression Profile of the Heart Subsequent to Chronic Doxorubicin Treatment. *Cardiovasc. Toxicol.* **2007**, *7*, 178–191. [CrossRef] [PubMed]
134. Bansal, N.; Amdani, S.M.; Hutchins, K.K.; Lipshultz, S.E. Cardiovascular disease in survivors of childhood cancer. *Curr. Opin. Pediatr.* **2018**, *30*, 628–638. [CrossRef] [PubMed]
135. Tang, H.; Tao, A.; Song, J.; Liu, Q.; Wang, H.; Rui, T. Doxorubicin-induced cardiomyocyte apoptosis: Role of mitofusin 2. *Int. J. Biochem. Cell Biol.* **2017**, *88*, 55–59. [CrossRef]
136. Samant, S.A.; Zhang, H.J.; Hong, Z.; Pillai, V.B.; Sundaresan, N.R.; Wolfgeher, D.; Archer, S.L.; Chan, D.C.; Gupta, M.P. SIRT3 Deacetylates and Activates OPA1 To Regulate Mitochondrial Dynamics during Stress. *Mol. Cell. Biol.* **2013**, *34*, 807–819. [CrossRef]
137. Aung, L.H.H.; Li, R.; Prabhakar, B.S.; Li, P. Knockdown of Mtfp1 can minimize doxorubicin cardiotoxicity by inhibiting Dnm1l-mediated mitochondrial fission. *J. Cell. Mol. Med.* **2017**, *21*, 3394–3404. [CrossRef]
138. Marques-Aleixo, I.; Alves, E.; Torrella, J.R.; Oliveira, P.; Magalhães, J.; Ascensao, A. Exercise and Doxorubicin Treatment Modulate Cardiac Mitochondrial Quality Control Signaling. *Cardiovasc. Toxicol.* **2018**, *18*, 43–55. [CrossRef]
139. Xia, Y.; Chen, Z.; Chen, A.; Fu, M.; Dong, Z.; Hu, K.; Yang, X.; Zou, Y.; Sun, A.; Qian, J.; et al. LCZ696 improves cardiac function via alleviating Drp1-mediated mitochondrial dysfunction in mice with doxorubicin-induced dilated cardiomyopathy. *J. Mol. Cell. Cardiol.* **2017**, *108*, 138–148. [CrossRef]
140. Zhuang, X.; Sun, X.; Zhou, H.; Zhang, S.; Zhong, X.; Xu, X.; Guo, Y.; Xiong, Z.; Liu, M.; Lin, Y.; et al. Klotho attenuated Doxorubicin-induced cardiomyopathy by alleviating Dynamin-related protein 1-mediated mitochondrial dysfunction. *Mech. Ageing Dev.* **2021**, *195*, 111442. [CrossRef]
141. Lim, K.; Lu, T.-S.; Molostvov, G.; Lee, C.; Lam, F.; Zehnder, D.; Hsiao, L.-L. Vascular Klotho Deficiency Potentiates the Development of Human Artery Calcification and Mediates Resistance to Fibroblast Growth Factor 23. *Circulation* **2012**, *125*, 2243–2255. [CrossRef] [PubMed]
142. Xie, J.; Cha, S.-K.; An, S.-W.; Kuro-O, M.; Birnbaumer, L.; Huang, C.-L. Cardioprotection by Klotho through downregulation of TRPC6 channels in the mouse heart. *Nat. Commun.* **2012**, *3*, 1238. [CrossRef] [PubMed]
143. Catanzaro, M.P.; Weiner, A.; Kaminaris, A.; Li, C.; Cai, F.; Zhao, F.; Kobayashi, S.; Kobayashi, T.; Huang, Y.; Sesaki, H.; et al. Doxorubicin-induced cardiomyocyte death is mediated by unchecked mitochondrial fission and mitophagy. *FASEB J.* **2019**, *33*, 11096–11108. [CrossRef] [PubMed]
144. Taguchi, N.; Ishihara, N.; Jofuku, A.; Oka, T.; Mihara, K. Mitotic Phosphorylation of Dynamin-related GTPase Drp1 Participates in Mitochondrial Fission. *J. Biol. Chem.* **2007**, *282*, 11521–11529. [CrossRef]
145. Cereghetti, G.M.; Stangherlin, A.; de Brito, O.M.; Chang, C.-R.; Blackstone, C.; Bernardi, P.; Scorrano, L. Dephosphorylation by calcineurin regulates translocation of Drp1 to mitochondria. *Proc. Natl. Acad. Sci. USA* **2008**, *105*, 15803–15808. [CrossRef]
146. Aung, L.H.H.; Chen, X.; Jumbo, J.C.C.; Li, Z.; Wang, S.-Y.; Zhao, C.; Liu, Z.; Wang, Y.; Li, P. Cardiomyocyte mitochondrial dynamic-related lncRNA 1 (CMDL-1) may serve as a potential therapeutic target in doxorubicin cardiotoxicity. *Mol. Ther. Nucleic Acids* **2021**, *25*, 638–651. [CrossRef]
147. Zhou, L.; Li, R.; Liu, C.; Sun, T.; Aung, L.H.H.; Chen, C.; Gao, J.; Zhao, Y.; Wang, K. Foxo3a inhibits mitochondrial fission and protects against doxorubicin-induced cardiotoxicity by suppressing MIEF2. *Free Radic. Biol. Med.* **2017**, *104*, 360–370. [CrossRef]
148. Chen, L.; Gong, Q.; Stice, J.P.; Knowlton, A.A. Mitochondrial OPA1, apoptosis, and heart failure. *Cardiovasc. Res.* **2009**, *84*, 91–99. [CrossRef]
149. Lemmon, M.A.; Schlessinger, J. Cell Signaling by Receptor Tyrosine Kinases. *Cell* **2010**, *141*, 1117–1134. [CrossRef]
150. Saraon, P.; Pathmanathan, S.; Snider, J.; Lyakisheva, A.; Wong, V.; Stagljar, I. Receptor tyrosine kinases and cancer: Oncogenic mechanisms and therapeutic approaches. *Oncogene* **2021**, *40*, 4079–4093. [CrossRef]
151. Hubbard, S.R. Structural analysis of receptor tyrosine kinases. *Prog. Biophys. Mol. Biol.* **1999**, *71*, 343–358. [CrossRef]
152. Robinson, D.R.; Wu, Y.-M.; Lin, S.-F. The protein tyrosine kinase family of the human genome. *Oncogene* **2000**, *19*, 5548–5557. [CrossRef]
153. Roberti, M.; Bottegoni, G. Non-ATP Competitive Protein Kinase Inhibitors. *Curr. Med. Chem.* **2010**, *17*, 2804–2821. [CrossRef]
154. Force, T.; Krause, D.S.; Van Etten, R.A. Molecular mechanisms of cardiotoxicity of tyrosine kinase inhibition. *Nat. Cancer* **2007**, *7*, 332–344. [CrossRef] [PubMed]
155. Varricchi, G.; Ameri, P.; Cadeddu, C.; Ghigo, A.; Madonna, R.; Marone, G.; Mercurio, V.; Monte, I.; Novo, G.; Parrella, P.; et al. Antineoplastic Drug-Induced Cardiotoxicity: A Redox Perspective. *Front. Physiol.* **2018**, *9*, 167. [CrossRef] [PubMed]
156. Di Lisi, D.; Madonna, R.; Zito, C.; Bronte, E.; Badalamenti, G.; Parrella, P.; Monte, I.; Tocchetti, C.G.; Russo, A.; Novo, G. Anticancer therapy-induced vascular toxicity: VEGF inhibition and beyond. *Int. J. Cardiol.* **2017**, *227*, 11–17. [CrossRef] [PubMed]
157. Chu, T.F.; Rupnick, M.A.; Kerkela, R.; Dallabrida, S.M.; Zurakowski, D.; Nguyen, L.; Woulfe, K.; Pravda, E.; Cassiola, F.; Desai, J.; et al. Cardiotoxicity associated with tyrosine kinase inhibitor sunitinib. *Lancet* **2007**, *370*, 2011–2019. [CrossRef]
158. Wang, Z. ErbB Receptors and Cancer. *Methods Mol. Biol.* **2017**, *1652*, 3–35.

159. Kenigsberg, B.; Jain, V.; Barac, A. Cardio-oncology Related to Heart Failure: Epidermal Growth Factor Receptor Target-Based Therapy. *Heart Fail Clin.* **2017**, *13*, 297–309. [CrossRef]
160. Grazette, L.P.; Boecker, W.; Matsui, T.; Semigran, M.; Force, T.; Hajjar, R.J.; Rosenzweig, A. Inhibition of ErbB2 causes mitochondrial dysfunction in cardiomyocytes: Implications for herceptin-induced cardiomyopathy. *J. Am. Coll. Cardiol.* **2004**, *44*, 2231–2238. [CrossRef]
161. Crone, S.A.; Zhao, Y.Y.; Fan, L.; Gu, Y.; Minamisawa, S.; Liu, Y.; Peterson, K.L.; Chen, J.; Kahn, R.; Condorelli, G.; et al. ErbB2 is essential in the prevention of dilated cardiomyopathy. *Nat. Med.* **2002**, *8*, 459–465. [CrossRef] [PubMed]
162. Kerkelä, R.; Grazette, L.; Yacobi, R.; Iliescu, C.; Patten, R.; Beahm, C.; Walters, B.; Shevtsov, S.; Pesant, S.; Clubb, F.J.; et al. Cardiotoxicity of the cancer therapeutic agent imatinib mesylate. *Nat. Med.* **2006**, *12*, 908–916. [CrossRef] [PubMed]
163. Kerkela, R.; Woulfe, K.C.; Durand, J.B.; Vagnozzi, R.; Kramer, D.; Chu, T.F.; Beahm, C.; Chen, M.H.; Force, T. Sunitinib-induced cardiotoxicity is mediated by off-target inhibition of AMP-activated protein kinase. *Clin. Transl. Sci.* **2009**, *2*, 15–25. [CrossRef] [PubMed]
164. Will, Y.; Dykens, J.A.; Nadanaciva, S.; Hirakawa, B.; Jamieson, J.; Marroquin, L.D.; Hynes, J.; Patyna, S.; Jessen, B.A. Effect of the Multitargeted Tyrosine Kinase Inhibitors Imatinib, Dasatinib, Sunitinib, and Sorafenib on Mitochondrial Function in Isolated Rat Heart Mitochondria and H9c2 Cells. *Toxicol. Sci.* **2008**, *106*, 153–161. [CrossRef] [PubMed]
165. Mellor, H.R.; Bell, A.R.; Valentin, J.-P.; Roberts, R.R.A. Cardiotoxicity Associated with Targeting Kinase Pathways in Cancer. *Toxicol. Sci.* **2010**, *120*, 14–32. [CrossRef]
166. Huang, W.S.; Metcalf, C.A.; Sundaramoorthi, R.; Wang, Y.; Zou, D.; Thomas, R.M.; Zhu, X.; Cai, L.; Wen, D.; Liu, S.; et al. Discovery of 3-[2-(imidazo [1,2-b]pyridazin-3-yl)ethynyl]-4-methyl-N-{4-[(4-methylpiperazin-1-yl)methyl]-3-(trifluoromethyl)phenyl}benzamide (AP24534), a potent, orally active pan-inhibitor of breakpoint cluster region-abelson (BCR-ABL) kinase including the T315I gatekeeper mutant. *J. Med. Chem.* **2010**, *53*, 4701–4719.
167. Talbert, D.R.; Doherty, K.R.; Trusk, P.B.; Moran, D.M.; Shell, S.A.; Bacus, S. A Multi-parameter In Vitro Screen in Human Stem Cell-Derived Cardiomyocytes Identifies Ponatinib-Induced Structural and Functional Cardiac Toxicity. *Toxicol. Sci.* **2015**, *143*, 147–155. [CrossRef]
168. Weng, Z.; Luo, Y.; Yang, X.; Greenhaw, J.J.; Li, H.; Xie, L.; Mattes, W.; Shi, Q. Regorafenib impairs mitochondrial functions, activates AMP-activated protein kinase, induces autophagy, and causes rat hepatocyte necrosis. *Toxicology* **2015**, *327*, 10–21. [CrossRef]
169. Thomson, M. Evidence of undiscovered cell regulatory mechanisms: Phosphoproteins and protein kinases in mitochondria. *Cell. Mol. Life Sci.* **2002**, *59*, 213–219. [CrossRef]
170. Wang, X.; Shen, X.; Yan, Y.; Li, H. Pyruvate dehydrogenase kinases (PDKs): An overview toward clinical applications. *Biosci. Rep.* **2021**, *41*, 20204402. [CrossRef]
171. Chaar, M.; Kamta, J.; Ait-Oudhia, S. Mechanisms, monitoring, and management of tyrosine kinase inhibitors-associated cardiovascular toxicities. *OncoTargets Ther.* **2018**, *11*, 6227–6237. [CrossRef] [PubMed]
172. Wang, H.; Sheehan, R.P.; Palmer, A.C.; Everley, R.A.; Boswell, S.A.; Ron-Harel, N.; Ringel, A.E.; Holton, K.M.; Jacobson, C.A.; Erickson, A.R.; et al. Adaptation of Human iPSC-Derived Cardiomyocytes to Tyrosine Kinase Inhibitors Reduces Acute Cardiotoxicity via Metabolic Reprogramming. *Cell Syst.* **2019**, *8*, 412–426.e7. [CrossRef] [PubMed]
173. Doenst, T.; Nguyen, T.D.; Abel, E.D. Cardiac metabolism in heart failure: Implications beyond ATP production. *Circ. Res.* **2013**, *113*, 709–724. [CrossRef] [PubMed]
174. Alhoshani, A.; Alanazi, F.E.; Alotaibi, M.R.; Attwa, M.W.; Kadi, A.A.; Aldhfyan, A.; Akhtar, S.; Hourani, S.; Agouni, A.; Zeidan, A.; et al. EGFR Inhibitor Gefitinib Induces Cardiotoxicity through the Modulation of Cardiac PTEN/Akt/FoxO3a Pathway and Reactive Metabolites Formation: In Vivo and in Vitro Rat Studies. *Chem. Res. Toxicol.* **2020**, *33*, 1719–1728. [CrossRef]
175. Barton, M.; Filardo, E.; Lolait, S.J.; Thomas, P.; Maggiolini, M.; Prossnitz, E.R. Twenty years of the G protein-coupled estrogen receptor GPER: Historical and personal perspectives. *J. Steroid Biochem. Mol. Biol.* **2018**, *176*, 4–15. [CrossRef]
176. Lappano, R.; De Marco, P.; De Francesco, E.M.; Chimento, A.; Pezzi, V.; Maggiolini, M. Cross-talk between GPER and growth factor signaling. *J. Steroid Biochem. Mol. Biol.* **2013**, *137*, 50–56. [CrossRef]
177. Rocca, C.; Femminò, S.; Aquila, G.; Granieri, M.C.; De Francesco, E.M.; Pasqua, T.; Rigiracciolo, D.C.; Fortini, F.; Cerra, M.C.; Maggiolini, M.; et al. Notch1 Mediates Preconditioning Protection Induced by GPER in Normotensive and Hypertensive Female Rat Hearts. *Front. Physiol.* **2018**, *9*, 521. [CrossRef]
178. Recchia, A.G.; De Francesco, E.M.; Vivacqua, A.; Sisci, D.; Panno, M.L.; Andò, S.; Maggiolini, M. The G Protein-coupled Receptor 30 Is Up-regulated by Hypoxia-inducible Factor-1α (HIF-1α) in Breast Cancer Cells and Cardiomyocytes. *J. Biol. Chem.* **2011**, *286*, 10773–10782. [CrossRef]
179. De Francesco, E.M.; Rocca, C.; Scavello, F.; Amelio, D.; Pasqua, T.; Rigiracciolo, D.C.; Scarpelli, A.; Avino, S.; Cirillo, F.; Amodio, N.; et al. Protective Role of GPER Agonist G-1 on Cardiotoxicity Induced by Doxorubicin. *J. Cell. Physiol.* **2017**, *232*, 1640–1649. [CrossRef]
180. Narayan, V.; Keefe, S.; Haas, N.; Wang, L.; Puzanov, I.; Putt, M.; Catino, A.; Fang, J.; Agarwal, N.; Hyman, D.; et al. Prospective Evaluation of Sunitinib-Induced Cardiotoxicity in Patients with Metastatic Renal Cell Carcinoma. *Clin. Cancer Res.* **2017**, *23*, 3601–3609. [CrossRef]

181. Heath, E.I.; Infante, J.; Lewis, L.D.; Luu, T.; Stephenson, J.; Tan, A.R.; Kasubhai, S.; Lorusso, P.; Ma, B.; Suttle, A.B.; et al. A randomized, double-blind, placebo-controlled study to evaluate the effect of repeated oral doses of pazopanib on cardiac conduction in patients with solid tumors. *Cancer Chemother. Pharmacol.* **2013**, *71*, 565–573. [CrossRef] [PubMed]
182. Abdel-Rahman, O.; Fouad, M. Risk of cardiovascular toxicities in patients with solid tumors treated with sorafenib: An updated systematic review and meta-analysis. *Future Oncol.* **2014**, *10*, 1981–1992. [CrossRef] [PubMed]
183. Hou, W.; Ding, M.; Li, X.; Zhou, X.; Zhu, Q.; Varela-Ramirez, A.; Yi, C. Comparative evaluation of cardiovascular risks among nine FDA-approved VEGFR-TKIs in patients with solid tumors: A Bayesian network analysis of randomized controlled trials. *J. Cancer Res. Clin. Oncol.* **2021**, *147*, 2407–2420. [CrossRef]
184. Casavecchia, G.; Galderisi, M.; Novo, G.; Gravina, M.; Santoro, C.; Agricola, E.; Capalbo, S.; Zicchino, S.; Cameli, M.; De Gennaro, L.; et al. Early diagnosis, clinical management, and follow-up of cardiovascular events with ponatinib. *Heart Fail. Rev.* **2020**, *25*, 447–456. [CrossRef] [PubMed]
185. Cortes, J.E.; Kim, D.-W.; Pinilla-Ibarz, J.; Le Coutre, P.; Paquette, R.; Chuah, C.; Nicolini, F.E.; Apperley, J.F.; Khoury, H.J.; Talpaz, M.; et al. A Phase 2 Trial of Ponatinib in Philadelphia Chromosome–Positive Leukemias. *N. Engl. J. Med.* **2013**, *369*, 1783–1796. [CrossRef]
186. Iacovelli, R.; Ciccarese, C.; Fornarini, G.; Massari, F.; Bimbatti, D.; Mosillo, C.; Rebuzzi, S.E.; DI Nunno, V.; Grassi, M.; Fantinel, E.; et al. Cabozantinib-related cardiotoxicity: A prospective analysis in a real-world cohort of metastatic renal cell carcinoma patients. *Br. J. Clin. Pharmacol.* **2019**, *85*, 1283–1289. [CrossRef]
187. Milling, R.V.; Grimm, D.; Krüger, M.; Grosse, J.; Kopp, S.; Bauer, J.; Infanger, M.; Wehland, M. Pazopanib, Cabozantinib, and Vandetanib in the Treatment of Progressive Medullary Thyroid Cancer with a Special Focus on the Adverse Effects on Hypertension. *Int. J. Mol. Sci.* **2018**, *19*, 3258. [CrossRef]
188. Kim, T.D.; le Coutre, P.; Schwarz, M.; Grille, P.; Levitin, M.; Fateh-Moghadam, S.; Giles, F.J.; Dörken, B.; Haverkamp, W.; Köhncke, C. Clinical cardiac safety profile of nilotinib. *Haematologica* **2012**, *97*, 883–889. [CrossRef]
189. Motzer, R.J.; Escudier, B.; Tomczak, P.; Hutson, T.E.; Michaelson, D.; Negrier, S.; Oudard, S.; Gore, M.E.; Tarazi, J.; Hariharan, S.; et al. Axitinib versus sorafenib as second-line treatment for advanced renal cell carcinoma: Overall survival analysis and updated results from a randomised phase 3 trial. *Lancet Oncol.* **2013**, *14*, 552–562. [CrossRef]
190. Rousseau, A.; Bertolotti, A. Regulation of proteasome assembly and activity in health and disease. *Nat. Rev. Mol. Cell Biol.* **2018**, *19*, 697–712. [CrossRef]
191. Yamamoto, L.; Amodio, N.; Gulla, A.; Anderson, K.C. Harnessing the Immune System against Multiple Myeloma: Challenges and Opportunities. *Front. Oncol.* **2021**, *10*. [CrossRef] [PubMed]
192. Taiana, E.; Cantafio, M.G.; Favasuli, V.; Bandini, C.; Viglietto, G.; Piva, R.; Neri, A.; Amodio, N. Genomic Instability in Multiple Myeloma: A "Non-Coding RNA" Perspective. *Cancers* **2021**, *13*, 2127. [CrossRef]
193. Annunziata, C.M.; Davis, R.E.; Demchenko, Y.; Bellamy, W.; Gabrea, A.; Zhan, F.; Lenz, G.; Hanamura, I.; Wright, G.; Xiao, W.; et al. Frequent engagement of the classical and alternative NF-kappaB pathways by diverse genetic abnormalities in multiple myeloma. *Cancer Cell* **2007**, *12*, 115–130. [CrossRef]
194. Motegi, A.; Murakawa, Y.; Takeda, S. The vital link between the ubiquitin–proteasome pathway and DNA repair: Impact on cancer therapy. *Cancer Lett.* **2009**, *283*, 1–9. [CrossRef]
195. Amodio, N.; Bellizzi, D.; Leotta, M.; Raimondi, L.; Biamonte, L.; D'Aquila, P.; Di Martino, M.T.; Calimeri, T.; Rossi, M.; Lionetti, M.; et al. miR-29b induces SOCS-1 expression by promoter demethylation and negatively regulates migration of multiple myeloma and endothelial cells. *Cell Cycle* **2013**, *12*, 3650–3662. [CrossRef] [PubMed]
196. Amodio, N.; D'Aquila, P.; Passarino, G.; Tassone, P.; Bellizzi, D. Epigenetic modifications in multiple myeloma: Recent advances on the role of DNA and histone methylation. *Expert Opin. Ther. Targets* **2016**, *21*, 91–101. [CrossRef] [PubMed]
197. Gandolfi, S.; Laubach, J.P.; Hideshima, T.; Chauhan, D.; Anderson, K.C.; Richardson, P.G. The proteasome and proteasome inhibitors in multiple myeloma. *Cancer Metastasis Rev.* **2017**, *36*, 561–584. [CrossRef] [PubMed]
198. Rajkumar, S.V.; Kumar, S. Multiple myeloma current treatment algorithms. *Blood Cancer J.* **2020**, *10*, 94. [CrossRef] [PubMed]
199. Ito, S. Proteasome Inhibitors for the Treatment of Multiple Myeloma. *Cancers* **2020**, *12*, 265. [CrossRef]
200. Cole, D.C.; Frishman, W.H. Cardiovascular Complications of Proteasome Inhibitors Used in Multiple Myeloma. *Cardiol. Rev.* **2018**, *26*, 122–129. [CrossRef]
201. Nowis, D.; Mączewski, M.; Mackiewicz, U.; Kujawa, M.; Ratajska, A.; Wieckowski, M.; Wilczynski, G.; Malinowska, M.; Bil, J.; Salwa, P.; et al. Cardiotoxicity of the Anticancer Therapeutic Agent Bortezomib. *Am. J. Pathol.* **2010**, *176*, 2658–2668. [CrossRef] [PubMed]
202. Tang, M.; Li, J.; Huang, W.; Su, H.; Liang, Q.; Tian, Z.; Horak, K.M.; Molkentin, J.D.; Wang, X. Proteasome functional insufficiency activates the calcineurin–NFAT pathway in cardiomyocytes and promotes maladaptive remodelling of stressed mouse hearts. *Cardiovasc. Res.* **2010**, *88*, 424–433. [CrossRef]
203. Hasinoff, B.B.; Patel, D.; Wu, X. Molecular Mechanisms of the Cardiotoxicity of the Proteasomal-Targeted Drugs Bortezomib and Carfilzomib. *Cardiovasc. Toxicol.* **2017**, *17*, 237–250. [CrossRef]
204. Kisselev, A.F.; Goldberg, A.L. Proteasome inhibitors: From research tools to drug candidates. *Chem. Biol.* **2001**, *8*, 739–758. [CrossRef]
205. Mårtensson, C.U.; Priesnitz, C.; Song, J.; Ellenrieder, L.; Doan, K.N.; Boos, F.; Floerchinger, A.; Zufall, N.; Oeljeklaus, S.; Warscheid, B.; et al. Mitochondrial protein translocation-associated degradation. *Nature* **2019**, *569*, 679–683. [CrossRef]

206. Pokorna, Z.; Jirkovsky, E.; Hlavackova, M.; Jansova, H.; Jirkovska, A.; Lencova-Popelova, O.; Brazdova, P.; Kubes, J.; Sotakova-Kasparova, D.; Mazurova, Y.; et al. In vitro and in vivo investigation of cardiotoxicity associated with anticancer proteasome inhibitors and their combination with anthracycline. *Clin. Sci.* **2019**, *133*, 1827–1844. [CrossRef] [PubMed]
207. Pancheri, E.; Guglielmi, V.; Wilczynski, G.M.; Malatesta, M.; Tonin, P.; Tomelleri, G.; Nowis, D.; Vattemi, G. Non-Hematologic Toxicity of Bortezomib in Multiple Myeloma: The Neuromuscular and Cardiovascular Adverse Effects. *Cancers* **2020**, *12*, 2540. [CrossRef] [PubMed]
208. Xiao, Y.; Yin, J.; Wei, J.; Shang, Z. Incidence and Risk of Cardiotoxicity Associated with Bortezomib in the Treatment of Cancer: A Systematic Review and Meta-Analysis. *PLoS ONE* **2014**, *9*, e87671. [CrossRef] [PubMed]
209. Laubach, J.P.; Moslehi, J.J.; Francis, S.A.; San Miguel, J.F.; Sonneveld, P.; Orlowski, R.Z.; Moreau, P.; Rosiñol, L.; Faber, E.A., Jr.; Voorhees, P.; et al. A retrospective analysis of 3954 patients in phase 2/3 trials of bortezomib for the treatment of multiple myeloma: Towards providing a benchmark for the cardiac safety profile of proteasome inhibition in multiple myeloma. *Br. J. Haematol.* **2017**, *178*, 547–560. [CrossRef] [PubMed]
210. Kuhn, D.J.; Chen, Q.; Voorhees, P.M.; Strader, J.S.; Shenk, K.D.; Sun, C.M.; Demo, S.D.; Bennett, M.K.; van Leeuwen, F.; Chanan-Khan, A.A.; et al. Potent activity of carfilzomib, a novel, irreversible inhibitor of the ubiquitin-proteasome pathway, against preclinical models of multiple myeloma. *Blood* **2007**, *110*, 3281–3290. [CrossRef]
211. Demo, S.D.; Kirk, C.J.; Aujay, M.A.; Buchholz, T.J.; Dajee, M.; Ho, M.N.; Jiang, J.; Laidig, G.J.; Lewis, E.R.; Parlati, F.; et al. Antitumor Activity of PR-171, a Novel Irreversible Inhibitor of the Proteasome. *Cancer Res.* **2007**, *67*, 6383–6391. [CrossRef] [PubMed]
212. Siegel, D.; Martin, T.; Nooka, A.; Harvey, R.D.; Vij, R.; Niesvizky, R.; Badros, A.Z.; Jagannath, S.; McCulloch, L.; Rajangam, K.; et al. Integrated safety profile of single-agent carfilzomib: Experience from 526 patients enrolled in 4 phase II clinical studies. *Haematology* **2013**, *98*, 1753–1761. [CrossRef] [PubMed]
213. Stewart, A.K.; Rajkumar, S.V.; Dimopoulos, M.A.; Masszi, T.; Špička, I.; Oriol, A.; Hájek, R.; Rosiñol, L.; Siegel, D.S.; Mihaylov, G.G.; et al. Carfilzomib, lenalidomide, and dexamethasone for relapsed multiple myeloma. *N. Engl. J. Med.* **2015**, *372*, 142–152. [CrossRef] [PubMed]
214. Siegel, D.S.; Dimopoulos, M.A.; Ludwig, H.; Facon, T.; Goldschmidt, H.; Jakubowiak, A.; Miguel, J.S.; Obreja, M.; Blaedel, J.; Stewart, A.K. Improvement in Overall Survival with Carfilzomib, Lenalidomide, and Dexamethasone in Patients with Relapsed or Refractory Multiple Myeloma. *J. Clin. Oncol.* **2018**, *36*, 728–734. [CrossRef] [PubMed]
215. Atrash, S.; Tullos, A.; Panozzo, S.; Bhutani, M.S.; Van Rhee, F.; Barlogie, B.; Usmani, S.Z. Cardiac complications in relapsed and refractory multiple myeloma patients treated with carfilzomib. *Blood Cancer J.* **2015**, *5*, e272. [CrossRef] [PubMed]
216. Waxman, A.J.; Clasen, S.C.; Garfall, A.L.; Carver, J.R.; Vogl, D.T.; O'Quinn, R.; Cohen, A.D.; Stadtmauer, E.A.; Ky, B.; Weiss, B.M. Carfilzomib-associated cardiovascular adverse events: A systematic review and meta-analysis. *J. Clin. Oncol.* **2017**, *35*, 8018. [CrossRef]
217. Cornell, R.F.; Ky, B.; Weiss, B.M.; Dahm, C.N.; Gupta, D.K.; Du, L.; Carver, J.R.; Cohen, A.D.; Engelhardt, B.G.; Garfall, A.; et al. Prospective Study of Cardiac Events During Proteasome Inhibitor Therapy for Relapsed Multiple Myeloma. *J. Clin. Oncol.* **2019**, *37*, 1946–1955. [CrossRef]
218. Chauhan, D.; Tian, Z.; Zhou, B.; Kuhn, D.; Orlowski, R.; Raje, N.; Richardson, P.; Anderson, K.C. In Vitro and In Vivo Selective Antitumor Activity of a Novel Orally Bioavailable Proteasome Inhibitor MLN9708 against Multiple Myeloma Cells. *Clin. Cancer Res.* **2011**, *17*, 5311–5321. [CrossRef]
219. Kupperman, E.; Lee, E.C.; Cao, Y.; Bannerman, B.; Fitzgerald, M.; Berger, A.; Yu, J.; Yang, Y.; Hales, P.; Bruzzese, F.; et al. Evaluation of the Proteasome Inhibitor MLN9708 in Preclinical Models of Human Cancer. *Cancer Res.* **2010**, *70*, 1970–1980. [CrossRef]
220. Moreau, P.; Masszi, T.; Grzasko, N.; Bahlis, N.J.; Hansson, M.; Pour, L.; Sandhu, I.; Ganly, P.; Baker, B.W.; Jackson, S.R.; et al. Oral Ixazomib, Lenalidomide, and Dexamethasone for Multiple Myeloma. *N. Engl. J. Med.* **2016**, *374*, 1621–1634. [CrossRef]
221. Wu, P.; Oren, O.; Gertz, M.A.; Yang, E.H. Proteasome Inhibitor-Related Cardiotoxicity: Mechanisms, Diagnosis, and Management. *Curr. Oncol. Rep.* **2020**, *22*, 66. [CrossRef] [PubMed]
222. Shen, X.; Wu, C.; Lei, M.; Yan, Q.; Zhang, H.; Zhang, L.; Wang, X.; Yang, Y.; Li, J.; Zhu, Y.; et al. Anti-tumor activity of a novel proteasome inhibitor D395 against multiple myeloma and its lower cardiotoxicity compared with carfilzomib. *Cell Death Dis.* **2021**, *12*, 429. [CrossRef]
223. Paradzik, T.; Bandini, C.; Mereu, E.; Labrador, M.; Taiana, E.; Amodio, N.; Neri, A.; Piva, R. The Landscape of Signaling Pathways and Proteasome Inhibitors Combinations in Multiple Myeloma. *Cancers* **2021**, *13*, 1235. [CrossRef] [PubMed]
224. Diwadkar, S.; Patel, A.A.; Fradley, M.G. Bortezomib-Induced Complete Heart Block and Myocardial Scar: The Potential Role of Cardiac Biomarkers in Monitoring Cardiotoxicity. *Case Rep. Cardiol.* **2016**, *2016*, 3456287. [CrossRef] [PubMed]
225. Meseeha, M.G.; Kolade, V.O.; Attia, M.N. Partially reversible bortezomib-induced cardiotoxicity: An unusual cause of acute cardiomyopathy. *J. Community Hosp. Intern. Med. Persp.* **2015**, *5*, 28982. [CrossRef]
226. Voortman, J.; Giaccone, G. Severe reversible cardiac failure after bortezomib treatment combined with chemotherapy in a non-small cell lung cancer patient: A case report. *BMC Cancer* **2006**, *6*, 129. [CrossRef]
227. Danhof, S.; Schreder, M.; Rasche, L.; Strifler, S.; Einsele, H.; Knop, S. 'Real-life' experience of reapproval carfilzomib-based therapy in myeloma—Analysis of cardiac toxicity and predisposing factors. *Eur. J. Haematol.* **2016**, *97*, 25–32. [CrossRef]

228. Hahn, V.S.; Zhang, K.W.; Sun, L.; Narayan, V.; Lenihan, D.J.; Ky, B. Heart Failure with Targeted Cancer Therapies: Mechanisms and Cardioprotection. *Circ. Res.* **2021**, *128*, 1576–1593. [CrossRef]
229. Jouni, H.; Aubry, M.C.; Lacy, M.Q.; Rajkumar, S.V.; Kumar, S.K.; Frye, R.L.; Herrmann, J. Ixazomib cardiotoxicity: A possible class effect of proteasome inhibitors. *Am. J. Hematol.* **2017**, *92*, 220–221. [CrossRef]

biomedicines

MDPI

Article

Saxagliptin Cardiotoxicity in Chronic Heart Failure: The Role of DPP4 in the Regulation of Neuropeptide Tone

Imre Vörös [1,2,3], Zsófia Onódi [1,2,3], Viktória Éva Tóth [1,2,3], Tamás G. Gergely [1,2,3], Éva Sághy [1], Anikó Görbe [1,4], Ágnes Kemény [5,6,7], Przemyslaw Leszek [8], Zsuzsanna Helyes [5,6,9], Péter Ferdinandy [1,4] and Zoltán V. Varga [1,2,3,*]

1 Cardiometabolic Research Group and MTA-SE System Pharmacology Research Group, Department of Pharmacology and Pharmacotherapy, Semmelweis University, 1085 Budapest, Hungary; voros.imre@med.semmelweis-univ.hu (I.V.); onodi.zsofia@med.semmelweis-univ.hu (Z.O.); toth.viktoria@med.semmelweis-univ.hu (V.É.T.); gergely.tamas@med.semmelweis-univ.hu (T.G.G.); saghy.eva@med.semmelweis-univ.hu (É.S.); gorbe.aniko@med.semmelweis-univ.hu (A.G.); peter.ferdinandy@pharmahungary.com (P.F.)
2 HCEMM-SU Cardiometabolic Immunology Research Group, Semmelweis University, 1085 Budapest, Hungary
3 MTA-SE Momentum Cardio-Oncology and Cardioimmunology Research Group, Semmelweis University, 1085 Budapest, Hungary
4 Pharmahungary Group, 6722 Szeged, Hungary
5 Szentágothai János Research Centre, University of Pécs, 7624 Pécs, Hungary; kemeny.agnes@pte.hu (Á.K.); zsuzsanna.helyes@aok.pte.hu (Z.H.)
6 Department of Pharmacology and Pharmacotherapy, Medical School, University of Pécs, 7624 Pécs, Hungary
7 Department of Medical Biology, University of Pécs, 7624 Pécs, Hungary
8 Department of Heart Failure and Transplantology, Cardinal Stefan Wyszyński National Institute of Cardiology, 04-628 Warszawa, Poland; przemyslaw.leszek@ikard.pl
9 PharmInVivo Ltd., 7629 Pécs, Hungary
* Correspondence: varga.zoltan@med.semmelweis-univ.hu; Tel.: +36-1-2104416

Citation: Vörös, I.; Onódi, Z.; Tóth, V.É.; Gergely, T.G.; Sághy, É.; Görbe, A.; Kemény, Á.; Leszek, P.; Helyes, Z.; Ferdinandy, P.; et al. Saxagliptin Cardiotoxicity in Chronic Heart Failure: The Role of DPP4 in the Regulation of Neuropeptide Tone. *Biomedicines* **2022**, *10*, 1573. https://doi.org/10.3390/biomedicines10071573

Academic Editors: Tânia Martins-Marques, Gonçalo F. Coutinho and Attila Kiss

Received: 11 May 2022
Accepted: 23 June 2022
Published: 1 July 2022

Publisher's Note: MDPI stays neutral with regard to jurisdictional claims in published maps and institutional affiliations.

Copyright: © 2022 by the authors. Licensee MDPI, Basel, Switzerland. This article is an open access article distributed under the terms and conditions of the Creative Commons Attribution (CC BY) license (https://creativecommons.org/licenses/by/4.0/).

Abstract: Dipeptidyl-peptidase-4 (DPP4) inhibitors are novel medicines for diabetes. The SAVOR-TIMI-53 clinical trial revealed increased heart-failure-associated hospitalization in saxagliptin-treated patients. Although this side effect could limit therapeutic use, the mechanism of this potential cardiotoxicity is unclear. We aimed to establish a cellular platform to investigate DPP4 inhibition and the role of its neuropeptide substrates substance P (SP) and neuropeptide Y (NPY), and to determine the expression of DDP4 and its neuropeptide substrates in the human heart. Western blot, radio-, enzyme-linked immuno-, and RNA scope assays were performed to investigate the expression of DPP4 and its substrates in human hearts. Calcein-based viability measurements and scratch assays were used to test the potential toxicity of DPP4 inhibitors. Cardiac expression of DPP4 and NPY decreased in heart failure patients. In human hearts, DPP4 mRNA is detectable mainly in cardiomyocytes and endothelium. Treatment with DPP4 inhibitors alone/in combination with neuropeptides did not affect viability but in scratch assays neuropeptides decreased, while saxagliptin co-administration increased fibroblast migration in isolated neonatal rat cardiomyocyte-fibroblast co-culture. Decreased DPP4 activity takes part in the pathophysiology of end-stage heart failure. DPP4 compensates against the elevated sympathetic activity and altered neuropeptide tone. Its inhibition decreases this adaptive mechanism, thereby exacerbating myocardial damage.

Keywords: saxagliptin; neuropeptide Y; substance P; neuropeptides; diabetes; heart failure; cardiotoxicity; cardiomyopathy

1. Introduction

Heart failure is a complex cardiovascular disease with high mortality rate and increasing prevalence [1]. Decreased cardiac output, when the heart is unable to provide the required perfusion to the peripheral tissues, is a critical consequence of heart failure. This event induces various compensatory mechanisms in order to maintain sufficient perfusion of vital organs. Important aspects of these compensatory processes are increased activity of the sympathetic nervous system and the consequently increased release of the sympathetic transmitters and co-transmitters (e.g., neuropeptide Y (NPY) and substance P (SP)), besides activation of the renin-angiotensin-aldosterone system, endothelin release, and inflammatory mechanisms [1]. In the short term, these processes will maintain tissue perfusion; however, in the long term, abnormalities in cellular signaling lead to inflammation, fibrosis, and cell death, which strongly promote the remodeling process, thus creating a vicious cycle to trigger the progression of heart failure.

The neuropeptide SP (acting through the neurokinin 1 receptor) is mainly expressed in the heart by C-fiber sensory nerves, innervating coronary arteries, intrinsic nerve bundles, coronary endothelial cells, and cardiomyocytes [2–7]. SP has been shown to play a detrimental role in various cardiovascular events, e.g., adverse cardiac remodeling, inflammation, necrosis, and fibrosis [8–12]. NPY is the most abundant neuropeptide in the heart, which exerts its actions through five known receptors (Y1 to Y5) [13]. NPY is expressed in the intracardiac ganglia, sympathetic nerves projecting to blood vessels, intrinsic parasympathetic cardiac neurons, and cardiomyocytes [14]. NPY was identified as a prognostic indicator of mortality in myocardial infarct patients [15].

The management of heart failure is still a challenging task for healthcare professionals, especially when co-morbidities are present in patients, e.g., type 2 diabetes [16]. Dipeptidyl-peptidase IV (DPP4) inhibitors are in general well-tolerated and relatively new options in the pharmacotherapy of type 2 diabetes [17,18]. It was discovered in the 1990s that the inactivation of the DPP-4 enzyme raised the levels of the incretin hormones, which in turn lowered the circulating glucose concentration and improved glycemia in animals [19]. In the following two decades, various small molecule DPP4 inhibitor substances have been developed and tested in subsequent clinical studies for the therapy of type 2 diabetes [19] and for other cardiovascular comorbidities [20–22]. The regulatory authorities have approved five of the DPP-4 inhibitors so far. Currently, these drugs are used as second- or third-line medications in the treatment of type 2 diabetes after metformin and sulphonylureas. These substances facilitate insulin secretion through increasing the level of glucagon-like peptide-1 indirectly by inhibiting DPP4, an enzyme responsible for the degradation of incretins (glucose-dependent insulinotropic peptide and glucagon-like peptide-1 [17,23]). DPP4 is a transmembrane exopeptidase, which cleaves dipeptides from the N-terminal site of its targets [24,25]. The cleavage results in the inactivation of the substrates in most cases, but sometimes changes in receptor affinity of certain substrates can also happen [25]. DPP4 has various types of substrates, e.g., incretins (glucagon-like peptide-1, glucose-dependent insulinotropic polypeptide), chemokines (e.g., stromal cell-derived factor 1), and neuropeptides (NPY, SP, peptide YY, and pancreatic polypeptide) [24,25]. Interestingly, it was recently identified that decreased circulating DPP4 activity in patients is associated with severe COVID-19 disease and is a strong prognostic biomarker of mortality [26]. The large randomized placebo-controlled phase 4 clinical trial (SAVOR-TIMI 53) on the cardiovascular safety of the DPP4-inhibitor saxagliptin surprisingly revealed that the heart-failure-associated hospitalization rate increased in the saxagliptin-treated group compared to controls [27]. The possible dangerous cardiovascular side effects of DPP4 inhibitors are included in the latest ESC heart failure guideline, which suggests avoiding the administration of gliptins in diabetic patients with heart failure [28]. Although this potentially cardiotoxic side effect of saxagliptin could considerably limit therapeutic use, the detailed mechanism by which DPP4 inhibitors may damage the heart is still unclear [27]. Therefore, we aimed to set up a relevant cell culture platform to mechanistically investigate the effect of DPP4 inhibition and the role of potentially important neuropeptide substrates such as SP and NPY. Moreover, we aimed to determine the expression of DPP4 and its substrates in the human heart and cell culture samples both at protein and mRNA levels.

2. Materials and Methods

2.1. Cell Culture

Human cardiac myocyte AC16 cell line was obtained from Merck (SCC109). Cells were plated on 6-well or 24-well plates (Thermo Fisher Scientific, Waltham, MA, USA) and maintained in growth medium (Dulbecco's Modified Eagle's Medium and Ham's F-12 Nutrient Mixture-DMEM/F12; Thermo Fisher Scientific, Waltham, MA, USA) supplemented with 10% fetal bovine serum (FBS; EuroClone, Pero MI, Italy) and antibiotic (100 U/mL penicillin and 100 µg/mL streptomycin; Thermo Fisher Scientific, Waltham, MA, USA) at 37 °C in a humidified atmosphere of 5% CO_2. Cells were cultured until 80–90% confluence, then used in experiments. Cells were kept in FBS-free DMEM/F12 medium during the whole experimental protocol.

2.2. Viability Measurements

In order to test the effect of different DDP4 inhibitors (alogliptin, linagliptin, saxagliptin, or vildagliptin) on cell viability, AC16 cells were treated with 500 nM of each gliptin in separate series for 24 h. Control group received DMSO vehicle.

In a separate experiment, cells were treated with 500 nM saxagliptin in the presence of different doses (5, 20, 50, and 100 nM) of NPY or SP. To achieve sufficient inhibition of DPP4 enzyme, saxagliptin was administered 1 h prior to neuropeptide treatment. Cell-culture-grade water containing DMSO and pure cell-culture-grade water were used as vehicles for saxagliptin and neuropeptides, respectively.

Cell viability was assessed by calcein viability assay [29]. After rinsing the plates with Dulbecco's phosphate-buffered saline (D-PBS), cells were incubated with calcein AM solution (1 µM) for 30 min at room temperature, protected from light. Fluorescence intensity was detected by Varioskan Lux multimode microplate reader (Thermo Fisher Scientific, Waltham, MA, USA) at room temperature, by using excitation wavelength: 490 nm and emission wavelength: 520 nm. Results are shown as RFU [30].

2.3. Co-Culture of Neonatal Rat Cardiac Fibroblasts and Myocytes

Primary neonatal rat hearts were isolated from 1–3-day-old Wistar rats (~3–4 animal/24-well plate) as previously described [31]. Briefly, rats were disinfected with 70% ethanol and sacrificed by cervical dislocation. Hearts were rapidly removed from an abdominal approach and placed in ice-cold PBS then washed with fresh volumes of ice-cold PBS three times under laminar hood. Ventricles were separated from atria in a sterile Petri dish containing ice-cold PBS. Afterwards, ventricles were minced properly with fine forceps and collected in 0.25% trypsin (5 mL/heart) containing Falcon tubes. Tissue fragments were digested by trypsin for 25 min in 37 °C water bath. The cell suspension was resuspended every 5 min by a 5 mL pipette. After digestion, the cell suspension was centrifuged for 15 min at 250 rcf at 4 °C and the supernatant was removed gently. Cell pellets were resuspended in 10 mL of Gibco Dulbecco's Modified Eagle Medium (DMEM) growth medium supplemented with 20% FBS. Cells were counted manually by hemocytometer then gently resuspended, divided equally into fibronectin-coated wells. Cells were maintained at 37 °C in a humidified atmosphere of 5% CO_2 and 95% air in a CO_2 incubator until ~100% confluence was reached. The medium was changed to DMEM growth medium supplemented with 10% FBS the day after preparation. The medium was changed again 6–8 h after the previous one to DMEM growth medium supplemented with 1% FBS. Cells were scored after the medium change. DMEM medium supplemented with 1% FBS was refreshed every 2–3 days until reaching the desired confluence and was used in scratch assay experiments.

2.4. Scratch Assay

Scratch assays were performed as reported previously [31]. Briefly, cells were seeded on 24-well plates, and were pretreated with 500 nM saxagliptin or its solvent for 1 h. After pretreatment, the cell monolayer was scratched with a 200 µL pipette tip in a 30° angle.

The medium was discarded to remove cell debris from wells, and cells were washed two times with Hank's Balanced Salt Solution (HBSS, Corning Inc., Somerville, MA, USA). Then, the cells were treated with a combination of saxagliptin and NPY or SP for 24 h in the same manner as described above. Cells were incubated for 24 h with the treatment solution at 37 °C in a humidified atmosphere of 5% CO_2 and 95% air in a CO_2 incubator. Images were taken at the start of the treatment (0 h) and 24 h after the scratch. The wounding area (unoccupied surface by cells) of the images was measured by ImageJ software [32] and expressed as percentage of baseline value (0 h). Data were collected from four independent experiments.

2.5. Human Heart Tissue Collection

The experiments were designed and implemented according to the ethical standards of the Declaration of Helsinki (1975). Patients gave their written informed consent to be involved in the study. The protocol was approved by the Polish Local Ethics Committee of the National Institute of Cardiology in Warsaw with the identification code IK-NPIA-0021-14/1426/18. Human tissue samples were collected in the Department of Heart Failure and Transplantology, Cardinal Stefan Wyszyński National Institute of Cardiology, Warszawa, Poland, as previously described [33]. Human hearts obtained from organ donors that were excluded from transplantation for various reasons were used as control (CON) samples. The donors with any relevant cardiovascular history or abnormalities were excluded in the present study. Failing hearts were obtained from patients suffering from end-stage heart failure of ischemic cardiomyopathy (ICM) or dilated cardiomyopathy (DCM). Clinical parameters of the human tissue samples used in Western blot (Supplementary Table S1) and in radioimmunoassay/ELISA (Supplementary Table S2) experiments are summarized in the supplementary files. Interventricular septum samples were collected during heart explantation, avoiding the inclusion of non-cardiac tissues, e.g., scar, adipose tissue, endocardium, epicardium, or coronary vessels. The samples were rinsed immediately in saline solution, blotted dry, frozen in liquid nitrogen, and kept at −80 °C until processing for further molecular assays. Another series of left ventricle samples was fixed in neutral buffered formalin and embedded in paraffin for histological assays.

2.6. RNA Scope® In Situ Hybridization

In situ hybridization of the DPP4 enzyme mRNA was performed on tissue slides of interventricular septum harvested from human control hearts using RNA Scope® Multiplex Fluorescent Kit v2 according to the manufacturer's instructions (Advanced Cell Diagnostics, Newark, CA, USA). Briefly, as reported previously [34], 4 μm formalin-fixed paraffin-embedded tissue sections were pretreated with heat, H_2O_2, and protease prior to hybridization with the following target oligo probes: 3plex-Hs-Positive Control Probe (catalog number: 320861), 3plex-Negative Control Probe (catalog number: 320871), Hs-DPP4 (catalog number: 477541, accession no.: NM_001935.3), Hs-VIM-C2 (catalog number: 310441-C2, accession no.: NM_003380.3), Hs-CD68-C2 (catalog number: 560591-C2, accession no.: NM_001040059.1), Hs-PECAM1-O1-C3 (catalog number: 487381-C3, accession no.: NM_000442.4), and Hs-RYR2-C2 (catalog number: 415831-C2, accession no.: NM_001035.2). Cell-type-specific markers were used to identify cardiomyocytes with a probe recognizing the mRNA of ryanodine receptor 2 (RYR2) [35], endothelial cells with a probe recognizing the mRNA of platelet endothelial cell adhesion molecule 1 (PECAM-1) [36], fibroblast cells with a probe recognizing the mRNA of vimentin (VIM) [37,38], and macrophages with a probe recognizing the mRNA of cluster of differentiation 68 (CD68) [39], respectively. Next, preamplifier, amplifier, and HRP-labeled oligo probes were then hybridized sequentially, followed by signal development with TSA fluorophores (TSA-Cy3, TSA-FITC, Akoya Biosciences, Marlborough, MA, USA). Each sample was quality-controlled for RNA integrity with a positive control probe specific to the housekeeping genes and with a negative control probe. The specific RNA staining signal was identified as red/green dots. Nuclei were stained with 4′,6-diamidino-2-phenylindole (DAPI). Imaging was performed with a Leica DMI8 Confocal microscope (Leica, Wetzlar, Germany).

2.7. DPP4 Protein Expression in the Heart

In order to investigate whether DPP4 expression was altered at the protein level in the human heart samples, Western blot was performed. Frozen tissue samples from the interventricular septum were homogenized with a TissueLyser (Hilden, Germany) in 1× radio immunoprecipitation assay buffer (RIPA; Cell Signaling Technology, Danvers, MA, USA) supplemented with 1× HALT Protease and Phosphatase Inhibitor cocktail (Thermo Scientific, Waltham, MA, USA). Protein concentration of the samples was determined by bicinchoninic acid assay kit (Thermo Scientific, Waltham, MA, USA). Equal amounts of protein (25 μg) from each sample were mixed with 1/4 volume of Laemmli buffer containing β-mercaptoethanol (Thermo Scientific, Waltham, MA, USA) and were loaded on 4–20% Tris-glycine sodium dodecyl sulfate-polyacrylamide gels (Bio-Rad, Hercules, CA, USA). After separation by electrophoresis, proteins were transferred onto polyvinylidene difluoride membrane (Bio-Rad, Hercules, CA, USA) with Trans-Blot® Turbo™ Transfer System (Bio-Rad, Hercules, CA, USA). Membranes were blocked in 5% bovine serum albumin (Bio-Rad, Hercules, CA, USA) in Tris-buffered saline containing 0.05% Tween-20 (0.05% TBS-T; Sigma, St. Louis, MO, USA) for 2 h at room temperature. Afterwards, membranes were incubated with anti-DPP-4 primary antibody overnight at 4 °C (1:1000 dilution). After three washes in TBS-T, membranes were incubated with HRP-conjugated anti-rabbit secondary antibodies (Cell Signaling, Danvers, MA, USA) for 2 h and washed in TBS-T. Signals were visualized after incubation with enhanced chemiluminescence kit (Bio-Rad, Hercules, CA, USA) by Chemidoc XRS+ (Bio-Rad, Hercules, CA, USA). Image analysis was performed using Image Lab™ 6.0 software (Bio-Rad, Hercules, CA, USA). Measurement of GAPDH content was used as loading control. Briefly, membranes were incubated with Restore Stripping Buffer for 15 min at room temperature (Thermo Scientific, Waltham, MA, USA), followed by incubation with anti-GAPDH primary antibody (1:5000 dilution, overnight at 4 °C), HRP-conjugated anti-rabbit secondary antibody (1:2000 dilution, 2 h at room temperature), and signal detection, as described previously.

2.8. SP-like Immunoreactivity

Human interventricular septum samples were homogenized in 1 mL of 20 mM phosphate buffer (K_2HPO_4 and KH_2PO_4, pH: 7.2) and 10 μL protease inhibitor (Gordox, 10,000 KIE/mL, Gedeon Richter Plc, Budapest, Hungary) with tissue homogenizer device (IKA T25 Digital ULTRA TURRAX). This was followed by centrifugation at 10,000 rpm at 4 °C for 15 min. The supernatant was collected and pooled at −80 °C.

Substance-P-like immunoreactivity was measured by the specific and sensitive radioimmunoassay method as described in detail in earlier publications [40] using the substance P competitive radioimmunoassay kit (cat.nr. RK-061-05, Phoenix Pharmaceuticals, Inc., Burlingame, CA, USA). Here, we only summarize the protocol briefly. Reconstituted positive controls and standards, 100 mL of tissue homogenates in duplicates, and 100 μL antiserum were incubated overnight at 4 °C in test tubes. Then, 100 μL 125I-labelled SP as tracer was added to the tubes on the consecutive day and additional overnight incubation was performed at 4 °C. Goat anti-rabbit IgG serum and normal rabbit serum were added to the designated tubes on the following day with a 90 min incubation period. Immunocomplexes were collected with centrifugation at 3000 rpm for at least 20 min at 4 °C, supernatant was carefully discarded, and pellet cpm was measured by g-counter (Gamma NZ-310, Budapest, Hungary). The results were expressed as fmol SP-like immunoreactivity per mg total protein weight.

2.9. NPY-like Immunoreactivity

Determination of NPY concentration in human interventricular septum samples was performed with RayBio® Human/Mouse/Rat Neuropeptide Y competitive Enzyme Immunoassay Kit (cat.nr.: EIA-NPY, RayBiotech Life Inc., Peachtree Corners, GA, USA) according to the manufacturer's instructions. We describe here the brief summary of the assay protocol. First, 100 µL of anti-NPY antibody solution was added to each well and 1.5 h of incubation with gentle shaking was performed at room temperature. After the incubation, the solution was removed and the wells were washed with 200–300 µL of 1× wash buffer solution, repeated four times. After the last washing step, the wash buffer was completely removed and the plate was inverted and blotted against clean paper towels. Next, 100 µL of standard reagents, positive control, and samples were added to the appropriate wells of the plate. The wells were covered and incubated at 4 °C overnight with gentle shaking. The solutions were removed, and the wells were washed four times with 1× wash buffer solution. To each well, 100 µL of prepared HRP-streptavidin solution was added and the plate was incubated for 45 min at room temperature with gentle shaking. After the next washing step, 100 µL of TMB One-Step Substrate Reagent was added to each well and incubation was performed for 30 min at room temperature, protecting the assay from light, with gentle shaking. At the end of the incubation period, 50 µL of stop solution was added to each well and color intensity measurement was performed at 450 nm with Labsystems DC plate reader immediately. Results were expressed as ng of NPY/mg of total protein content of the samples. The total protein concentration was measured with a bicinchoninic acid assay kit (Thermo Scientific Pierce Protein Research Products, Rockford, IL, USA) using bovine serum albumin as a standard.

2.10. Statistical Analysis

Statistical analyses were performed and graphs were created using GraphPad Prism version 8 (GraphPad Software, San Diego, CA, USA). One-way analysis of variance (ANOVA), two-way ANOVA, unpaired *t*-test, and Mann–Whitney test were used to find statistically significant differences. Differences were considered significant at values of $p < 0.05$. Unless noted otherwise, all data represent the mean ± SEM.

3. Results

A large-scale, multicenter, randomized, double-blind, and placebo-controlled phase 4 clinical trial called SAVOR-TIMI 53 aimed to assess the cardiovascular safety, efficacy, and potential cardioprotective benefits of saxagliptin, a DPP4 inhibitor [27]. According to the conclusions of that study, saxagliptin was found to be neutral in most of the primary and secondary end-points except that the heart-failure-associated hospitalization rate increased in the saxagliptin-treated patients compared to the control group. This result raised the question whether saxagliptin exerts any potentially harmful effect on the cardiovascular system.

3.1. Protein Expression of DPP4 and NPY Decreased in Human Failing Heart Samples

Substance P and neuropeptide Y are important substrates of DPP4 that can mediate various detrimental effects in cardiovascular diseases [41–43]. Therefore, here we performed Western blot, ELISA, and radioimmunoassay experiments in order to measure the expression of DPP4, NPY, and SP at the protein level in interventricular septum samples from failing human hearts (patients with ischemic cardiomyopathy (ICM) or dilated cardiomyopathy (DCM)) and also from healthy control (CON) hearts (detailed patients' characteristics in Supplementary Tables S1 and S2). We have found that the expression of DPP4 decreased significantly both in the ICM and DCM groups compared to the control (Figure 1A,B). NPY showed a significantly decreased expression in the DCM group and a decreasing tendency in the ICM group compared to the control (Figure 1C). The expression of SP remained around a similar level in each heart sample (Figure 1D).

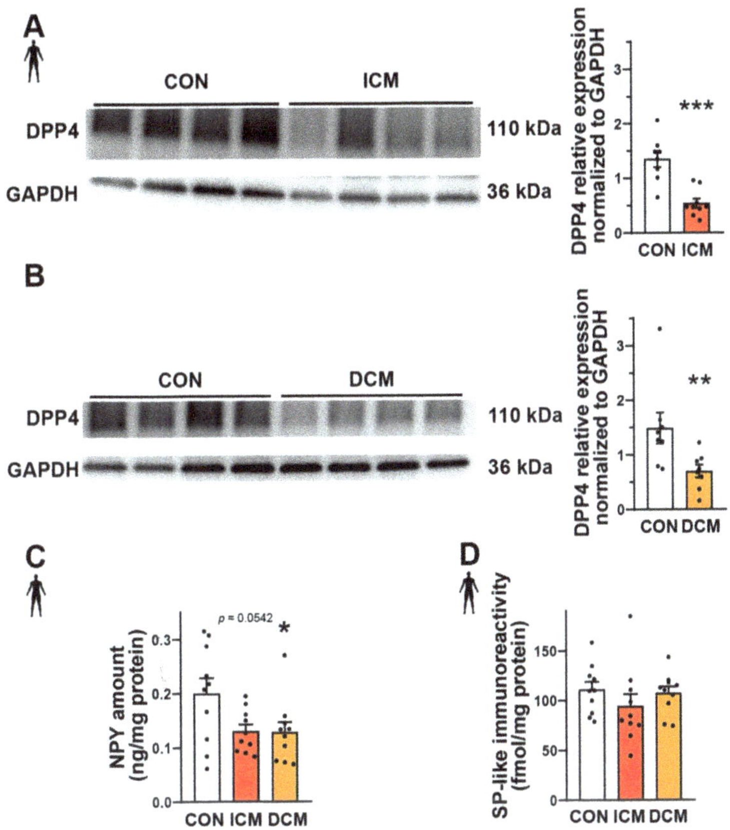

Figure 1. Protein expression of DPP4 and neuropeptides in failing human heart samples. Western blot analysis of DPP4 enzyme (**A**,**B**), ELISA (**C**), and radioimmunoassay (**D**). Quantification (**A**–**D**) of the DPP4 and neuropeptide substrates (NPY and SP) content in interventricular septum samples of healthy patients (CON) or patients with ischemic (ICM) or dilated cardiomyopathy (DCM). One-way ANOVA with Tukey's post hoc test, unpaired t-test, and Mann–Whitney test * $p < 0.05$, ** $p < 0.01$, *** $p < 0.001$ vs. CON, group size: n = 8–10. Data are expressed as mean ± SEM.

After the successful measurements of the protein expressions, our following aim was to assess which cell types express the DPP4 enzyme.

3.2. DPP4 mRNA Is Primarily Localized in Cardiomyocytes of the Human Left Ventricle

In order to clarify the cell-type-specific expression of DPP4, RNA Scope® in situ hybridization assay was performed on left ventricular tissue slides of healthy humans. Expression of the mRNA of DPP4 was shown primarily in RYR2 mRNA-positive cardiomyocytes (Figure 2A) and it was also detected to some extent in PECAM-1 mRNA-positive endothelial cells (Figure 2B), but not in VIM mRNA-positive fibroblasts (Figure 3A) and CD68 mRNA-positive macrophages (Figure 3B).

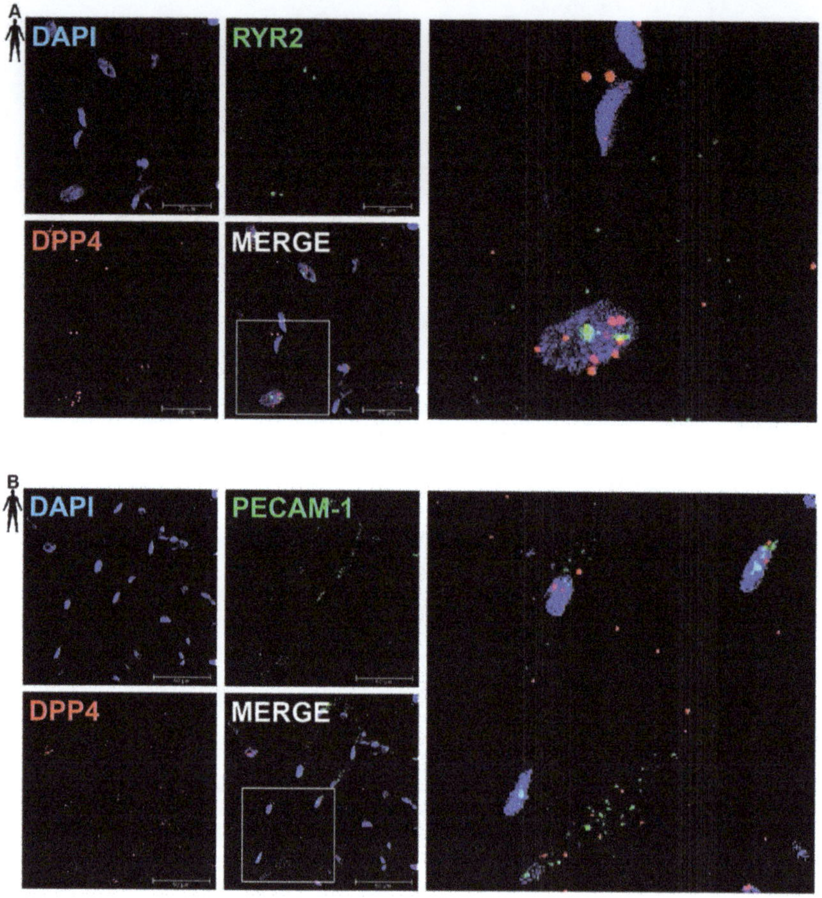

Figure 2. Representative confocal microscopy images of RNA Scope®-DPP4 mRNA expression in human control left ventricle. Nuclei were stained with DAPI (blue). Fluorescein-labeled tyramide (green) was used to visualize mRNA of RYR2 (cardiomyocyte marker, (**A**)) or PECAM-1 (endothelium marker, (**B**)) and Cy3-labeled tyramide (red) was used to visualize mRNA of DPP4, respectively. Scale bar represents 25 or 50 µm. (RYR2: ryanodine receptor 2, PECAM-1: platelet endothelial cell adhesion molecule-1).

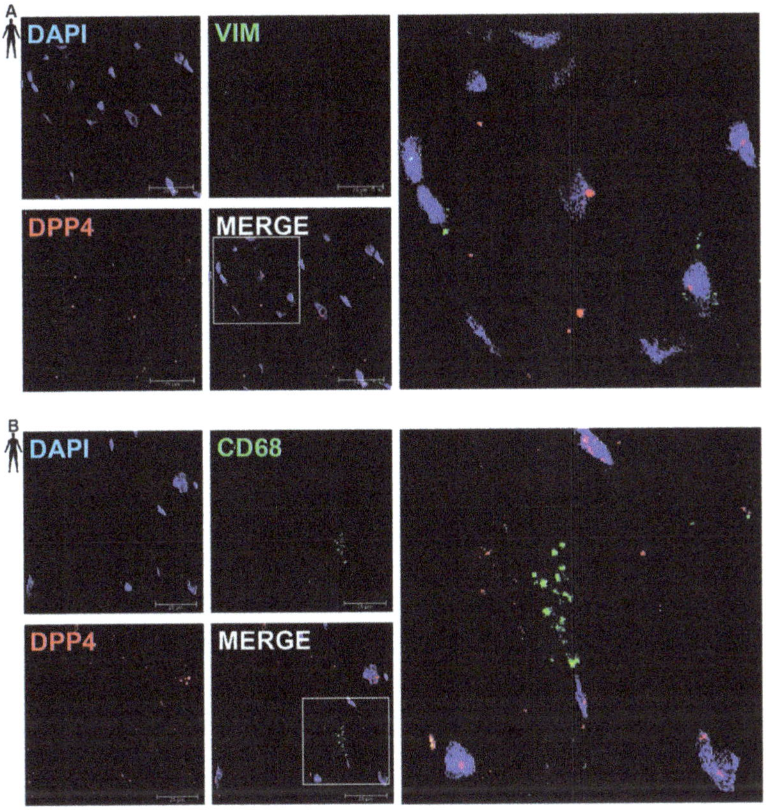

Figure 3. Representative confocal microscopy images of RNA Scope®-DPP4 mRNA expression in human control left ventricle. Nuclei were stained with DAPI (blue). Fluorescein-labeled tyramide (green) was used to visualize mRNA of VIM (fibroblast marker, (**A**)) or CD68 (macrophage marker, (**B**)) and Cy3-labeled tyramide (red) was used to visualize mRNA of DPP4, respectively. Scale bar represents 25 μm. (VIM: vimentin, CD68: cluster of differentiation 68).

There was no detectable signal on the negative control slides (Supplementary Figure S1). These observations indicate that DPP4 is primarily expressed in cardiac myocytes and endothelial cells in the heart tissue.

3.3. DPP4 Inhibition and Neuropeptide Substrates Do Not Affect Cell Viability

After successful determination of DPP4 and its substrates' expression profile, we tested the potential detrimental effects of DPP4 inhibition and/or its substrates (SP, NPY) on cardiomyocytes. In order to begin the in vitro cell culture experiments, a proper cell culture model was required. We chose the AC16 cell line because these cells express the DPP4 enzyme at the protein level according to the results of the Western blot measurements (Figure 4A), also confirming our RNA Scope® results. First, we hypothesized that direct cytotoxic effect of gliptins might be responsible for the slightly increased cardiovascular risk observed in SAVOR-TIMI, thus we performed the cell viability test on AC16 cells treated with saxagliptin or other clinically used gliptins. We have found that DPP4 inhibition by any of the tested gliptins (saxagliptin, vildagliptin, linagliptin, alogliptin) alone, at a concentration of 500 nM, does not have a toxic effect on cell viability of AC16 cells (Figure 4B). Next, we shifted our focus to investigating whether reduced DPP4 enzyme activity by gliptins increases the potential cytotoxic effects of its substrates NPY and SP. Thus, we treated the cells with NPY or SP or in combination with saxagliptin. We have found that neither neuropeptides (NPY or SP) alone nor in combination with saxagliptin can reduce the viability of AC16 cells (Figure 4C). These results indicate that saxagliptin and the neuropeptide substrates of DPP4 do not influence the viability of healthy cells; therefore, a different approach and model are required to test the potential cardiotoxic effect of saxagliptin.

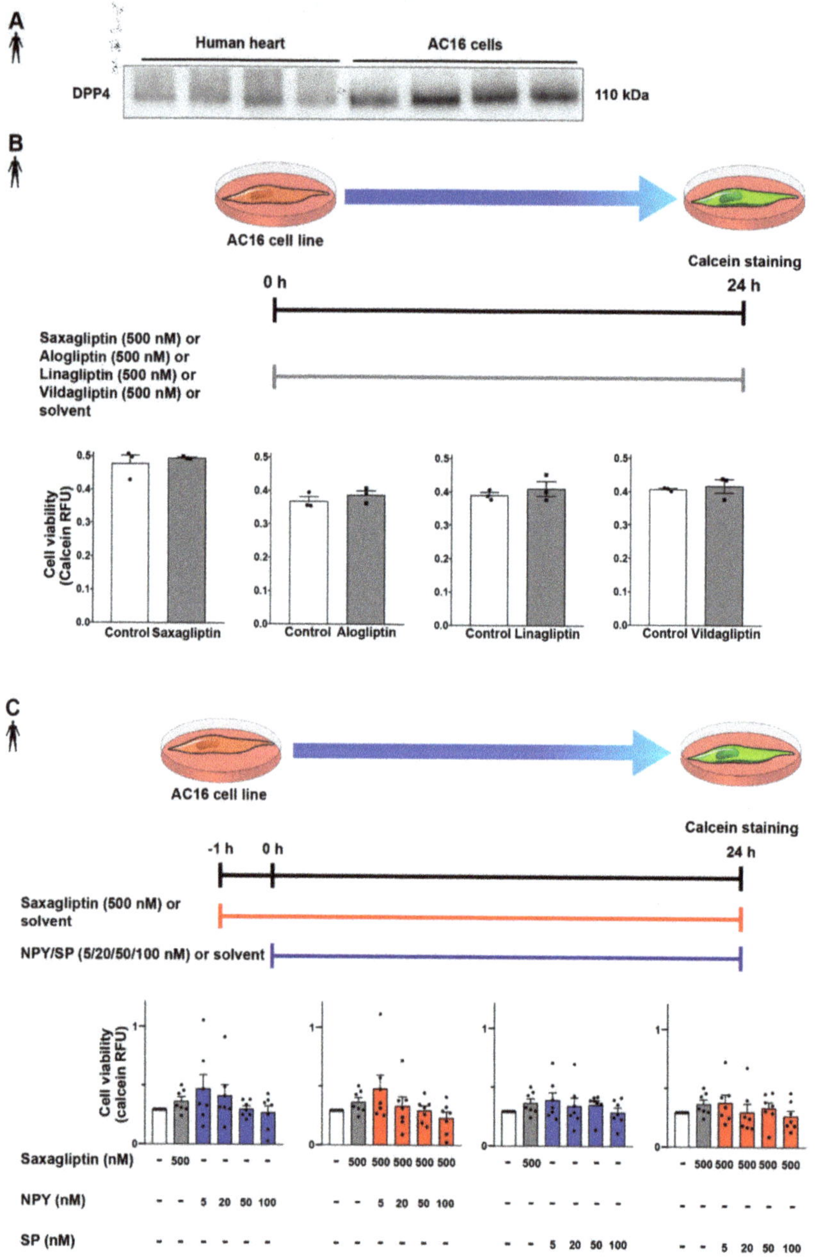

Figure 4. Effect of DPP4 inhibition and/or neuropeptide substrates on the viability of AC16 cells. Western blot analysis of DPP4 (**A**) in healthy human left ventricle samples and AC16 cells. In vitro

treatment protocol with various gliptins on AC16 cell line and cell viability (calcein assay) results (**B**). In vitro treatment protocol with neuropeptides and their combined administration with saxagliptin and their effect on the viability of AC16 cells (**C**). One-way ANOVA, Tukey's post hoc test, and unpaired *t*-test. Data are presented as mean ± SEM. Group sizes: (**B**) n = 3 from 1 independent experiment, (**C**) n = 7 from 7 independent experiments. RFU: relative fluorescence unit.

3.4. Both Saxagliptin and the Neuropeptides Alter the Migration Capacity of Cardiac Fibroblasts

Literature data suggest that SP and NPY may exert their harmful cardiovascular effects by the modulation of fibroblast activities [11,12,44,45]. Accordingly, we continued our experiments with a co-culture model of primary neonatal rat cardiomyocytes and cardiac fibroblasts, and performed scratch assay (which is widely used to model in vitro cell migration and wound healing) experiments (Figure 5A,B) with the same treatment protocol (neuropeptides and/or saxagliptin) as described above (Figure 5A,C).

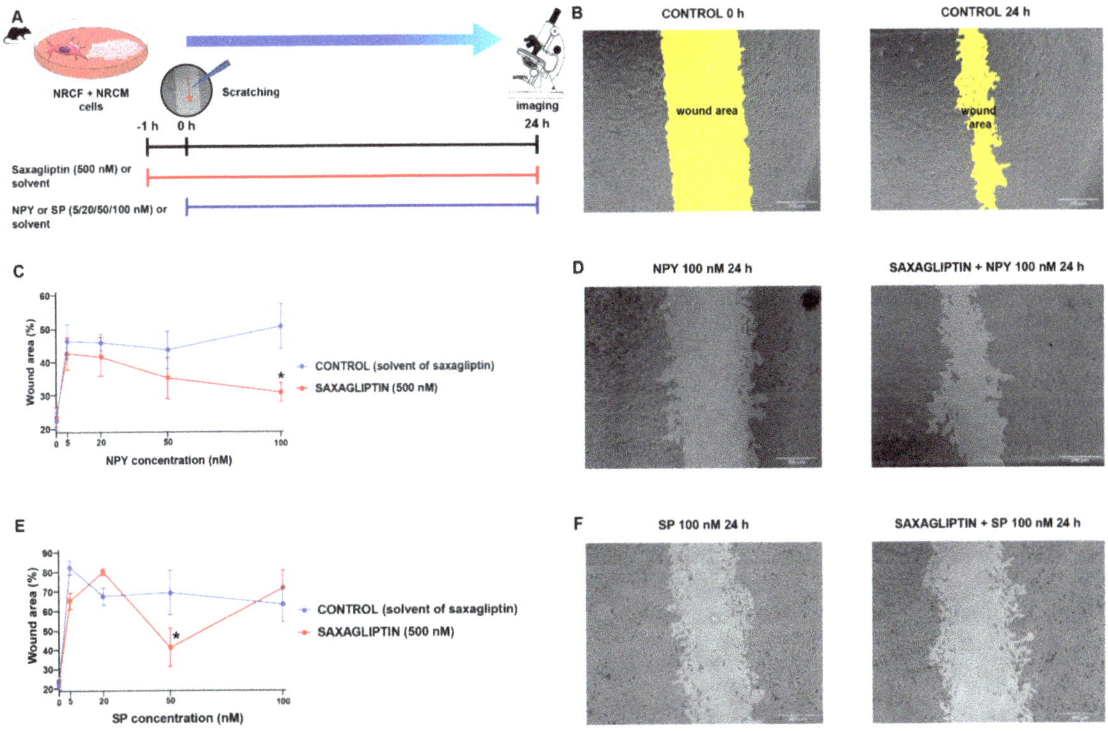

Figure 5. The cell migration speed in saxagliptin- and neuropeptide-treated cardiomyocyte/fibroblast co-culture model. Scratch assay treatment protocol (**A**) and the results of the treatment with NPY (**C**) and SP (**E**). Representative bright field microscope images of control (**B**), NPY-treated (**D**), and SP-treated groups (**F**). Two-way ANOVA, * $p < 0.05$ vs. CON. Group sizes: n = 8–22 from 3–5 independent experiments. Data are presented as mean ± SEM in percentage of the wound area compared to each corresponding 0 h (baseline) values.

We have found that administration of either NPY (Figure 5C,D) or SP (Figure 5E,F) reduces the migration speed of the cells significantly compared to the control groups (CONTROL and SAXAGLIPTIN). At the highest concentration of NPY (100 nM), the administration of saxagliptin (500 nM) restored the migration capacity of fibroblasts. In the case of the SP administration, there was no meaningful difference caused by the administration of saxagliptin.

4. Discussion

The surprising results of the SAVOR-TIMI 53 [27] clinical trial revealed that the heart-failure-associated hospitalization rate is increased in the saxagliptin-treated patients. These results point to the importance of cardiovascular safety testing in preclinical disease models to reveal the mechanisms of this clinically relevant cardiotoxicity. Therefore, in our present study, we focused on the investigation of the potential role of saxagliptin and the neuropeptide substrates of DPP4 in this phenomenon. Here, we showed that the protein expression of DPP4 and NPY decreased in the cardiac tissue of heart failure patients. Moreover, we also demonstrated by in situ hybridization that DPP4 mRNA is expressed mainly by cardiomyocytes. Interestingly, we have found that neither DPP4 inhibition alone nor in combination with neuropeptides such as NPY and SP affected the viability of AC16 cells, suggesting that the direct cytotoxic effect might not be primarily responsible for the observed adverse outcome in the clinical scenario [27]. In contrast, neuropeptides decreased cell migration speed in the co-culture of primary neonatal rat cardiomyocytes and cardiac fibroblasts. However, saxagliptin co-administration restored fibroblast migration speed in comparison to NPY. These results showed that DPP4 inhibition by saxagliptin and the increased level of neuropeptides in heart failure can modulate fibroblast migration; therefore, this may interfere with adaptive cardiac remodeling in heart failure, indicating a potential hidden drug cardiotoxicity mechanism [46].

Dipeptidyl-peptidase-4 is a pivotal enzyme in the degradation of various substances such as incretins, neuropeptides, and chemokines, among others. DPP4 inhibitors have become important drugs for the treatment of type 2 diabetes due to the improved regulation of glycemia through the increased glucagon-like peptide-1 levels. In addition, expression of DPP4 in the vascular endothelial cells raised its potential role in regulating vascular functions as well. Experimental results with various DPP4 inhibitors proved that DPP4 may play a role in the modulation of nitric oxide release [47] or reduce the severity of atherosclerotic lesions [48]. However, its role in regulating cardiac myocyte and fibroblast functions has not been investigated so far. We provide here the first evidence that the protein expression of DPP4 and NPY decreased in cardiac tissue samples of heart failure patients. This decrease in DPP4 expression was not related to the etiology of heart failure (ischemic or non-ischemic origin). Our findings are in line with previous data showing that cardiac tissue NPY protein expression is decreased in rats with volume-overload-induced heart failure, while in parallel, these studies revealed that the circulating level of NPY is increased due to heart failure [13]. Further supporting our results, Ajijola et al. have found that NPY immunoreactivity is decreased, but its mRNA expression did not change in the stellate ganglia of heart failure patients compared to healthy controls, suggesting increased release of NPY from the stellate ganglia [49].

We identified here that the DPP4 mRNA is localized mainly in cardiomyocytes and endothelial cells of the healthy human left ventricle. Previous studies revealed that DPP4 is widely expressed on the surface of various cell types, including leukocytes and epithelial or endothelial cell populations in several organs [50–54]. In line with these findings, we have also confirmed the presence of DPP4 mRNA in endothelial cells in the human heart as well. Our data highlight that cardiomyocytes could be a potential target cell type for saxagliptin, thus DPP4 inhibition in cardiomyocytes may play a role in mediating harmful effects.

The cardiovascular safety of saxagliptin in heart failure is quite controversial according to the currently available, conflicting literature data. Most of the clinical investigations and meta-analyses suggest that saxagliptin is harmful or at least shows controversial cardiovascular safety [55–58], while others found saxagliptin to be neutral [59–61], and there are several preclinical results on cardioprotective properties of DPP4 inhibitors [62–64]. Therefore, we investigated whether DPP4 inhibition in the presence of neuropeptide substrates could affect cardiomyocyte viability in vitro. We found that neither DPP4 inhibition alone nor in combination with neuropeptides exerts any cytotoxic effect on the human cardiomyocyte-like AC16 cells. According to the literature data, effects of saxagliptin treatment were investigated in various models of diabetes and/or acute cardiac damages. It was found that saxagliptin treatment exerted a direct cytoprotective effect in vitro against glucose and oxygen depletion and reoxygenation in cultured primary human brain microvascular endothelial cells [65]. Additionally, other groups demonstrated that saxagliptin was also able to decrease oxidative stress through regulation of endothelial nitric oxide synthase in a Goto-Kakizaki rat model of non-obese type 2 diabetes [66]. Saxagliptin also ameliorated cardiac diastolic dysfunction in isoproterenol-treated rats [63] and reduced cardiac ischemia-reperfusion injury in an ex vivo rat model of I/R injury with type 2 diabetes [64]. The discrepancy between the clinical and preclinical data is potentially due to differences in the mechanisms involved in the experimental models, since the majority of studies focus on diabetic models and acute cardiac damage, but not on chronic conditions such as heart failure.

Dipeptidyl-peptidase-4 inhibition by sitagliptin has been reported to decrease collagen deposition, and activation of pro-fibrotic signaling in rat hearts potentially improves fibrosis in heart failure [67]. These data suggest that the DPP4 enzyme and its substrates might be important in cardiac fibrosis, cell migration, and remodeling. Therefore, next, we investigated their potential effect on the migration speed of fibroblast cells and their potential interfering effect with adverse cardiac remodeling in a cardiomyocyte-fibroblast co-culture scratch assay model. We have found that administration of either NPY or SP significantly reduces the migration speed of the cells compared to the vehicle-treated controls. At the highest concentration of NPY, saxagliptin administration restored the migration capacity of fibroblasts. Interestingly, in other cell types, NPY has been shown to stimulate migration of human umbilical endothelial cells [68], to promote ischemic angiogenesis and vascularization in the rat ischemic hind limb [69], and to increase motility of neuroblastoma cells [70]. Our finding suggests that NPY might act differently on cardiac fibroblasts, reducing their migration and likely preventing cardiac fibrosis. Interestingly, saxagliptin reverses the effect of NPY on fibroblast migration; therefore, saxagliptin in the presence of NPY may exert harmful profibrotic effects.

In conclusion, we have shown that the expression of DPP4 enzyme and the neuropeptide tone is altered in human failing hearts. Although DPP4 inhibition neither alone nor in combination with neuropeptides has any detectable cytotoxic effect in vitro, neuropeptides inhibited cell migration in primary neonatal rat cardiomyocyte/fibroblast co-cultures, which is reversed by saxagliptin co-administration. Our results highlight that saxagliptin and NPY may interfere with cardiac tissue remodeling and thus play a role in the pathophysiological mechanisms of end-stage chronic heart failure. We believe that the DPP4 enzyme could exert a compensatory function against the altered neuropeptide tone caused by the elevated sympathetic activity in heart failure. Inhibition of DPP4 by saxagliptin could impair this adaptive mechanism, and thereby exacerbate myocardial damage, although further experiments are required to more deeply understand the potential role of other DPP4 substrates and the possible direct toxic effects of saxagliptin.

Supplementary Materials: The following supporting information can be downloaded at: https://www.mdpi.com/article/10.3390/biomedicines10071573/s1, Table S1: Patient characteristics in Western blot experiments; Table S2: Patient characteristics in radioimmunoassay and ELISA experiments; Figure S1: Representative confocal microscopy images of RNA Scope® negative control in histological samples of human control left ventricle.

Author Contributions: Conceptualization, Z.V.V., Z.O., I.V., Z.H. and P.F.; methodology, Z.O., I.V., V.É.T., T.G.G., É.S., Z.V.V., A.G. and Á.K.; investigation, I.V., Z.O., V.É.T., T.G.G. and Á.K.; resources, P.L.; data curation, Z.V.V.; writing—original draft preparation, I.V., Z.O., Z.V.V., T.G.G. and Á.K.; visualization, I.V. and Z.V.V.; supervision, Z.V.V., Z.H. and P.F.; funding acquisition, Z.V.V. and P.F. All authors have read and agreed to the published version of the manuscript.

Funding: This study was supported by the National Research, Development, and Innovation Office of Hungary (NVKP-16-1-2016-0017 National Heart Program and 2017-1.2.1-NKP-2017-00002 the National Brain Research Program-Zs. Helyes, University of Pécs) and by the Higher Education Institutional Excellence Program of the Ministry of Human Capacities in Hungary, within the framework of the Therapeutic Development Thematic Program of the Semmelweis University, Research Excellence Programme of the National Research, Development and Innovation Office of the Ministry of Innovation and Technology in Hungary (TKP/ITM/NKFIH), National Research, Development, and Innovation Office of Hungary (NKFIA; FK134751 to Z.V.V., K139237 to A.G.). Project no. RRF-2.3.1-21-2022-00003 has been implemented with the support provided by the European Union. I.V. was supported by the New National Excellence Program of the Ministry for Innovation and Technology from the source of the National Research, Development and Innovation Fund (ÚNKP-21-3-II-SE-14). T.G.G., Z.O., and I.V. were supported by EFOP-3.6.3-VEKOP-16-2017-00009. T.G.G was supported by Gedeon Richter Talentum Foundation's scholarship. The work was supported by the European Union's Horizon 2020 Research and Innovation Programme under grant agreement no. 739593 and by a Momentum Research Grant from the Hungarian Academy of Sciences (LP-2021-14 to Z.V.V.). É.S. and Z.V.V. were supported by the János Bolyai Research Scholarship of the Hungarian Academy of Sciences. VEKOP-2.3.3-15-2017-00016 grant made confocal microscopy imaging possible. The research was financed by the Thematic Excellence Programme (2020-4.1.1.-TKP2020) of the Ministry for Innovation and Technology in Hungary, within the framework of the Therapeutic Development and Bioimaging Thematic Programs of the Semmelweis University. The project (TKP2021-EGA-23) has been implemented with the support provided by the Ministry of Innovation and Technology of Hungary from the National Research, Development and Innovation Fund, financed under the [TKP2021-EGA/TKP2021-NVA/TKP2021-NKTA] funding scheme (TKP2021-EGA-16 and TKP2021-EGA-13 for University of Pécs), and 2020-1.1.6-JÖVŐ-2021-00013 by NKFIA.

Institutional Review Board Statement: The experiments were designed and implemented according to the ethical standards of the Declaration of Helsinki (1975). The protocol was approved by the Polish Local Ethics Committee of the National Institute of Cardiology in Warsaw with the identification code IK-NPIA-0021-14/1426/18.

Informed Consent Statement: Patients gave their written informed consent to be involved in the study.

Data Availability Statement: The datasets used and/or analyzed during the current study are available from the corresponding author on reasonable request.

Acknowledgments: The authors thank Teréz Bagoly, Regina Nagy, Tünde Petrovics, and Krisztina Kecskés for their expert technical assistance in the experiments. We are grateful to Kornél Király for his professional advice, and for providing vildagliptin for our experiments. The authors wish to acknowledge SERVIER Medical Art (https://smart.servier.com/ (accessed on: 20 September 2021)) for use of their medical art kits when making the illustrations for the present article.

Conflicts of Interest: P.F. is the founder and CEO of Pharmahungary Group, a group of R & D companies (www.pharmahungary.com (accessed on 20 September 2021)). Z.H. is the founder and scientific director of PharmInVivo Ltd. (www.pharminvivo.hu (accessed on 20 September 2021)). The remaining authors declare that the research was conducted in the absence of any commercial or financial relationships that could be construed as a potential conflict of interest.

Abbreviations

ANOVA	analysis of variance
CD68	cluster of differentiation 68
CON	control
Cy3	cyanine 3
DAPI	4′,6-diamidino-2-phenylindole
DCM	dilated cardiomyopathy
DMEM	Dulbecco's modified eagle medium
DMSO	dimethyl sulfoxide
D-PBS	Dulbecco's phosphate-buffered saline
DPP4	dipeptidyl-peptidase-4

ELISA	enzyme-linked immunosorbent assay
FBS	fetal bovine serum
GAPDH	glyceraldehyde 3-phosphate dehydrogenase
HBSS	Hank's balanced salt solution
HRP	horseradish peroxidase
ICM	ischemic cardiomyopathy
mRNA	messenger RNA
PECAM-1	platelet endothelial cell adhesion molecule 1
RFU	relative fluorescent unit
RYR2	ryanodine receptor 2
SEM	standard error of the mean
VIM	vimentin

References

1. Braunwald, E. Heart failure. *JACC Heart Fail.* **2013**, *1*, 1–20. [CrossRef] [PubMed]
2. Dalsgaard, C.J.; Franco-Cereceda, A.; Saria, A.; Lundberg, J.M.; Theodorsson-Norheim, E.; Hökfelt, T. Distribution and origin of substance P- and neuropeptide Y-immunoreactive nerves in the guinea-pig heart. *Cell Tissue Res.* **1986**, *243*, 477–485. [CrossRef]
3. Papka, R.E.; Urban, L. Distribution, origin and sensitivity to capsaicin of primary afferent substance P-immunoreactive nerves in the heart. *Acta Physiol. Hung.* **1987**, *69*, 459–468. [PubMed]
4. Wharton, J.; Polak, J.M.; McGregor, G.P.; Bishop, A.E.; Bloom, S.R. The distribution of substrate P-like immunoreactive nerves in the guinea-pig heart. *Neuroscience* **1981**, *6*, 2193–2204. [CrossRef]
5. Rysevaite, K.; Saburkina, I.; Pauziene, N.; Vaitkevicius, R.; Noujaim, S.F.; Jalife, J.; Pauza, D.H. Immunohistochemical characterization of the intrinsic cardiac neural plexus in whole-mount mouse heart preparations. *Heart Rhythm.* **2011**, *8*, 731–738. [CrossRef]
6. Milner, P.; Ralevic, V.; Hopwood, A.M.; Fehér, E.; Lincoln, J.; Kirkpatrick, K.A.; Burnstock, G. Ultrastructural localisation of substance P and choline acetyltransferase in endothelial cells of rat coronary artery and release of substance P and acetylcholine during hypoxia. *Experientia* **1989**, *45*, 121–125. [CrossRef] [PubMed]
7. Pintér, E.; Pozsgai, G.; Hajna, Z.; Helyes, Z.; Szolcsányi, J. Neuropeptide receptors as potential drug targets in the treatment of inflammatory conditions. *Br. J. Clin. Pharmacol.* **2014**, *77*, 5–20. [CrossRef]
8. Mak, I.T.; Chmielinska, J.J.; Kramer, J.H.; Spurney, C.F.; Weglicki, W.B. Loss of neutral endopeptidase activity contributes to neutrophil activation and cardiac dysfunction during chronic hypomagnesemia: Protection by substance P receptor blockade. *Exp. Clin. Cardiol.* **2011**, *16*, 121–124.
9. Weglicki, W.B.; Phillips, T.M. Pathobiology of magnesium deficiency: A cytokine/neurogenic inflammation hypothesis. *Am. J. Physiol.* **1992**, *263*, R734–R737. [CrossRef]
10. Weglicki, W.B.; Mak, I.T.; Phillips, T.M. Blockade of cardiac inflammation in Mg^{2+} deficiency by substance P receptor inhibition. *Circ. Res.* **1994**, *74*, 1009–1013. [CrossRef]
11. Dehlin, H.M.; Manteufel, E.J.; Monroe, A.L.; Reimer, M.H., Jr.; Levick, S.P. Substance P acting via the neurokinin-1 receptor regulates adverse myocardial remodeling in a rat model of hypertension. *Int. J. Cardiol.* **2013**, *168*, 4643–4651. [CrossRef]
12. Kumaran, C.; Shivakumar, K. Calcium- and superoxide anion-mediated mitogenic action of substance P on cardiac fibroblasts. *Am. J. Physiol.-Heart Circ. Physiol.* **2002**, *282*, H1855–H1862. [CrossRef]
13. Callanan, E.Y.; Lee, E.W.; Tilan, J.U.; Winaver, J.; Haramati, A.; Mulroney, S.E.; Zukowska, Z. Renal and cardiac neuropeptide Y and NPY receptors in a rat model of congestive heart failure. *Am. J. Physiol. Ren. Physiol.* **2007**, *293*, F1811–F1817. [CrossRef]
14. McDermott, B.J.; Bell, D. NPY and cardiac diseases. *Curr. Top. Med. Chem.* **2007**, *7*, 1692–1703. [CrossRef]
15. Ullman, B.; Hulting, J.; Lundberg, J.M. Prognostic value of plasma neuropeptide-Y in coronary care unit patients with and without acute myocardial infarction. *Eur. Heart J.* **1994**, *15*, 454–461. [CrossRef]
16. McMurray, J.J.; Adamopoulos, S.; Anker, S.D.; Auricchio, A.; Bohm, M.; Dickstein, K.; Falk, V.; Filippatos, G.; Fonseca, C.; Gomez-Sanchez, M.A.; et al. ESC Guidelines for the diagnosis and treatment of acute and chronic heart failure 2012: The Task Force for the Diagnosis and Treatment of Acute and Chronic Heart Failure 2012 of the European Society of Cardiology. Developed in collaboration with the Heart Failure Association (HFA) of the ESC. *Eur. Heart J.* **2012**, *33*, 1787–1847. [CrossRef]
17. Thornberry, N.A.; Gallwitz, B. Mechanism of action of inhibitors of dipeptidyl-peptidase-4 (DPP-4). *Best Pract. Res. Clin. Endocrinol. Metab.* **2009**, *23*, 479–486. [CrossRef]
18. Deacon, C.F. Dipeptidyl peptidase 4 inhibitors in the treatment of type 2 diabetes mellitus. *Nat. Rev. Endocrinol.* **2020**, *16*, 642–653. [CrossRef]
19. Ahrén, B. DPP-4 Inhibition and the Path to Clinical Proof. *Front. Endocrinol.* **2019**, *10*, 376. [CrossRef]
20. White, W.B.; Cannon, C.P.; Heller, S.R.; Nissen, S.E.; Bergenstal, R.M.; Bakris, G.L.; Perez, A.T.; Fleck, P.R.; Mehta, C.R.; Kupfer, S.; et al. Alogliptin after acute coronary syndrome in patients with type 2 diabetes. *N. Engl. J. Med.* **2013**, *369*, 1327–1335. [CrossRef]

21. Green, J.B.; Bethel, M.A.; Armstrong, P.W.; Buse, J.B.; Engel, S.S.; Garg, J.; Josse, R.; Kaufman, K.D.; Koglin, J.; Korn, S.; et al. Effect of Sitagliptin on Cardiovascular Outcomes in Type 2 Diabetes. *N. Engl. J. Med.* **2015**, *373*, 232–242. [CrossRef]
22. Rosenstock, J.; Perkovic, V.; Johansen, O.E.; Cooper, M.E.; Kahn, S.E.; Marx, N.; Alexander, J.H.; Pencina, M.; Toto, R.D.; Wanner, C.; et al. Effect of Linagliptin vs Placebo on Major Cardiovascular Events in Adults with Type 2 Diabetes and High Cardiovascular and Renal Risk: The CARMELINA Randomized Clinical Trial. *JAMA* **2019**, *321*, 69–79. [CrossRef]
23. Mentlein, R. Dipeptidyl-peptidase IV (CD26)—Role in the inactivation of regulatory peptides. *Regul. Pept.* **1999**, *85*, 9–24. [CrossRef]
24. Zhu, L.; Tamvakopoulos, C.; Xie, D.; Dragovic, J.; Shen, X.; Fenyk-Melody, J.E.; Schmidt, K.; Bagchi, A.; Griffin, P.R.; Thornberry, N.A.; et al. The role of dipeptidyl peptidase IV in the cleavage of glucagon family peptides: In vivo metabolism of pituitary adenylate cyclase activating polypeptide-(1-38). *J. Biol. Chem.* **2003**, *278*, 22418–22423. [CrossRef]
25. Ussher, J.R.; Drucker, D.J. Cardiovascular biology of the incretin system. *Endocr. Rev.* **2012**, *33*, 187–215. [CrossRef]
26. Nadasdi, A.; Sinkovits, G.; Bobek, I.; Lakatos, B.; Forhecz, Z.; Prohaszka, Z.Z.; Reti, M.; Arato, M.; Cseh, G.; Masszi, T.; et al. Decreased circulating dipeptidyl peptidase-4 enzyme activity is prognostic for severe outcomes in COVID-19 inpatients. *Biomark. Med.* **2022**, *16*, 317–330. [CrossRef]
27. Scirica, B.M.; Bhatt, D.L.; Braunwald, E.; Steg, P.G.; Davidson, J.; Hirshberg, B.; Ohman, P.; Frederich, R.; Wiviott, S.D.; Hoffman, E.B.; et al. Saxagliptin and cardiovascular outcomes in patients with type 2 diabetes mellitus. *N. Engl. J. Med.* **2013**, *369*, 1317–1326. [CrossRef]
28. McDonagh, T.A.; Metra, M.; Adamo, M.; Gardner, R.S.; Baumbach, A.; Böhm, M.; Burri, H.; Butler, J.; Čelutkienė, J.; Chioncel, O.; et al. 2021 ESC Guidelines for the diagnosis and treatment of acute and chronic heart failure: Developed by the Task Force for the diagnosis and treatment of acute and chronic heart failure of the European Society of Cardiology (ESC) With the special contribution of the Heart Failure Association (HFA) of the ESC. *Eur. Heart J.* **2021**, *42*, 3599–3726. [CrossRef]
29. Makkos, A.; Szantai, A.; Paloczi, J.; Pipis, J.; Kiss, B.; Poggi, P.; Ferdinandy, P.; Chatgilialoglu, A.; Gorbe, A. A Comorbidity Model of Myocardial Ischemia/Reperfusion Injury and Hypercholesterolemia in Rat Cardiac Myocyte Cultures. *Front. Physiol.* **2019**, *10*, 1564. [CrossRef]
30. Brenner, G.B.; Makkos, A.; Nagy, C.T.; Onodi, Z.; Sayour, N.V.; Gergely, T.G.; Kiss, B.; Gorbe, A.; Saghy, E.; Zadori, Z.S.; et al. Hidden Cardiotoxicity of Rofecoxib Can be Revealed in Experimental Models of Ischemia/Reperfusion. *Cells* **2020**, *9*, 551. [CrossRef]
31. Jelemensky, M.; Kovacshazi, C.; Ferenczyova, K.; Hofbauerova, M.; Kiss, B.; Pallinger, E.; Kittel, A.; Sayour, V.N.; Gorbe, A.; Pelyhe, C.; et al. Helium Conditioning Increases Cardiac Fibroblast Migration Which Effect Is Not Propagated via Soluble Factors or Extracellular Vesicles. *Int. J. Mol. Sci.* **2021**, *22*, 10504. [CrossRef] [PubMed]
32. Schneider, C.A.; Rasband, W.S.; Eliceiri, K.W. NIH Image to ImageJ: 25 years of image analysis. *Nat. Methods* **2012**, *9*, 671–675. [CrossRef] [PubMed]
33. Varga, Z.V.; Pipicz, M.; Baan, J.A.; Baranyai, T.; Koncsos, G.; Leszek, P.; Kusmierczyk, M.; Sanchez-Cabo, F.; Garcia-Pavia, P.; Brenner, G.J.; et al. Alternative Splicing of NOX4 in the Failing Human Heart. *Front. Physiol.* **2017**, *8*, 935. [CrossRef] [PubMed]
34. Voros, I.; Saghy, E.; Pohoczky, K.; Makkos, A.; Onodi, Z.; Brenner, G.B.; Baranyai, T.; Agg, B.; Varadi, B.; Kemeny, A.; et al. Somatostatin and Its Receptors in Myocardial Ischemia/Reperfusion Injury and Cardioprotection. *Front. Pharmacol.* **2021**, *12*, 663655. [CrossRef]
35. Lanner, J.T.; Georgiou, D.K.; Joshi, A.D.; Hamilton, S.L. Ryanodine receptors: Structure, expression, molecular details, and function in calcium release. *Cold Spring Harb. Perspect. Biol.* **2010**, *2*, a003996. [CrossRef]
36. Feng, D.; Nagy, J.A.; Pyne, K.; Dvorak, H.F.; Dvorak, A.M. Ultrastructural localization of platelet endothelial cell adhesion molecule (PECAM-1, CD31) in vascular endothelium. *J. Histochem. Cytochem.* **2004**, *52*, 87–101. [CrossRef]
37. Lawson, J.S.; Syme, H.M.; Wheeler-Jones, C.P.D.; Elliott, J. Characterisation of feline renal cortical fibroblast cultures and their transcriptional response to transforming growth factor beta1. *BMC Vet. Res.* **2018**, *14*, 76. [CrossRef]
38. Kim, H.D. Expression of intermediate filament desmin and vimentin in the human fetal heart. *Anat. Rec. Off. Publ. Am. Assoc. Anat.* **1996**, *246*, 271–278. [CrossRef]
39. Greaves, D.R.; Gordon, S. Macrophage-specific gene expression: Current paradigms and future challenges. *Int. J. Hematol.* **2002**, *76*, 6–15. [CrossRef]
40. Nemeth, J.; Helyes, Z.; Gorcs, T.; Gardi, J.; Pinter, E.; Szolcsanyi, J. Development of somatostatin radioimmunoassay for the measurement of plasma and tissue contents of hormone. *Acta Physiol. Hung.* **1996**, *84*, 313–315.
41. Wang, J.; Hao, D.; Zeng, L.; Zhang, Q.; Huang, W. Neuropeptide Y mediates cardiac hypertrophy through microRNA-216b/FoxO4 signaling pathway. *Int. J. Med. Sci.* **2021**, *18*, 18–28. [CrossRef]
42. Dehlin, H.M.; Levick, S.P. Substance P in heart failure: The good and the bad. *Int. J. Cardiol.* **2014**, *170*, 270–277. [CrossRef]
43. Tan, C.M.J.; Green, P.; Tapoulal, N.; Lewandowski, A.J.; Leeson, P.; Herring, N. The Role of Neuropeptide Y in Cardiovascular Health and Disease. *Front. Physiol.* **2018**, *9*, 1281. [CrossRef]
44. Widiapradja, A.; Chunduri, P.; Levick, S.P. The role of neuropeptides in adverse myocardial remodeling and heart failure. *Cell Mol. Life Sci.* **2017**, *74*, 2019–2038. [CrossRef]
45. Costoli, T.; Sgoifo, A.; Stilli, D.; Flugge, G.; Adriani, W.; Laviola, G.; Fuchs, E.; Pedrazzini, T.; Musso, E. Behavioural, neural and cardiovascular adaptations in mice lacking the NPY Y1 receptor. *Neurosci. Biobehav. Rev.* **2005**, *29*, 113–123. [CrossRef]

46. Ferdinandy, P.; Baczkó, I.; Bencsik, P.; Giricz, Z.; Görbe, A.; Pacher, P.; Varga, Z.V.; Varró, A.; Schulz, R. Definition of hidden drug cardiotoxicity: Paradigm change in cardiac safety testing and its clinical implications. *Eur. Heart J.* **2019**, *40*, 1771–1777. [CrossRef]
47. Mason, R.P.; Jacob, R.F.; Kubant, R.; Walter, M.F.; Bellamine, A.; Jacoby, A.; Mizuno, Y.; Malinski, T. Effect of enhanced glycemic control with saxagliptin on endothelial nitric oxide release and CD40 levels in obese rats. *J. Atheroscler. Thromb.* **2011**, *18*, 774–783. [CrossRef]
48. Ta, N.N.; Schuyler, C.A.; Li, Y.; Lopes-Virella, M.F.; Huang, Y. DPP-4 (CD26) inhibitor alogliptin inhibits atherosclerosis in diabetic apolipoprotein E-deficient mice. *J. Cardiovasc. Pharmacol.* **2011**, *58*, 157–166. [CrossRef]
49. Ajijola, O.A.; Chatterjee, N.A.; Gonzales, M.J.; Gornbein, J.; Liu, K.; Li, D.; Paterson, D.J.; Shivkumar, K.; Singh, J.P.; Herring, N. Coronary Sinus Neuropeptide Y Levels and Adverse Outcomes in Patients With Stable Chronic Heart Failure. *JAMA Cardiol.* **2020**, *5*, 318–325. [CrossRef]
50. Zhong, J.; Rao, X.; Rajagopalan, S. An emerging role of dipeptidyl peptidase 4 (DPP4) beyond glucose control: Potential implications in cardiovascular disease. *Atherosclerosis* **2013**, *226*, 305–314. [CrossRef]
51. Gorrell, M.D. Dipeptidyl peptidase IV and related enzymes in cell biology and liver disorders. *Clin. Sci.* **2005**, *108*, 277–292. [CrossRef]
52. Heike, M.; Mobius, U.; Knuth, A.; Meuer, S.; Meyer zum Buschenfelde, K.H. Tissue distribution of the T cell activation antigen Ta1. Serological, immunohistochemical and biochemical investigations. *Clin. Exp. Immunol.* **1988**, *74*, 431–434.
53. Dinjens, W.N.; ten Kate, J.; van der Linden, E.P.; Wijnen, J.T.; Khan, P.M.; Bosman, F.T. Distribution of adenosine deaminase complexing protein (ADCP) in human tissues. *J. Histochem. Cytochem.* **1989**, *37*, 1869–1875. [CrossRef]
54. Shigeta, T.; Aoyama, M.; Bando, Y.K.; Monji, A.; Mitsui, T.; Takatsu, M.; Cheng, X.W.; Okumura, T.; Hirashiki, A.; Nagata, K.; et al. Dipeptidyl peptidase-4 modulates left ventricular dysfunction in chronic heart failure via angiogenesis-dependent and -independent actions. *Circulation* **2012**, *126*, 1838–1851. [CrossRef]
55. Seferovic, P.M.; Coats, A.J.S.; Ponikowski, P.; Filippatos, G.; Huelsmann, M.; Jhund, P.S.; Polovina, M.M.; Komajda, M.; Seferovic, J.; Sari, I.; et al. European Society of Cardiology/Heart Failure Association position paper on the role and safety of new glucose-lowering drugs in patients with heart failure. *Eur. J. Heart Fail.* **2020**, *22*, 196–213. [CrossRef]
56. Kongwatcharapong, J.; Dilokthornsakul, P.; Nathisuwan, S.; Phrommintikul, A.; Chaiyakunapruk, N. Effect of dipeptidyl peptidase-4 inhibitors on heart failure: A meta-analysis of randomized clinical trials. *Int. J. Cardiol.* **2016**, *211*, 88–95. [CrossRef]
57. Scirica, B.M.; Braunwald, E.; Raz, I.; Cavender, M.A.; Morrow, D.A.; Jarolim, P.; Udell, J.A.; Mosenzon, O.; Im, K.; Umez-Eronini, A.A.; et al. Heart failure, saxagliptin, and diabetes mellitus: Observations from the SAVOR-TIMI 53 randomized trial. *Circulation* **2014**, *130*, 1579–1588. [CrossRef]
58. Xia, C.; Goud, A.; D'Souza, J.; Dahagam, C.; Rao, X.; Rajagopalan, S.; Zhong, J. DPP4 inhibitors and cardiovascular outcomes: Safety on heart failure. *Heart Fail. Rev.* **2017**, *22*, 299–304. [CrossRef]
59. Iqbal, N.; Parker, A.; Frederich, R.; Donovan, M.; Hirshberg, B. Assessment of the cardiovascular safety of saxagliptin in patients with type 2 diabetes mellitus: Pooled analysis of 20 clinical trials. *Cardiovasc. Diabetol.* **2014**, *13*, 33. [CrossRef]
60. Men, P.; Li, X.T.; Tang, H.L.; Zhai, S.D. Efficacy and safety of saxagliptin in patients with type 2 diabetes: A systematic review and meta-analysis. *PLoS ONE* **2018**, *13*, e0197321. [CrossRef]
61. Pollack, P.S.; Chadwick, K.D.; Smith, D.M.; Billger, M.; Hirshberg, B.; Iqbal, N.; Boulton, D.W. Nonclinical and clinical pharmacology evidence for cardiovascular safety of saxagliptin. *Cardiovasc. Diabetol.* **2017**, *16*, 113. [CrossRef] [PubMed]
62. Fleenor, B.S.; Ouyang, A.; Olver, T.D.; Hiemstra, J.A.; Cobb, M.S.; Minervini, G.; Emter, C.A. Saxagliptin Prevents Increased Coronary Vascular Stiffness in Aortic-Banded Mini Swine. *Hypertension* **2018**, *72*, 466–475. [CrossRef] [PubMed]
63. Ikeda, J.; Kimoto, N.; Kitayama, T.; Kunori, S. Cardiac DPP-4 inhibition by saxagliptin ameliorates isoproterenol-induced myocardial remodeling and cardiac diastolic dysfunction in rats. *J. Pharmacol. Sci.* **2016**, *132*, 65–70. [CrossRef] [PubMed]
64. Bradic, J.; Milosavljevic, I.; Bolevich, S.; Litvitskiy, P.F.; Jeremic, N.; Bolevich, S.; Zivkovic, V.; Srejovic, I.; Jeremic, J.; Jovicic, N.; et al. Dipeptidyl peptidase 4 inhibitors attenuate cardiac ischaemia-reperfusion injury in rats with diabetes mellitus type 2. *Clin. Exp. Pharmacol. Physiol.* **2021**, *48*, 575–584. [CrossRef]
65. Zeng, X.; Li, X.; Chen, Z.; Yao, Q. DPP-4 inhibitor saxagliptin ameliorates oxygen deprivation/reoxygenation-induced brain endothelial injury. *Am. J. Transl. Res.* **2019**, *11*, 6316–6325.
66. Keller, A.C.; Knaub, L.A.; Miller, M.W.; Birdsey, N.; Klemm, D.J.; Reusch, J.E. Saxagliptin restores vascular mitochondrial exercise response in the Goto-Kakizaki rat. *J. Cardiovasc. Pharmacol.* **2015**, *65*, 137–147. [CrossRef]
67. Esposito, G.; Cappetta, D.; Russo, R.; Rivellino, A.; Ciuffreda, L.P.; Roviezzo, F.; Piegari, E.; Berrino, L.; Rossi, F.; De Angelis, A.; et al. Sitagliptin reduces inflammation, fibrosis and preserves diastolic function in a rat model of heart failure with preserved ejection fraction. *Br. J. Pharmacol.* **2017**, *174*, 4070–4086. [CrossRef]
68. Ghersi, G.; Chen, W.; Lee, E.W.; Zukowska, Z. Critical role of dipeptidyl peptidase IV in neuropeptide Y-mediated endothelial cell migration in response to wounding. *Peptides* **2001**, *22*, 453–458. [CrossRef]
69. Lee, E.W.; Michalkiewicz, M.; Kitlinska, J.; Kalezic, I.; Switalska, H.; Yoo, P.; Sangkharat, A.; Ji, H.; Li, L.; Michalkiewicz, T.; et al. Neuropeptide Y induces ischemic angiogenesis and restores function of ischemic skeletal muscles. *J. Clin. Investig.* **2003**, *111*, 1853–1862. [CrossRef]
70. Abualsaud, N.; Caprio, L.; Galli, S.; Krawczyk, E.; Alamri, L.; Zhu, S.; Gallicano, G.I.; Kitlinska, J. Neuropeptide Y/Y5 Receptor Pathway Stimulates Neuroblastoma Cell Motility Through RhoA Activation. *Front. Cell Dev. Biol.* **2020**, *8*, 627090. [CrossRef]

MDPI
St. Alban-Anlage 66
4052 Basel
Switzerland
www.mdpi.com

Biomedicines Editorial Office
E-mail: biomedicines@mdpi.com
www.mdpi.com/journal/biomedicines

Disclaimer/Publisher's Note: The statements, opinions and data contained in all publications are solely those of the individual author(s) and contributor(s) and not of MDPI and/or the editor(s). MDPI and/or the editor(s) disclaim responsibility for any injury to people or property resulting from any ideas, methods, instructions or products referred to in the content.

www.ingramcontent.com/pod-product-compliance
Lightning Source LLC
Chambersburg PA
CBHW061816210926
51850CB00006B/720
9783036589886